DEUTSCHE FORSCHUNGS- UND VERSUCHSANSTALT
FÜR LUFT- UND RAUMFAHRT E.V. (DFVLR)

SOLAR THERMAL ENERGY UTILIZATION

German Studies on Technology and Application

Editor: M. Becker

Volume 3:
Solar Thermal Energy for Chemical Processes

Springer-Verlag
Berlin Heidelberg GmbH 1987

Dr.-Ing. Manfred Becker ·

Hauptabteilung Energietechnik der
Deutschen Forschungs- und Versuchsanstalt für Luft- und Raumfahrt e. V. (DFVLR), Köln

Additional material to this book can be downloaded from http://extras.springer.com

ISBN 978-3-540-18032-6   ISBN 978-3-662-09933-9 (eBook)
DOI 10.1007/978-3-662-09933-9

2362/3020-543210

Preface

The energy crisis in 1973 and 1979 initiated a great number of
activities and programs for low and high temperature application of solar energy. Synthetic fuels and chemicals produced by
solar energy is one of them, where temperatures in the range of
600-1000°C or even higher are needed. In principle such high
temperatures can be produced in solar towers. For electricity
production, the feasibility and operation of solar tower plants
has been examined during the SSPS - project (Small Solar Power
System) in Almeria, Spain.

The objective of Solar Thermal Energy Utilization is to extend
the experience from the former SSPS - program in to the field
of solar produced synthetic fuels. New materials and technologies have to be developed in order to research this goal.
Metallic components now in use for solar receivers need to be
improved with respect to transient operation or possibly
replaced by ceramics. High temperature processes, like
steam-methane reforming, coal conversion and hydrogen production need to be developed or at least adapted for the unconven-
tional solar operation. Therefore Solar Thermal Energy Utilization is a long term program, which needs time for its development much more time than the intervals expected in between
further energy crisis. The "Studies on Technology and Application on Solar Energy Utilization" is a necessary step in the
right direction in order to prepare for the energy problems in
the future.

Prof. Dr. H. F. Knoche

Rheinisch-Westfälische Technische Hochschule Aachen,
Federal Republic of Germany

PART I

LURGI

STEAM REFORMING OF METHANE UTILIZING
SOLAR HEAT

W.D. MÜLLER

LURGI, FRANKFURT

Contents

Steam Reforming of Methan Utilizing Solar Heat (Part I)

1. PURPOSE OF STUDY AND FIRST CONCLUSIONS

There is a worldwide interest to use solar energy
to save or substitute fossil material, which is taken
as fuel or chemical feedstock in present technologies.
Among the possibilities, which are studied in detail for
such substitution of fossil energy by solar energy,
the reforming of methane with water vapour may have
a good chance for application as this process predominantly is used for the production of several basic
materials in chemistry, as ammonia, methanol and hydrogen. This process needs as endothermic reaction a high
portion of energy for firing, which may be substituted
by solar energy.

Among the possibilities to introduce such solar
radiation to the process there are two basic modes:
an integrated and a separated solution.

In the <u>integrated solution</u> catalyst containing reforming tubes are directly radiated by the (reflected)
solar energy. Solar receiver and chemical reformer
form an integrated unit.

In the <u>separated solution</u> the (reflected) solar energy
is heating up in the solar receiver any heat carrying
medium, which is recycled, and which on its part is
heating the chemical reactor, which may be arranged
separate from the receiver.

It was the purpose of this study to generate data that
would allow a preselection of a system, i.e. integrated
or separated solution, which is being thought to be
superior to the other and which should be further

examined in future studies.

As the time for the preparation of this study was very short, only results, limited in assertion, could be expected, especially as the different lines had to be followed up in some details without being sure about the final conclusions and because obviously different chemical products influence the efficiency and applicability of one system or the other.

As will be shown in course of the following statements, none of the systems, which were studied in this study, could be selected off-hands as being superior to others as either unsolved technical problems raise uncertainties about the general realization of such systems or the economic outlook seems questionable. It is therefore the strong recommendation of the investigators to continue with the variety of the studied systems - or even more, which were laid aside in an early state of investigations without justified judgement - and to deepen the preliminary results in such a way that general assertions can be made and proven prior to the selection of any system, which shall be studied in future experiments.

This study was prepared by LURGI GMBH, Division Gas-technology, Frankfurt/Main and MAN AG - Neue Technologie, München to equal parts.

MAN's constributions mainly were referring to the part of the solar receivers with varying tasks and conditions as well as investigations of gaseous and liquid heat transfer systems.
LURGI's constributions cover the field of process arrangements and selections as well as investigations on overall economy of the different studied systems.

Considering the short period of study time with its need of parallel work at MAN and Lurgi it has to be accepted that some discrepancies at accompanying conditions could not be eliminated without redoing all work. Therefore these small differences, which are not felt to be essentail, have been accepted within this study.

2. INVESTIGATIONS OF OVERALL SYSTEMS

Steam reforming of methane for the generation of
synthesis gas is being applied worldwide for the production of the following basic chemicals
- ammonia
- methanol
- hydrogen
- oxo-alcohols

These products are gained either each by itself or in
combination of each other. Present technologies have
reached a high standard, which allows the use of any
residue gases to make the overall processes most
economic. Economic optimizations have influenced
the choice of processes even in such a manner that
residue gases are accepted or willingly generated
with regard to low investments, if a low specific total
energy consumption for the product will finally show up.

This demonstrates that there is a high dependency of
several process steps from each other and that it is
very difficult to compare such optimized overall systems
with any new system, depending not only from sunshine
but also from available technologies and specific conditions at the site , where such plants shall be built
and which dictate to design each of them as "tailor-
made".

In spite of such limitations there has been made the
attempt to compare the overall process sequences for
the production of

- methanol + hydrogen
- hydrogen
- ammonia

with similar accompanying conditions. Thus the specific steam generation and consumption of the follower
processes have been taken into consideration, when
designing the steam reforming plant, but no further
integration and optimization has been tried. Thus
the direct comparison with modern plants is not recommended; this comparison would have to be made specifically for selected systems in future investigations,
where exactly the same conditions were chosen for a
conventional and a solar system. The aim of this study
was to show the influence of different products in
steam reforming only.

The study could show - as was to be expected - that
amongst the indicated products only the production of

- methanol in combination with hydrogen and
- ammonia

is interesting for a solar energy coupling.

In hydrogen production so much offgas is available
already in conventional plants that such plants are
designed to export steam as normally no other use can
be found. With solar energy addition only the export
of steam would be enlarged, what cannot be the purpose
of such technology.

The production of methanol as single product also
is not feasible for the same reason as the available
purgegas quantities from the synthesis plant nearly
cover all firing needs. With solar energy addition
there would be a surplus of purge gas. Therefore the
additional production of hydrogen by Pressure-Swing-
Adsorption techniques of the purgegas is envisaged to
make such systems meaningful.

The attached figures 1 to 3 show the overall arrangements, which were the basis for the comparisons. Results of the integration of an integrated or separated
system in the different production sequences will be
shown together with the concepts of energy coupling.

Besides the two already mentioned concepts of solar
combination of the reformer - integrated or separated -
there are two further differentiations - low temperature
and high temperature reactions. Any integrated or
separated solution may be combined with a low temperature or high temperature reaction stage. A further
differentiation may be possible for the separated
concepts as the heat carrying media may be gaseous,
liquid or solid.

Solid heat carriers have not been investigated within
the scope of this study. From the other possible
combinations about 20 solutions have been calculated
in some detail. The systems of highest interest
according to the judgement of LURGI and MAN will be
presented hereafter. It should be noted that within
the scope of this study all calculations in the reforming plant have been made under the assumption of
no thermal losses in a first approach. In a second
step estimated thermal losses were introduced and
the results may be compared. This would have to be
detailed in continuing studies.

Some essential results may be listed in advance to
support and prepare some final conclusions:

-    There cannot be seen an important advantage in
     using high temperature reforming reactors. The
     efficiency of the solar receiver will be higher
     with lower system temperatures. Additionally
     material problems will be more stringent with
     higher temperatures.

The amount of fossil fuel saved will not depend
on the temperature of the reactor. No economic
and simple method could be found to operate an
integrated high temperature reactor during sunless
times as any flue gases cannot be used in such a
concept, all duties are already covered from other
waste heat sources.

- For a given thermal output of the solar receiver
  the throughput of natural gas for reforming will
  depend on the system chosen with respect to all
  parameters (integrated or separated, high
  temperature or "low" temperature reaction, with
  storage system(s) or without).
  The smallest throughput of natural gas can be
  achieved with storage systems and/or when running
  the reforming plant with variable load. In such
  cases the specific consumption figures per unit
  of product or the specific part of solar energy
  in the product become most favourable though the
  absolute amount of used solar heat is the same in
  all considerations. By the storage systems it is
  possible to egalize the fossil fuel from the
  process with the needs of the firing in such a
  manner that the natural gas for firing tends
  towards zero.

- By using the two stage concepts with solar heated
  prereactor and smaller (conventional) fired
  reformer it is possible to achieve higher temperatures in the reformer (which are essential for
  high efficiency) due to available experience,
  especially because of the higher homogenity of
  heat fluxes in standard reformers in comparison
  with solar heated reformers.

If the same high reforming temperature should be achieved with a separated system, very high maximum temperatures of the heat carrying material would be necessary, which at present still are not possible with availalbe technology.

- The separated solutions suffer from additional energy requirements for the recycle. Especially air as heat carrying substance does not seem adequate. Molten salts would obviously be most advisable.

4. DESCRIPTION OF VARIOUS REFORMING SYSTEMS

The statements of the preceding chapter shall be proven by the explanation of the following process flow sketches for the reforming units of various types.

The classification letters shall indicate:

Type A: The solar energy is directed to a high temperature reformer. Only a separated solution seems possible, as $^+$wall temperatures of the reforming tubes will be too high, if exposed to solar irradiation, and service time of such tubes would be too short.

Type B: The solar energy is directed to a prereactor at moderate reforming temperatures, separated solution with gas circuit as heat carrying system.

Type C: The solar energy is directed to a prereactor at moderate reforming temperatures, integrated solution, fossil firing of the cavity during sunless times.

Type D: The solar energy is directed to a prereactor at moderate reforming temperatures, separated solution with a molten salt circuit as heat carrying system.

Type E: The solar energy is directed to a reformer, which is integrated in a receiver and which achieves about 800°C equilibrium temperature. The cavity shall be fired with fossil material during sunless times.

$^+$otherwise

The Roman numbers following the letters indicate:

I   : Syngas production for methanol and hydrogen
              as product, steam systems of synthesis and
              reforming plant being tied together.

II   : Syngas production for ammonia as product,
              steam systems of synthesis and reforming
              plant being tied together. The syngas pro-
              duction differs from (I), as shift conver-
              sion steps additionally have been introduced
              to enlarge the amount of hydrogen.

Small letters, following the Roman numbers, indicate
subsystems, for instance for steady load of the plant
for a full day or for a varying load with varying
energy supply from the sun.

For the different reforming concepts, the following
explanations apply:

Type A I, Fig. 4

This is merely a reference concept to show the possibilities of a high temperature system in a separated
mode. It is the preferred choice amongst others that
were investigated, which use available temperatures and
duties in different manner but with no better results.
Due to a relatively big amount of solar energy, which
can be introduced in relation to the overall heat
balance, (realtively) small quantities of natural gas
can be handled with a given amount of solar energy.
But the overall efficiency is not better than for
other cases because:

- The residual gas of the synthesis can be fired in
  a directly fired heater only, which will have an
  overall efficiency of about 65 % only due to exergetic
  reasons as otherwise the heating tubes will be
  destroyed by too high temperatures

- even with the high temperature of $900^{\circ}C$ for the
  heat carrying medium, which cannot be achieved with
  available receiver technologies at present, the reforming rate in the main reactor is rather poor.
  Though no natural gas is taken for firing (residual
  gas only!) the specific quantity of natural gas per
  products is high in comparison to other routes.

Type B I, Fig. 5

This is a reference concept for a separated solution
with prereactor and gas circuit as heat carrying medium.
The available energy from the residual gas is not
sufficient to maintain reaction temperature in the main
reformer, i.e. during any time and even with partial
load there is a need for additional natural gas for
firing.

With the standard conditions, which have been chosen
throughout all comparison until now (i.e. 50,0 MW thermal
max. output of receiver, 43,0 MW thermal medium output of receiver for about 8,5 hours, which amounts to
a total solar energy flow of 366 MW/d) the amount of
methanol produced and of natural gas consumed is much
bigger than for A I.

Type C I, Fig. 6

This is the reference concept for an integrated solution
It seems inevitable that a chemical production plant
has to be operated continuously throughout day and
night. Therefore it is proposed that the solar receiver cavity shall be fired by fossil material during
night or when the sun is hidden by clouds. As the
portion of energy, which can be taken by the integrated prereactor is limited, the throughput of natural
gas is high for the standard conditions. But it was
excluded from the investigations that a steam generator
may be included also in the cavity.

This was thought to be too complicated in control of
the plant and in operation. At the same time it would
have taken load from the fossil heated part of the
plant, while during night time the flue gas quantity
would have become very big and demand and offer of
energy would have become different.

Type C I is thought to be ideal, if the fossil firing
of the cavity can be realized within the scope of
assumptions. It is quite acceptable also if the high
throughput of natural gas and the relevant high quantity of product can be allowed.

For balancing of energy streams it has to be accepted
that a prereactor has to be placed upstream of the
solar heated reactor.

Type D I, Fig. 8

This is an alternate concept for a separate solution,
where the heat carrier is a molten salt. Process-
wise the concept is very similar to Type B I, Fig. 5.

There are some features that make this concept very
interesting:

- the efficiency of the solar receiver will be high
  due to the moderate temperatures;

- the energy requirements for recycling the heat
  carrier are negligible, they are only a small
  portion of those of the reference concepts "B"
  with gaseous heat carriers;

- the concept offers the possibility of heat storage
  by the liquid in a high energy density. This concept allows the egalization of all available energy
  from the sun on a total day of 24 hours. Thus
  fossil firing may be excluded entirely and the
  specific natural gas consumption may be reduced to
  the theoretical minimum, which is needed for methanol
  production.

- this concept offers great flexibility by adjusting the heat storage. It can be designed in such a way that the maximum design throughput (solar maximum energy input) may be operated all time with additional natural gas firing, or it may be operated in following the solar energy input, so that no fluid storage is necessary and that a fossil operated plant will be used at any load, which is wanted within the scope of the technical possibilities, for instance between 15 and 100 % of design. Therefore it will only be necessary to design the heat exchanger within the flue gas duct, which has to heat up the molten salt, for the changing load;

- the concept allows small lines as connection between solar receiver and chemical plant. It is expected that all chemical reactors. can be placed on ground and that the problems of connecting hot piping is minimum for such systems.

But there have been found also great difficulties for designing such plants:

- It could be expected from normal experience that the heat transfer coefficients of a liquid phase are much better than for a gaseous phase and that therefore smaller temperature approaches between heat carrier and reacting agent in the catalyst could be chosen. This did not turn out as realistic, because the massflow of the liquid phase would then be reduced so drastically that no velocities for turbulent flow can be achieved within the reactors.

This caused as consequences:
- The massflow had to be increased, thereby increasing the lower temperature of the molten salt from about 450°C (expected originally) to about 580°C.
- This massflow increase at the same time will increase the storage volumes drastically for any solution with storage systems.

- Due to high temperatures alloy material, which is expensive, has to be used predominantly within the reactors. Steel compensators for stress relief cannot be avoided either due to conditions during start-up, shut down and failures. To design those parts reasonably the differential pressures of molten salt and gaseous phase have to be kept small under all circumstances. No systems could be found yet that would automatically assume such small differential pressures also under dondition of failure of one side. For the judgement of this circumstance it has to be kept in mind that

- a tower of about 200 m height will generate a pressure of about 35 bar at ground level with molten salts.
- the pressure of the reactants shall be in the order of magnitude of about 20 bar
- even with depressurizing the salt system or keeping it at 20 bar, there is still a big problem, if one side loses pressure as the liquid and the gas will depressurize at very different rates so that any small design differential pressure will not be kept under all circumstances.

The solution of such problems may be in a design
pressure difference of about 10 bar by artificial
means, but the problems and related costs could
not yet be solved.

- According to the heat transfer problems and the
  related temperature levels the minimum temperature
  in the cycle will be about 580°C. No pumps are
  available for molten salts at such high temperatures.

Therefore, though the initially enumerated advantages
make system D desirable, the appropriate technical
solutions have not yet been found.

Type E I, Fig. 11

As it is known that studies of different origin favour
a concept, where high temperature reforming is performed,
a similar concept has been studied with same accompanying
conditions by Lurgi. The same comments as for system
C I do apply. The comparing tables do not show a
preference for plants of this type.

Table I allows to compare the different types as
they were presented above.

Similar concepts are offered for the production of
ammonia. Changes are dependent from different need
of steam in the ammonia synthesis and the introduction
of a high and low temperature shift conversion step to generate
more hydrogen. The concepts evaluated were the types
B II, C II and D II, Fig. 8, 9 and 10. Type E II

(fig. 12) was added later. The figures for these types of process arrangement can be compared by <u>table II</u>.

<u>Tables III and IV</u> allow to compare the introduced systems on a uniform basis, i.e. per ton of product.

TABLE I: Main Data $CH_3OH + H_2$

Basis: Ntherm solar max = 50,0 MW

N therm solar total = 366 MWh/d   43 MW for 8,5 h

Type / Figure	A Ia /4		A Ib /4		B I / 5		C I / 6		D Ia / 7		D Ib / 7		E I	
losses	no	yes	no	yes	no	yes	no	yes	no	yes	no	yes	no	yes
CH_3-OH-production t/d	310,645	310,645	166,080	149,927	1128,3	1128,3	1.422,8	1.422,8	348,53	348,53	1.142,7	1.142,7	568,791	568,791
H_2-production m^3_N/d	201.733	201.733	100,684	97.393	732.732	732.732	923.979	923.979	226,335	226,335	742.093	742.093	369.385	369.385
Natural Gas Process m^3_N/d	290.000	290.000	155.041	139.962	996.914	996.914	1257.114	1257.114	307.943	307.943	1009.650	1009.650	507.396	507.396
Natural Gas Firing m^3_N/d	45.950	54.690	–	–	147.279	161.098	195.975	211.149	18,341	25,157	149.663	164.614	89.211	96.345
Natural Gas Total m^3_N/d	335.950	344.690	155.041	139.962	1144.193	1158.012	1453.089	1468.263	326.284	333.100	1159.313	1174.264	596.607	603.741
Natural Gas Saving by Sun m^3_N/d	52.788	52.788	(52.788)	(52.788)	39.290	39.290	39.290	39.290	39.290	39.290	39.290	39.290	39.290	39.290
Residual Gas Equivalent as Natural Gas m^3_N/d	75.166	75.166	40.186	36.277	212.255	212.255	267.655	267.655	65.565	65.565	214.967	214.967	112.373	112.373
Energy Requirement for Heat Carrier kWh/d as Air	30.480	30.480	16.195	14.620	52.800	52.800	–	–	–	–	–	–	–	–
as He	9.600	9.600	5.006	4.519	20.400	20.400	–	–	8.400	8.400	–	–	–	–
as H_2	4.560	4.560	2.356	2.127	8.880	8.880	–	–	–	–	11.600	11.600	–	–
Natural Gas Equivalent for Heat Carrier Transport m^3_N/d as Air	8.600	8.600	4.569	4.125	14.910	14.910	–	–	–	–	–	–	–	–
as He	2.700	2.700	1.412	1.275	5.760	5.760	–	–	2.360	2.360	–	–	–	–
as H_2	1.270	1.270	665	600	2.510	2.510	–	–	–	–	3.260	3.260	–	–
Heat storage Melting for NS-salt m^3	–	–	–	–	–	–	–	–	2x 1319	2x1319	–	–	–	–
Residual Gas Storage m^3_N	20.468	18.310	41.905	39.478	–	–	–	–	–	–	–	–	–	–

TABLE II: Main Data HN_3-Production
Basis: Ntherm Solar max = 50,0 MW Ntherm Solar total= 366 MWh/d 43 MW for 8,5 h

Type / Figure		B II / 8		C II / 9		D II / 10		E II /12	
losses		no	yes	no	yes	no	yes	no	yes
NH_3-Production	t/d	1.318,8	1.318,8	1.647,0	1.647,0	927.7	647,5	677.176	677.176
Natural Gas to Process	m^3_N/d	976.847	976.847	1.219.937	1.219.937	687.118	479.600	479.558	479.558
Natural Gas Firing	m^3_N/d	16.567	34.088	30.420	48.979	-	-	16.378	23.429
Natural Gas Total	m^3_N/d	993.414	1.010.935	1.250.357	1.268.916	687.118	479.600	513.936	520.987
Natural Gas Saving by Sun	m^3_N/d	39.290	39.290	39.290	39.290	39.290	39.290	39.290	39.290
Residual Gas Equivalent as Natural Gas	m^3_N/d	371.707	371.707	464.207	464.207	261.460	182.496	175.632	175.632
Energy Requirement for Heat Carrier kWh/d as Air		60.635	60.635	-	-	8.400	8.400	-	-
as He		25.360	25.360	-	-	-	-	-	-
as H_2		12.130	12.130	-	-	-	-	-	-
Natural Gas Equivalent for Heat Carrier Transport m^3_N/d as Air		17.100	17.100	-	-	2.360	2.360	-	-
as He		7.120	7.120	-	-	-	-	-	-
as H_2		3.410	3.410	-	-	-	-	-	-
Heat Storage Melting m^3 for NS-salt		-	-	-	-	2x1319	2x 1319	-	-
Residual Gas Storage m^3_N		90.277	64.897	67.510	40.530	-	-	92.572	84.332

TABLE III: Specific Data for 1 t CH_3OH + 649 $m^3_N H_2$

No.	Type / Figure	A Ia/4 no	A Ia/4 yes	A Ib/4 no	A Ib/4 yes	B I/5 no	B I/5 yes	C I/6 no	C I/6 yes	D Ia/7 no	D Ia/7 yes	D Ib/7 no	D Ib/7 yes	E I no	E I yes
1	Process Feed N.G. m^3_N	933,5	933,5	933,5	933,5	883,6	883,6	883,6	883,6	883,6	883,6	883,6	883,6	892,1	892,1
2	Fossil Firing N.G. m^3_N	147,9	176,1	-	-	130,5	142,8	137,8	148,4	52,6	72,2	131,0	144,1	156,8	169,4
3	Fossil Firing Resid. Gas as N.G. m^3_N	242,0	242,0	242,0	242,0	188,1	188,1	188,1	188,1	188,1	188,1	188,1	188,1	197,6	197,6
4	N.G. need for Energy Production (max.) m^3_N	27,7	27,7	27,5	27,5	13,2	13,2	-	-	6,8	6,8	2,9	2,9	-	-
5	Solar Energy Equivalent as N.G. m^3_N	169,9	169,9	317,8	352,1	34,8	34,8	27,6	27,6	112,7	112,7	34,4	34,4	69,1	69,1
6	Firing Requirement Total as N.G.=(2)+(3)+(4)+(5) m^3_N	587,5	615,7	587,7	621,6	366,6	378,9	353,5	364,1	360,2	379,8	356,4	369,5	423,5	436,1
7	Specific NG-consumption with Solar Energy m^3_N	1109,1	1137,3	961,0	961,0	1027,3	1029,5	1021,3	1032,0	943,0	962,6	1017,5	1030,6	1048,9	1061,5
8	Specific NG-consumption without Solar Energy m^3_N	1279,0	1307,2	1278,8	1313,1	1062,1	1074,3	1048,9	1059,6	1055,7	1075,3	1051,9	1065,0	1118,0	1130,6
9	Relative Saving by Sun = $\frac{(5)}{(2)+(4)+(5)}$ %	49,2	45,5	92,0	92,8	19,5	18,2	16,7	15,7	65,5	58,8	20,4	19,0	30,6	29,0

TABLE IV: Specific Data for 1 t NH_3-Production

No.	Type / Figure	B II/8		C II/9		D II/10		E II/12	
	losses	no	yes	no	yes	no	yes	no	yes
1	Process Feed N.G. m^3_N	740,7	740,7	740,7	740,7	740,7	740,7	734,8	734,8
2	Fossil Firing N.G. m^3_N	12,5	25,8	18,4	29,7	–	–	24,2	34,6
3	Fossil Firing Res. Gas as N.G. m^3_N	281,9	281,9	281,9	281,9	281,9	281,9	259,4	259,4
4	N.G. Need for Energy Production (max.) m^3_N	13,0	13,0	–	–	2,5	3,6	–	–
5	Solar Energy Equivalent as N.G. m^3_N	29,8	29,8	23,9	23,9	42,4	60,7	58,0	58,0
6	Firing Requirement Total as N.G. m^3_N	337,2	350,5	324,2	335,5	326,8	346,2	341,6	352,0
7	Specific N.G.-consumption with Solar Energy m^3_N	766,3	779,6	759,2	770,4	743,2	744,3	759,0	769,4
8	Specific N.G.-Consumption without Solar Energy m^3_N	796,1	809,4	783,1	794,3	785,6	805,0	817,0	827,4
9	Relative Saving by Sun = (5)/((2)+(4)+(5)) %	53,9	43,4	56,5	44,6	94,4	94,4	70,6	62,6

COMMENTS to tables I to IV

1) Each system was evaluated first on a basis without
   losses. Then losses (in the reforming part only!)
   were introduced, differentiated in height depending
   from the temperature levels of the equipment.

   All losses were compensated by additional fossil firing,
   the throughput of the plant was kept constant by this
   prodeedings. Only in systems with storage installations
   the losses may be compensated by different operation of
   the solar heated reactors. In these cases the throughput will be smaller "with losses". Exact calculation
   would have to be done for these cases to verify that
   gaseous storage systems really can be left out, when
   thermal storage already is provided.

   The losses generally were estimated on basis of values
   of experience. The losses would have to be calculated
   more accurately in course of detailed design of such
   systems. Figures may slightly vary by such conditions.

2) For all calculation the same daily thermal output from
   the solar receiver was assumed though the detailed calculations of MAN show some differences due to varying
   efficiencies with changing temperatures.

   The "Natural Gas Saving by Sun" was calculated as
   366 MHh/d, divided by the achievable efficiency of

the fossil furnace.   As for systems
B to D the efficiency of the fossil part was assumed
(and controlled) to be constant, this leads to the
same amount of natural gas saving throughout the systems.
Such good efficiency cannot be obtained for systems A,
which is the reason that the solar saving is bigger
there though the system is inadequate.

3) The energy requirements for the circulation of a heat
   carrier have been calculated for each system. It has
   been assumed that the efficiency of power generation,
   transmission and motors will be around 32 % only.

4) As Natural Gas pure methane with 0,5 % by volume of
   nitrogen has been used throughout this study.

5) For systems A to D a molar ratio of steam to carbon
   of 2,6 has been taken. Systems E I and E II have been
   calculated with $H_2O/C = 4,2$. Thereby it was possible
   to produce the same amount of methanol + hydrogen or
   ammonia respectively from the same amount of natural
   gas with the lower reforming temperature. Systems
   E I and E II are penalized by the higher steam consumption
   the specific natural gas consumption is higher than for
   comparable systems.

6) Operation modes of different systems:

A I a: The plant is constantly run at maximum rate,
        being set by 50 MW thermal energy being supplied
        by the sun.

A I b: The plant is run at maximum rate only at noon
        with solar maximum energy input. At other
        times the throughput is throttled. During
        solar times an excess of residual gas is gene-
        rated and stored, which is used at times with-
        out solar energy input. At proper egalization
        of throughput and storage the need for additional
        natural gas firing can be put to zero.

        The specific consumptions and saving for A I b
        are better than for A I a.

B I, C I
and DIb: The plant is constantly run at maximum rate,
        being set by 50 MW thermal energy being applied
        by the sun. The load during sunless times is
        being maintained by firing of additional natural
        gas.

D I a: The plant is constantly run at a capacity,
        which corresponds to the average of daily solar
        energy input. The plant is equipped with storage
        systems for molten salts at high temperatures.
        The surplus of solar energy input above the
        daily average is lead to the storage and is
        regained from there at times, when the solar
        energy is below average or zero.

B II, C II

and E II: The plant is constantly run at maximum load,
            equivalent to 50 MW thermal solar energy
            input, the load during sunless times to be
            maintained by firing of additional natural
            gas.
            Residual gas storage systems are necessary
            for these systems, as during times with sun
            there will exist a surplus of residual gas,
            which shall be used, when the sun is down.

D II :   This is a system equivalent to DI a. The
            same comments are applicable.
            The necessity of gaseous storage could not
            be quantified.

For the systems B I and D I the possible design of
the pertinent prereactors was studied in detail.
Thermodynamic calculations have been performed to
homogenize heat, mass and transfer balances.

Much investigation has been done, from MAN as well
as from LURGI, to preselect proper heat carrying molten
salt systems. Experience for the temperature range,
where such salts shall be used obviously is not
available, some risks will stay with such systems, even
if all other problems which are enumerated hereafter,
can be solved.

The results are presented in the accompanying copies
of drawings. For the standard plant size 2 reactors
will be needed for D I b, but only 1 reactor would
be needed for D I a, as the average load is much
smaller. System B I will also need 2 reactors of the
shown design.

As was described already in chapter 4 there have been
found a lot of problems, which could not yet be solved
by shortage of time. They refer mainly to

- heat insulation problems and
- differential pressure in independent systems.

without going into details these problems may shortly
be explained as follows:

- The prereactor has to tolerate heat carrying
  media of 700 to 800 $^{\circ}$C at the inlet. There are
  only few materials available that can withstand
  these temperatures for prolonged times, but
  - as Incoloy 800 H - they are rather expensive.
  It must be the aim of construction to limit the
  wall thickness of the vessels to save money.

- It is known from normal chemical plant design
  that the pressure of reforming should be in the
  range of about 20 bar to operate such plants
  economically. It was not possible up to now to
  study the effects of altered pressure conditions
  in a solar heated system.

- For reasons of contruction elements (expansion
  joints and for instance the thickness of the tube
  sheet, which separates pressurewise the heat
  carrier and the working gas) the pressure of the
  heat carrier has to be in the same magnitude.

- With high temperatures and about 20 bar the
  wallthickness becomes that big that with present
  manufacturing devices - even apart from price -
  such vessels cannot be manufactured.

- The logical escape then calls for internal insulation to reduce metal temperatures and to make
  the use of other, cheaper construction material
  possible.

- While such measures are deemed possible for
  gas as heat carrier, unsolved problems are
  envisaged for molten salts as heat carriers,
  because the heat carrier - also for design reasons -
  normally has to penetrate the insulation material
  for pressure egalization.

- An often used solution, to cool the pressure carrying walls with the heat carrier medium of lowest
  temperature of the cycle is no adequate design
  strategy in this case, as this lowest design
  temperature will be in the range of 600 °C, which
  would not allow the use of normal high temperature
  steel. Only marginal improvemtns would be realized,
  because of the improved stress allowance for any
  metal at lower temperature.

- With gas on one side and a liquid (molten salt)
  on the other side considerations also must be given
  to the different behaviour of these media, when
  depressurizing by any reason. While the gaseous
  medium will depressurize at a limited rate, the
  depressurization of a liquid may take place rapidly.
  There may be caused differential pressure much
  beyond allowed limits during such instabilities,
  which could lead to destruction of the vessel.

While such difficulties may be overcome by proper
design and selection of systems, it must be stated again
that the solutions are not contained in this phase of
the studies.

Table of Contents

1. Introduction

2. Receiver Conceptual Design

   2.1 Integrated Reactor Cavity (IRC)

   2.2 Separated Reactor Cavity - Gas-cooled (SRC-G)

   2.3 Separated Reactor Cavity - Liquid-cooled (SRC-L)

3. Discussion of Meltings

4. Assessment of Different Concepts

5. Proposed Scope of Work for Study Phase of the STAP-Project

   5.1 Concept Study

   5.2 Preliminary Design of the Experimental Plant at Almeria

6. General Aspects of Experiments at Almeria

Introduction

The study aims at the utilization of solar energy for the reforming process of methane. The solar related part of such a plant is investigated under considering the following principle aspects:

- solar radiation heats up the process gas itself in a combined receiver/reactor (integrated solution)

- solar radiation heats up a heat transfer medium which then will be used to heat up the process gas in a separated reactor (separated solution)

To compare the different concepts also under commercial scale relations a solar thermal input of 50 MW for each concept was assumed. The design work was concentrated at the receiver assuming that a corresponding heliostatfield is possible to design in every case.

2. Receiver Conceptual Design

For the integrated solution an "Integrated Reactor Cavity" (IRC) was
predesigned and for the separated solution two different types of heat
transfer media have been investigated. According to this a "Separated
Reactor Cavity-Gas-cooled" (SRC-G) and a "Separated Reactor Cavity-
Liquid-cooled" (SRC-L) were predesigned.

2.1 Integrated Reactor Cavity (IRC)

The total thermal power of 50 MW is divided up on two reactor/receiver
modules each with a designed power of 25 MW. Both modules have to be
oriented symmetrically to geographic north.
The coolant of the receiver tubes is the process gas itself which reacts
in the catalyst filled tubes. The necessary heat for this reaction is
transferred in the receiver.
Due to limitations of the catalyst the tubes shall not exceed a length
of aprr. 4 m. For this reason the tubes were arranged in three floors.
To keep small the deformation of the receiver/reformer tubes which occurs
due to a strong unhomogeneous radiation all tubes are located at only
indirect irradiated surfaces of the cavity (see fig. 1). The tubes are
collected to panels each of them with it's own inlet and outlet header.
Because only short tube length are allowed, a relative high effort is
necessary to provide the procress gas to each nanel and also to collect
the gas again.
For the exchange of dead catalyst each tube has a feed-in and removal
flange.
The operation of the prereactor/receiver shall be at constant thermal
load as far as possible. Therefore burners shall be installed in the
cavity itself. If the solar radiation decreases, the lacking power is
substituted by ignition of the burners. Also cloud passes shall be
equalized by these burners. Because only a part of the total burner
power can be transferred to the heating surfaces, the flue gas must be
extracted and used in a waste-heat boiler.

This combination of solar and fossil fuel heating of a cavity is not realized up to now. Therefore some open questions exist which will also require development effort:

- gas exchange of flue gas and ambient air via the aperture

- The CO_2- and H_2O-contents of flue gas will increase the infrared absoptivity of the gas bulk in the cavity. So the cavity losses could have a considerable part which arises from the hot gas bulk inside the cavity.

- Exchange of radiation between the flame and the cavity walls as well as its combination at simultaneously solar operation.

A design sketch of this receiver is shown in fig. 2.

2.2 Separated Reactor Cavity - Gas-cooled (SRC-G)

The coolant for the receiver tubes is a gas which circulates in a separate primary loop. In the receiver the gas is heated up to 800 ^{O}C and the heat is extracted in the prereactor.
For the design estimations air was assumed to be the coolant, but also other gases could be of interest like helium or also hydrogen. The total thermal power under design point conditions which is delivered to the process shall be again 50 MW. These power is drawn out of two receiver modules, each of them delivers 25 MW. The apertures of both cavities are faced to the north direction symmetrically.
The heat exchanger tubes are arranged in the direct irradiated areas of the receiver which leads to lower thermal losses. The increased loads due to a higher ununiformity of irradiation are well investigated in the GAST receiver design. The arrangement of tubes is shown in fig. 3. The incoming gas is heated up in a low-temperature part and enters then the high-temperature part to reach the final temperature. The low- and the high-temperature part are mechanical decoupled by a compensation system. Thermally both are connected in series and are mass-flow controlled together.
A design sketch of this receiver is shown in fig. 4.

2.3 Separated Reactor Cavity - Liquid-cooled (SRC-L)

The coolant for the receiver tubes is a liquid preferably molten
salt or molten sodium. Due to the better heat transfer properties
of these media higher heat flux densities can be accepted and so
the receiver size can be reduced. Additionally the flow velocities
are low so that large tube lengths can be used without high pressure
losses. These all result in a more flexible tubing of the receiver
so that nearly all receiver walls can be equipped with heat exchanging tubes and only one cavity is needed for a thermal power of 50 MW.
The arrangement of tubes is shown in fig. 5, and fig. 6 shows a design sketch.

The main features of the investigated receiver concepts are compiled in
table 1. Table 2 shows the main process data as well as cavity data and
data of the tubes.
The time dependent thermal power of the receivers delivered to the plant
under the assumption of constant outlet temperature are shown in fig. 7
to fig. 9 for the March 21st.
These curves consider the heliostat, aperture and receiver efficiencies
and base upon a direct solar insolation as shown in fig. 10. Table 3 compares the energy delivery for operation at design day conditions and the
needed heliostatfield area.

3. Discussion of Meltings

The use of meltings in a solar plant could have some advantages like

- high possible heat flux density in the receiver

- no phase change during normal operation

- decoupling of solar dynamics and process dynamics by a small storage capacity

- potential of larger storage capacity

- low power consumption for circulation

as well as some disadvantages like

- possibility of freezing of the medium in lines, armatures and vessels

- several special aspects of corrosion, purification and safety have to be considered

Some mixture of salts ("Hitec", draw salt) and also sodium are well
.investigated and are used in some well established technologies, but
only up to temperatures of 550 ^{O}C in maximum.
The proposal made in the STAP-project, to use a prereactor for the
steam reforming of methane, with upper process temperatures of about
680 ^{O}C, makes it possible to think about the use of meltings because
the materials for the containments as well as some stable meltings
seems to be available. Some properties for meltings of interest are
compiled in table 4. Table 5 shows a rough assessment of these meltings
under an operational point of view.
The salt mixture KCl-CuCl* and sodium seem to be possible candidates
for heating this prereactor. The use of a sodium-potassium mixture has
the potential to decrease the melting point to -14 ^{O}C, but a real advantage - which can compensate the high price - by omitting any trace
heating must be proven.

*Like reported by: Etter, D.E., C.J. Wiedenheft, Solar Energy Materials 2
(1980) 423

For long term and large scale application an extensive program for
the investigation of corrosion of sodium as well as of KCl-CuCl must
be started. In the case of sodium well proven data are available only
up to temperatures of 600 ^{O}C.
With sodium some special hazards are combined. The safety aspects of
a sodium/water reaction are well investigated during the fast breeder
programs. Some new aspects occur through the fact that now instead of
water a mixture of CH_4, H_2O, CO, CO_2 and H_2 would have to be heated.

Further some development effort will be necessary if valves or pumps
in the hot leg of the loop will be necessary. Today they are available
normally only up to temperatures of 550 ^{O}C.
Some of these open questions may lose importance if only a short test
operation as in the STAP-project is planned.
Nevertheless some development effort will be necessary if the decision
is made to use anv melting as heat transfer medium during the STAP-
project.
The emphasis of this effort will mainly be

- corrosion problems

- development of precautions to prevent any hazard in the case of sodium

- adaption of valves or pumps if necessary in the hot leg

4. Assessment of Different Concepts

Table 6 compares the different concepts of solar energy integration.
The solution with an integrated prereactor/receiver requires to install some main components on the tower. These are the integrated
prereactor/receiver, the main reactor and the exhaust heat boiler
where the flue gas of both the main reactor and in the case of additional firing that of the prereactor/receiver has to be utilized. The connections to the plant on groundlevel then mainly are supply lines for
water, methane, product gas and fuel.
In the case of the separated concepts it is assumed to locate only the
receiver on the top. For the gas-cooled receiver this requires hot gas
lines (800 ^{o}C down-stream, 400 ^{o}C up-stream, 20 bar) with considerable
diameter. For the liquid-cooled receiver these lines reach appr. 700 ^{o}C
at lower diameters.
The required heliostatfield area corresponds to the efficiencies at design point (march 21st, 12.00 o'clock) conditions. The liquid-cooled
receiver which produces simultaneously the most energy on the march 21st
requires the lowest area.
The energy, which is necessary to circulate the heat transfer media, is
also shown in the table 6. The circulation of air consumes a considerable amount of the gathered solar energy so that for a commercial plant
helium or also hydrogen - which is produced in these plants - have to be
taken in consideration strongly.
One further figure which can be used to compare the different concepts
is the relative amount of fuel which can be saved in relation to a pure
fossil fuel fired plant.
These figures have to be considered together with the flow schemes and
give an overview about the potential of fuel saving of the different
concepts. The concept using a melting as coolant has a good storage capability which was not considered in table 6. Using of storages can of
course increase the amount of fuel which can be replaced for this concept.

The cost relation between the concepts so far as they are connected
to the solar related part of the plant might be approximately the
following: Gas-cooled separated receiver 100 %, liquid-cooled separated receiver 95 % and integrated prereactor/receiver 90 %. These preliminary figures are results of a very rough estimation.
Table 7 gives an overview about general advantages and disadvantages
of the investigated concepts.

5. Proposed Scope of Work for Study Phase of the STAP-Project

(Time schedule see fig. 11)

5.1 Concept Study (solar related part of the plant)

- investigation of integrated receiver (burner), melting cooled
  receiver
- energy transport system (gas, liquid) for indirect coupling concept
- concept decision
- definition of an experimental plant at Almeria (flow diagram, heat
  balances and preliminary sizing of main equipment, plot plan)
- cost and time estimation of an experimental plant at Almeria

5.2 Preliminary Design of the Experimental Plant at Almeria

- preliminary design of the receiver
  . cavity shape
  . arrangement of heating areas
  . materials selection
  . rough stress estimations
  . description of main operating modes
- preliminary design of primary loop (if any)
- preliminary design of secondary energy source (if any)
- identification of technical problems
  (accompanying research and development)
- clarifying of and adaptions to site conditions (data acquisition
  system, field, tower etc.)

6. General Aspects of Experiments at Almeria

Main aim of testing must be the experimental demonstration of the steam reforming of methane using solar-process heat, e.g. replacing of fuel in the combustion section of a steam/methane reforming unit still maintaining the product conditions typical of fossil-fired reformers (reducing the risks in the users' eyes).

Depending on the extension of the heliostatfield and on the available budget about 3 MW_{th} receiver size seems to be a proper size to show the above mentioned feasibility. For saving money it is desirable to use the possibilities of the Almeria plant as far as possible. Adaptions to any site requirements have not been considered yet, but serious principal problems are not expected.

The system will be based on a cavity receiver, a scaled version of the 50 MW_{th} conceptual design. Process parameters as

. volume ratio steam/methane

. temperatures

. pressures

. heat fluxes (tubes, rear wall, aperture) etc.

should be taken over from the outlined plant as far as possible and necessary to assess technical risks of an upscaled plant.

The items to be investigated may be

. function of receiver and steam reformer

. check of design values of components

. gathering operational experiences under a variety of solar conditions

It would be advisable to build the chemical experimental plant in such a way that the experimental results can be utilized for future applications. This might result in slightly higher investment costs, but would result in more reliable lay-out data for future plants. Critical scaling relationship with respect to the full-size commercial plant and smaller-scale experiments should be assessed. Until now the proposer does not see problems in upgrading.

The experimental work in Almeria, the measurements and their analysis should provide the following information:

- absorbed heat load on tubes

- thermal efficiency of receiver

- extent of methane conversion and reaction efficiency

- effect of flux and temperature variation on the reaction

- variation of the process and thermal efficiency as a function of daily hours

- gain of experiences for using high temperature solar heat

Prereactor Integrated IRC	Gas-Cooled Separated Receiver SRC - G	Liquid-Cooled Separated Receiver SRC - L
2 receiver modules	2 receiver modules	1 receiver module
coolant process gas	coolant gas (air or helium)	coolant liquid (salt or metal)
heating surfaces	heating surfaces	heating surfaces
indirect irradiated	direct irradiated	direct irradiated
arranged in 3 floors (in each the gas is heated up from 525 °C to 680 °C)	devided in a low temperature (LT) and high temperature (HT) section	all cavity surfaces are covered with panels
consisting of 26 panels all panels in parallel	consisting of 18 panels (1 LT panel and 1 HT panel are combined in serie)	consisting of 8 panels
panels consist of vertical tubes with considerable pitch		panels consist of horizontal loops without remarkable pitch
control	control	control
	module and panel have to be massflow controlled	
additional combustion		
burner are integrated to equalize a lack of solar irradiation		
development effort	development effort	development effort
detailed design	only adaptions from GAST-project necessary	detailed design
integration of burner has to be investigated (it affects losses and radiation exchange)		investigation of corrosion

Table 1: Main features of the investigated receiver concepts

	IRC	SRC-G	SRC-L
Symbol:	□	◺	◿
Type:	Prereactor integrated	Prereactor	separated
Coolant:	Process Gas	Gas (air or helium)	Liquid (salt or metal)

Design Point Operating Data

		IRC	SRC-G	SRC-L
thermal power	MW	2 · 25	2 · 25	50
inlet pressure	bar	22	20	5
inlet temperature	°C	525	442	430
outlet temperature	°C	680	800	700
heat flux density appr.				
av. in tubes	kW/m²	30	26	60

Cavity Geometry

		IRC	SRC-G	SRC-L
number of cavities		2	2	1
height	m	21,0	22,0	10,7
depth	m	16,5	21,0	13,0
width	m	23,2	24,0	13,0
inner surface	m²	1710	1880	720
inner volume	m³	5240	6250	1430

Cooling Tubes Geometry

		IRC	SRC-G lower panel	SRC-G upper panel	SRC-L
irradiated tube length	m	4,0	12,0	8,0	54 ÷ 97
outer diameter d_o	mm	60,3	50,0		40
wall thickness	mm	3,6	5,0		2,0
relative pitch t/d_o		1,5	2,0		1,2
number of tubes/cavity		1100	300	300	96

		IRC	SRC-G	SRC-L
av.heat transfer coefficient	W/(m² K)	600	450	8000
max.tube temp.	°C	750	850	710

Table 2: Main data of the investigated receiver concepts

	IRC	SRC-G	SRC-L
Symbol:	□	◣	◿
Type:	Prereactor integrated	Prereactor	separated
Coolant:	Process Gas	Gas (air or helium)	Liquid (salt or metal)
Receiver operation	constant gas temperatures, constant total thermal power (foss.burners)	constant coolant temperatures, thermal power dependent on solar radiation	
Daily thermal energy delivery from solar radiation (march 21., Almeria, no clouds), $TJ = 10^{12}$ J	1,30	1,32	1,36
Total installed mirror area m^2 (for 50 MW thermal power from solar input at march 21, 12 h)	97.700	98.100	90.400
$\dfrac{\text{Daily thermal energy}}{\text{Installed mirror area}}$, $\frac{MJ}{m^2}$	13,3	13,5	15,0

Table 3: Thermal energy delivery for operation on March 21st

Material	Melting-temp. /°C/	Boiling-temp. /°C/	Density /kg m⁻³/	Heat-cond. /Wm⁻¹ K⁻¹/	Spec. heat cap. /kWh kg⁻¹K⁻¹/	Dyn. visc. /g m⁻¹ s⁻¹/	Price /DM t⁻¹/	Remarks
Na	98	887	900	70	$3,6 \cdot 10^{-4}$	0,28	5.000,--	
Mass-% Na-K 40 - 60	5	Na: 887 K: 774	877	26	$2,4 \cdot 10^{-4}$	0,24	50.000,--	
Mass-% CuCl-KCl 72 - 28	150	CuCl: 1490 KCl: 1411	2900	**	$2 \cdot 10^{-4}$	1,4	5.000,--	
GS 250 Alkali Hydroxide (≈30 %) Alkali Chloride	250	Alk-OH: ≈1300 Alk-Cl: ≈1400	1700	**	*	*	1.900,--	Hot quenching furnace salt, manufacturer: Houghton, exact mixture unknown
GS 345 Alkali Hydroxide (≈30 %) Alkali Chloride Alkali Carbonate (few)	345	Alk-CO3: decomposes before boiling	1900	**	*	*	1.600,--	- dito -

* no data available until now, should be similar to CuCl-KCl

** no data available until now, some data of other chloride salts show a somewhat reduced heat conductivity in comparison to nitrate salts, about 0,3 W m⁻¹ K⁻¹ are expected

Table 4: Compilation of properties and prices of meltings

Criteria under consideration	Na	Na-K	KCl-CuCl	GS 250, GS 345
Stability of medium up to temperatures of 750 °C	stable	stable, boiling point (BP) of potassium: BP = 774 °C nearly reached	stable	stable at least up to 700 °C
Corrosion	Long term experiences at operational temperatures above 600 °C are not available			
	some investigations at elevated temperatures are available further investigations of corrosion will be necessary			Hydroxides have a serious corrosion effect
Hazards	Explosive reaction with water Reaction of Na, K with a mixture of CH_4, H_2O, CO, CO_2, H_2 must be expected to be very violent. Precautions to limit the hazard must be developed		no hazards are known	
Effects to SSPS experimental plant	Heat transfer system could be used partially (must be further checked)		A new heat transfer system is required, but only a small storage seems to be necessary.	

Table 5: Rough assessment of meltings from an operational point of view

- 49 -

Type of receiver	Prereactor integrated	Prereactor separated	
	process gas	gas (air or helium)	liquid (molten salt or metal)
Coolant	process gas	gas (air or helium)	liquid (molten salt or metal)
Main components to be arranged at the tower	Integrated prereactor/receiver burner-system, main reactor, exhaust-heat boiler	Receiver	Receiver
Connections to the plant located on ground level	supply lines (water, CH_4, fuel)	heat transfer gas lines: 440 °C up-stream, 800 °C down-stream	lines for melting transport: 430 °C up-stream, 700 °C down-stream
Required heliostat area	97.700 m²	98.100 m²	90.400 m²
Mean solar energy gathered per clear day	361 MWh	366 MWh	378 MWh
Energy equivalent of fuel to be consumed for circulation of coolant		Air: 137 MWh Helium: 31 MWh	30 MWh
Mean fossil energy saved per clear day	361 MWh	Air: 229 MWh Helium: 335 MWh	348 MWh
Fuel saving related to the fuel consumption of a pure fossil fuel fired plant	CH_3OH 16.5 % NH_3 28.6 %	CH_3OH Air 16.6 % Helium 24.1 % NH_3 Air 21.5 % Helium 31.5 %	CH_3OH 19.8 % NH_3 33.2 %

Table 6: Comparison of different solar integration schemes

Table 7 compares three solar integration schemes. The table is laid out with three main columns.

Prereactor Integrated IRC	Gas-Cooled Separated Receiver SRC - G	Liquid-Cooled Separated Receiver SRC - L
Advantages	**Advantages**	**Advantages**
· no energy need for circulation of a primary loop coolant · constant load operation by additional combustion directly in the cavity · low material and cavity temperatures by heating up the process gas itself	· technology developed in the GAST-project can be used · Relative low risk concerned with component realization · only the receiver has to be located at the tower	· low energy need for coolant circulation · only the receiver has to be located at the tower · small connection lines between receiver and plant · required heliostat area rel. small · good storage potential
Disadvantages	**Disadvantages**	**Disadvantages**
· no storage potential · relative high effort for control · main components to be located on tower	· high energy consumption for coolant circulation in the case of air as coolant (no serious disadvantage in the case of the He and H_2 (which is produced in the plant)) · high material and cavity temperatures · long hot gas lines between receiver and plant	· trace heating necessary (not regarded in energy considerations) · high hazard potential (in case of Na or Na-K)
Risks/Development Effort	**Risks/Development Effort**	**Risks/Development Effort**
· combination of solar and fossil fuel heating within the same cavity not realized before	· low, adaption from GAST-project	· corrosion investigations · safety investigations · adaptions of valves and pumps for the hot leg of the loop

Table 7: Discussion of the different solar integration schemes

- 51 -

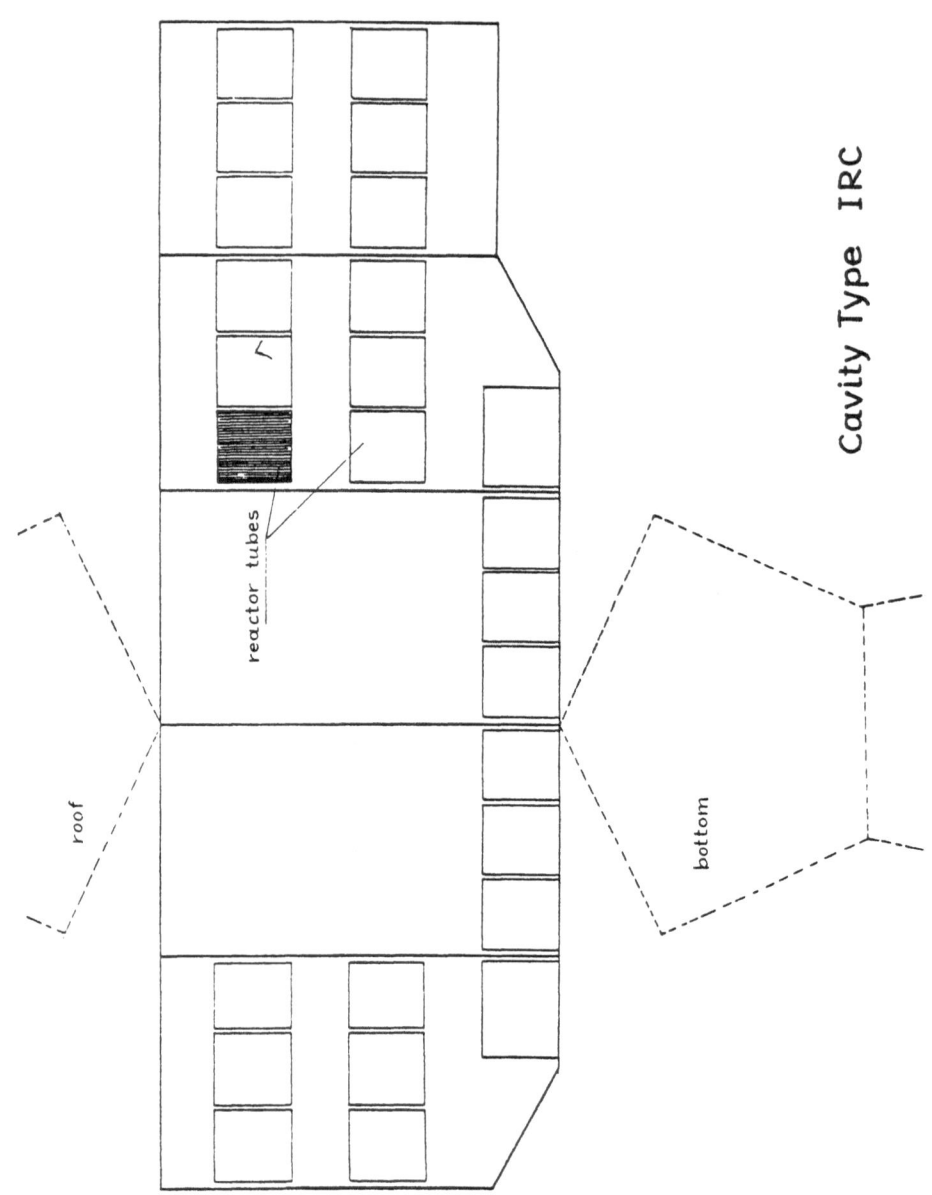

Cavity Type IRC

Fig. 1: Arrangement of tubes in the IRC-type receiver

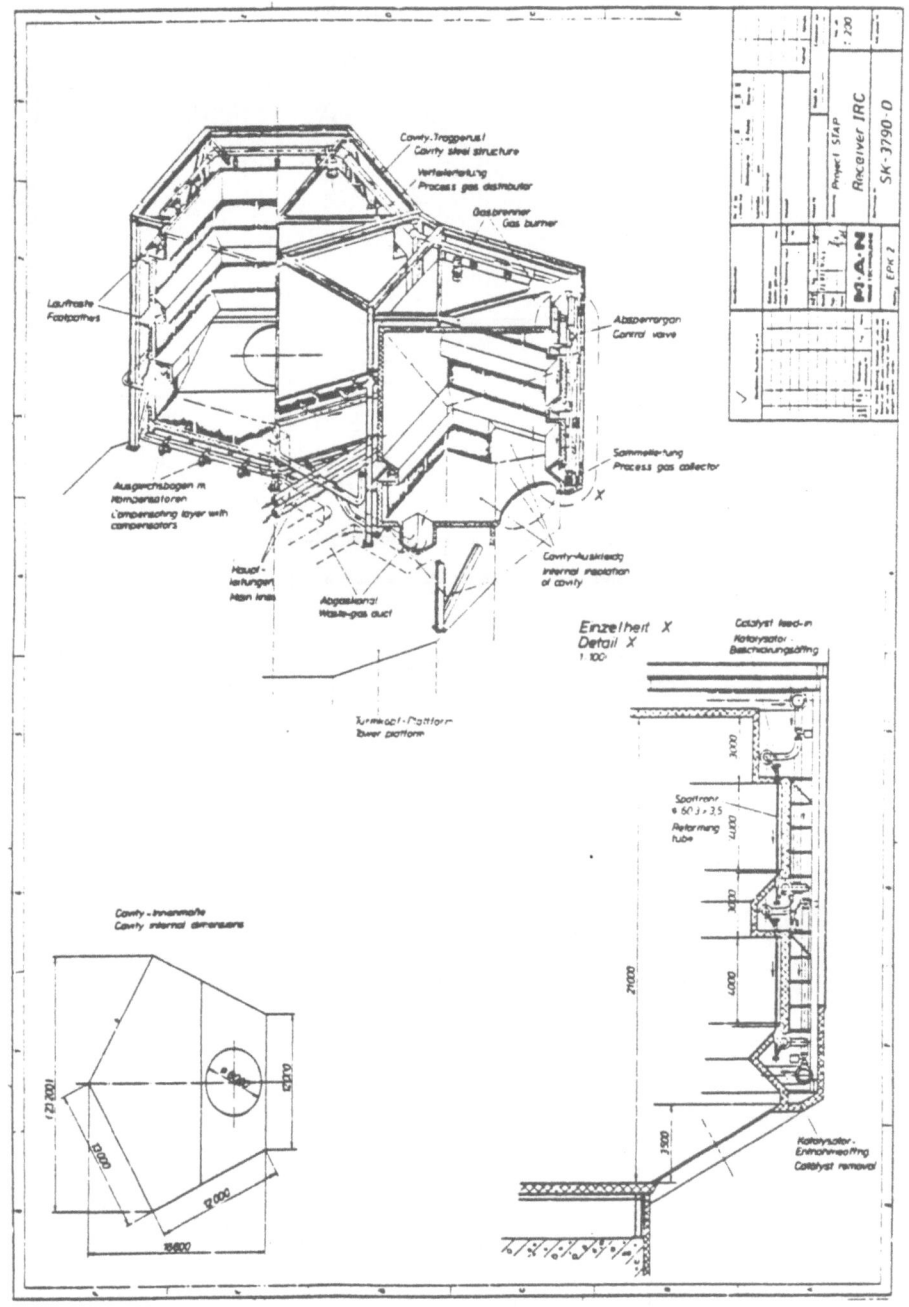

Fig. 2: Sketch of the IRC-type receiver

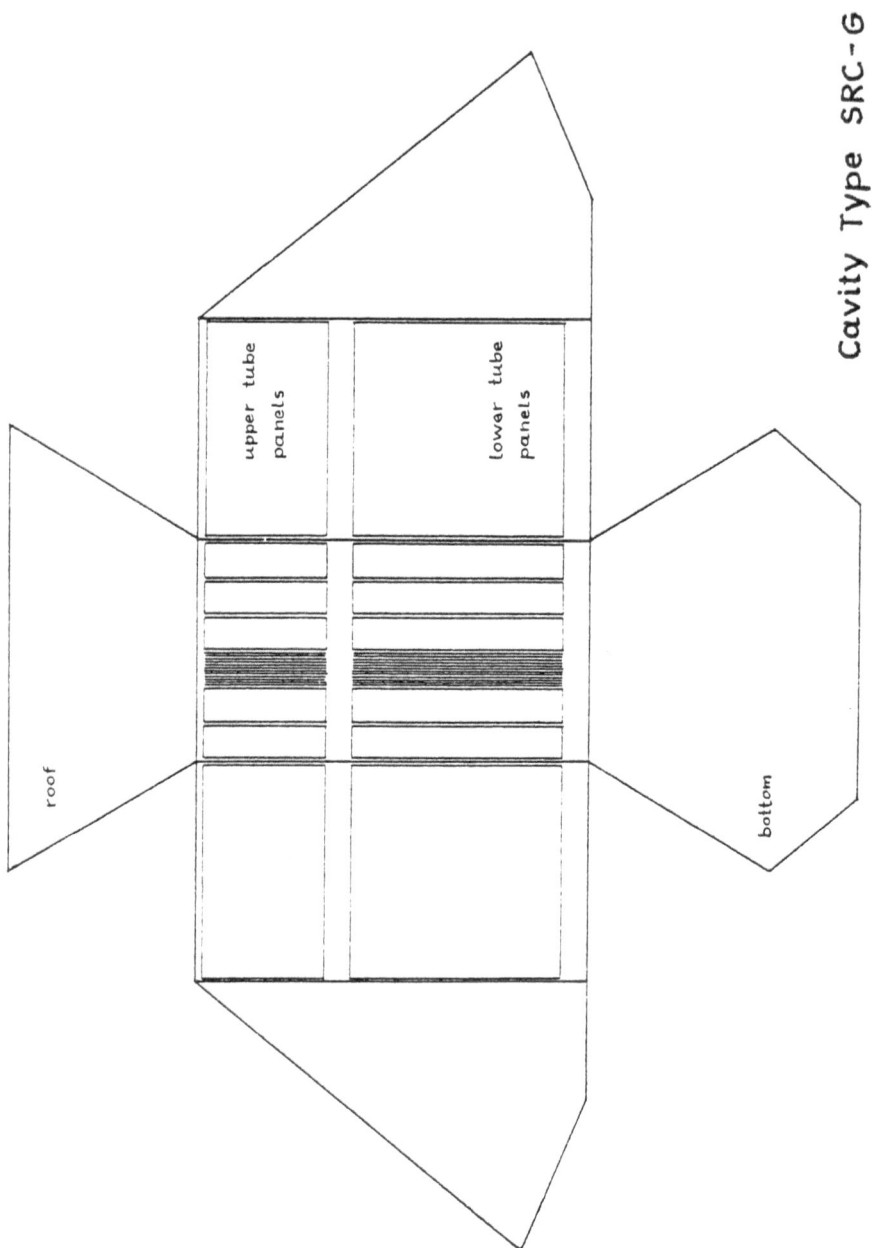

Fig. 3: Arrangement of tubes in the SRC-G type receiver

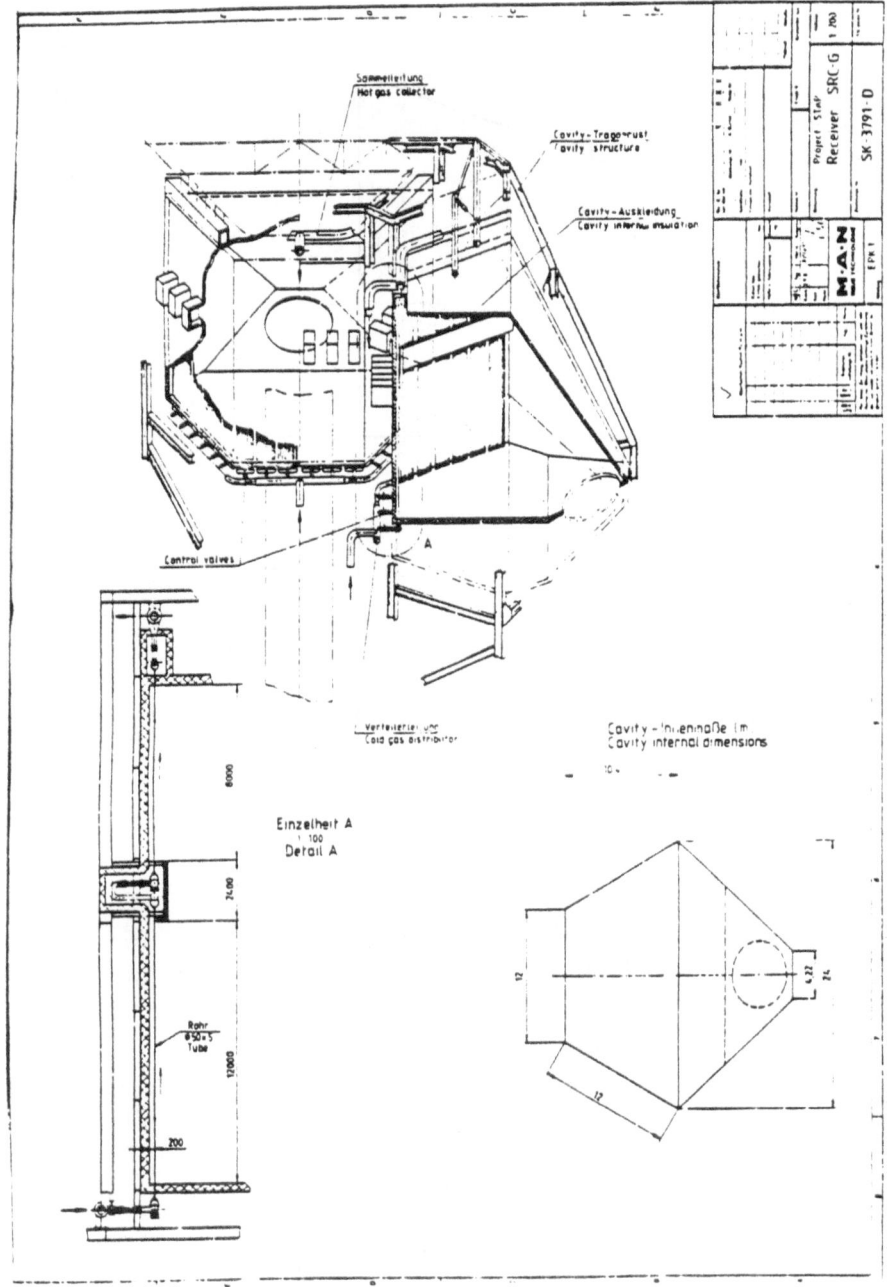

Fig. 4: Sketch of the SRC-G type receiver

- 55 -

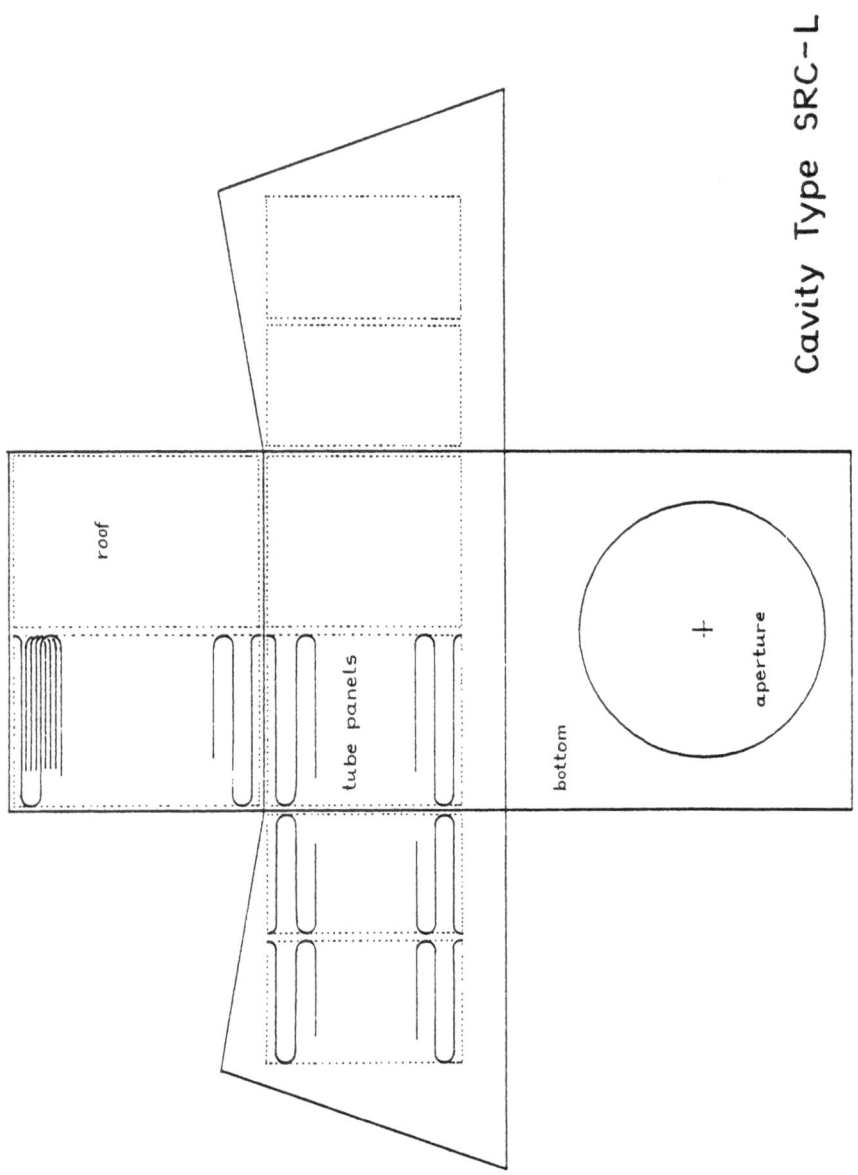

Cavity Type SRC-L

Fig. 5: Arrangement of tubes in the SRC-L type receiver

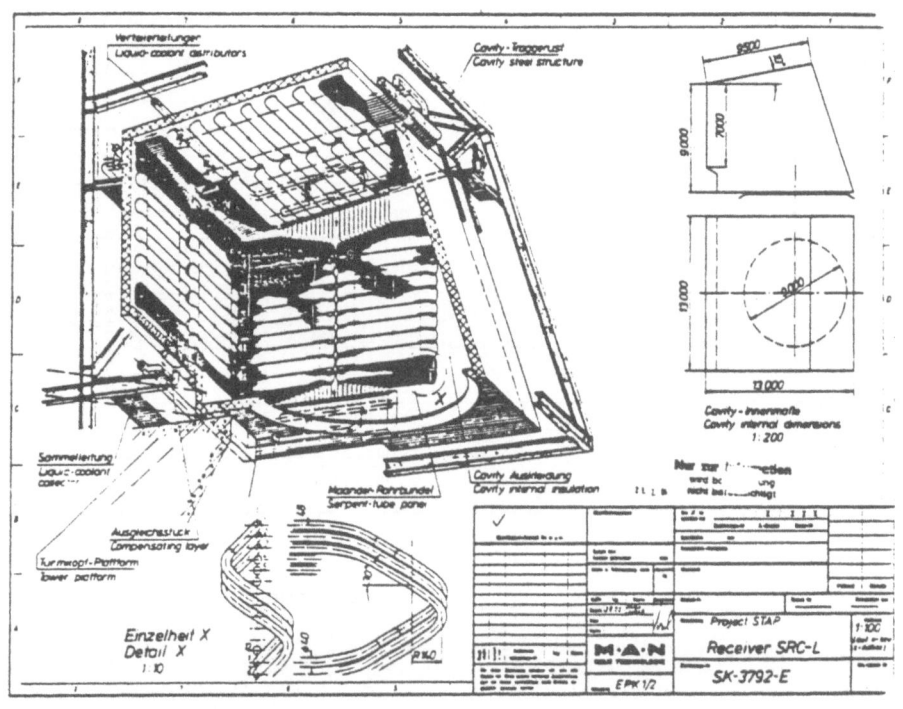

Fig. 6: Sketch of the SRC-L type receiver

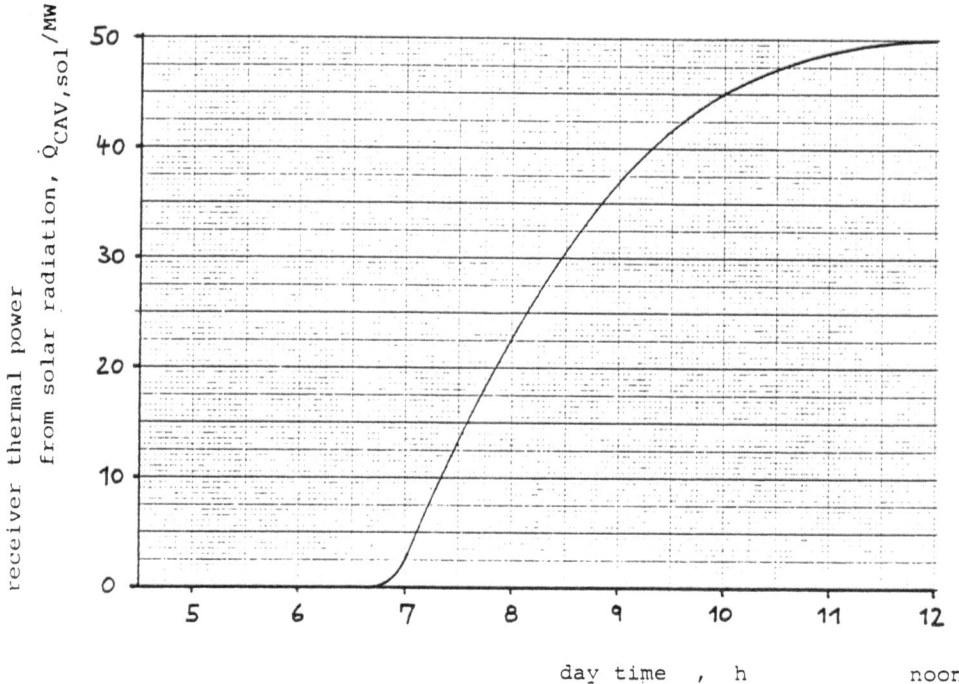

Fig. 7: Cavity Type IRC (integrated reactor)

Daily thermal energy delivery from solar radiation:

$$Q = 1,30 \text{ TJ} = 361 \text{ MWh}$$

(for march 21st, Almeria, no clouds)
Installed mirror area in heliostat field: 97.700 m²

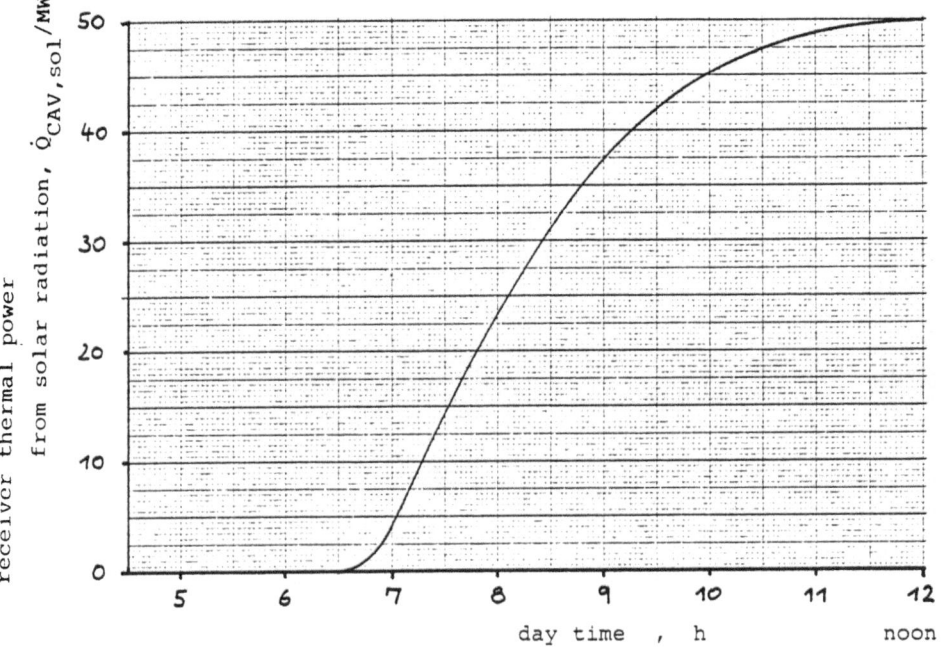

Fig. 8 Cavity Type SRC-G (separated reactor, gas cooled)

Daily thermal energy delivery from solar radiation:

$$Q = 1,32 \text{ TJ} = 366 \text{ MWh}$$

(for march 21st, Almeria, no clouds)
Installed mirror area in heliostat field: 98.100 m²

- 59 -

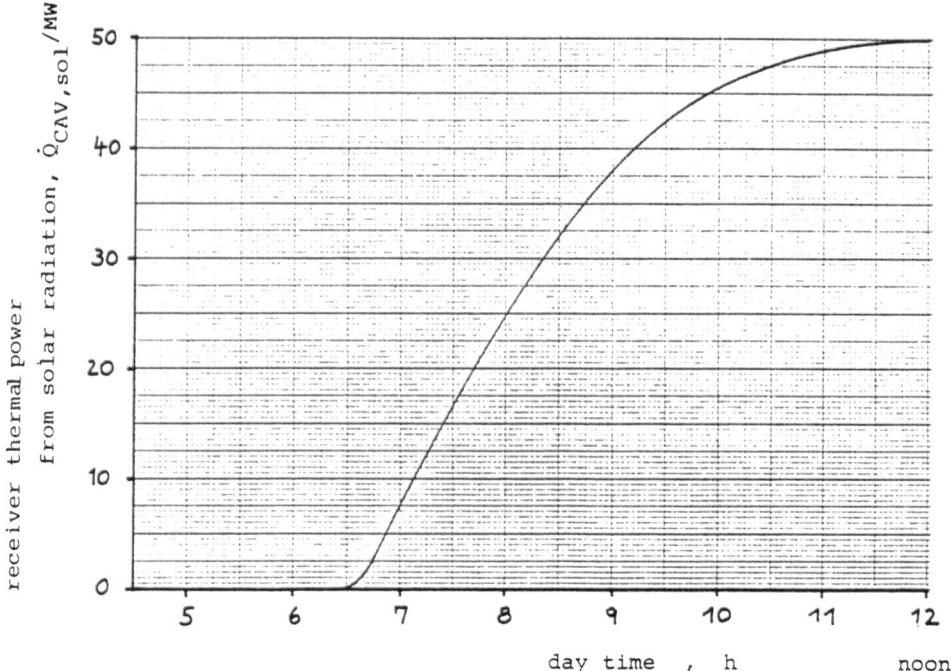

Fig. 9: Cavity Type SRC-L (separated reactor, liquid cooled)

Daily thermal energy delivery from solar radiation:

$$Q = 1,36 \text{ TJ} = 378 \text{ MWh}$$

(for march 21st, Almeria, no clouds)
Installed mirror area in heliostat field: 90.400 m²

- 60 -

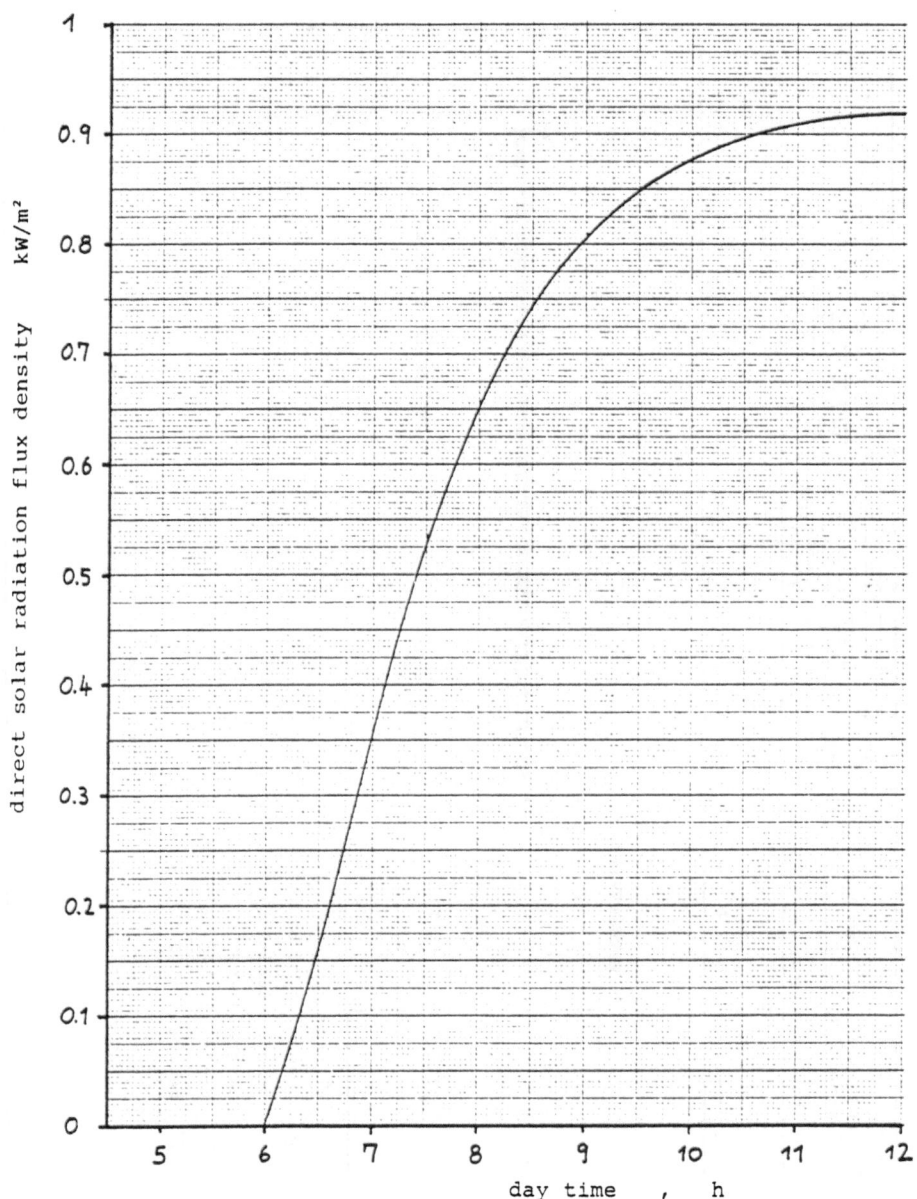

Fig. 10: Insolation on a surface normal to direct radiation direction at the 21st of march, lattitude of Almeria, Spain

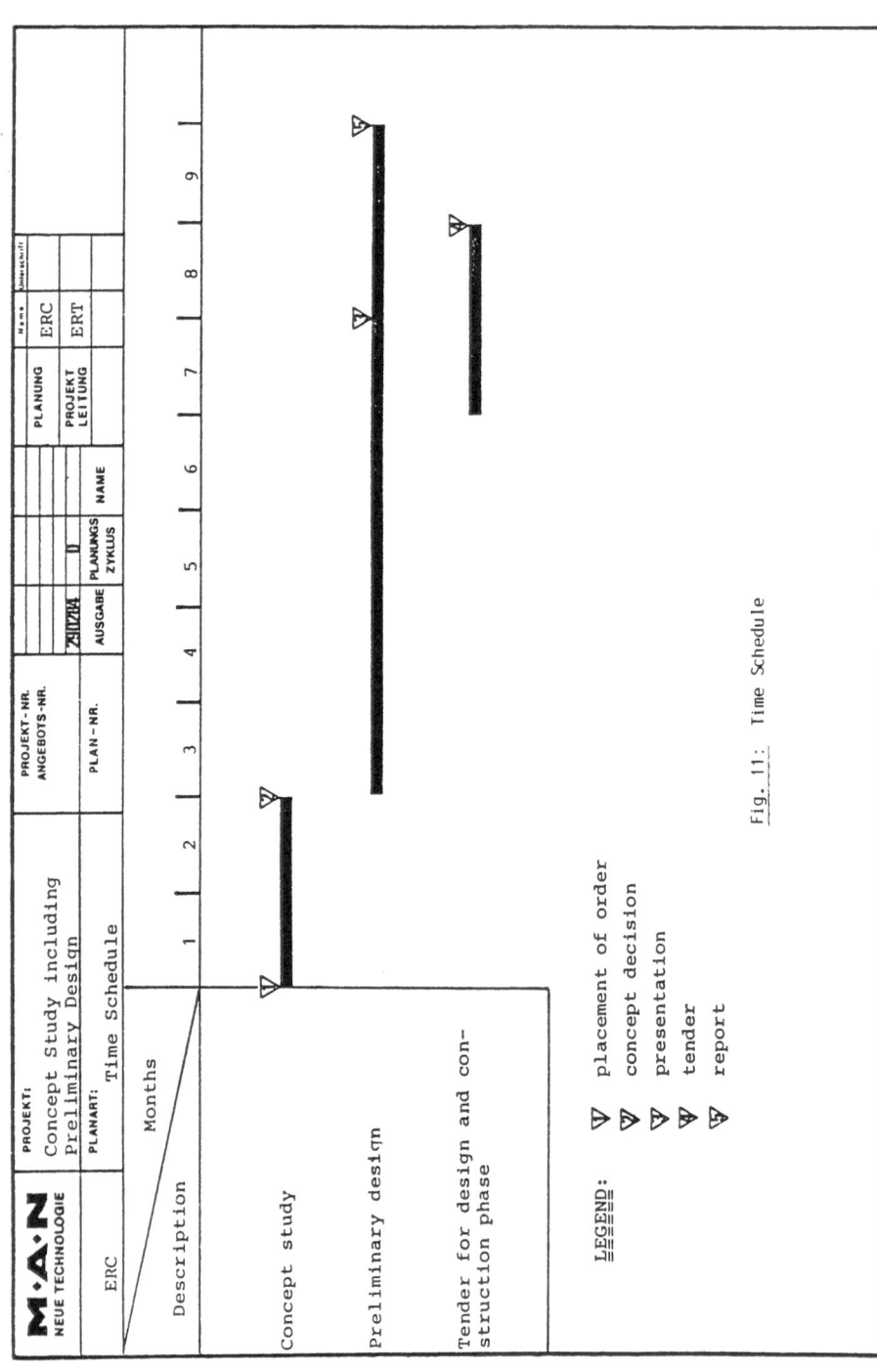

Fig. 11: Time Schedule

As will be substantiated in chapter 8, MAN and LURGI propose to base continuing studies on system B. Figure 13 shows the elements, which would have to be built in a testing plant - as at Almeria, Spain - to receive meaningful results for extrapolation for future plants.

As the reformer in concept B is of conventional design, no need can be seen to demonstrate its performance with additional energy expense. If the prereactor can be run continuously with a variation of solar and fossil energy supply, the behaviour of the fossil fired reformer can be extrapolated due to available experience.

Even with a design of about 3 MW thermal energy only to the prereactor about 60.000 m^3_N/d natural gas will be used by the process on continuous operation. This gas could be flared, but the expense for the supply of natural gas will be high and it even may be critical to get such quantities at the test location.

It therefore will be recommended to the customer to install additionally a methanation plant, which will allow to synthesize again "natural gas" from the prereactor's effluent gas. This plant of course will increase the investment cost but will save about 50.000 m^3_N/d of natural gas. Even other fuel than natural gas may then be taken for the supply of fossil firing. Thus the investment will pay out in a very short time. Further details may ve developped during continuations of this study.

a) The comparative tables III and IV with their
    specific data would favour a system as D Ia or D II,
    which are those with molten salt storage systems.
    The other systems with acceptable specific con-
    sumption figures differ by less than 2 % only
    amongst each other.

b) It has been shown in the preceeding chapters that
    these molten salt systems include the highest
    portion of unsolved problems and risks.

c) It has to be recognized that the absolute amount
    of energy, which is substituted by the sun, is
    equal in all discussed systems. The gain in costs
    therefore will be equal in all cases, independent
    from the absolute amount of product, which is
    manufactured in the plant.
    Therefore the absolute amount of profit will be
    the highest for the lowest investment for a given
    solar energy rate, but storage systems in any case
    will sensibly raise the costs of the solar part.

d) The comparable costs of a solar heated plant must
    include all equipment, which is necessary to couple
    the solar energy into the process. Thus the pre-
    reactor and the heat carrier system has to be in-
    cluded in the scope of calculations. It can be
    shown that the costs of these items will amount
    to only about 20 % of the other costs of the
    solar part (mirror field and receiver).

The type of prereactor as studied in figures
14 to 17 therefore hardly influences the comparable costs of a solar heated plant, only the
technical feasibility should guide a selection.

e) The energetically most interesting systems A I b,
   D I a and D II need storage systems but depend on
   small chemical plants as following plants as well.
   The small chemical plants suffer from specific
   relatively high investment costs. If one adds the
   cost of the solar storage system to the cost of
   the chemical plant one probably will reach a total
   investment nearly equal to that one, which is
   necessary for the systems B and C without storage
   but with bigger chemical plants. Thus it can be
   expected that the apparent energetic advantage is
   no advantage economically, the attraction of the
   storage systems is lost then immediately.

f) With all the above reasons in mind, MAN and LURGI
   propose to investigate predimonantly

                     System B

in continuing studies.
This system promises enough specific solar saving
on one side and contains little risks on the other
side as it seems to be studied most diligently and
comparably to other projects. A benefit for further
evaluations of this type of reforming will be that
experience is available from the plant ADAM II/EVA II
at Kernforschungsanlage Jülich, which has been
built by LURGI and which has proven the reliability
of such a design.

The receiver type, which will be necessary, is
well studied within the GAST-project and its
integration into the overall system should be
possible in a shorter period than for other
receiver type.

g) As integrated solution also claims high interest,
it should be studied in continuing investigations,
whether the fossil firing of the cavity should be
included in the experiment at Almeria. This would
mean to have the options that the heat carrying
air would be heated up either in the receiver by
solar radiation, or by indirect heat exchange with
flue gas in the flue gas duct or in the receiver by
heat exchange with fossil generated flames/flue
gases, which would be sent to the stack without
heat exchange in such special experiment.
If the general behaviour of such a system could be
demonstrated and well proven in this scale, future
plants could then be based also on variant C. In
the judgement of the investigators, system C could
be more attractive only with regard to simpler
operating. Specific cost or energy advantages
cannot be seen for system C or any other directly
insolated reformer.

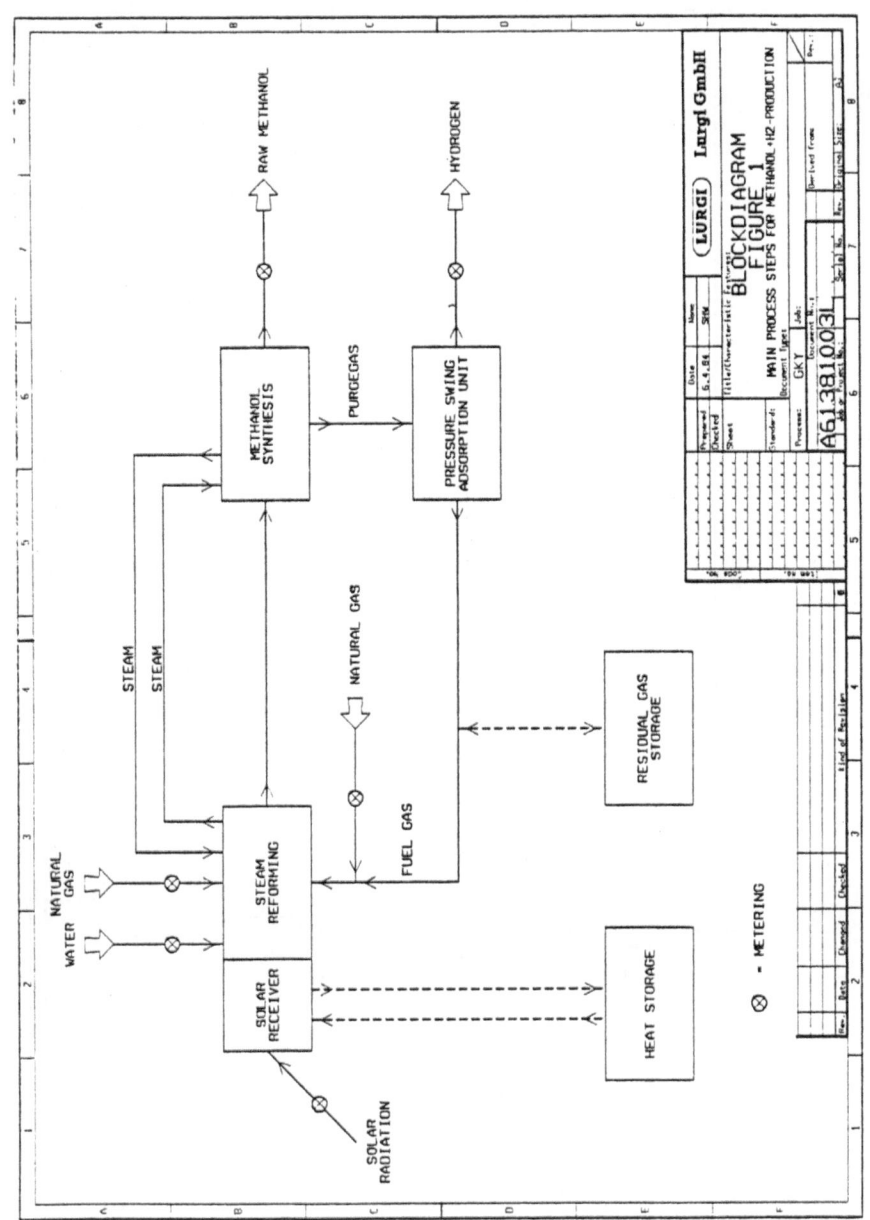

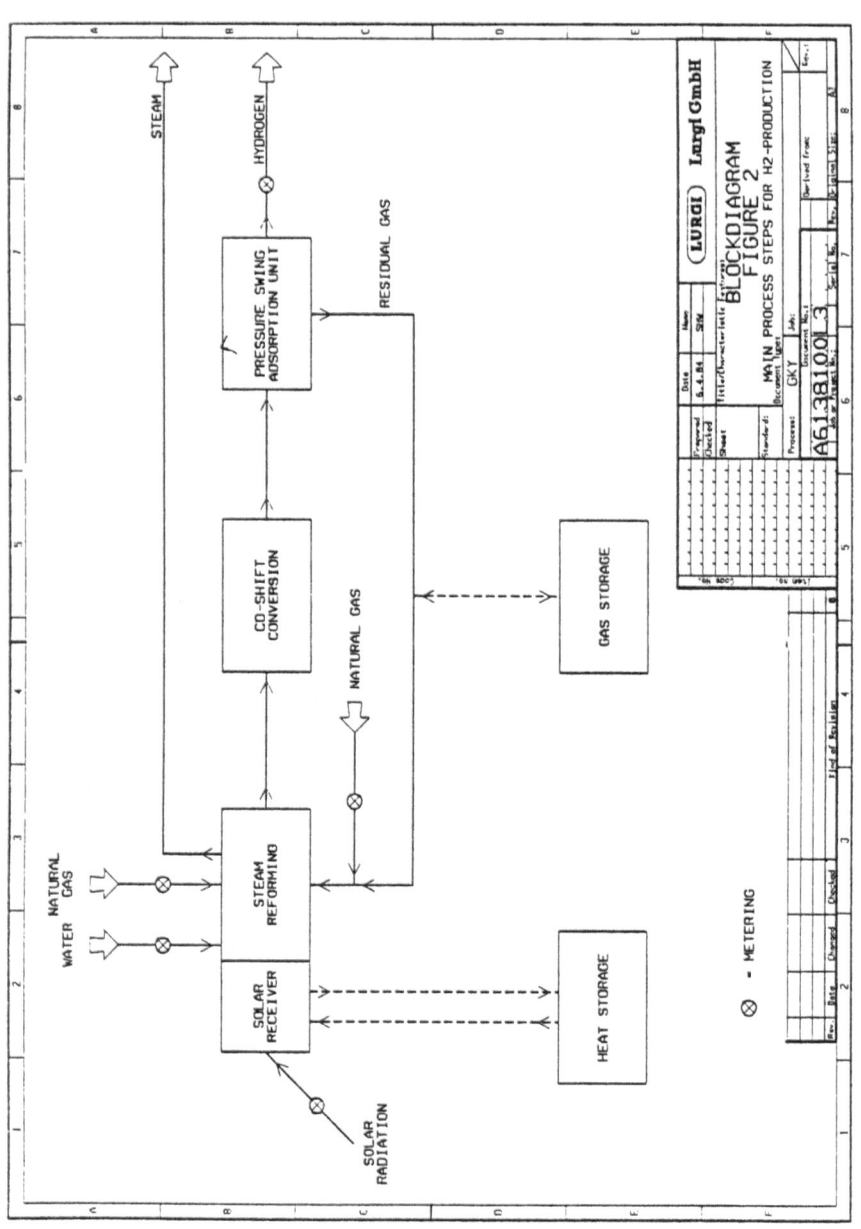

STEAM

HYDROGEN

RESIDUAL GAS

PRESSURE SWING ADSORPTION UNIT

CO-SHIFT CONVERSION

STEAM REFORMING

SOLAR RECEIVER

NATURAL GAS

WATER

NATURAL GAS

SOLAR RADIATION

GAS STORAGE

HEAT STORAGE

⊗ - METERING

LURGI   Lurgi GmbH

BLOCKDIAGRAM
FIGURE 2

MAIN PROCESS STEPS FOR H2-PRODUCTION

A6138100L.3

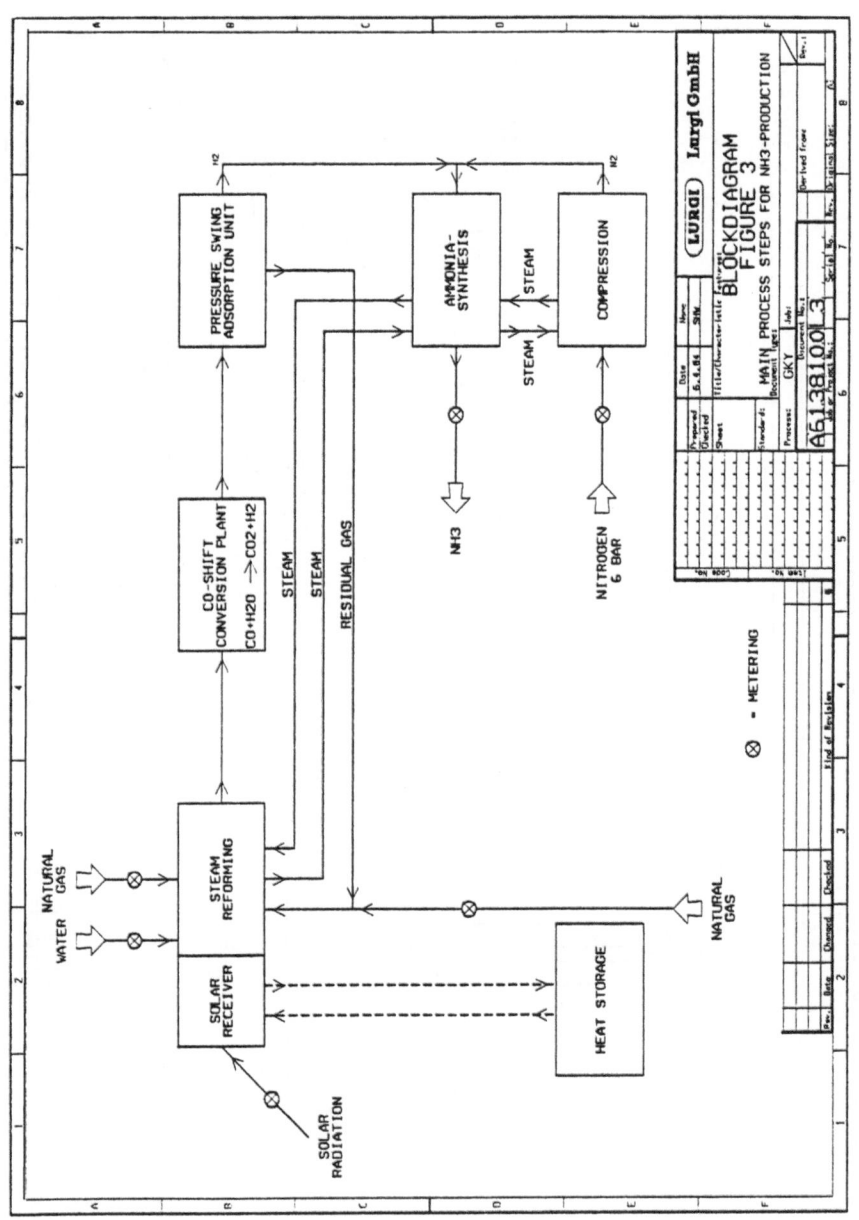

BLOCKDIAGRAM
FIGURE 3

MAIN PROCESS STEPS FOR NH3-PRODUCTION

LURGI Lurgi GmbH

A61.38100.3

- 69 -

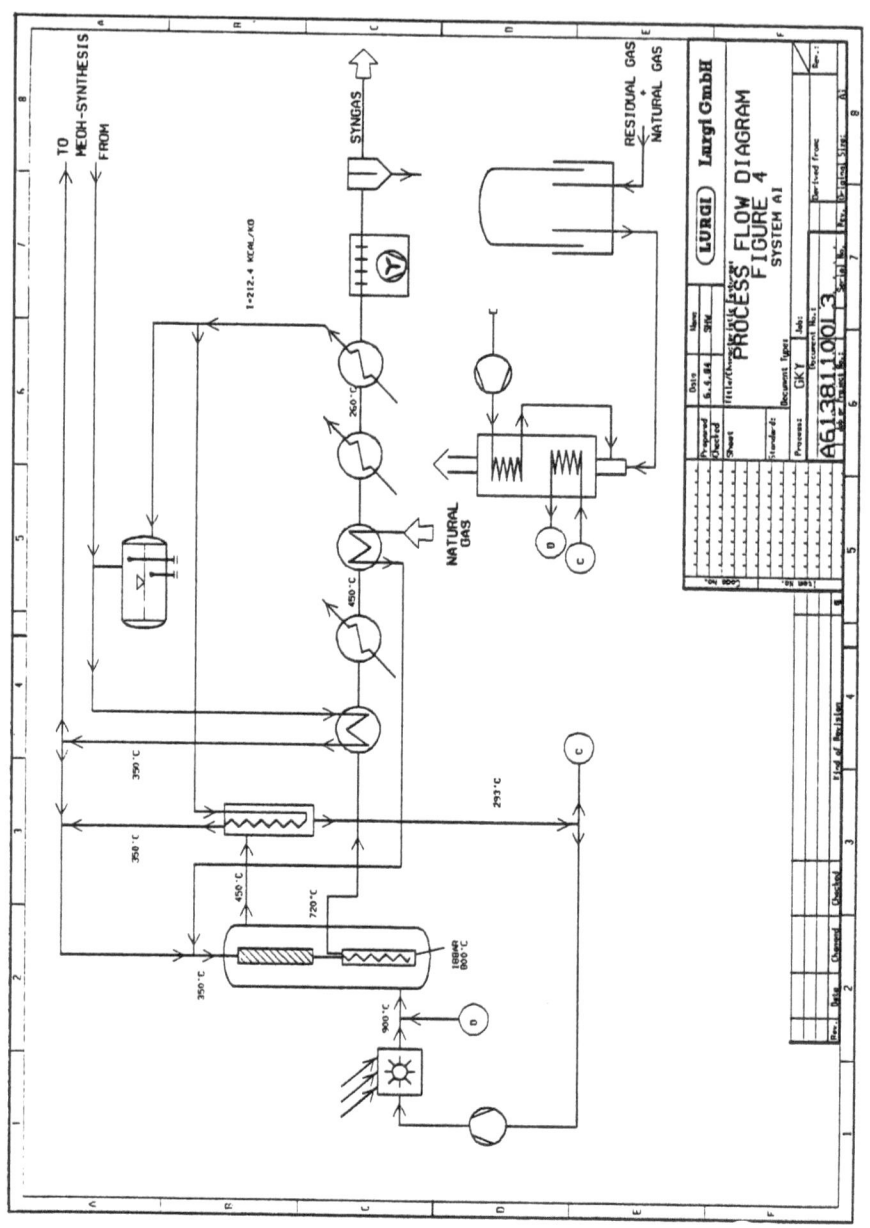

TO MECH-SYNTHESIS FROM

SYNGAS

1=212,4 KCAL/KG

NATURAL GAS

RESIDUAL GAS + NATURAL GAS

250°C

450°C

350°C

350°C

293°C

450°C

720°C

350°C

188BAR 800°C

900°C

LURGI Lurgi GmbH

PROCESS FLOW DIAGRAM
FIGURE 4
SYSTEM AI

A61.381.1001.3

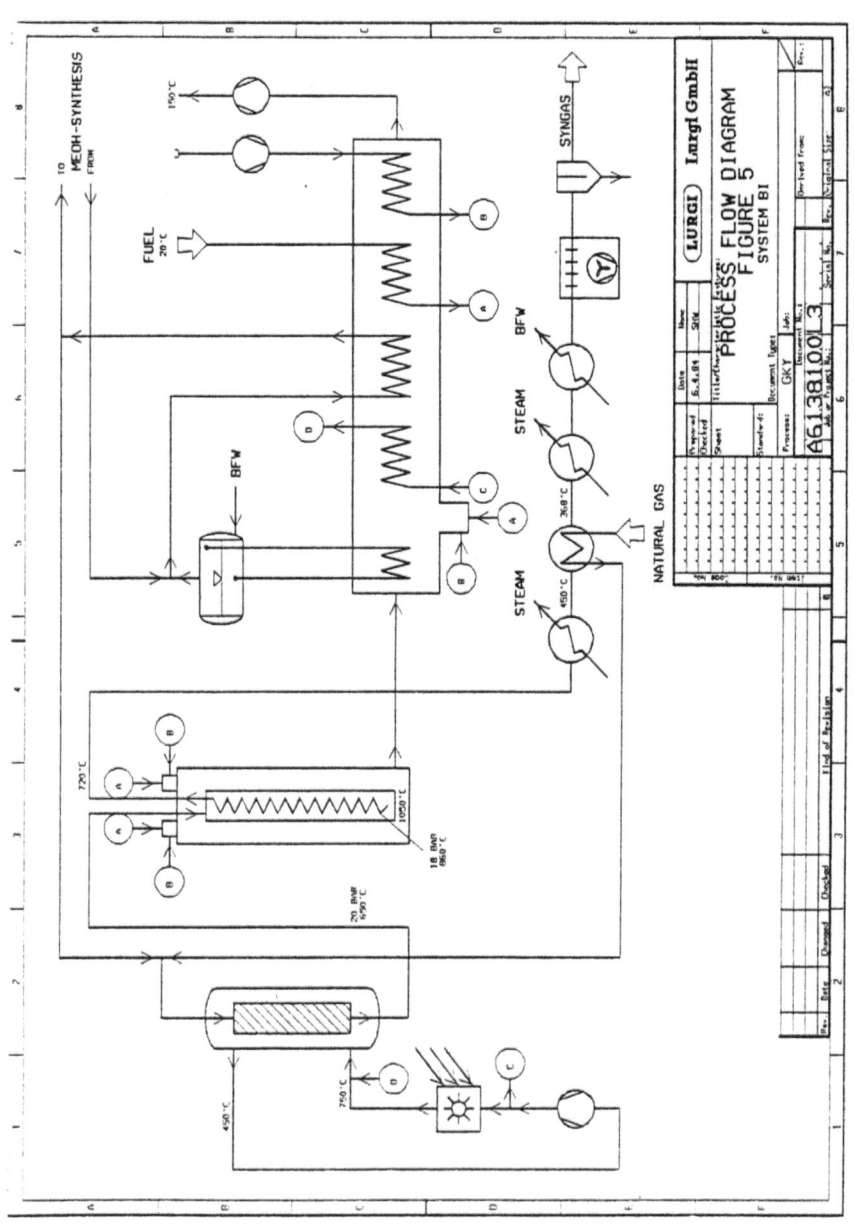

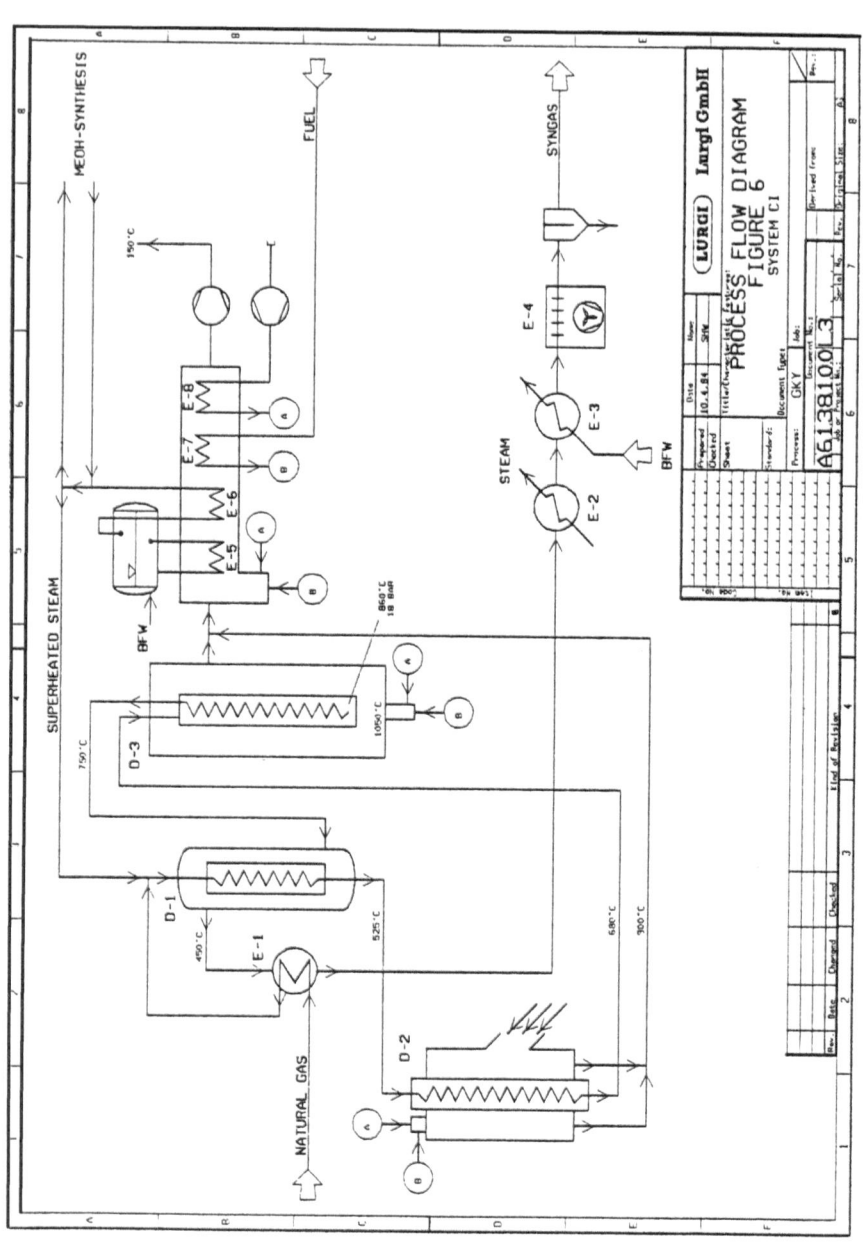

MEOH-SYNTHESIS

FUEL

SYNGAS

SUPERHEATED STEAM

NATURAL GAS

BFW

STEAM

BFW

E-8
E-7
E-6
E-5

E-4

E-3

E-2

E-1

D-1

D-2

D-3

150°C

750°C

490°C

525°C

680°C

900°C

1050°C

860°C
18 BAR

LURGI Lurgi GmbH

PROCESS FLOW DIAGRAM
FIGURE 6
SYSTEM C1

A6138100L3

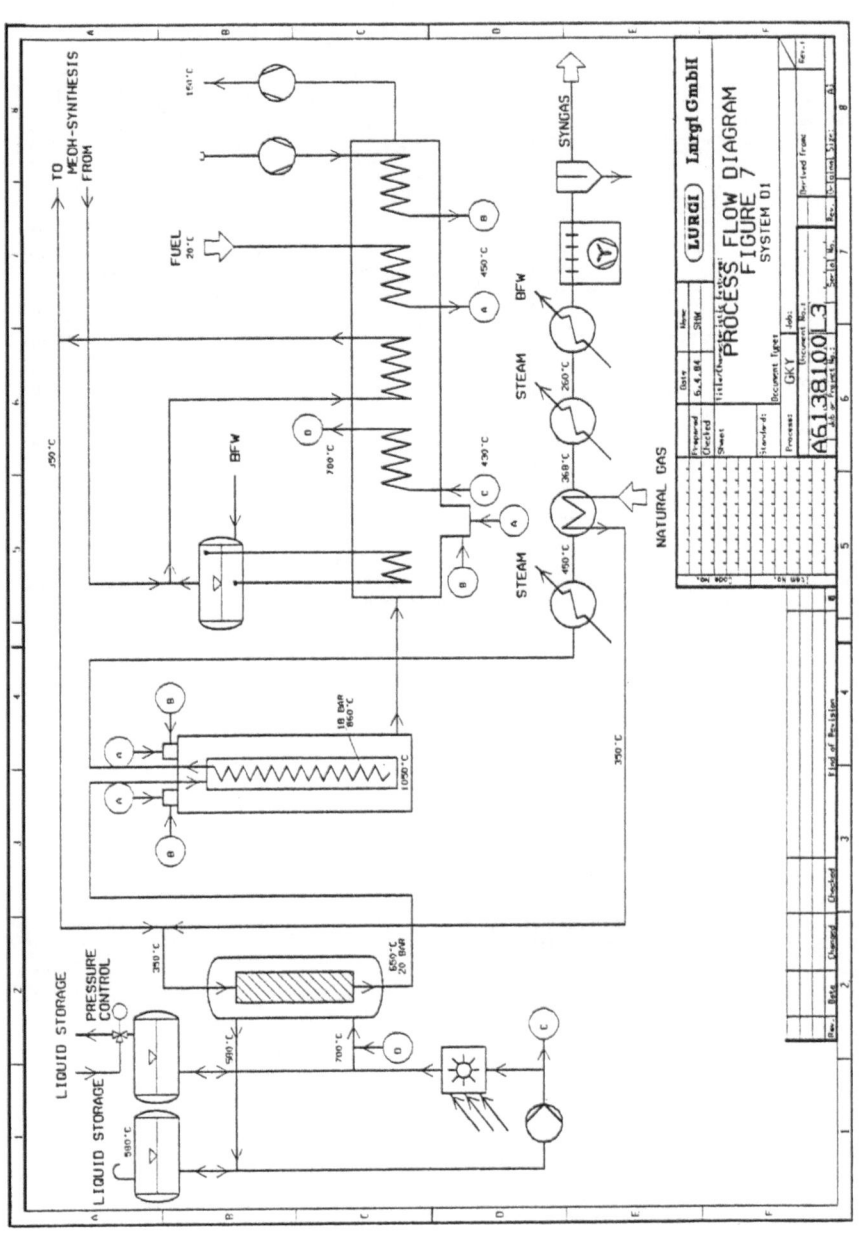

PROCESS FLOW DIAGRAM
FIGURE 7
SYSTEM 01

LURGI   Lurgi GmbH

A61.38100L3

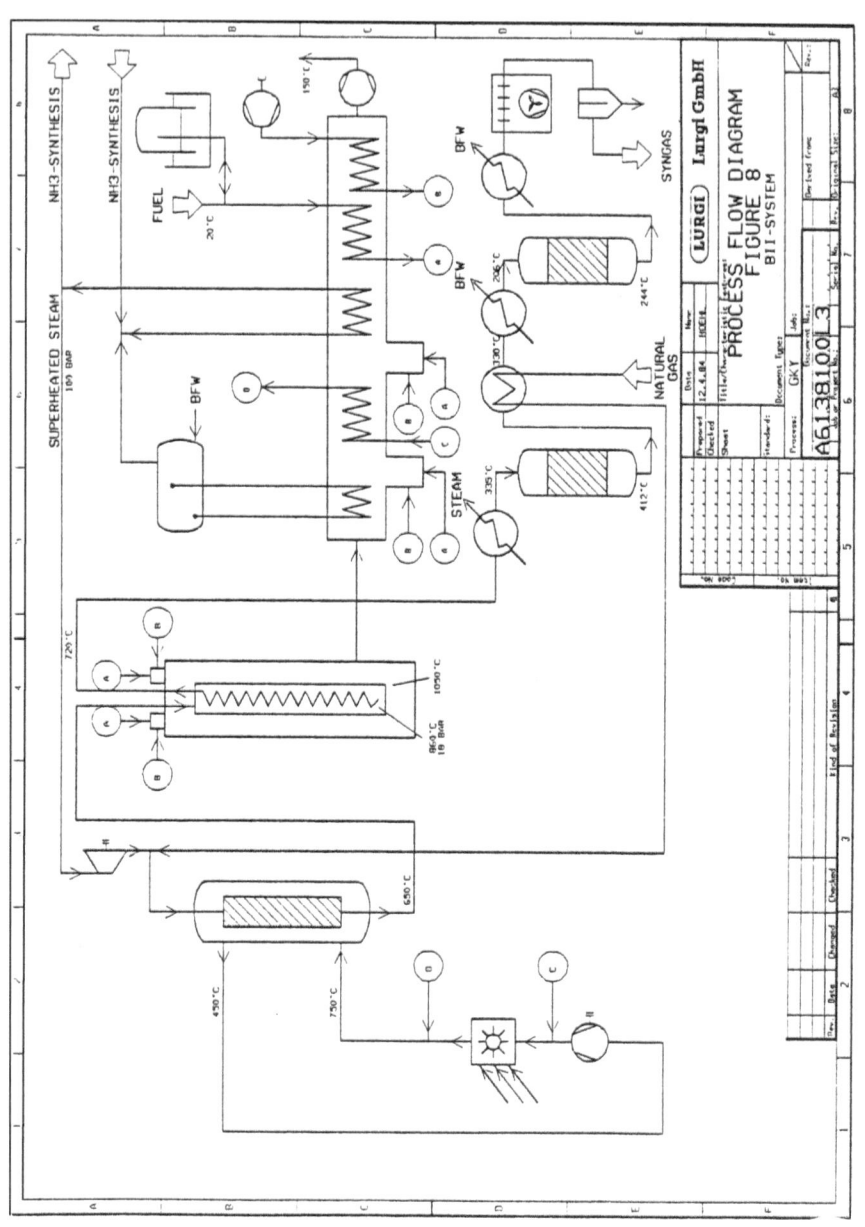

PROCESS FLOW DIAGRAM
FIGURE 8
BII-SYSTEM

LURGI GmbH

- 74 -

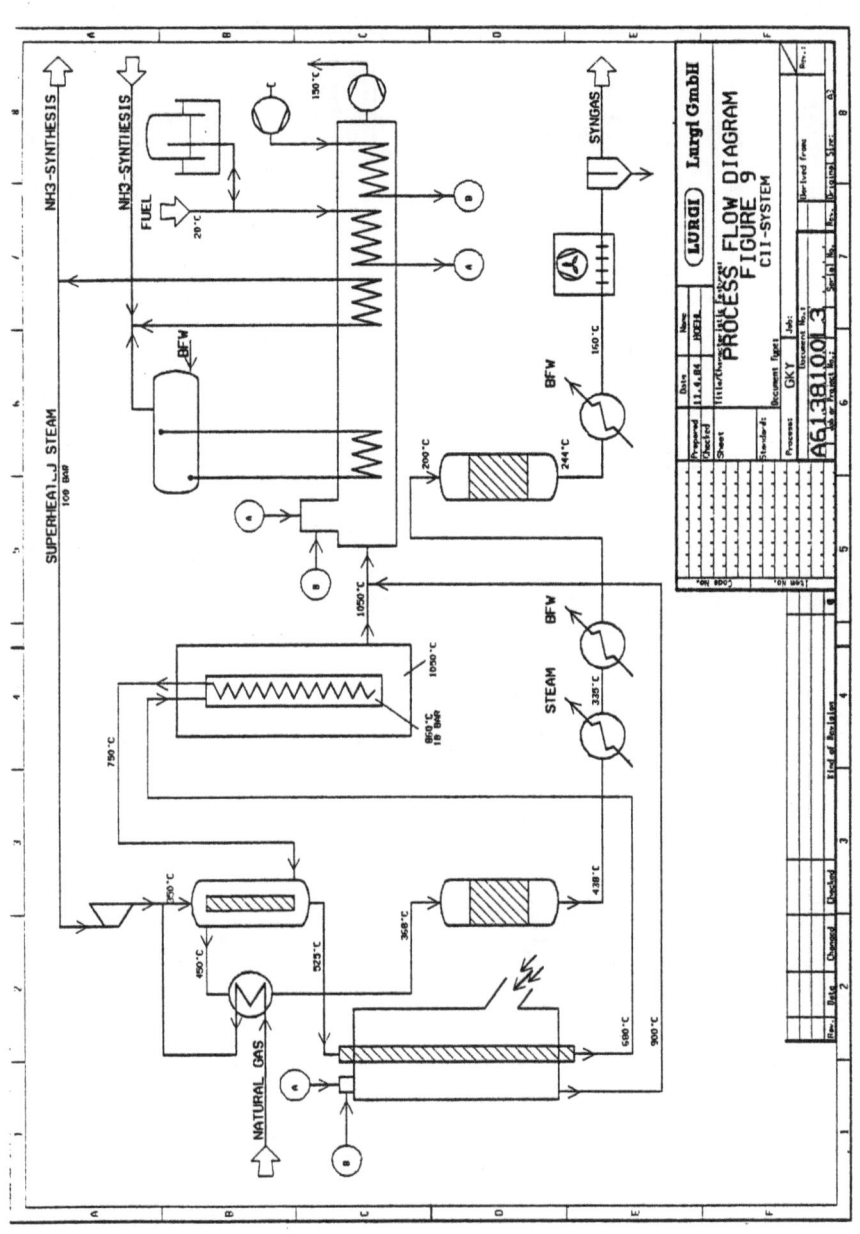

SUPERHEATED STEAM
100 BAR

NH3-SYNTHESIS

NH3-SYNTHESIS

FUEL

BFW

SYNGAS

NATURAL GAS

STEAM

BFW

150 °C
20 °C
200 °C
244 °C
160 °C
1050 °C
1050 °C
860 °C
10 BAR
790 °C
335 °C
438 °C
750 °C
450 °C
525 °C
368 °C
500 °C
900 °C

LURGI Lurgi GmbH

PROCESS FLOW DIAGRAM
FIGURE 9
C11-SYSTEM

A61.38100.3

- 75 -

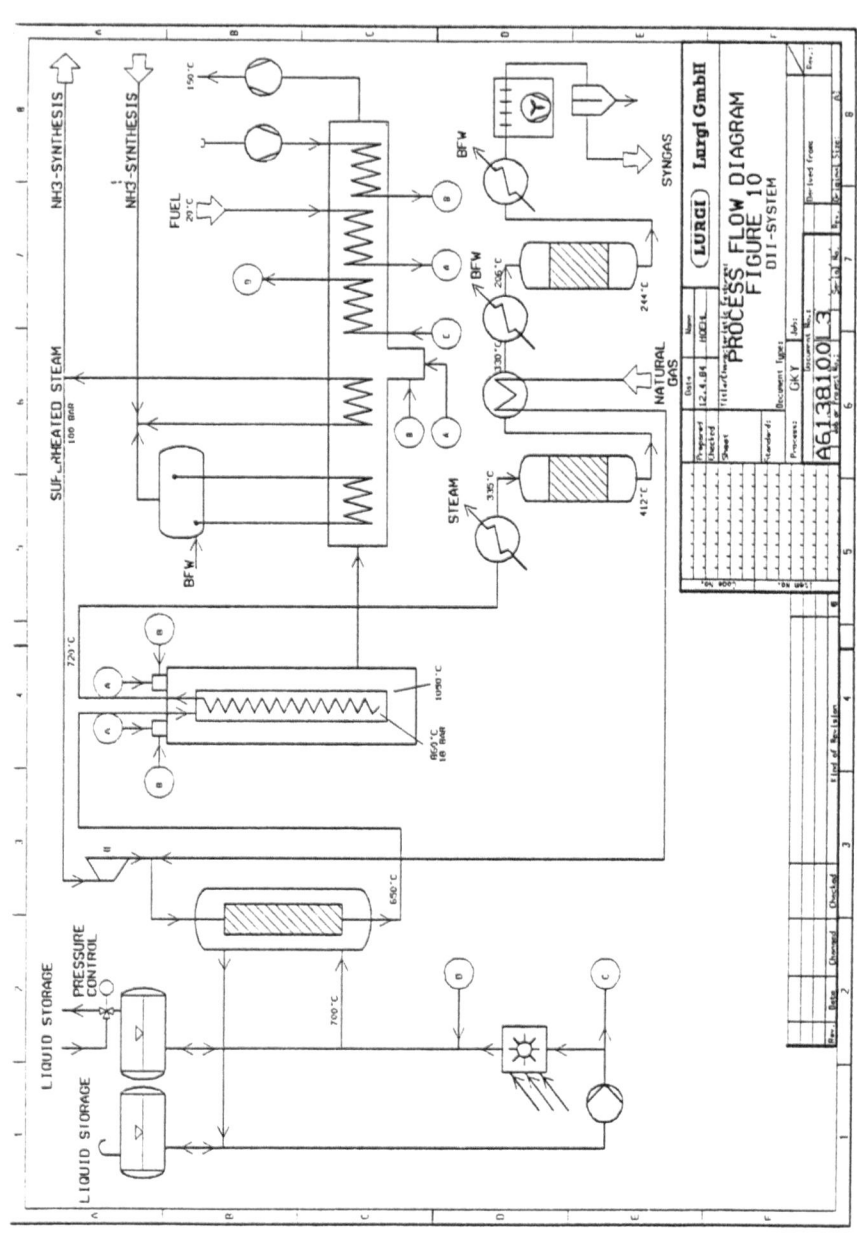

PROCESS FLOW DIAGRAM
FIGURE 10
D11-SYSTEM

LURGI Lurgi GmbH

A6138100 3

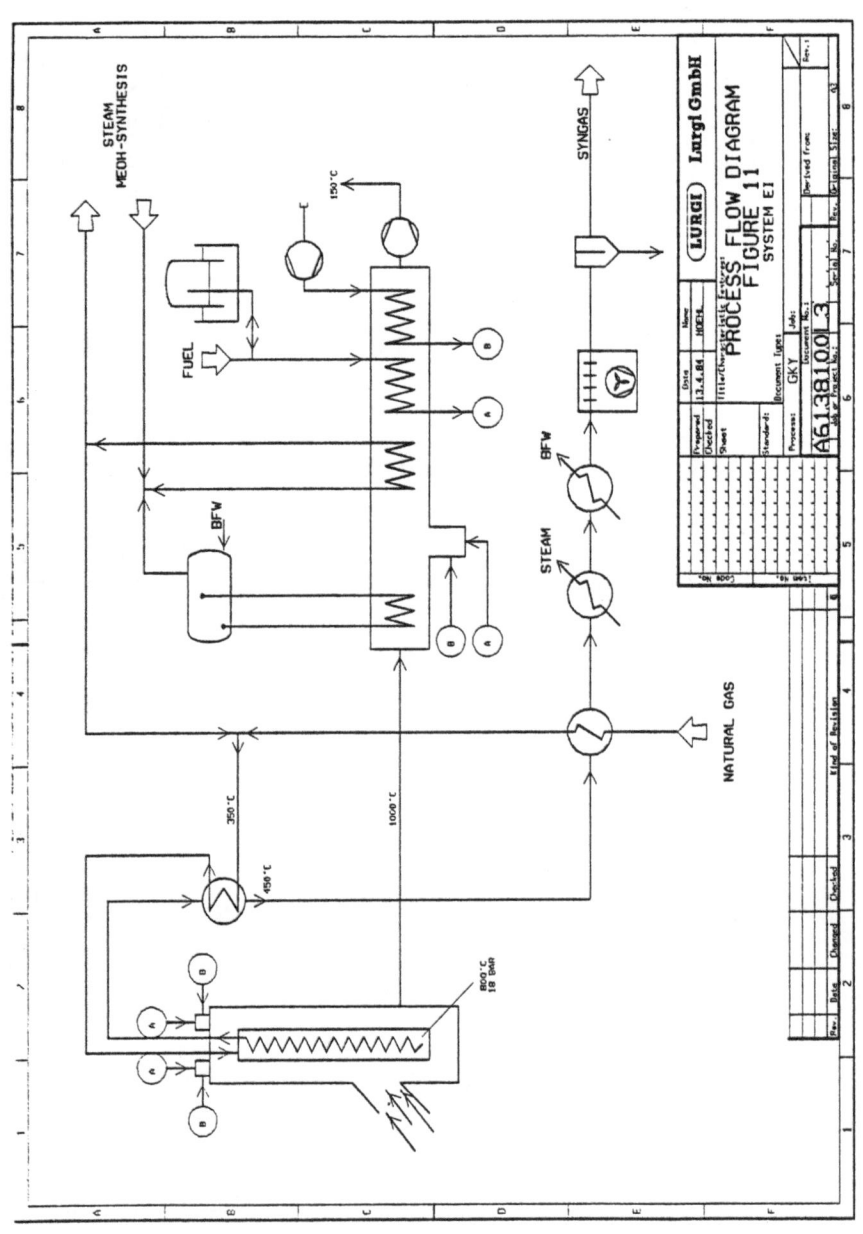

STEAM
MEOH-SYNTHESIS

FUEL

BFW

150 °C

SYNGAS

BFW

STEAM

NATURAL GAS

350 °C

450 °C

1000 °C

800 °C
18 BAR

- 77 -

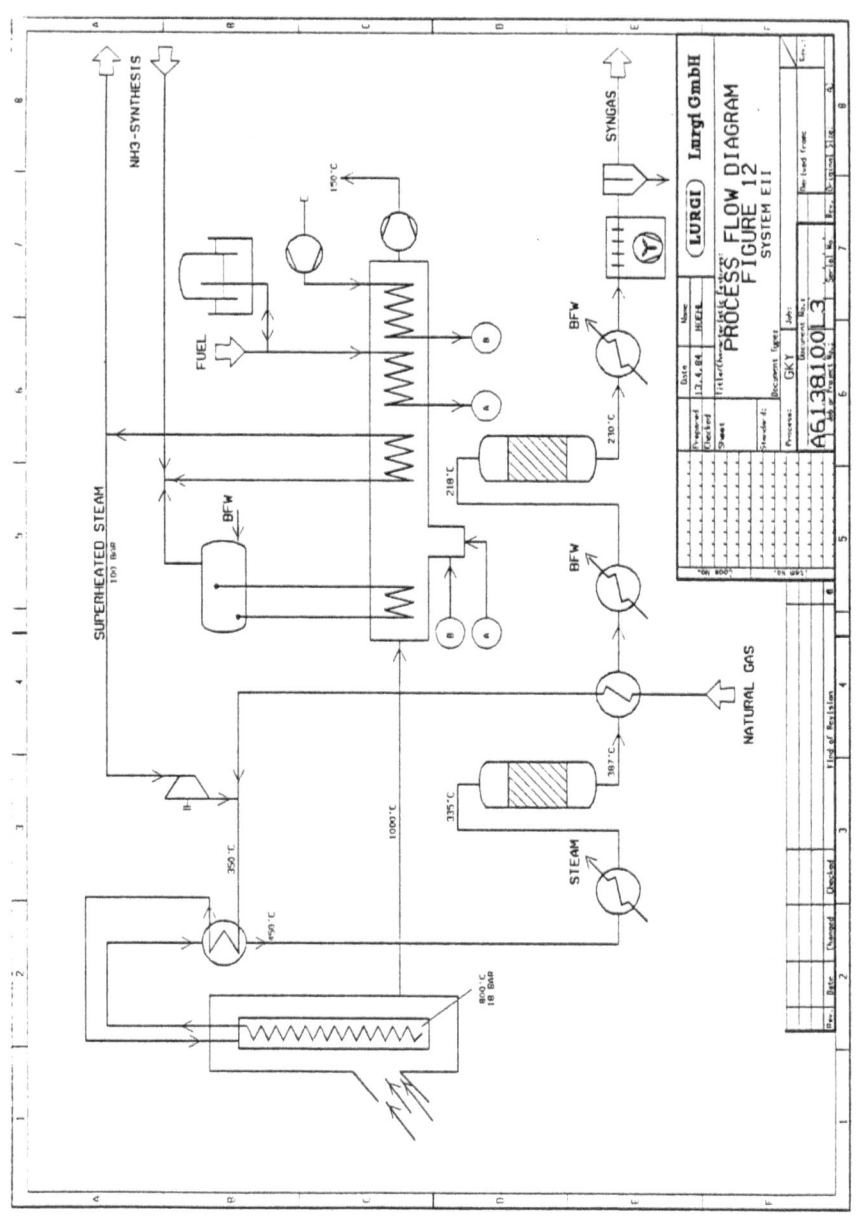

PROCESS FLOW DIAGRAM
FIGURE 12
SYSTEM E11

LURGI   Lurgi GmbH

NH3-SYNTHESIS

SYNGAS

SUPERHEATED STEAM
100 Bar

FUEL

BFW

BFW

BFW

NATURAL GAS

STEAM

350 °C

1000 °C

150 °C

210 °C

270 °C

335 °C

387 °C

650 °C

800 °C
18 Bar

A6138100L3

GKY

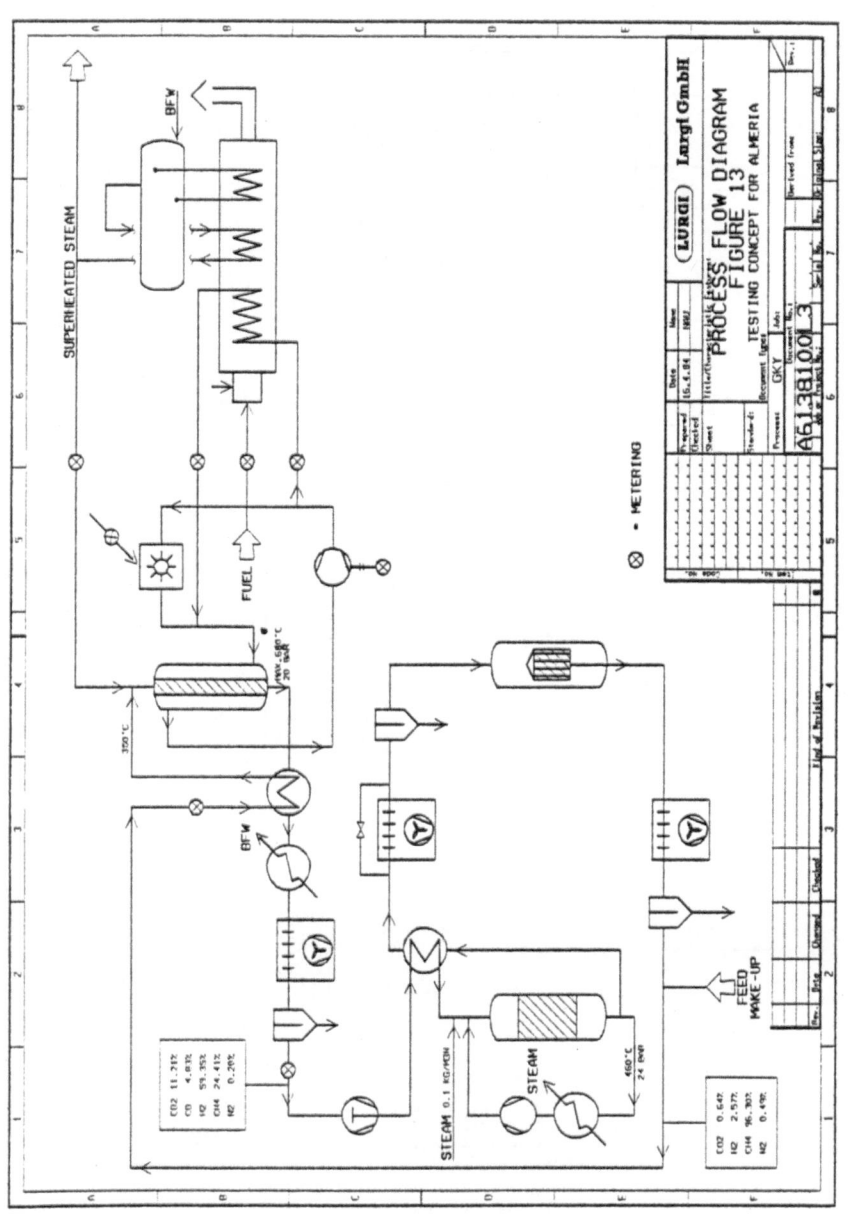

⊗ = METERING

LURGI GmbH

PROCESS FLOW DIAGRAM
FIGURE 13
TESTING CONCEPT FOR ALMERIA

A6138100 3

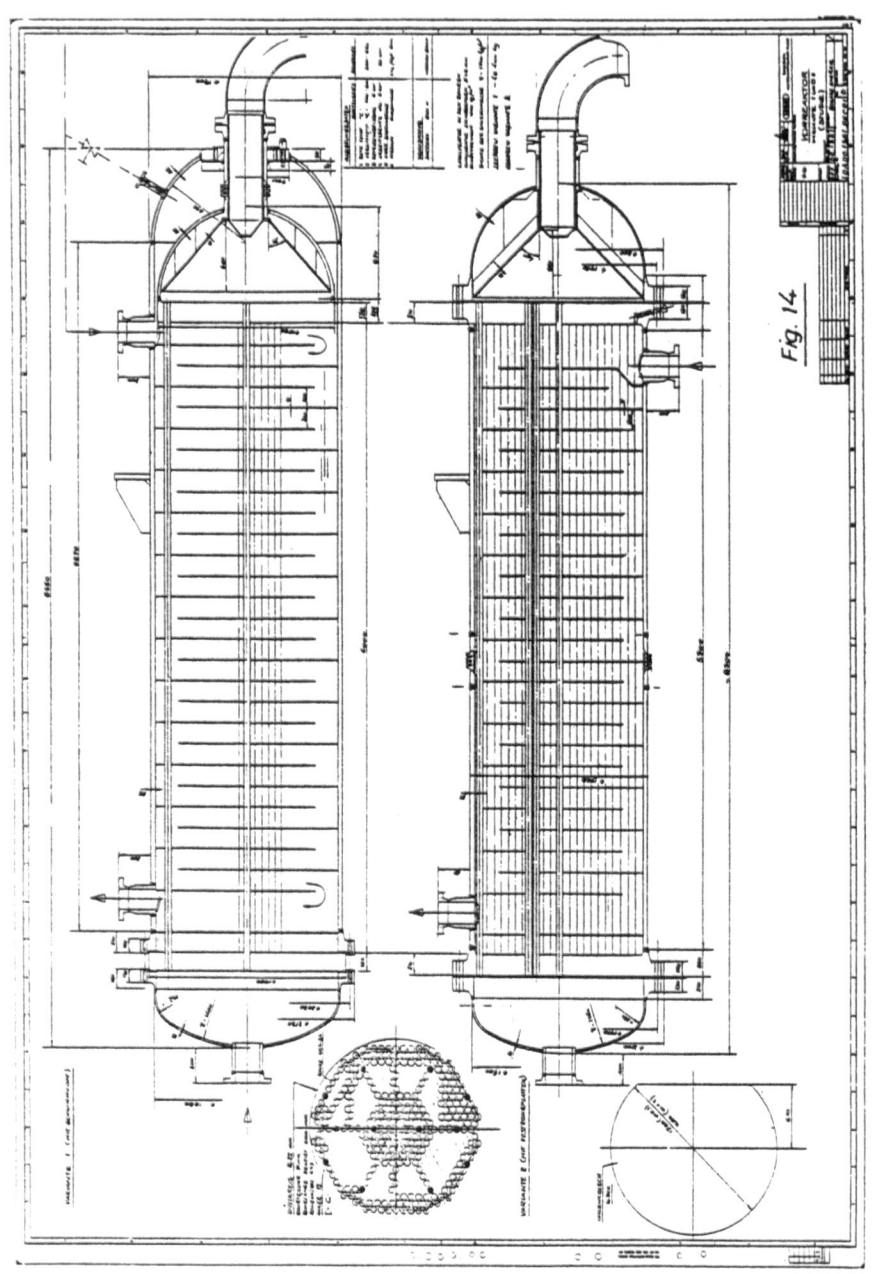

Fig. 14

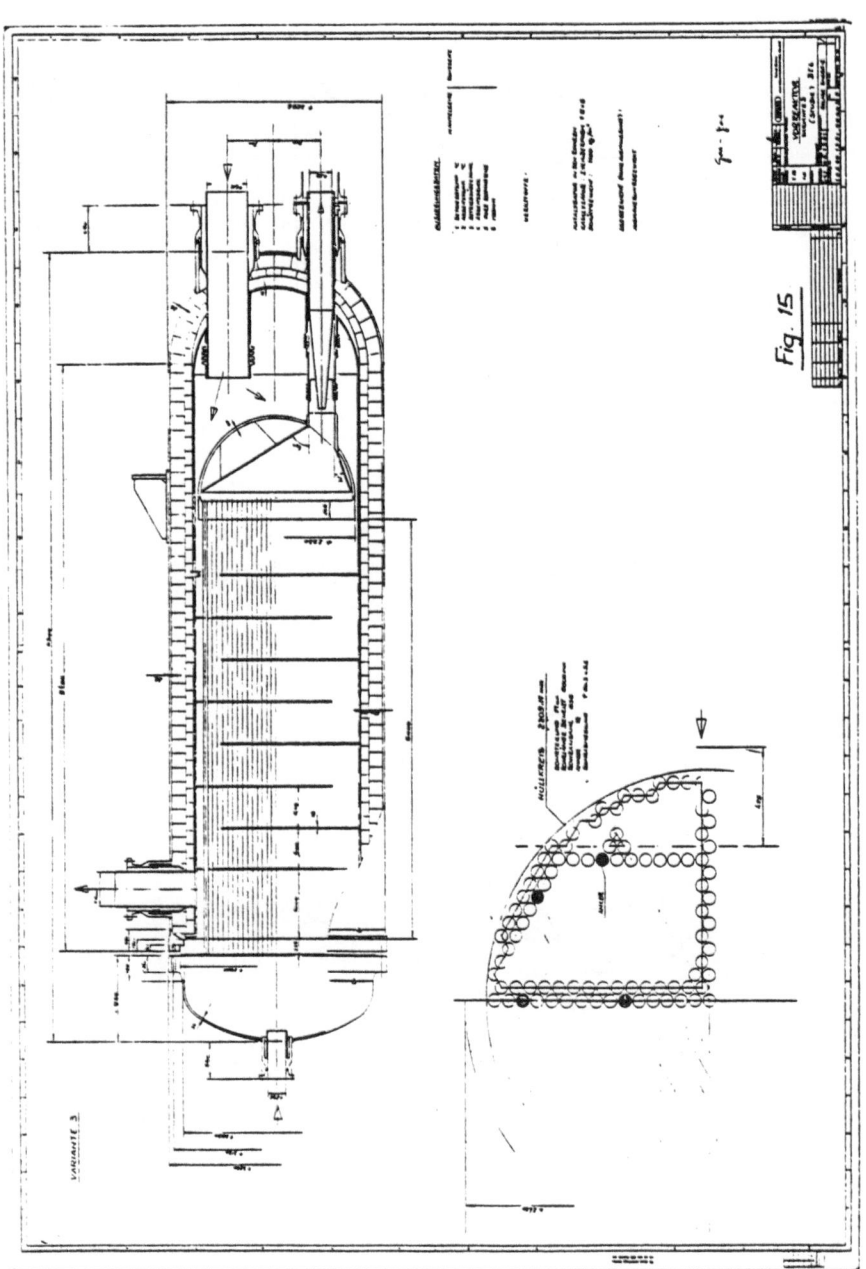

Fig. 15

- 81 -

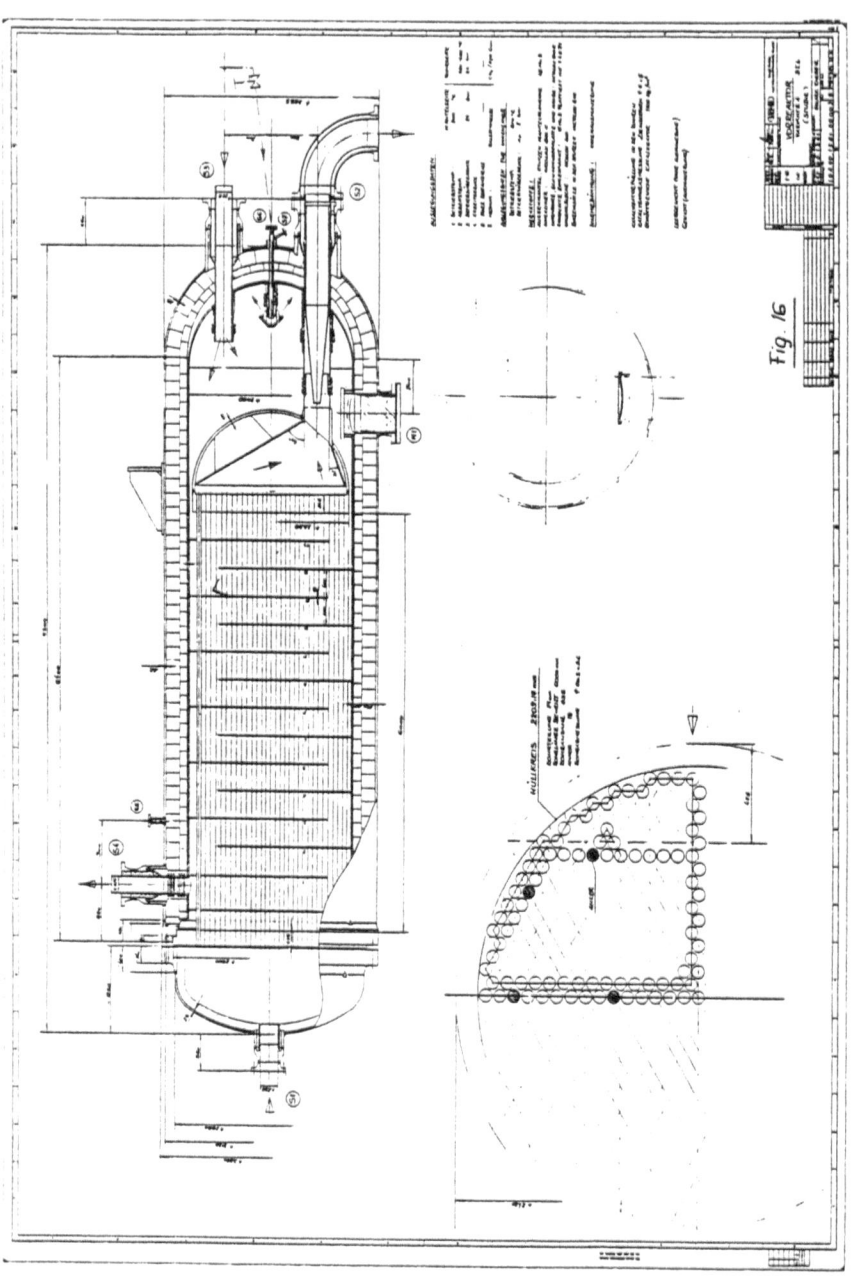

Fig. 16

- 82 -

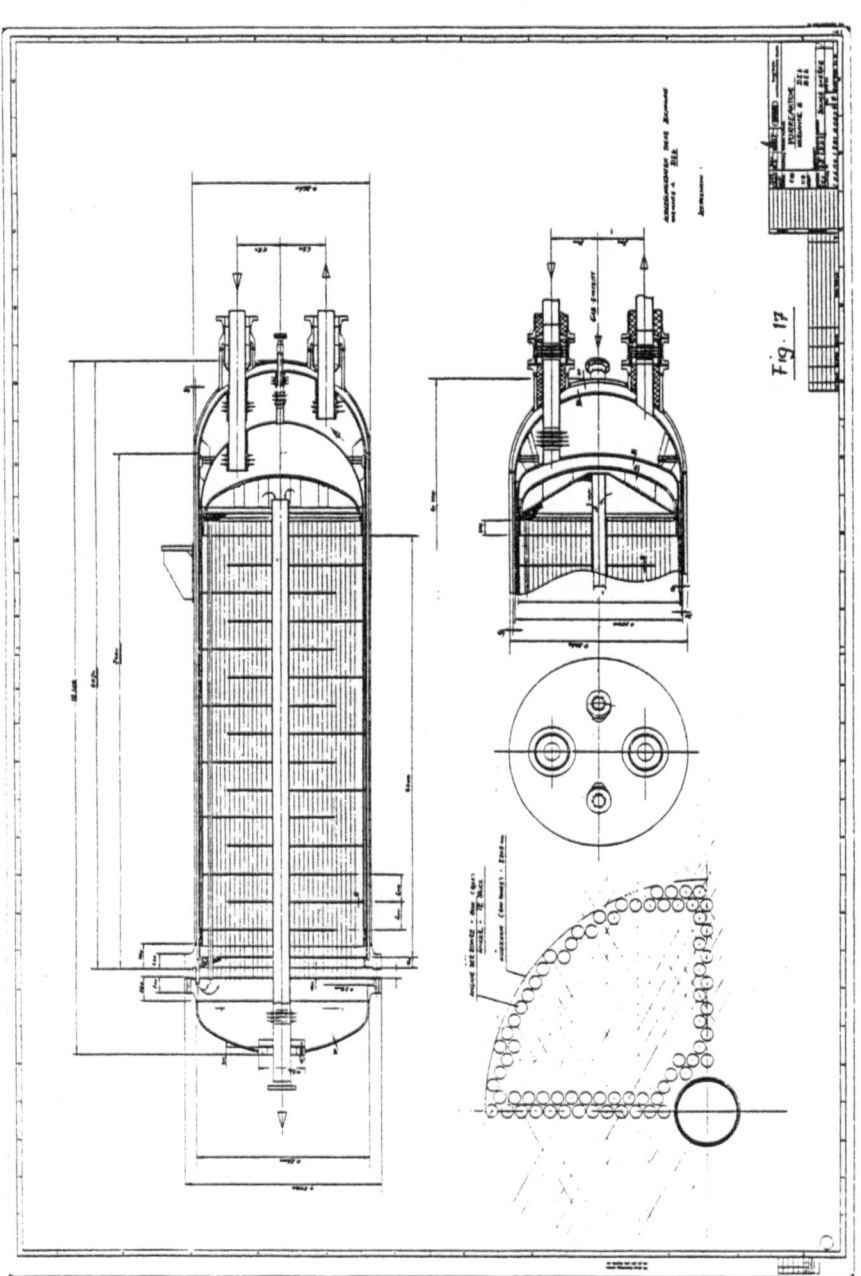

Fig. 17

- 83 -

PART II

STEAM REFORMING OF METHANE UTILIZING
SOLAR HEAT

W.D. MÜLLER

LURGI, FRANKFURT

Steam Reforming of Methane Utilizing Solar Heat (Part II)

1.0   INTRODUCTION

1.1   Purpose of the study and relation of this study with earlier investigations

In a preceding study GKY-61381-00 several different schemes have been evaluated, how solar energy can be used to generate synthesis gas from natural gas, the synthesis gas to be used for the production of ammonia or methanol and hydrogen.

In the first study it could be shown that it is advisable to introduce a "heated prereactor" with limited temperatures to achieve with at present available technologies a reliable plant layout.

When making the conclusions of the afore mentioned study it became evident that the overall efficiency of the investigated systems would not vary very much with the principle of the process configuration but with the layout and construction of the solar receiver, though the specific amount of solar energy in the product was varying with different process arrangements. The final recommendation was to furher investigate a system, which would be readily available with little risk.

Such a system had been presented as "System B" (new figure 1), which included a gaseous heat transfer system. The heat carrier gas was being heated in the solar receiver or/and in a fossil fired combustion chamber to about 800 $^{\circ}C$ and was then ·transmitting the accepted energy in a heated prereactor to the natural gas/steam-mixture passing a catalyst, thereby generating in an endothermic reaction hydrogen and carbonmonoxide.

This prereacted gas was then passed to a fossil steam
reformer which generated with less fossil energy than
in a nonsolar plant a suitable synthesis gas with little
unreacted methane.

When checking the proposed "System B" for such elements which are specific for the steam reforming in a
solar plant and where direct experience from conventional plants will not be availabel, one will find
- the gas cooled solar receiver
- the carrier gas heater in a fire box
  for high outlet temperatures
- the gas heated prereactor
- the compressor for hot carrier gas

though for all of these elements theoretical investigations or some similar but not identic references are
available. Therefore any testing of a facility to prove
the viability of such a concept should at least include
these enumerated items.

Consequently a proposal had been included in the first
study, how such plant configuration could look like
(new figure 2) , when at the same time the investment
should be low but also the feedstock consumption should
still be acceptable for remote locations as for instance
the solar testing plant at Almeria/Spain which is not
connected to any industrial complex in its vicinity.
The proposed system includes the afore mentioned items
and as addition - as a kind of utility plant - a methanation plant, which allows to recombine methane from
synthesis gas, generated in the solar reforming plant,
thereby saving nearly all feedstock, as this is being
turned around in a closed cycle and only using small
amounts of fuel, electricity and demineralized water
for operation, those quantities varying with specific
loads and intentsity of sunshine.

This proposal principally had been accepted by the customer, DFVLR, and as in the first study,

LURGI GMBH, Frankfurt am Main
and
MAN A.G.-Neue Technologie, München

were entrusted with a basic layout of such a testing facility. It should be assumed that this facility was to be placed at Almeria/Spain and that the solar reflectors, which are at present available at this site, shall be used only.

The task of this study shall be to determine the proper layout and to develop the investment of such an installation.

As in the first study a partition of investigations between LURGI and MAN had been found to be economic and useful, it was proposed and accepted that the study was divided into a solar part and a reaction part, the first to be handled by MAN, the latter by LURGI.

1.2   External supplies to the plant

The present design of the pilot plant is based on the availability of the following feedstocks, energies and utilities.

1.2.1 Natural Gas Feedstock

At present no data are available on the quality of liquefied natural gas, which may be bought for Almeria. Therefore the same analysis as in the prestudy has been adopted, which is

$$CH_4 \qquad 99,5 \text{ mol-\%}$$

$$N_2 \qquad 0,5 \text{ mol-\%}$$

By addition of the methanation plant the total loop is
very similar to the ADAM-EVA-Plant at Jülich, with the
only difference that the reacted gas contains less hydrogen and carbon monoxide due to the lower reaction temperature. Only small amounts of natural gas have to be added
to the closed loop to compensate for losses.

1.2.2   Sunshine

The central receiver will make availabel a thermal output
of
          1,4 MW
for design conditions at March 21./Sept. 23 with a heat
carrier gas outlet temperature of
          800°C

The total load gathered during the equinoctial days will
amount to
          9,68 MWh

Details will be given in chapter (2).

1.2.3   Electric Energy

It is assumed that electricity is available with
          voltage   380/220 V
          frequency   50 Hz.

1.2.4   Demineralized Water

It is assumed that fully demineralized water will be made
available with quantities as listed in chapter (5).

If only raw water, for instance of quality of drinking
water, could be made available, boiler feed water in adequate quality could be produced by distillation of this

water with some excess heat still available within the proposed unit, but this is not considered in the present layout.

1.2.5 Fuel

Fuel will be needed to maintain production during non-solar times. As, different from later plants with chemical products, no byproduct fuel is available this has to be taken from external sources.

For this study it has been assumed that the fuel will be fuel oil with a lower heating value of
$$H_u = 11,86 \text{ kWh/kg.}$$

It is further anticipated that no treatment of the generated flue gas will be necessary (pollution control), beqfore they are released to the atmosphere.

1.2.6 Diesel Oil

Diesel oil will be needed to operate the driver of the airblower K-3, as this driver, a Diesel engine will allow to vary the speed and thereby the recycle rate of air within the primary loop very effectively.

1.2.7 Nitrogen

For purging purposes nitrogen is needed, when the plant will be started or shut down.

Nitrogen will have to be delivered in liquefied form and will be stored on site. It will be evaporated for use and will be available with high purity at a pressure of about 12 bar gauge.

1.2.8 Hydrogen

Hydrogen will be needed for the reduction of catalysts
for the first start up and in limited amounts for each
shut down and start up of the plant.

The hydrogen will be stored as compressed gas in commercially available steel cylinders.

1.2.9 Carbondioxide

The experience of the ADAM-EVA-Plant at Jülich shows that
by preferred solution of carbondioxide and methane in the
gascondensate and loss of these components during water
treatment the loop always is enriched with hydrogen. The
loss of carbon from the loop must therefore be corrected
by the addition of carbondioxide to the loop. Carbondioxide shall also be stored in liquefied form in a
sphere at ambient temperature and introduced into the
loop over a dosing station by its own available vapour
pressure.

1.3   Internally generated utilities

Cooling water in limited amounts and instrument air will
be needed for the operation of the test facility. Their
generatioin will be described in chapter (3).

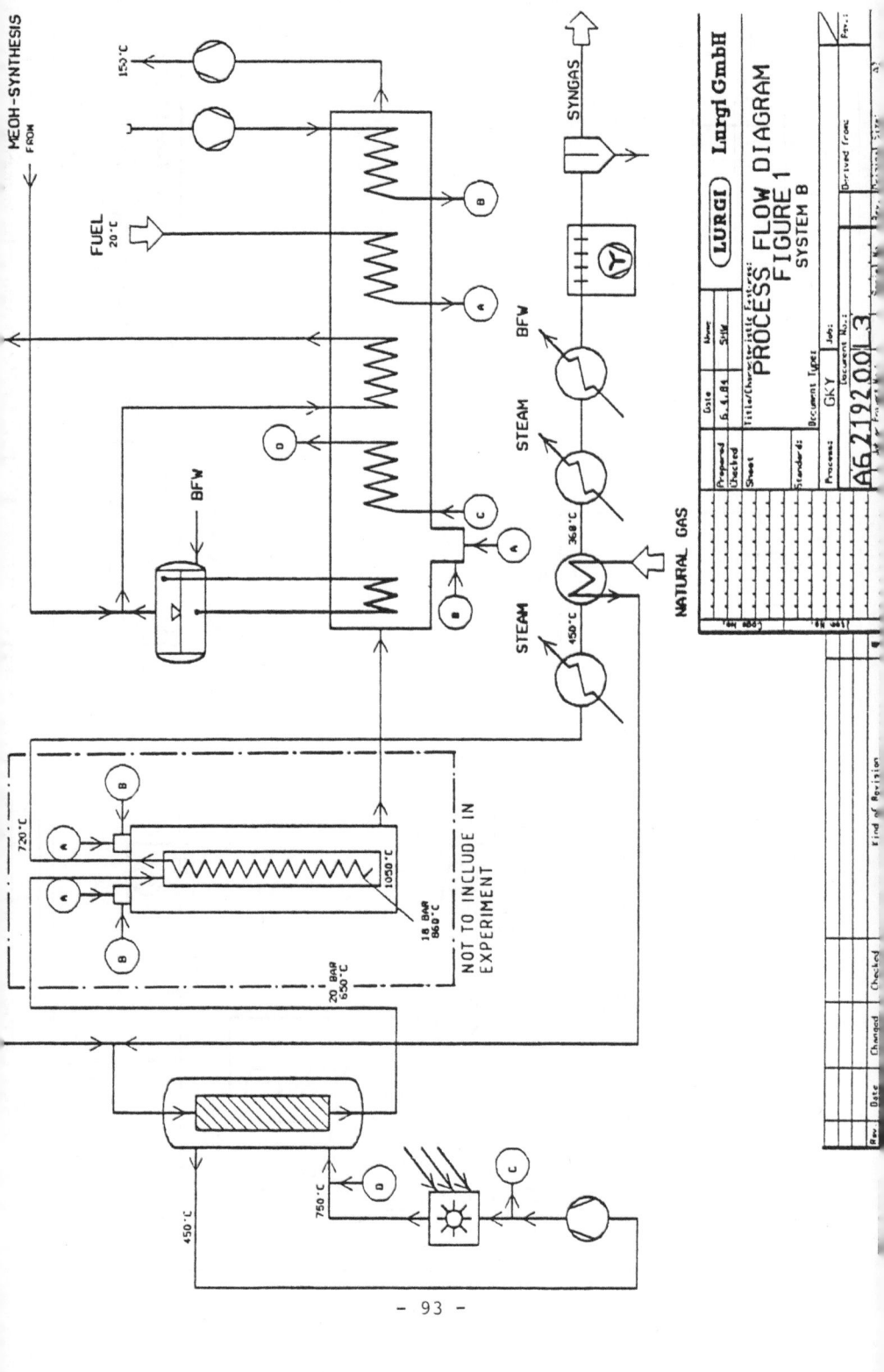

PROCESS FLOW DIAGRAM
FIGURE 1
SYSTEM B

LURGI Lurgi GmbH

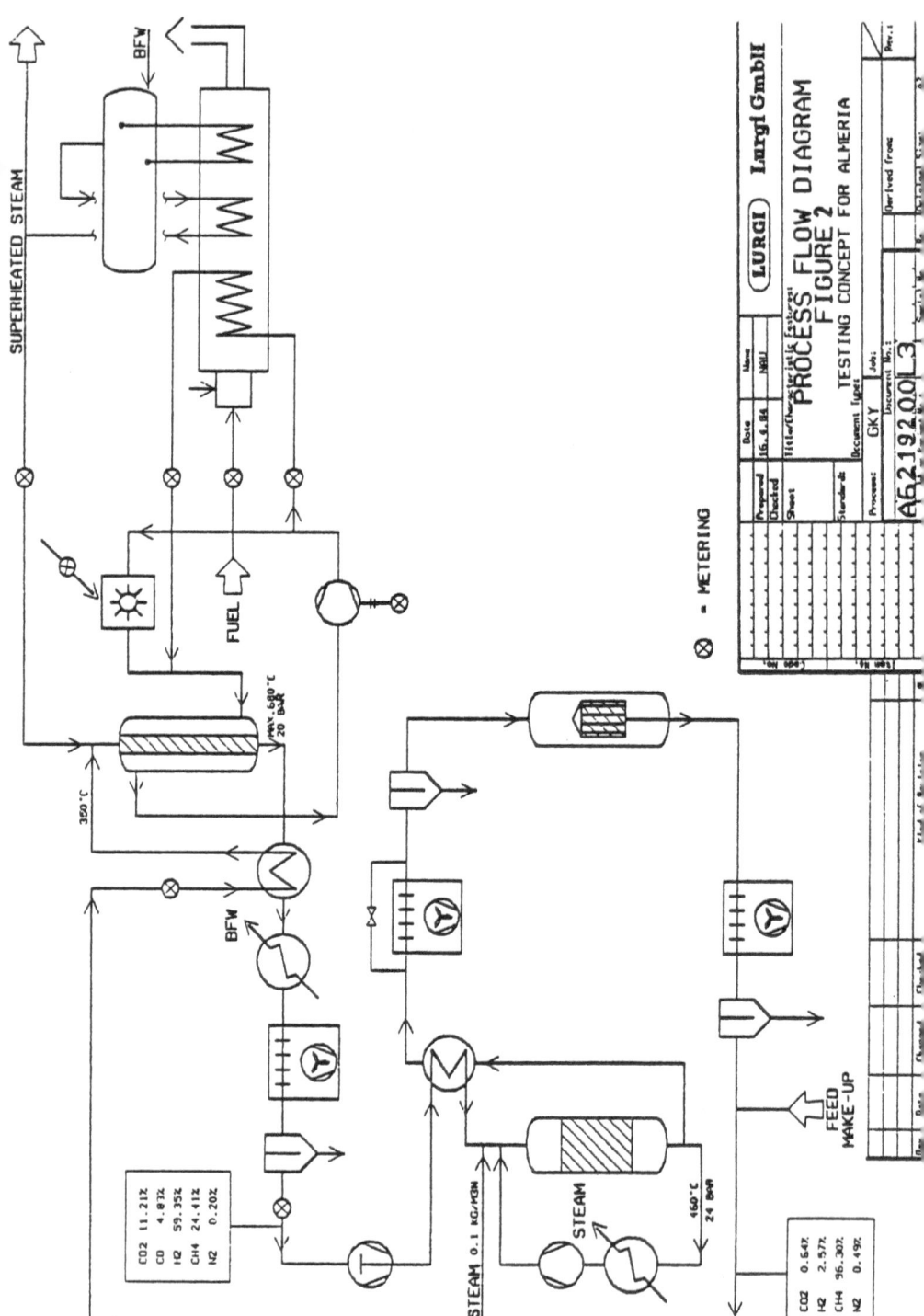

SUPERHEATED STEAM

BFW

FUEL

BFW

350°C

MAX. 680°C
20 BAR

STEAM 0.1 KG/NM3

STEAM

160°C
24 BAR

FEED
MAKE-UP

CO2	11.21%
CO	4.87%
H2	59.35%
CH4	24.41%
N2	0.20%

CO2	0.64%
H2	2.57%
CH4	96.30%
N2	0.49%

⊗ = METERING

LURGI   Lurgi GmbH

Date	Name	Title/Charge Einheiten
16.4.84	NHU	

PROCESS FLOW DIAGRAM
FIGURE 2
TESTING CONCEPT FOR ALMERIA

Document No.:

GKY   Job:

Prepared		Derived from
Checked		
Sheet		
Standards		
Process:		

A6219200l3   Document No.:

2.0 SYSTEM DESCRIPTION

2.1 Basic Data

By appointment this study was to be handled for a plant
which should be located at Almeria/Spain and which should
use the solar collectors as being available at present
at this location with the Small Solar Power System (SSPS)
of the International Energy Agency.

When having made a preliminary investigation of the possibilities of installed and of new equipment with the raised
temperature level, it was fixed by the partners, MAN and
LURGI, that the following parameters should be guiding for
the current investigations.

Number of available heliostats	93
Installed max. active heliostat area	about 3900 m^2
Design point	21st of March
	10 : 30 o'clock
	solar time
Approx. insolations at design point	800 W/m^2
Resulting thermal power between	
inlet and outlet of receiver	1400 kW
for	
inlet temperature of receiver	450 ^{o}C
outlet temperature of receiver	800 ^{o}C
Temperature loss between receiver EO and	
prereactor R 1	20 K

heat carrier medium	air
heat carrier pressure at inlet of prereactor R 1	22 bar (absolute)
max.allowed pressure drop for heat carrier cycle	3 bar
minimum heat load of fossil fired heater E 10	230 kW
maximum heat load of fossil fired heater E 10	1630 kW
Outlet/Equilibrium conditions of hydrocarbon/steam reforming cycle at preheater R 1	
temperature	680 ^{o}C
pressure	20 bar (absolute)

Feed analysis of hydrocarbon

CH_4 99,5 % (molar)

N_2 0,5 % (molar)

On the basis of these figures the design and layout of the relevant parts of each partner, MAN and LURGI, was started. It was accepted that minor deviations would result in each partner's portion by progress of the work but that such slight deviations especially at interfaces between the partners' portions, would be acceptable in the scope of this study.

2.2 Energy Collecting System (ECS)

The Energy Collecting System (ECS) has to produce
process heat with a temperature of 800 °C which
shall be used in the reactor to operate the Methane
reforming process.

The preliminary design has been done for an experimental plant to be erected at the Plataforma Solar
de Almeria in the South of Spain. Since 1981 a
solar power plant is already operated there under
contract of the International Energy Agency. It is
planned to continue the operation of this plant as
an experimental one for applications of high temperature process heat.

The ECS described in this chapter will use the
existing facilities as far as possible especially
the existing heliostat field.

The ECS consists of the heliostat field which fo-
cuses the solar radiation onto an air cooled cavity
receiver arranged on the existing steel structure
tower. The pressurized air which circulates in a
closed loop is heated up to 800 °C in the receiver
and then it is led to the reactor. A turbocompressor
circulates the air and a fossil fired heater in
parallel to the receiver can cover gaps in insolation and can maintain the over-night operation.
The flow scheme of the ECS is shown in fig. 2.2-4.

2.2.1 Design Assumptions

The preliminary design on the ECS is based on the
following assumptions and boundary conditions.

Location
  Plataforma Solar de Almeria
  37° 06' North
   2° 23' West
  500 m above sea leavel

Insolation

. Direct solar insolation over
  the day at 3-21/9-21, 6-21, 12-21   see fig. 2.2.-1

. Sunshape (acc. to Kuiper) /2.2-2/

$$f(\alpha) = \frac{I(\alpha)}{I_{dir}} = \begin{cases} \dfrac{1 + \beta\sqrt{1 - \dfrac{\tan^2\alpha}{\tan^2\delta}}}{\pi \tan^2\delta(1 + \frac{2\beta}{3})} & \alpha \leq \delta \\[4mm] 0 & \alpha > \delta \end{cases}$$

δ = 5,58 mrad, β = 2,2 (acc. to /2.2-2/)

$I(\alpha)$ = Power coming from the sun per solid angle
       unit and area $[W \cdot (m^2\ sr)^{-1}]$

I_{dir} = Insolation (acc. to fig. 2.2-1) $[W \cdot m^{-2}]$

As design value for the insolation 800 W/m² has
been chosen. This value is reached on 10:30 o'clock
at equinox (fig. 2.2-1). In accordance to the daily
insolation curves shown in fig. 2.2-1 all the year
the heliostat field will have a reserve in power.
It is believed that this reserve improves the operational availibility and reduces load restrictions
due to bad insolation conditions.

Heliostats

. Heliostat geometry   see fig. 2.2-2

. Reflectivity
 (acc. to fig. 76 in /2.2-3/   85 %

. Beam quality /2.2-3/   2,7 mrad

. Tracking /2.2-3/   1,5 mrad

. Total error (rss) /2.2-3/   3,1 mrad

. Heliostat field layout
 and definition of nominal
 focal zones   see fig. 2.2-3

. Actual focal length per zone   see tab. 2.2-1

. Field coordination system
 /2.2-3/
 positive x-axis   east
 positive y-axis   geographic north
 positive z-axis   height
 origin: projection of the aperture center point
      of the CRS-cavity receiver at ground level

. Heliostat coordinates   see tab. 2.2-2

Aperture

. Coordinates of aperture   x-value: 0
 center point (same system   y-value: 2 m
 of coordinates as above)
               z-value: 42,95 m

. Orientation of aperture plane
 angle between normal direction
 and y-axis $\varphi = 0°$

Tower
. height   40,75 m
. top width   6 m
. bottom width   7,5 m
. max. weight on top   30 t

2.2.2   Primary Loop Design

2.2.2.1   Description of the ECS

The solar energy is coupled in via the receiver E-0 (see fig. 2.2-4). The pressurized air which acts as coolant for the receiver is heated up to 800 °C and is then fed into the reactor together with the hot air stream coming from the fossil fired heater E-10. The fossil fired heater will be operated continiuously on its lowest load level which is necessary to keep the process running also when no insolation is available. If the process requires more power then the missing portion may be delivered by the heater E-10.

The air circulating in the primary loop leaves the reactor with 430 °C in maximum. If this temperature is exceeded the massflow can be splitted flowing through the cooler E-16 partly. The cooler E-16 is designed to handle about 50 K excess temperature to protect the compressor.
The pressure losses of the loop are covered by the compressor K-3. A single-stage, radial-flow turbo-compressor shall be used which is designed for the high intake temperature. The compressor shall be driven by a Diesel engine so that the massflow can be controlled by the speed of the Diesel-engine and - so far necessary - by the by-pass via valve V5.

Behind the compressor the gasflow is splitted up into the receiver branch and into the heater branch. The reactor R-1 and the fossil fired heater E-10 are arranged in a steel structure which is located near the tower on ground level (see plotplan). Because the lines between these components can be made short no thermal or pressure losses were assumed and all the line losses are considered to be in the riser and the downcomer which connect the receiver on the appr. 40 m high tower with the plant.

An auxiliary part of the primary loop is the charging
and pressure control system which wasn't considered
during the preliminary design study.
The daily thermal energy (21st of March) which can
be gathered from solar radiation is shown in fig. 2.2-5
considering the field and receiver losses.
For the design point the following steps of solar
power occure in the system:

Date: 21.3.

Time: 10:30

	Power	Efficiency
Insolation: 800 W/m²		
Potential power to system:	2924 kW	
Power in aperture plane:	2275 kW	0,778
Power into receiver:	2077 kW	0,913
Thermal output of receiver:	1462 kW	0,703
Total efficiency:	-	0,5

The main consumer in the primary loop is the turbo-
compressor. It needs at design conditions appr.
150 kW to be supplied by the Diesel engine. The specific consumption of the Diesel engine is appr.
210 g/kWh.

2.2.2.2   Description of Main Operational Modes

Reactor load has to be constant:

The explanation shall be started at time τ_0
(s. fig. 2.2-6), it could be for instance at
equinox noon.
The receiver is working under its maximum load.
Some of the heliostats are not used because the
insolation exceeds the design value (The dashed
line in fig. 2.2-6 represents the solar power
which could be delivered by the heliostat field if
all heliostats are used).

The heater E-10 is working at its minimum load. The
throttel D1 is opened as far as possible and the
throttel D2 is closed so far that the massflow
through the receiver branch and the heater branch
is adjusted to that value which is given in fig. 2.2-4.
If there are any fluctuations of insolation the receiver outlet temperature will be held at 800 °C by
splitting the massflow between the receiver and the
heater E-10.
Assuming the time proceeds. The receiver power decreases continuously due to the decreasing insolation.
If the operator notices the reduction of the receiver
power (measuring point 20, fig. 2.2-7) he can get
some more heliostats in operation.
If the time τ_1 is reached all heliostats are focused
onto the receiver. Because the insolation decreases
further the receiver power decreases continuously
from now on too. The power will be shifted more and
more to the heater E-10 by opening D2. If both
throttle valves are opened as far as possible about
50 % of the reactor power will be covered by the
heater E-10. The further reduction in receiver power requires to close the throttle D1 while the
throttle D2 is opened fully.
At the time τ_2 the minimum admissible power level
of the receiver is reached and the shut down procedure of the receiver has to be started.

Reactor load follows the solar insolation (cloudless day):

In this case the fossil fired heater E-10 is running with its minimum load constantly except the
equalizing of quick insolation changes to maintain
the reactor inlet temperature.
Using fig. 2.2-6 again - after the time τ_1 the
receiver power decreases. Because the mass flow in
the primary loop didn't change yet the above explained temperature control shifts the power to the

fossil fired heater. In this operational mode the power of the heater is controlled. If the set point of the heater load is exceeded a controller which isn't shown yet in fig. 2.2-7 will reduce the total mass flow in the primary loop by decreasing the compressor speed. The consequence is that the receiver outlet temperature increases and the mass flow will be shifted to the receiver back and the heater load E-10 reaches its setpoint again.
The reduction of massflow in the primary loop and the constant inlet and outlet temperatures of the reactor mean a reduction in reactor power. To maintain the process temperatures the massflow on the process side on the reactor has to be adjusted so that the set point of the massflow relation (measuring point 19, fig. 2.2-7) is reached again.

Start up and shut down procedures:

Both procedures have to consider to use the solar energy in the maximum extent. It is assumed that always if the solar receiver is started-up the plant is already running with the fossil fired heater E-10 at least under part load. During the study phase these operation modes haven't been investigated in detail. In general the following sequences are necessary.

Start up procedure:
If the massflow through the receiver is zero yet the aperture will be opened and some heliostats must be focused on the receiver (depending on the present insolation). The receiver is warmed up to a temperature somewhat above 430 °C then the massflow through the receiver starts. The irradiation of the receiver increases (by increasing insolation or by increasing the number of focused heliostats) and also the massflow is increased slowly so the required receiver outlet temperature of

800 °C will be reached at minimum admissible massflow. Now the power of the receiver can be increased
and the automatic temperature control will shift
the massflow from the heater E-10 to the receiver.

Shut down procedure:
If the minimum power of the receiver is reached
the number of heliostats must be reduced so that
the cavity temperature will decrease. Simultaneously the massflow is reduced.
If the final cavity temperature is reached (It may
be different for the over-night shut-down and cloud
interuptions of limited duration.) all heliostats
will be defocused and the aperture will be closed.

During the start-up and shut-down operation the receiver outlet temperature could be lower than 800 °C
over a short period of time. Due to the fact that the
reactor requires only 780 °C at the inlet and the
massflow is low through the receiver in this case
it might be possible to maintain the reactor inlet
temperatures also during receiver start up and shut
down procedures by the heater E-10.

References chapter 2.2

/2.2-1/   Becker, M., A. Kalt, W. Grasse
          The SSPS-Plants and their Potential for
          the Production of Process Heat and Solar
          Fuels and Chemicals"
          in the proceedings of the
          "International Seminar on Solar Thermal
          Heat Production and Solar Fuels and
          Chemicals"
          Stuttgart, October '83
          Edited by INSOLAR

/2.2-2/   F. Biggs, C.N. Vittitoe
          "The Helios Model for the Optical Behavior
          of Reflecting Solar Concentrators"
          SAND 76-0347
          March '79

/2.2-3/   Becker, M., H. Ellgering, D. Stahl
          "Construction Experience Report for the
          Central Receiver System (CRS) of the
          International Energy Agency (IEA)
          Small Solar Power Systems (SSPS) Project"
          Edited by IEA-SSPS Operating Agent DFVLR
          March '83

ROW	ZONE	ACTUAL AVERAGE MIRROR FOCAL LENGTH, M	AVERAGE PER ZONE, M
1		72.5	
2	1	75.2	76
3		78.4	
4		90.5	
5	2	96.4	97
6		99.3	
7		102.4	
8		114.4	
9	3	123.3	122
10		127.5	
11	4	131.0	137
12		140.7	

Table 2.2 - 1: Focal length of the different rows resp. zones/ 2.2-3

| HELIOSTAT COORDINATES | | | HELIOSTAT | | FOCAL |
X	Y	Z	ADDRESS	ROW	ZONE
-.001	42.208	.840	1	1	1
9.754	41.154	.870	2	1	1
18.902	37.795	.883	18	1	1
-.001	52.854	1.052	5	2	1
9.910	51.835	1.073	6	2	1
19.457	49.060	1.073	12	2	1
28.200	44.648	1.042	20	2	1
.000	65.111	1.280	9	3	1
9.905	64.008	1.334	10	3	1
19.510	62.040	1.340	14	3	1
28.616	58.357	1.297	16	3	1
37.187	53.349	1.261	22	3	1
.001	79.263	1.580	33	4	2
14.016	78.026	1.628	34	4	2
27.736	74.217	1.617	44	4	2
40.687	68.273	1.572	50	4	2
52.149	59.718	1.463	54	4	2
7.931	87.331	1.783	36	5	2
23.483	84.274	1.792	42	5	·2
37.806	78.943	1.773	48	5	2
51.225	70.555	1.679	56	5	2
.009	95.970	1.898	39	6	2
17.090	94.181	1.972	40	6	2
33.537	89.620	1.957	46	6	2
48.934	82.302	1.868	52	6	2
62.797	72.381	1.776	58	6	2
9.447	104.849	2.130	66	7	2
28.035	101.493	2.166	74	7	2
45.570	94.942	2.132	82	7	2
61.870	85.339	2.033	90	7	2
-.001	114.920	2.305	67	8	3
20.726	113.086	2.363	68	8	3
40.226	107.756	2.369	80	8	3
58.837	98.758	2.267	88	8	3
11.709	125.431	2.563	70	9	3
33.521	121.305	2.593	78	9	3
54.567	113.686	2.555	84	9	3
73.769	102.260	2.428	94	9	3
-.006	137.359	2.753	73	10	3
24.547	135.040	2.834	76	10	3
48.160	128.630	2.833	86	10	3
70.111	117.648	2.740	92	10	3
13.410	149.633	3.055	98	11	4
40.229	144.779	3.118	104	11	4
64.933	135.465	3.054	110	11	4
7.483	163.058	3.288	100	12	4
22.564	161.987	3.352	102	12	4
36.585	159.110	3.375	106	12	4
50.600	155.455	3.376	108	12	4
64.038	150.270	3.338	112	12	4
-9.750	41.152	.768	3	1	1
-18.892	37.800	.700	19	1	1
-9.909	51.820	.968	7	2	1
-19.440	49.064	.876	13	2	1
-28.167	44.641	.750	21	2	1
-9.898	64.010	1.203	11	3	1
-19.511	62.026	1.122	15	3	1
-28.617	58.363	1.016	17	3	1
-37.186	53.337	.848	23	3	1
-14.034	78.034	1.464	35	4	2
-27.743	74.219	1.328	45	4	2
-40.691	68.278	1.118	51	4	2
-52.121	59.718	.922	55	4	2
-7.918	87.331	1.714	37	5	2
-23.465	84.284	1.545	43	.5	2
-37.800	78.940	1.370	49	5	2
-51.203	70.550	1.140	57	5	2
-17.053	94.170	1.772.	41	6	2
-33.514	89.598	1.613	47	6	2

Table 2.2 - 2:

Heliostat coordinates (field inclination of appr. 2,1°
from NE to SW), z-value indicates upper edge of foundation /2.2-3/
(continued next page)

X	Y	Z	HELIOSTAT ADDRESS	ROW	FOCAL ZONE
-48.911	82.296	1.390	53	6	2
-62.770	72.356	1.110	59	6	2
-9.449	104.847	2.026	65	7	2
-28.040	101.500	1.853	75	7	2
-45.558	94.951	1.653	83	7	2
-61.882	85.329	1.372	91	7	2
-20.731	113.085	2.148	69	8	3
-40.230	107.750	1.935	81	8	3
-58.840	98.741	1.644	89	8	3
-11.740	125.429	2.432	71	9	3
-33.530	121.311	2.246	79	9	3
-54.557	113.689	1.960	85	9	3
-73.760	102.253	1.632	95	9	3
-24.530	135.030	2.571	77	10	3
-48.157	128.634	2.316	87	10	3
-70.102	117.647	1.982	93	10	3
-13.411	149.652	2.928	97	11	4
-40.235	144.774	2.690	103	11	4
-64.930	135.458	2.378	109	11	4
-7.470	163.068	3.203	99	12	4
-22.565	161.984	3.107	101	12	4
-36.588	159.110	2.970	105	12	4
-50.603	155.443	2.827	107	12	4
-64.019	150.264	2.657	111	12	4

Table 2.2 - 2: continued

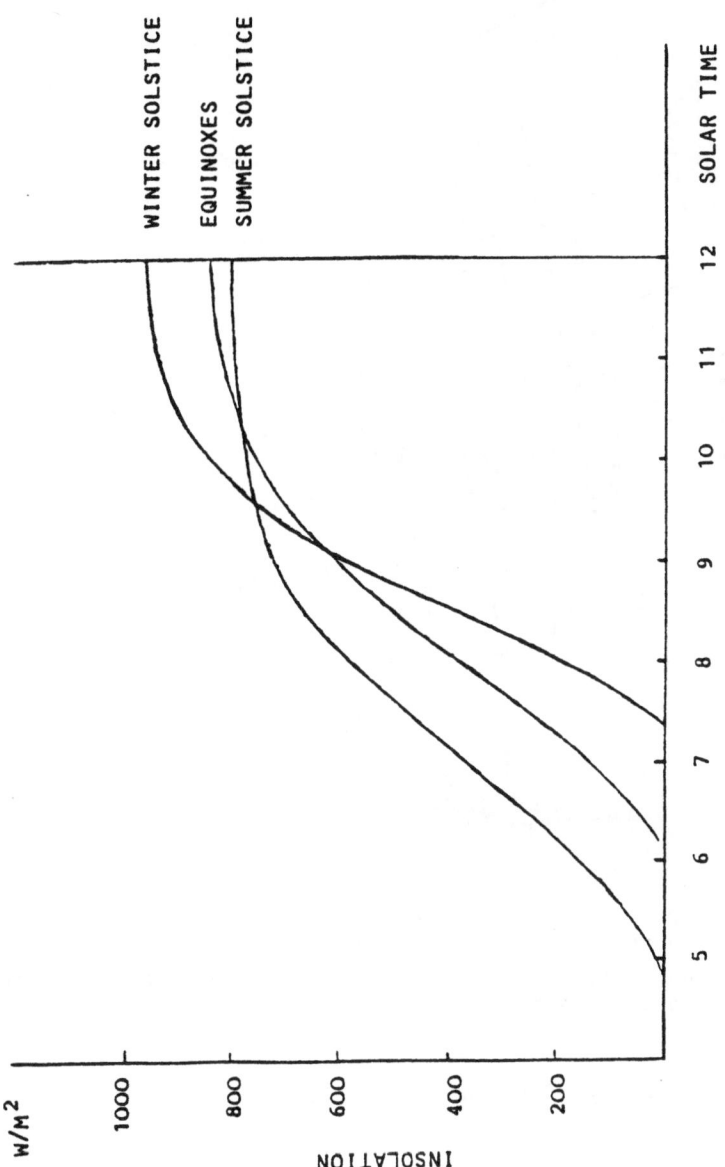

Fig. 2.2 - 1: Direct insolation over the day /2.2 - 1/

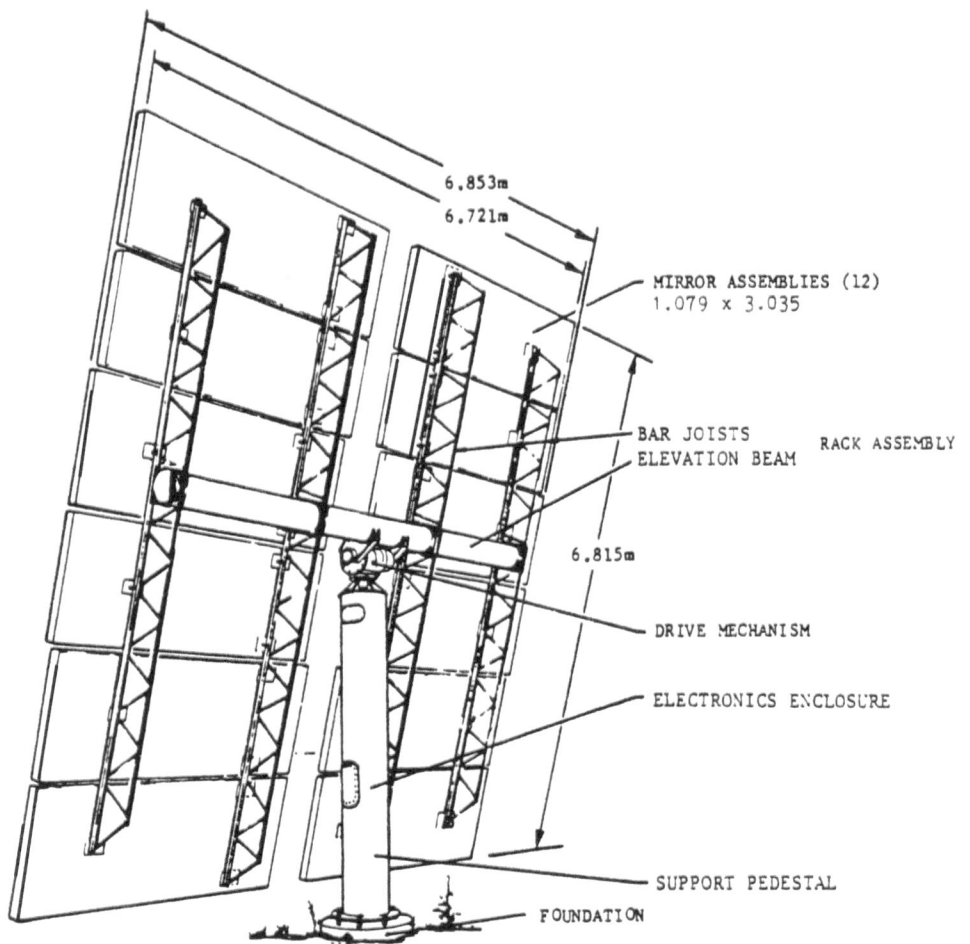

- Point of intersection of the elevation and azimuth axis':
  3,84 m above upper edge of foundation
- Normal distance between the center of the mirror plane and this poin
  of intersection: 0,114 m

Fig. 2.2 - 2: Geometry of the heliostat /2.2 - 3/

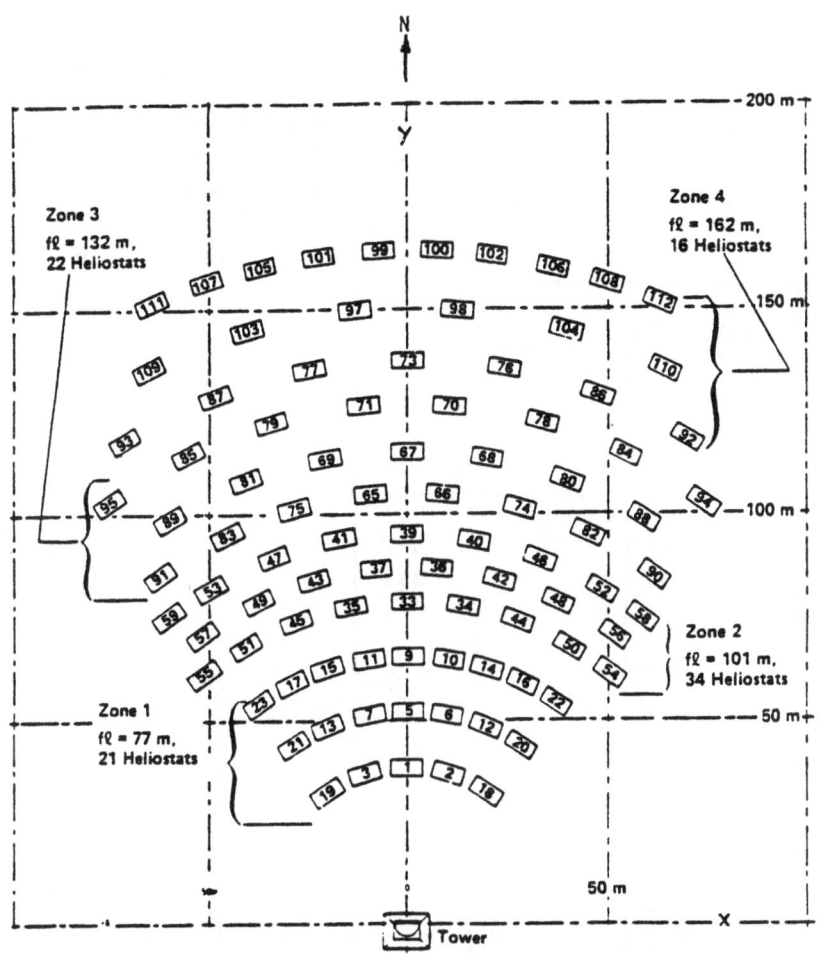

Fig. 2.2 - 3: Heliostatfield layout and definition of nominal focal zones /2.2 - 3/

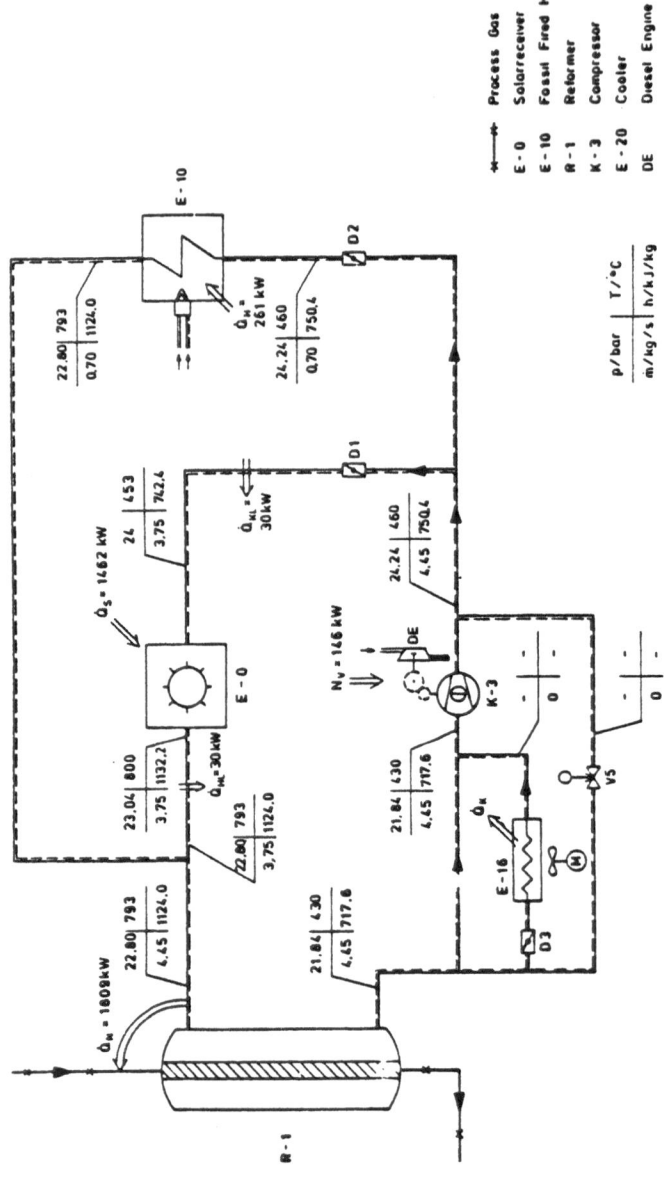

Fig. 2.2 - 4: Heat balance - primary loop at solar design point conditions

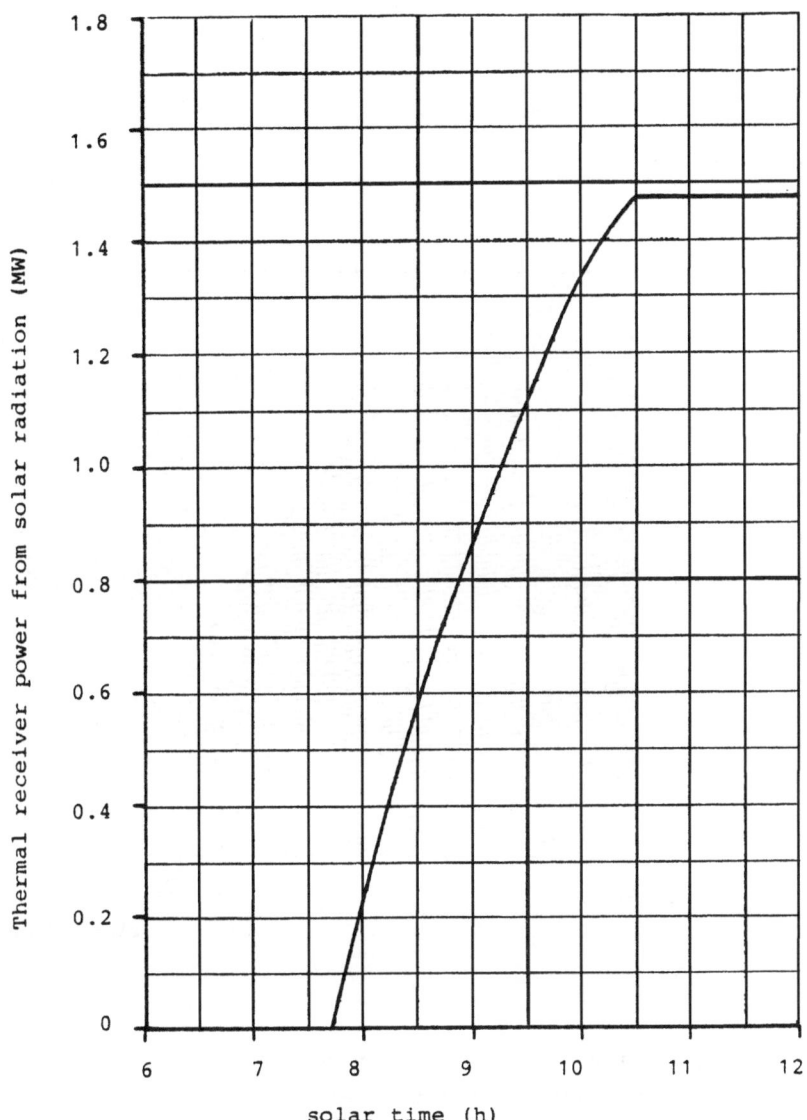

Fig. 2.2 - 5: Daily thermal energy delivery from solar radiation:
Q = 9.68 MWh
(Almeria, 21st. of March, no clouds)

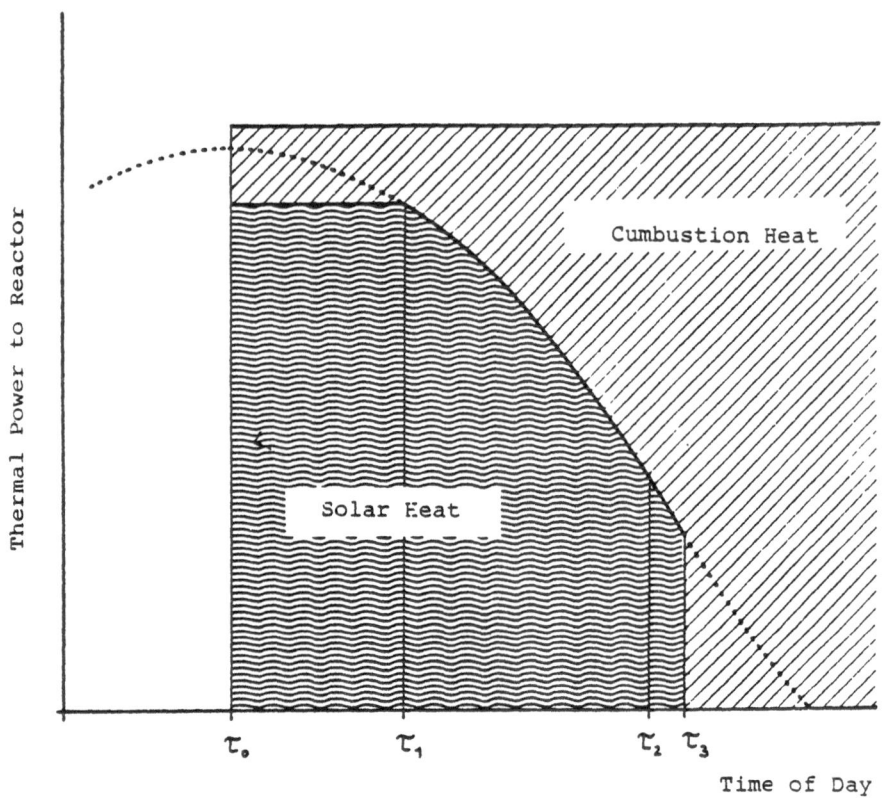

Fig. 2.2 - 6: Sharing of Thermal Power between Solarreceiver and Fossi
Fired Heater (qualitative)

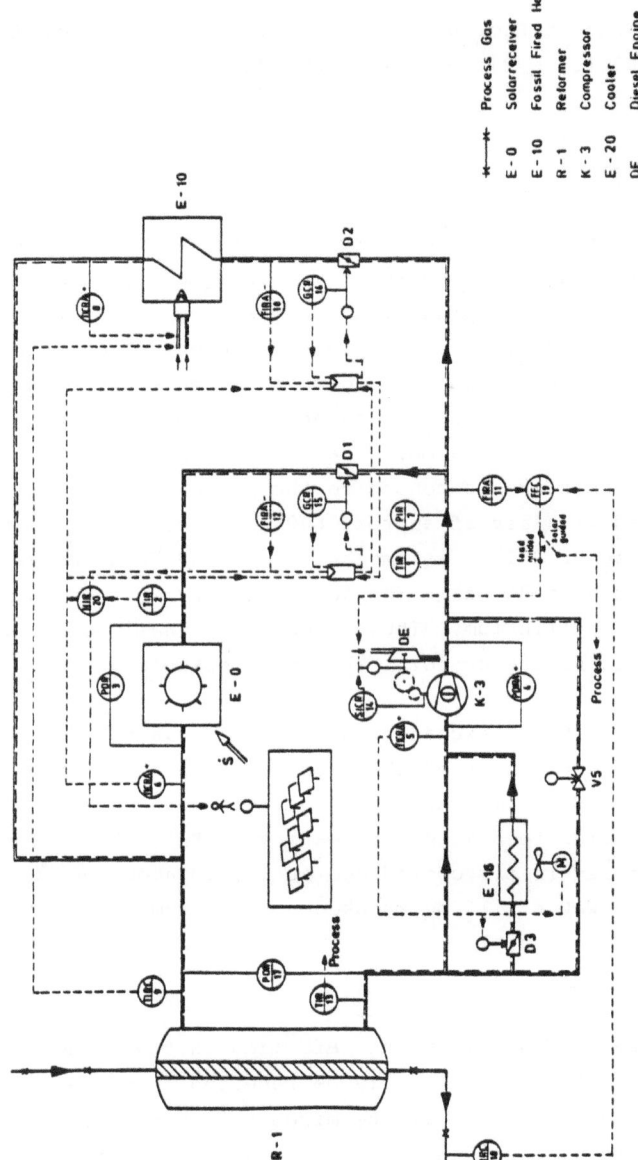

Fig. 2.2 - 7: Primary loop - instrumentation and control scheme

2.3   PROCESS DESCRIPTION OF THE ENERGY CONVERSION SYSTEM

See Process Flow Diagrams D 62192-00-00002 (Case A),
-00003 (Case B), -00004 (Case C)

2.3.1   Reforming

Methanated Gas, "SNG", from methanation unit is being
compressed by compressor K 1. The gas may be cooled direc-
ly after the compressor or only for bypass operation, depending from type of compressor chosen. The gas quantity
is flow controlled. The gas is preheated in exchanger E 1
to about 350 $^{\circ}$C and subsequently mixed with superheated
process steam at a molar ratio of H_2O/C = 2,6. The gas/
vapour mixture is routed to the prereactor R 1 which is
heated with air of approx. 800 $^{\circ}$C. The reacted gas leaves
prereactor R 1 at a temperature of 680 $^{\circ}$C at an absolute
pressure of 20 bar. The gas is supposed to be in thermodynamic equilibrium. These conditions shall be maintained
at this reference point whenever possible.

The outlet gas exchanges heat in E 1 against the feed gas.
The sensible heat of this gas stream further is used for
saturated steam generation in waste heat boiler E 2 and
for boiler feed water preheating in E 3. Finally the gas
is cooled in air cooled cooler E 4 to about 50 $^{\circ}$C. Condensed water will be knocked out by drums F 1 and F 13.

2.3.2   Methanation

The reacted gas from the reforming unit will not be compressed but will directly be reheated in exchanger E 5 to
about 240 $^{\circ}$C. It will be mixed with hot recycle gas. The
mixture will be lead to a methanation reactor R 2, which
contains a nickel catalyst. By the exothermic reaction
and the chosen recycle ratio the outlet temperature of the
reactor R 2 will be limited to 460 $^{\circ}$C. The larger portion
of this gas will be cooled in waste heat boiler E 6 to
about 260 $^{\circ}$C and forms the recycle gas. The separator

F 15 is a safety device only to protect the recycle compressor K 2 against water slags for the not yet observed case if a leak should occur in E 8 and release water to the gas stream. The remaining portion of methanated gas from reactor R 2, which is formed from the incoming reformed gas, preheats said incoming gas, thereby being cooled down to about 250 ^{o}C. Due to the composition of the reformed gas, which contains relatively little carbon monoxide but much carbon dioxide in relation to hydrogen, much water as vapour is formed in reactor R 2. To achieve a high quality product gas it is advisable to condense part of this vapour before submitting the intermediate product gas to final methanation. Therefore, a part of the intermediate product gas is cooled in air cooled cooler E 7 and the condensed water is separated in knock-out drum F 2. This portion is recombined with the hot part, resulting in a mixing temperature of about 175 ^{o}C. The mixed gas, lowered in water vapour content is reheated in the internal heat exchanger E 9 of reactor R 3 to an adequate catalyst inlet temperature. The catalyst is being cooled by this heat exchange, achieving favourable equilibrium conditions at the outlet of this reactor. The product gas is cooled in air cooled cooler E 8 and the achieved process condensate is separated in knock-out drum F 3. The cooled gas will be completed by the dissolved gas, regained from the process condensates and by the feed make up gas and will be lead to the compressor K 1, whereby the loop of the process gases is closed.

2.3.3   Steam Generation and Flue Gas Duct

The process unit "Reforming" is dependent from the reliable supply of superheated high pressure steam. This steam will be generated to a high portion by the use of the sensible heat of the reactants, i.e. by cooling the reformed gas in waste heat boiler E 2 (with appertaining feed water preheater E 3) and by cooling the methanated gas in methanation waste heat boiler E 6. The still mis-

sing portion of steam will be produced in the flue gas
waste heat boiler E 12, which is incorporated in the flue
gas duct.

The flue gas duct is required
a) to provide fossil energy for the heating of air as supplement or substitute of solar energy at times, if sunshine is not available

b) to egalize the energy balance of the process system or
the steam generation respectively.

As (a) and (b) are tasks, which are independent from each
other there is the need to apply two burner systems at the
flue gas duct: The first one will provide the energy for
(a) as required by the overall duty split of the energy
streams to achieve the desired throughput, while (b) will
add so much fossil fuel to the flue gas duct as will be required to have a positive balance of steam. As is shown
in case-B-flowsheet, the secondary burner can be shut off
under certain load conditions, for instance , if the maximum load shall be realized by fossil firing only. In this
case the sensible heat and the mass flow of the flue gas
from the first burner is that high, that excess steam will
be generated without the secondary burner in operation.

Superheating of the required process steam will be effected in superheater E 11. By the proper choice of the construction's material the temperature of the steam will be
allowed to rise to high temperatures, 700 $^{\circ}$C for instance,
at the outlet of E 11. This highly superheated steam then
will be quenched with boiler feed water to achieve the desired steam quality for the process.

Excess steam will be won as saturated steam and will be
condensed in air cooled cooler E 14. The condensate will
be separated in F 16 and will be reused after renewed
treatment in the water treatment plant.

The arrangement of a combustion air blower K 8 and of a
flue gas blower K 4 with their appertaining control
systems is standard praxis for such type of plants. The
flue gas stack C-1 will be a steel pipe which will be
fixed to the solar tower and which will end about 5 m
above the cavity receiver and which will therefore discard
the flue gases at the highest possible point of the area
whereby it is assumed that the flue gases will influence
solar irradiation to the reflectors only very little.

2.4   PROCESS DESCRIPTION OF THE AUXILIARY PLANTS

         See Process Flow-Diagrams D 62192-00-00005 Water Treatment
                          and   00010 Utilities

2.4.1   Water Treatment

         The proposed treatment is - except for sizes - identic
         with that one provided for KFA Jülich for ADAM-EVA-II-
         Plant. The collected gas condensates are filtered in F 9
         and depressurized into vessel F 10. The dissolved gases,
         mainly CO_2 and CH_4 are freed from the water by this de-
         pressurization. They are cooled down in E 13, condensed
         water is knocked out in F 11 and returned to vessel F 10
         and the freed gases are compressed by recompressor K 6 and
         returned to the gas loop upstream of main compressor K 1.

         The process condensate is pumped out of vessel F 10 to the
         predegaser F 6. Therein the water is stripped with excess
         steam in the upper part, thereby driving out nearly all re-
         sidual CO_2 and CH_4. In the lower part the water is cooled
         with air, compressed by blower K 5, returning the water va-
         pour containing air to the atmosphere.

         By stripping and cooling of the process condensate water
         is lost, which will be made up from an external source for
         demineralized water. This is supplied by tank car and
         pumped to storage vessel F 5, which also receives all

clean condensate, which was not in contact with process
media. The preheated process condensate and the make-up water together with any condensate is mixed and pumped
through mixed bed filters D 1/D 2 for fine polishing of
the water. The mixed bed filters are used alternately,
the active masses of one of these filters always to be regenerated outside of the plant, while the other filter is
in operation. The internal regeneration of the filter
masses has been excluded to avoid handling of chemicals
waste water problems for the proposed site location.

After this treatment the water only will contain dissolved
traces of oxygen which are stripped off with life steam,
which is blown into boiler feed water supply drum F 7, the
main degasser being directly placed on top of this drum.
The escaping excess steam is used as stripping vapour in
predegassing tower F 6, upper part. The boiler feed water
will be pumped to the heat exchanger E 3 and to the steam
drum F 4. Corrosion inhibitors are added from vessel F 8
by dosing pump G 5.

2.4.2   Natural Gas Supply

Natural Gas will be supplied in liquified form by road tanker at low temperatures. It will be pumped from the road
tanker to the storage pressure vessel F 17 at a reduced
pressure in relation to the normal operating mode. In normal operation a small portion of the liquefied gas will be
sent to the pressurization exchanger E 17. In heat exchanger with the atmosphere this portion will be evaporated and build up the desired pressure above the liquid,
around 15 bar. Liquid gas will be taken under this pressure to the main natural gas evaporator E 18, where also
this (main) portion will be evaporated and heated up to
temperatures around 0 $^{\circ}$C by heat exchange with atmosphere.
The low pressure storage vessel F 18 serves as a buffer
vessel. The gas is intermittently compressed to about 50
to 70 bar pressure by the natural gas compressor K 10.

The gas is stored under this pressure in the high pressure
storage vessel F 19 from where it is used to compensate
the losses of the process system only.

2.4.3   Nitrogen Supply

An identic system as described for the natural gas supply,
probably only different in size, will be installed for the
supply of nitrogen with regard to the low temperature
storage vessel F 20, pressurization evaporator E 19 and
main evaporator E 20.

Nitrogen will be used to inertize the process system for
start-up and after shut-down and for small purging streams
during normal operation. As there is no need for high
pressure nitrogen (the system pressure of the process can
be lowered for start-up and shut-down) there will be no installations for compression (equivalent to F 18, K 10,
F 19).

2.4.4   Carbon-Dioxide Storage

Carbondioxide (CO_2) will be needed as explained in chapter 1.2.9.

Carbondioxide will be stored in liquefied form in a
non-insulated tank F 21. As the amounts of CO_2, which will
have to be used, are very small there is no need for evapo-
rators as the exchange of heat with the surrounding atmosphere will be sufficient to evaporize the needed quantities in the tank at sufficiently high pressure. The Co_2
storage tank will be placed on a balance. The quantitiy
of carbondioxide, which was used by the process, is determined by building weight differences.

2.4.5 Hydrogen Storage

Hydrogen (H_2) will be needed for the first start-up of
the plant. It will be bought in steel cylinders in gaseous
form at about 200 bar pressure and will be stored on site
in a sheltered place. It will be ready for use without
further treatment.

2.4.6 Instrument Air Supply

Due to the considerable amount of instrumentation, which
will be required for the operation of the testing facility,
it is assumed that the available instrument air supply at
Almeria will not be sufficient and therefore a new installation has been foreseen. The atmospheric air is being
sucked in by piston compressor K 12 through filter and suction noise dampener F 24 and is being compressed to about
10 bar. The air is cooled in E 24 and any condensate is
knocked out in separator F 25.

Humid air is being stored in vessel F 26 in such a quantity that the demand of instrument air of the total plant
will be ensured for about 20 minutes' time, which will
allow a safe shut-down, if the instrument air supply cannot be restored in normal function again within a short
time.

The humid air is then passed over an adsorbent, which will
reduce its dew point to about - 20 °C, the drying plant
will be of the type "heat-les". Dry product instrument
air is being taken to regenerate one vessel, while the
drier is on stream, without the need of supplying heat to
this system. The switching of the driers F 27 A/B from ad-
sorbing duty to regeneration and vice versa is being done
automatically. Final drier F 28 is installed for safety
only as a guard filter; it may be regenerated only once
per year.

The dried instrument air will be available at ambient temperature with an overpressure of minimum 4 bar.

2.4.7   Cooling Water Supply

Cooling water will be needed in limited quantities only, mainly for internal cooling purposes of some compressors and for the coolers E 15 and E 24.

As raw water supply for normal, open cooling towers will be difficult at the envisaged location of Almeria, a closed circuit is being proposed.

The cooling water will be pumped to the consumers by cooling water pump G 9 and will return to the evaporator E 22 of a refrigeration unit. An expansion vessel F 23 will allow for specific volume changes in the cooling water circuit.

Vapours of the evaporated coolant will escape from evaporator E 22 and will be compressed by coolant compressor K 11. The vapours will be condensed in condenser E 23, the liquid will return to the evaporator E 22. The condenser E 23 will be cooled by circulating water, which on its part is cooled in air cooled cooler E 21. This concept of an additional water circuit is chosen as it tis supposed to be cheaper as if the coolant vapours were condensed in the air cooled cooler directly, as K 11, E 23, E 22 are normally supplied as package unit. The cheapest and most economic solution would have to be found out during detailed engineering.

2.5   DESCRIPTION OF OVERALL CONTROL SYSTEM

2.5.1   Principles
         The overall control system should allow
         a) to adjust itself in such a manner that a preset value
            for the power (or throughput) will automatically be
            achieved
         b) that equipment will be safeguarded against excessive
            operating conditions, i.e. too high temperatures
         c) that set-up conditions in one circuit will adjust other
            circuits, which normally are independent, in such a way
            to maintain production at a reduced capacity, if no
            general shut-down will be necessary
         d) to minimize energy requirements for the circulation of
            the heat carrier in the primary loop (and in the pro-
            cess circuit, if possible)

2.5.2   Description of Overall Control System during normal
         Operation
         The reference numbers of this chapter refer to the
         drawing "Schematic Overview of Overall Control System",
         Lurgi drawing No. D 6219200, IL00009.
         A value for the instant power shall be assumed. As this
         power will be linked with prefixed energy levels at inlet
         and outlet of reactor R 1 in the primary loop, this chosen
         power will correspond to a mass-flow of air in the primary
         loop. The following description will assume that a total
         power input of solar and fossil energy of 1,2 MW, corres-
         ponding to about 75 % of maximum throughput, will be
         required by the "master controller". It is further
         assumed that electronic controllers will be chosen, which
         take and send signals of 4 mA to 20 mA, corresponding to
         0 to 100 %; 75 % therefore correspond to a signal of
         16 mA (0,75 x (20-4)+4), if the full scales of (1) and (2)
         correspond to 1,6 MW.

The "master controller" is represented either by "Program controller" (1) or by the set value of the "Signal Generator" (2). By the selector (3) it is chosen, whether the automatic program or the manual fixing shall be guiding the total process sequence. The program would allow for instance in an optimum manner to follow the solar irradiation during daytime and to hold a constant fossil night load.

Under the guiding assumptions at the outlet of selector (3) a signal of 16 mA would be available and would pass selector (4) under normal conditions. The signal is split The full signal will go as set point signal to flow controller (6), while a signal multiplied by a constant (or preselectable) factor in multiplicator (5a) will go as set point signal to power controller (5). Under the assumption that 80 % of the load normally shall be borne by the solar part, 20 % by the fossil part the arriving signal should be $0,8 \times 0,75 \times (20-4)+4 = 13,6$ mA, corresponding to $0,8 \times 1,2 = 0,96$ MW (if full scale of XRC (P) (5) corresponds to 1,6 MW also).

Remark:
If scales of the cascaded controllers change, the signals within the loop have to be adjusted generally by factoring also!! Task of detailed engineering!

At flow controller FRC 6 the full scale for 1,6 MW thermal load would correspond to a circulation flow of 11700 m^3N/h the incoming signal of 16 mA therefore would demand 8775 m^3N/h flow rate. (if the final value of the scale would correspondend to 15000 m^3N/h the incoming signal would have to be altered to 14,692 mA by factoring). Flow controller FRC-6 therefore would alter its outgoing signal unless the circulation quantity of 8775 m^3N/h is reached. This is achieved by first closing the bypass around air recycle blower K 3 at minimum speed of K 3, and if the volume is not yet reached with fully closed bypass, the speed of the driver (Diesel-engine) will be increased afterwards to the required level.

The available air quantity of 8775 m^3N/h will be sent to the solar cavity receiver and/or to the fossil fired hea - ter E-10. The split will be made by three-way-control valve (10 a) in such a manner that the preset outlet temperature of TRC-10 (800 ^oC) will be reached. The quantity through E-0 should correspond to 80 % of total, i.e. 7020 m^3N/h, the rest of 1755 m^3N/h should be passed to E-10.

It was assumed that the solar power input would/should correspond to 0,96 MW. This has to be controlled and achieved. Control will be done in counting the quantity of mirrors, being focussed to the cavity, by XQRC-7 and multipliying the present value-signal with the signal of a radiation measuring instrument, RR-8, in multiplier (9).

The corresponding signal may say that only 0,85 MW are achieved at this time, controller XQRC-7 therefore will increase the number of focussed mirrors to such a quantity unless 0,96 MW are achieved as an outlet signal from multiplier (9). By this available energy input in cavity receiver E 0 the corresponding flow at an outlet temperature of 800 ^oC should amount to 7020 m^3N/h, (for instance to be measured at FIA-L-14). The air left for fossil fired heater E 10 will be heated in such a manner that the outlet temperature will normally be 800 ^oC, i.e. the set value of TRCACO-H-11 shall be 800 ^oC. The quantity of fuel will be established in such a way that this outlet temperature will be achieved. The following controls, for instance for burner air, are not shown in the attached sketch. As both temperature controllers, TRC-10 and TRC-11 are set to 800 ^oC also the corresponding mixture will have this temperature. The need of TRC-12 is given for insteady conditions only and will be dealt with later on.

The total air flow now will be available at the inlet of reactor R 1 with 800 ^oC or slightly lower due to thermal losses. If the throughput of natural gas and steam on the catalyst side of the reformer will properly be established

the air temperapure at the outlet of reactor R 1 will
necessarily reach 430 $^{\circ}$C, assuming that the system was designed adequately.

At the same time the goal of an outlet temperature of
680 $^{\circ}$C in the reacted gas must be achieved. Due to this
physical combinations the outlet temperature of the reacted gas can be taken to control the throughput of natural
gas (and steam), which shall be reformed. TRC-16 therefore will in steady state send a signal which will demand
for a natural gas quantity of about 772 m^{3}N/h (dry),-
corresponding to 75 % of maximum load. If the scale of
FRC-17 was 1200 m^{3}N/h the incoming signal should be
14,29mA. The proper steam quantity will be added by a
flow-ratio-controller FFRC-23.

Thus the total air stream (8775 m^{3}N/h) will be available
with a temperature of 430 $^{\circ}$C (maximum) at the outlet of
reactor R-1. If an inlet temperature of 435 $^{\circ}$C would be
allowed for air recycle blower K-3, the set value of-
TIC-22 should be 435 $^{\circ}$C, resulting in the lowest possible
outlet signal (4 mA) as this temperatures will never be
reached, with the consequence that three way valve 22 a
will direct all gas directly to the blower and, as by
assumption of the first parts of this description, the bypass valve (6 b) of the compressor is closed, no energy
has to be extracted at compressor cooler E-16, cooling air
compressor K-9 even could be shut-down. If, at a differ-
entlychosen load the bypass valve (6 b) would be open,
the temperature at the inlet of K-3 could rise by the increased temperature of the recycled air, if no cooling
air was passed through exchanger E-16, thereby initiating
temperature controller TIC-22 to raise its outlet signal,
starting compressor K-9 and opening valve 22 c. If for
any reason the inlet temperature for air recycle compressor K 3 would be limited to 380 $^{\circ}$C, the temperature con-
trollerwould send a signal of any value unless the set
temperature of 380 $^{\circ}$C would be reached. The consequence
would be:

a) the compressor K 9 will be started and valve (22 c)
will be opened entirely
b) three-way control valve (22 a) will change its position
so long until the cooled stream arriving from the outlet of compressor cooler E-16 and the hot gas stream
from the reactor R 1 (the part which goes directly to
the air recycle compressor K 3) result in a mixture of
the selected temperature. If the bypass of air recycle
compressor should be in operation, its cooling duty
will be handled automatically at the same time.

Thus the circuits are closed, the overall control functions
correspond to the principles as layed down above.

2.5.3   Description of Overall Control System for Disturbed Plant
Conditions

In the following paragraphs the principles, how to handle
disturbances of the independent systems, will be handled.
This shall be a guide only, how the overall control system
from principle may work. It is not assumed that the present version is a complete layout of the overall system,
which should be done in course of the detailed engineering
of such a plant.

2.5.3.1   Decreased Flow in Primary Loop
It is assumed that for any reason the air flow of the primary loop cannot be maintained at the desired rate, for
instance, because the bypass valve of compressor K 3 is
mechanically blocked in an open position.

As described in part (2) of this chapter the flow at FRC-6
should be 8775 m^3N/h. It is assumed that only 6150 m^3N/h
can be achieved by this failure. The actual flow is
measured by FFIC 20 and the arriving signal is compared
with the set value signal, which demands 8775 m^3N/h.
This controller builds the ratio of both signals and re-
cognizes that the actual ratio is 0,7 instead of set ratio of 1,0. As a consequence this controller FFIC-20 will

change its outlet signal. The function of this controller will be made thisway that a ratio lower than the set
ratio will result in decreasing outlet signals. The assumed failure therefore will result in a decreasing outlet signal which will pass the minimum selector relais
(21) (if conditions in naturalgas reforming circuit are
not disturbed) and will arrive at the next minimum selector relais (4). From the moment when this signal is smaller than the preselected signal of the master controller,
the "disturbance signal" will guide the overall process.
FFIC-20 will lower the set value of FRC-6 so long until
the ratio of 1,0 will be reached again at FFIC-20, of
course at a lower mass flow rate than originally foreseen. In consequence also the demand of solar energy w-
ill be decreased proportionally with the result, that
some mirros will be defocussed. The split of 80 % to solar cavity, 20 % to fossil fired heater(percentages related to overall, reduced mass flow of air)will be maintained, if sufficient sunshine is available. If not,
TRC-10 will let only the quantity to the solar cavity receiver, which can be heated to the desired temperature,
the rest to be independently heated to 800 $^{\circ}$C in the fossil part.

In consequence of the reduced mass flow of air the outlet
temperature of reformer gas would decrease with the effect
that a reduced mass flow also would result in this circuit

Thus the failure of the bypass valve (6 b) of the air recycle compressor K 3 is automatically compensated by the
overall control system.

2.5.3.2   Decreased Flow in Secondary Loop .
The same system as described for the primary loop shall
be installed for the secondary loop. Instead of the signal of FFIC-20 then the signal of FFIC-19 will be selected for overall control by the minimum selectors (21) and
(4). The only change is that the primary loop in this
case becomes a "secondary" loop as a follower of signals.

In both cases (3.1 and 3.2) the overall control system will return automatically to the master controller's guidance, if the upset condition in the loops is over.

2.5.3.3   Influence of Clouds

If clouds are passing the solar field this will be recognized by the radiation recorder(s) RR-8 with the consequence that more mirrors will be focussed to the cavity to maintain power input. If this stays without effect the temperature at the outlet of the cavity will decrease and TRC-10 will counteract by reducing the mass-flow through the solar receiver, leaving the remaining portion for fossil heating.

2.5.3.4   Discrepancy of Solar Demand and Input

There might be the effect that there is no good correspondence between the energy gathered by the solar receiver and the solar power calculated by the multiplication of signals from RR-8 and XQRC-7. As long as the actual thermal energy is smaller than the calculation this is not harmful as the flow through cavity E-0 will be reduced by TRC-10. If too much energy is collected this would tend to pass all air mass flow through the cavity leaving nothing for E-10, thereby risking a burnout. Therefore a limiting relais 10 d will ensure that always a minimum flow passes through both elements, E-0 and E-10.

With a mass flow of air through E-0 at the presently available maximum (three-way-valve 10 a at minimum passage to E-10) and too high radiation into the cavity the outlet temperature of TRC-10 will rise above the set value. In this moment the signal of TRC-10 will be split by relais (10 c) and an inverse function will be built. This signal will then be selected by a minimum signal selector (10 e) with the effect on the solar power controller XRC(P)-5 of a reduction of the set value, finally resulting in a defocussing of mirrors.

The rest of the control system will adjust itself properly.

2.5.3.5   Start-up/Shut-down of Solar Receiver

It is desirable to limit gradients within the solar cavity
during start-up and shut-down. For example it is sche-
duled that the air mass flow is shut down entirely only,
when the outlet temperature of E-0 has reached 700 ^{o}C
with a 20 % mass-flow of the maximum. This will demand
that the outlet temperature of the fossil fired heater
E-10 must be increased beyond 800 ^{o}C to maintain a mixed
temperature of 800 ^{o}C at the inlet of reactor R 1.

The not yet described items of the overall control system
shall serve to achieve this aims to the utmost possible
degree, which will be described by the case of shut-down
by gradually diminishing solar radiation. All mirrors
will be focussed to the cavity but nevertheless the required energy will not be made available. The throughput
through the cavity is gradually lowered unless three-way-
valve 10 a reaches its final position, governed by the
limiting relais (10 d). With further decreasing solar
energy supply the outlet temperature of E 0 will decrease.
When it reaches 700 ^{o}C the cut-out signal of TRCACOL10
will close valve (clapet)(10 f) and the energy supply will
be made on the fossil basis only.

In the period between 800 ^{o}C and 700 ^{o}C at the outlet of
the receiver E-0 the mixing temperature at the inlet to
reactor R 1 would be lower than 800 ^{o}C, if this temperature was maintained at the outlet of the fossil fired heater E-10. Therefore an additional temperature controller
TRC-12 shall be installed which will try to maintain 800 ^{o}C
as mixing temperature, increasing the set point of TRC-11.
As an outlet temperature of 1000 ^{o}C could be required under
special load cases to maintain 800 ^{o}C as mixing temperature, the set point for TRC-11 must be limited, for instance to 830 ^{o}C, which will be effected by the limiting
relais 13. It may happen now that the mixing temperature,
measured at TRC-12, cannot be maintained with the preset
mass-flow of air. This may result in a temperature drop
at TRC-16, too. The throughput of natural gas for reaction then will be reduced to stabilize the total system.

Should for any reason the mixing temperature decrease below a fixed value, for instance 720 $^\circ$C, it will not be
possible to make specification grade gas any more, the
plant then will be shut down by TIACO-L 23 (or a relevant
switch of TRC-12), cutting the signal for the mass flow
of the primary loop, putting it to zero. Other safety devices will follow which will have to be bridged for new
start-up.

For a smooth start-up of the solar receiver, when the
fossil part is already working, a program (XRC-10 b) may
be leading the set-point of TRC-10. Additional splittings
may also be introduced,for instance switching solar receiver E-0 and fossil fired heater E-10 in temporary
series with reduced throughput of E-0, on the same principle of control.

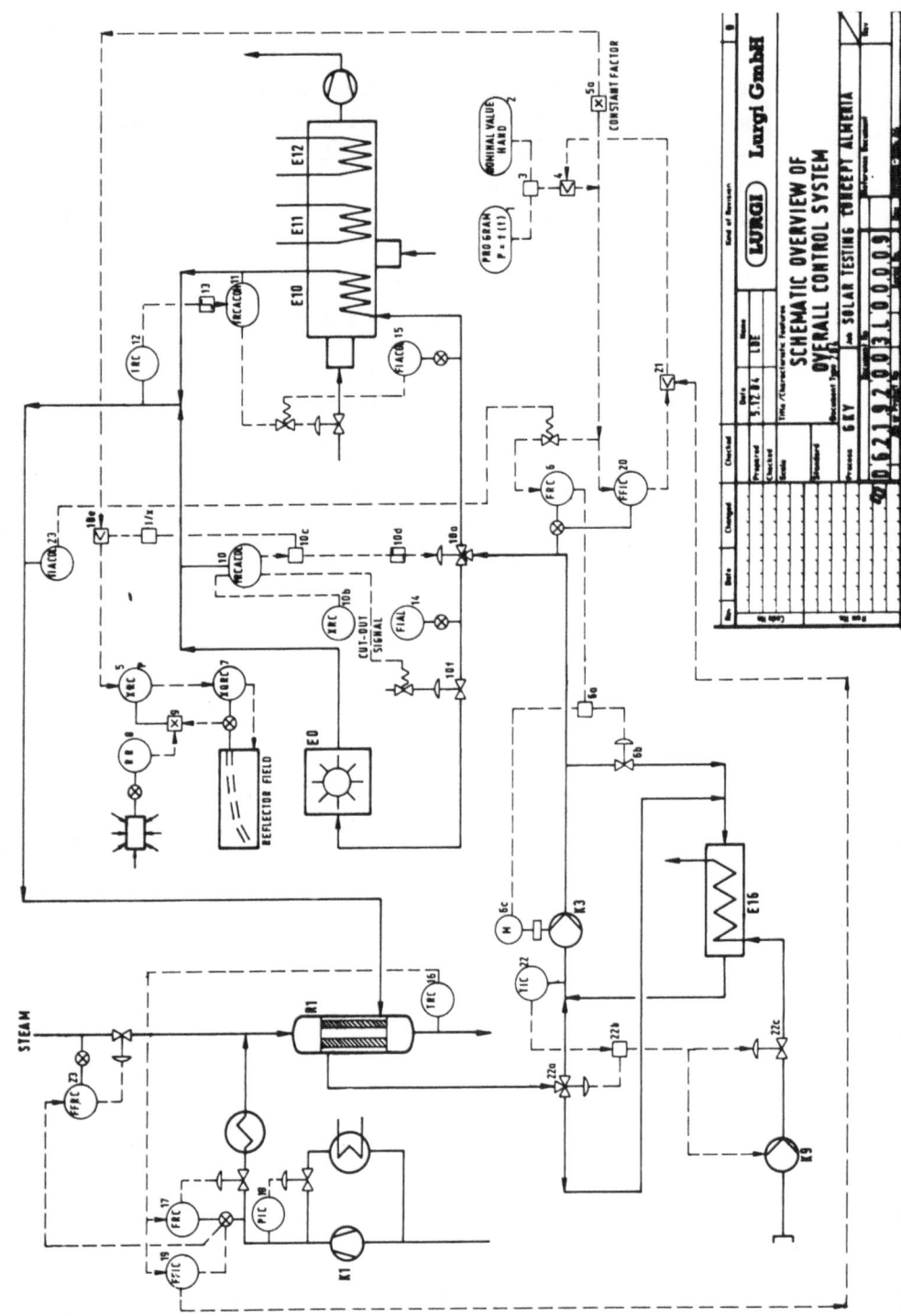

3.0　　DESCRIPTION OF COMPONENTS

The constructional design of the various elements of the plant need special attention due to the high temperatures which are required and the unequal distribution of energy flows at the same time, and due to the wide range of loads which are expected.

Therefore, the following items have been preliminarily designed according to these requirements.

3.1 Component Description of the Energy Collecting System (ECS)

3.1.1 Receiver E-0

3.1.1.1 Description of the Receiver

The receiver has to transfer the incomming radiation energy to the coolant air which circulates in the primary loop and transports the heat to the reactor. During this design study it was assumed that the receiver will be mounted on the existing tower of the SSPS-CRS plant.

Sketch SK-4565-B shows a general view of the receiver on the top of the tower. It is a cavity-type receiver with direct irradiated heater tubes. The heater tubes are arranged on the three rectangular rear walls of appr. 2,9 m width each.
The walls are insulated by ceramic fiber insulation. If the receiver is out of operation the cavity will be closed by an aperture flap gate with horizontal axis. For the receiver support structure a steel tube frame-work will be used.
The gas inlet is on the bottom of the cavity, the hot gas outlet is on the cavity top.
The main design data are compiled in the following. Some small differences between the figures indicated in fig. 2.2-4 and these design figures arised during the study. They will be adjusted by performing the detailed design.

Design requirements

Coolant	: air
Hours at 800 °C outlet temperature	: 20.000
Number of full load cycles	: 10.000

Main operational data at design
point conditions (21st of March,
10:30 solar time, SSPS-CRS heliostat field)

In the coolant absorbed thermal power   : 1475 kW
Inlet conditions
  Temperature   : 430 °C
  Pressure   : 24 bar
Outlet conditions
  Temperature   : 800 °C
Pressure drop (related to inlet pressure): < 4 %
Massflow   : 3,53 kg/s
Efficiency   : 71,2 %
  Reflection losses   : 1,5 %
  Emission losses   : 13,0 %
  Convection and conduction losses   : 14,3 %

Fig. 3.1-1 represents the cavity part load efficiency
at constant inlet and outlet temperatures and variable massflow.
The distribution of the direct solar irradiation
under design point conditions is shown in fig. 3.1-2
and the wall temperatures are shown in fig. 3.1-3.
The aperture of the cavity has been optimized for
the design point. The optimization has been done
under considering the dependence of the cavity losses and the intercept losses on the aperture diameter and inclination. The result is shown in
fig. 3.1-4. The distribution of the power hitting
the aperture plane at the surrounding surface is
shown in fig. 3.1-5. This irradiation results in
surface temperatures of 700 °C at the aperture rim.
The aperture is insulated by ceramic fiber boards
on the cavity inside as well as on the cavity outside
to protect the steel structure from inadmissible
temperatures. However on both surfaces the temperatures are appr. 700 °C (see also fig. 3.1-3) and the
square tube carrying the insulation (see sketch

SK-4565-B) will reach also temperatures near 700 °C.
A weak forced draft air cooling will solve this
problem because the heat flux densities to be handled
are low and no high temperature resistent steel must
be used.
The heating surface of the receiver consists of nine
identical panels. Three panels are located on each
of the three rectangular back walls. Sketch SK-4564-B
shows such a panel.
On the hot gas outlet all panels are connected to a
ring collector collecting the hot gas streams of all
panels. The reference point is located also there. The
thermal expansion of the panels is compensated on the
gas-inlet side by a lateral compensator.

The main data of the panels are compiled:

Number of tubes per panel: 22
Material of inlet header : 15Mo3
Material of tubes   : X10NiCrAlTi3220H
Diameter of tubes   : 22 mm
Relative pitch   : 2
Material of outlet header: X10NiCrAlTi3220H

The power which is absorbed by each of the panels
is different and depends on time. Clouds which cover
the heliostat field partly also affect the power distribution of the panels. Under design point conditions the power of each panel is shown in fig. 3.1-6.
The outlet temperature of each panel is 800 °C. The
different massflow through the panels is realized by
throttle valves shown in sketch SK-4564-B. The control concept of the receiver will be explained below.
The absorbed heat flux density (averaged over the
tube circumference in each cross section), the max.
tube wall temperature and the gas temperature as well
as the max. stress factor represents fig. 3.1-7 for
the maximum loaded panel (The stress factor is defined
as the relationship of operational stresses to the
permissible stresses).

The reached design of the receiver is the state after
a preliminary design phase. Some improvements might
be possible.

3.1.1.2   Control Scheme of the Receiver

The receiver control system must fulfill two tasks
 - maintain the outlet temperature at 800 °C,
 - to do that with the lowest possible additional
   pressure loss due to the throttle valves.
The receiver control scheme shall be explained with
the aid of sketch SK-4589-E.
The throttle valves D11 to D19 will adjust the massflow through each of the nine panels in such a way
that the outlet temperature of all panels is equal
to the rated value.

Because lowest possible pressure losses are required
at least one of the throttle valves is in the desired
position Y_{Set} and produces the lowest possible pressure drop.
Assuming further throttle D1 is totally opened and
throttle D2 is in a position between $Y_{D2\ max}$ (max.
flow section) and $Y_{D2\ min}$ (min. flow section). Now
the irradiation power into the receiver changes -
let's say it is reduced. The panel outlet temperatures decrease and the throttles D11 to D19 will close
accordingly. Then the pressure drop is increased in
the primary loop for the present. Y_{max} will be smaller
than Y_{Set} and $\Delta Y < 0$.
The switch is in the right hand position because also
the right hand condition is fulfilled. The signal ΔY
is negated, it means if ΔY represents a negative deviation the throttle D2 will increase the flow section
and reduce the pressure drop in the heater branch.
Therefore the massflow in the receiver branch will decrease and the panel outlet temperatures exceed the
rated value. This will be corrected by opening the
throttle valves again until $\Delta Y = 0$ and the new steady-
state is reached at the lowest possible pressure drop.

3.1.1.3   Auxiliary Cavity Firing

For using the solar energy in chemical plants special
problems arise to maintain the process conditions - at
least the process temperatures - constant over days
better weeks. Therefore any secondary fossil fired
energy source is always included in the plant concepts
proposed for using solar process heat in chemical plants.
The cavity of gas cooled receivers which mostly have
a big ratio of cavity surface area and aperture area
could be regarded as an radiation part of a heater
with the aperture as a disturbtion and the idea is
obvious to use the fossil fired cavity as a secondary
energy source. This solution could have the following
advantages

- it opens possibilities for integrated reactor/recei-
  ver concepts where the reaction is running in the
  receiver,
- it could replace an additional heater,
- it could smooth transient loads of the receiver.

Because such a cavity must be able to be operated in
a mixed operational mode - it means fossil fired and
solar "fired" simultaneously - the main open questions
are:

- What are the losses through the aperture?
- What enthalpy contents one will find in the flue
  gas?

To answer this questions theoretically it will be necessary a lot of effort and this effort can't replace
the experimental confirmation.

So it is proposed - fully separated from the Methane
reforming experiment - to install some burners in the
cavity and to investigate the above mentioned questions
in parallel.

Thermodynamic estimations have been already done to
get design ideas for such an auxiliary cavity firing.
For the present pure fuel fired operation is assumed
and the receiver inlet conditions (T_{air} = 430°,
p = 24 bar) and receiver outlet conditions (T_{air} =
800 °C) must be the same as at solar design point

conditions. It means 1475 kW must be transferred
to the heat carrier gas air of the primary loop.
The fuel to be used is natural gas L. The cavity
may be filled by combustion gases. The luminous
flame volume is negligible so it can be assumed
that the heat will be transferred to the tubes from
a gas bulk with a homogeneous temperature by convection and by radiation.
For the heat transfer to the tubes by convection a
heat-transfer coefficient of 20 $W/(m^2 \cdot K)$ is used,
the factors of radiation exchange have been determined in a complex manner described by Schupe and
Jeschar /3.1-1/. As a result has been found a temperature of the gas bulk of 840 °C which is also
nearly the flue gas temperature.
In this case a power of 1467 kW is transferred to
the cycle fluid - as required. About 20 % is transferred by convection and 80 % by radiation.
The total power of the auxiliary cavity firing to
be installed is about 3 MW which is splitted in

   power to be transferred into the cycle fluid:

                                            ≈ 1500 kW
   power leaving with flue gas*:   ≈ 1000 kW
   losses via an opened aperture: ≈ 600 kW
The indicated losses are the total receiver losses
which occure if the aperture is open and the thermal
power to the cycle fluid is supplied by the firing
system only. This is an operational mode of theoretical relevance because the receiver will be operated
in this mode for a short time only - if at all - and
then the aperture will be closed and only the conduction losses have to be covered.

*It has to be used in the process yet.

However for a mixed operation which is of most interest these losses will occure and must be covered
by the solar irradiation and the firing together.
Then the losses can be shared only theoretically
between both energy sources.
These losses are investigated for a pure solar
operation but the effect of an auxiliary cavity firing
like
  - changing the thermo-optical properties by a
    combustion gas filled cavity and
  - excitation of the flow in the cavity by bur-
    ners and flue gas exhaust
must be investigated additionally.
The influence of the changed infrared absorption
properties due to the combustion gas containing cavity has been estimated and drawn in fig. 3.1-8. It
leads to somewhat increased radiation losses.
The gas exchange via the aperture will mainly be
influenced by the location of burners and flue gas
exhaust. Some rough ideas of the flow pattern are
shown in fig. 3.1-10 for an arrangement of burners
and flue gas exhaust which is shown in fig. 3.1-9. For another arrangement using recuperative burners the flow
pattern are drawn in fig. 3.1-11.
Because it can't be prevented to suck ambient air it
must be avoided that these air can take energy with
it if it leaves the cavity again. Especially care
must be taken that this "by-padded" air can't be
mixed intensively with the hot gases inside the cavity and can't reach the receiver back walls where
it could also take away useful heat.
A solution could be to suck in a well defined massflow of ambient air and to suck off it at the cavity
bottom again (see fig. 3.1-10, 3.1-11). Due to big
density difference of the cold ambient air and the
hot cavity gas bulk mixing of both should not occure
intensively and only a few amount of heat can withdrawn from the cavity bottom.

The investigation of all these questions should be
done experimentally and could be a task of general
interest of using solar process heat in any process
beside the Methane reforming.

3.1.2   Auxiliary Loop Components

Some auxiliary components of the primary loop shall
be described briefly.

Compressor K-3:
type: 1-stage radial-flow turbocompressor

max. massflow	:	4,5 kg/s
max. intake temperature	:	430 °C
suction pressure	:	21 bar
discharge pressure	:	24,5 bar
isentropic efficiency	:	0,64
drive power to be installed:		250 kW
design point operation (acc. to fig. 2.2-4):		
massflow	:	4,45 kg/s
intake temperature	:	430 °C
suction pressure	:	21,84 bar
discharge pressure	:	24,24 bar
outlet temperature	:	460 °C
drive power	:	146 kW

To drive the compressor a Diesel engine is foreseen which can be operated speed controlled easily to realize different mass flows in the primary
loop.

Diesel engine:

type: D 2866 TE, 4-stroke 6 cylinder

power at 2200 rpm	:	250 kW
torque at 2200 rpm:		1085 Nm
consumption	:	210 g/kWh

Cooler E-16:

type: forced draft air cooler

tube side:
  inlet temperature   :   430 °C
  inlet pressure   :   22 bar
  outlet temperature   :   270 °C
  pressure drop   :   0,05 bar
  massflow   :   1,5 kg/s

cooling power   :   255 kW

shell side:
  suction air temperature   :   30 °C
  installed power of fan drive: 4,5 kW

Low temperature lines:

max. operational temperature: 430 °C
max. operational pressure   :   25 bar
nominal diameter   :   150 mm
material   :   15Mo3
thermal losses per length   :   ≈ 0,4 kW/m

High temperature lines:

max. operational temperature   800 °C
max. operational pressure   25 bar
nominal diameter   200 mm
material   X10NiCrAlTi3220H
thermal losses per length   ≈ 0,4 kW/m

The hot gas line will have an outer insulation too.
Reference points of the lines will be the connections
to the reactor (fixed) and to the receiver (fixed in
vertical movement only). The thermal expansion (about
450 mm for the hot line and 250 mm for the cold line)
will be compensated by angular compensators at the
reactor level. About in the middle of the tower the
lines will be supported by constant hangers additionally.

References chapter 3.1

/3.1-1/ Schupe, W., R. Jeschar
          Vereinfachte Berechnung des Strahlungs-
          wärmeüberganges in Industrieöfen und Ver-
          gleich mit Messungen in einer Versuchs-
          brennkammer
          Gas, Wärme international, Bd. 24 Nr. 2 (1975)

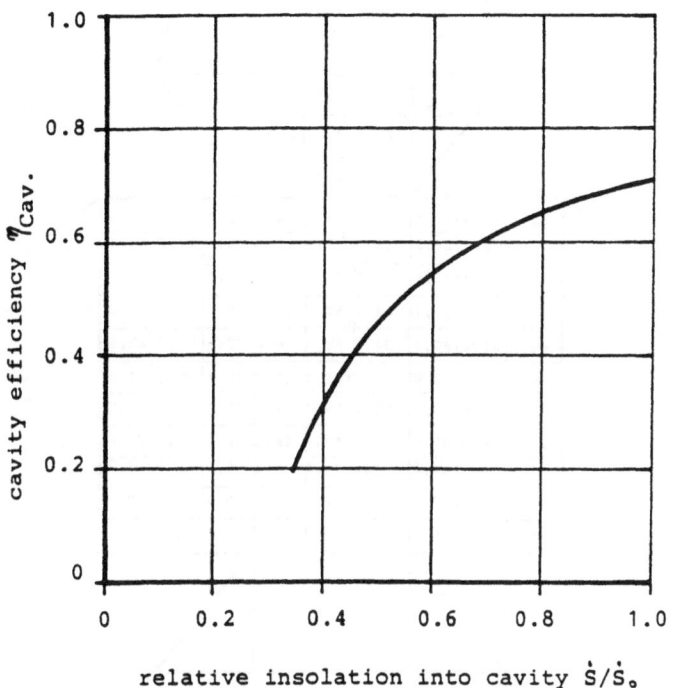

Fig.3.1-1: Cavity part load efficiency

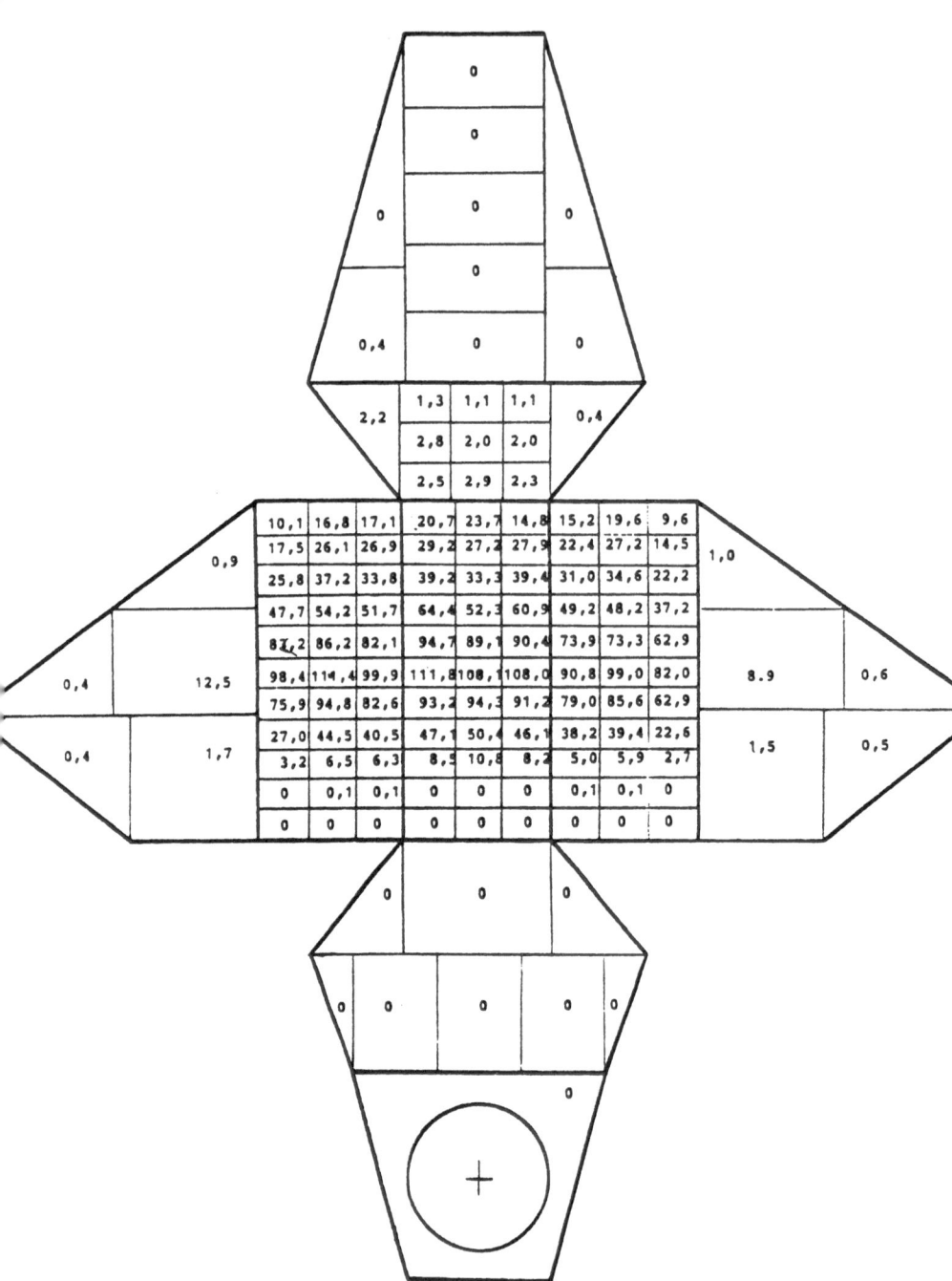

<u>Fig. 3.1 - 2:</u> Distribution of incident solar flux density on the
inside receiver surfaces in kW/m²
(design point: 21st of March, 10.30 o'clock)

- 146 -

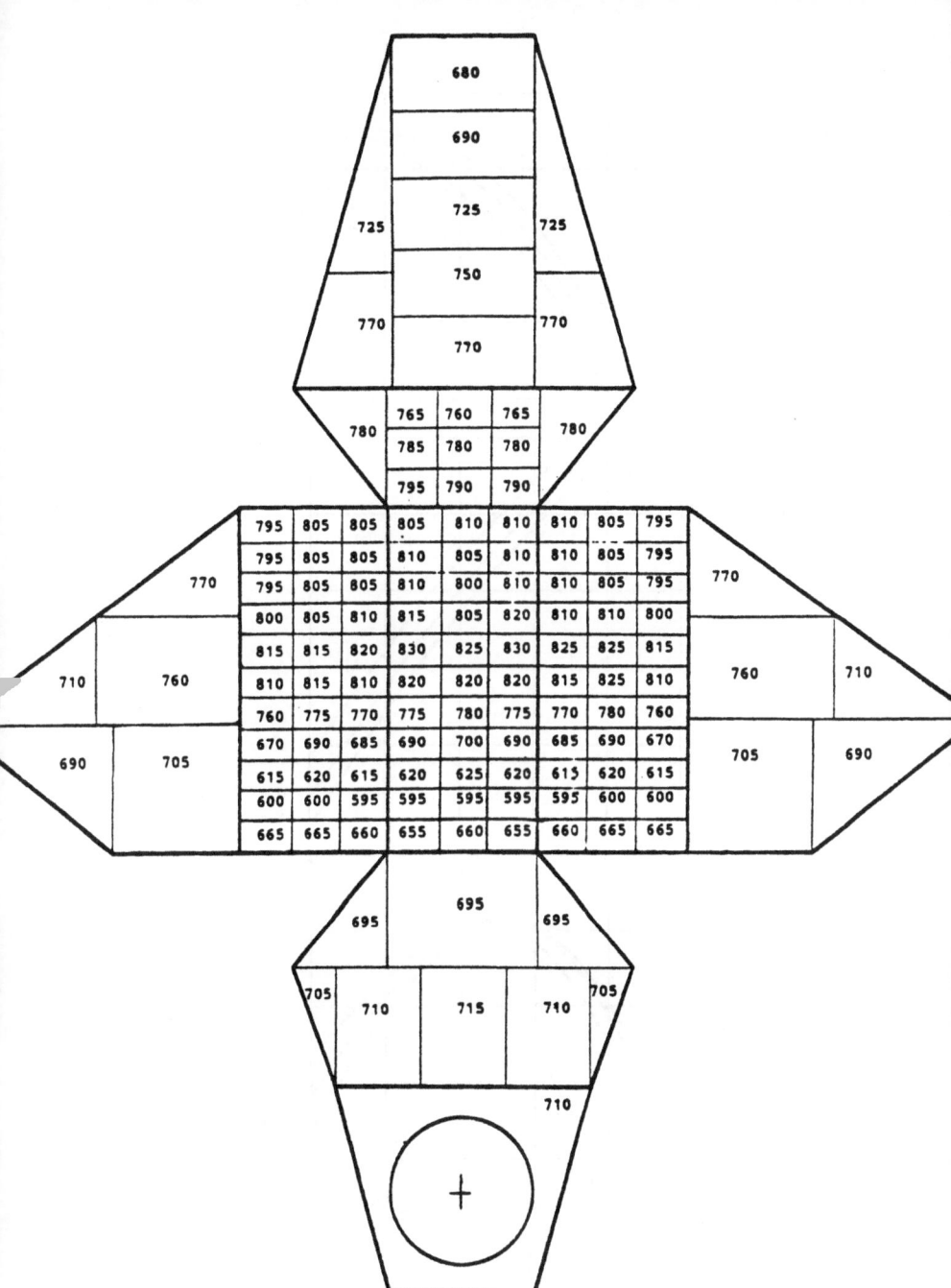

ig. 3.1 - 3: **Wall temperatures (°C) inside the receiver at design point**

- 147 -

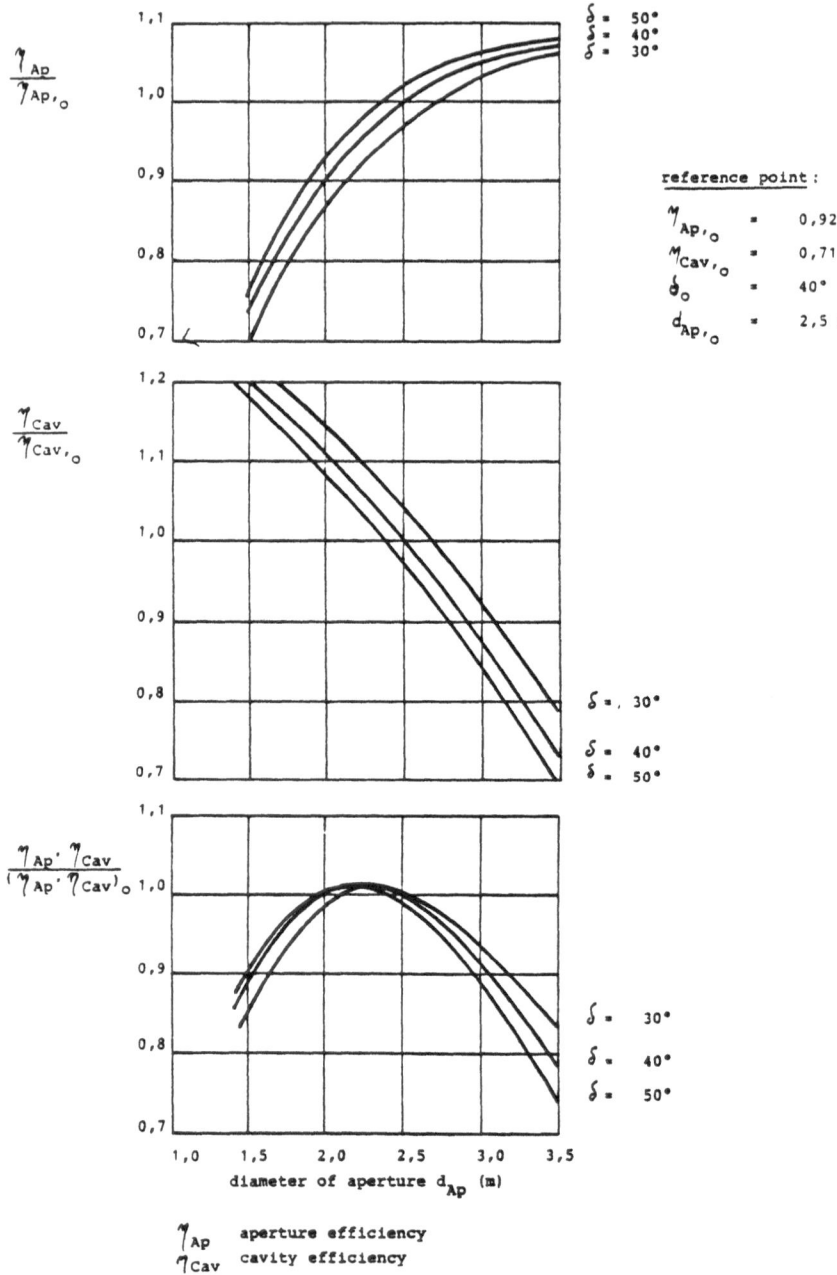

$\frac{\eta_{Ap}}{\eta_{Ap,o}}$

$\delta = 50°$
$\delta = 40°$
$\delta = 30°$

reference point :

$\eta_{Ap,o}$ = 0,92

$\eta_{Cav,o}$ = 0,71

δ_o = 40°

$d_{Ap,o}$ = 2,5 m

$\frac{\eta_{Cav}}{\eta_{Cav,o}}$

$\delta =, 30°$

$\delta = 40°$
$\delta = 50°$

$\frac{\eta_{Ap} \cdot \eta_{Cav}}{(\eta_{Ap} \cdot \eta_{Cav})_o}$

$\delta = 30°$

$\delta = 40°$

$\delta = 50°$

diameter of aperture d_{Ap} (m)

η_{Ap} aperture efficiency
η_{Cav} cavity efficiency
δ angle between aperture plane and horizontal plane

Fig. 3.1 - 4:   Optimum of aperture diameter dependent on aperture and cavity losses

- 148 -

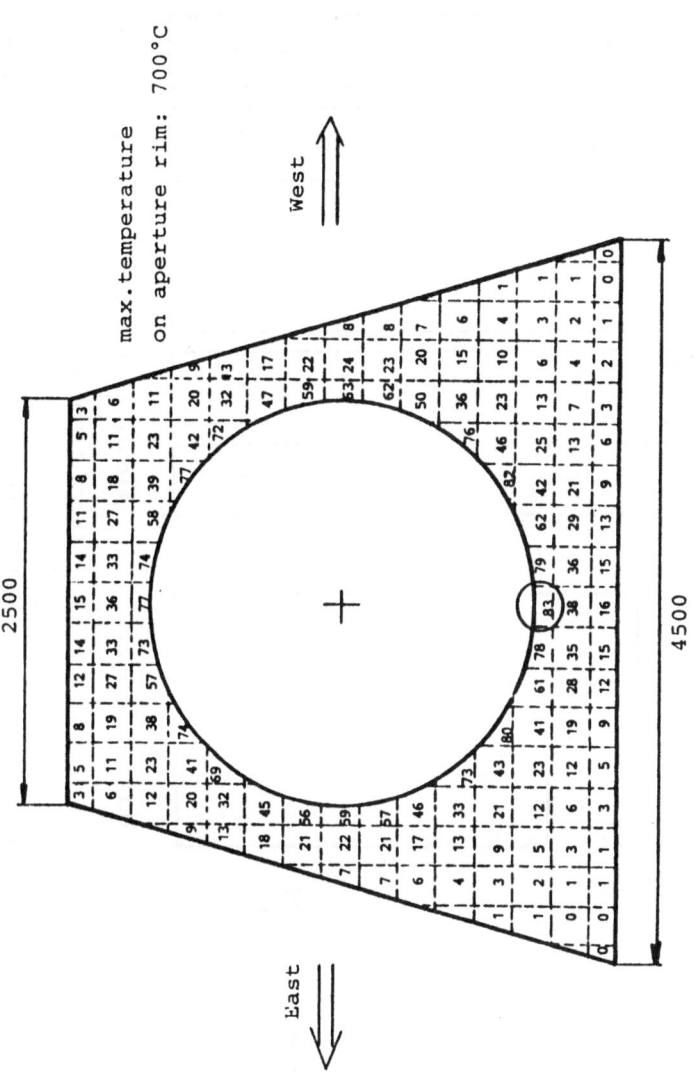

max.temperature
on aperture rim: 700°C

Fig. 3.1 - 5: Solar flux density in kW/m² on the outside aperture
surrounding receiver wall at design point

- 149 -

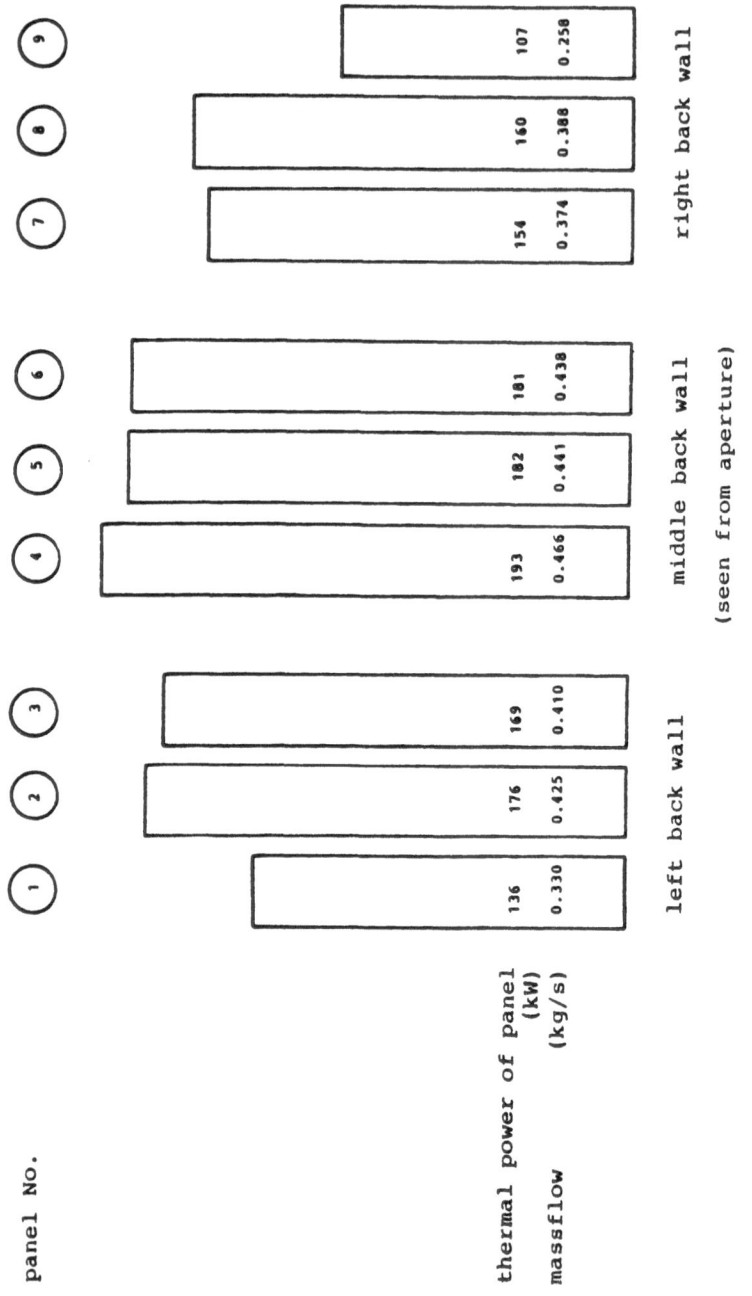

panel No.

① ② ③   ④ ⑤ ⑥   ⑦ ⑧ ⑨

thermal power of panel (kW)

massflow (kg/s)

	1	2	3	4	5	6	7	8	9
thermal power (kW)	136	176	169	193	182	181	154	160	107
massflow (kg/s)	0.330	0.425	0.410	0.466	0.441	0.438	0.374	0.388	0.258

left back wall   middle back wall (seen from aperture)   right back wall

Fig. 3.1 - 6: Distribution of thermal net power on the receiver back walls at design point

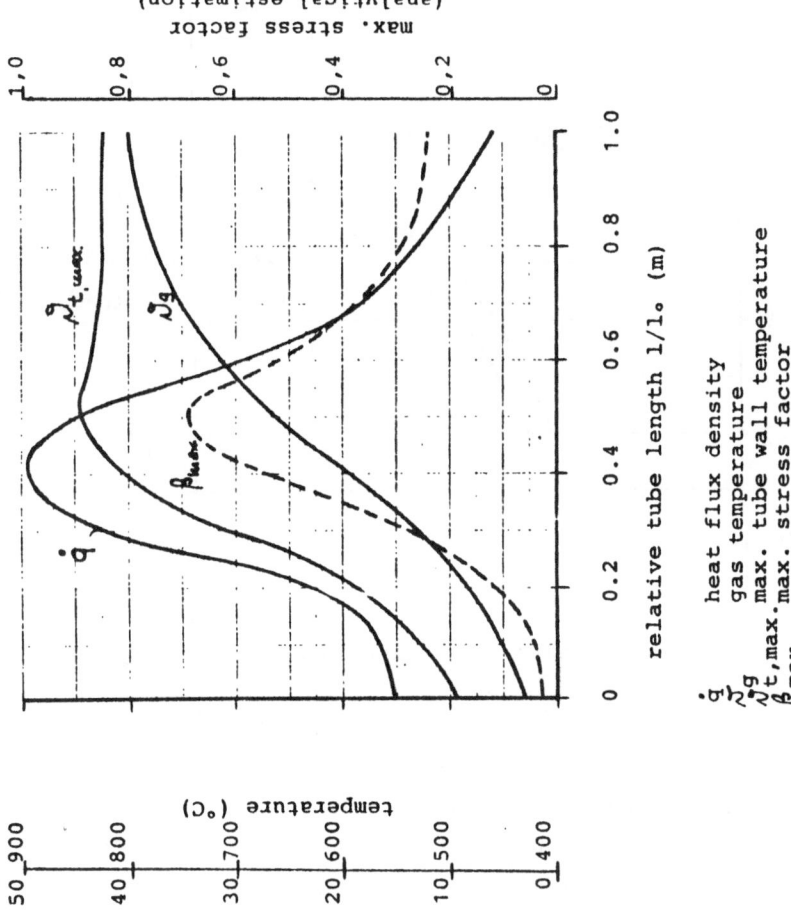

max. stress factor
(analytical estimation)

1,0   0,8   0,6   0,4   0,2   0

temperature (°C)

heat flux density (kW/m²)

relative tube length l/l_0 (m)

0   0.2   0.4   0.6   0.8   1.0

$\dot{q}$ heat flux density
ϑ_g gas temperature
$\vartheta_{t,max}$ max. tube wall temperature
β_{max} max. stress factor

Fig. 3.1 - 7:   Steady state conditions for max. loaded panel (No.4)
at design point

- 151 -

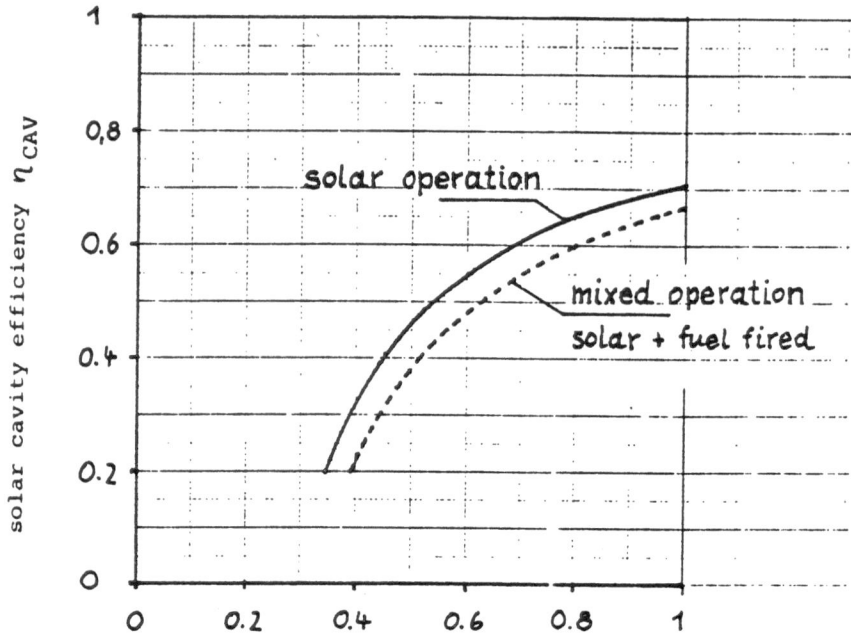

Fig. 3.1 - 8: Cavity part load efficiency.
Influence of hot combustion gas radiation loss
on solar efficiency.

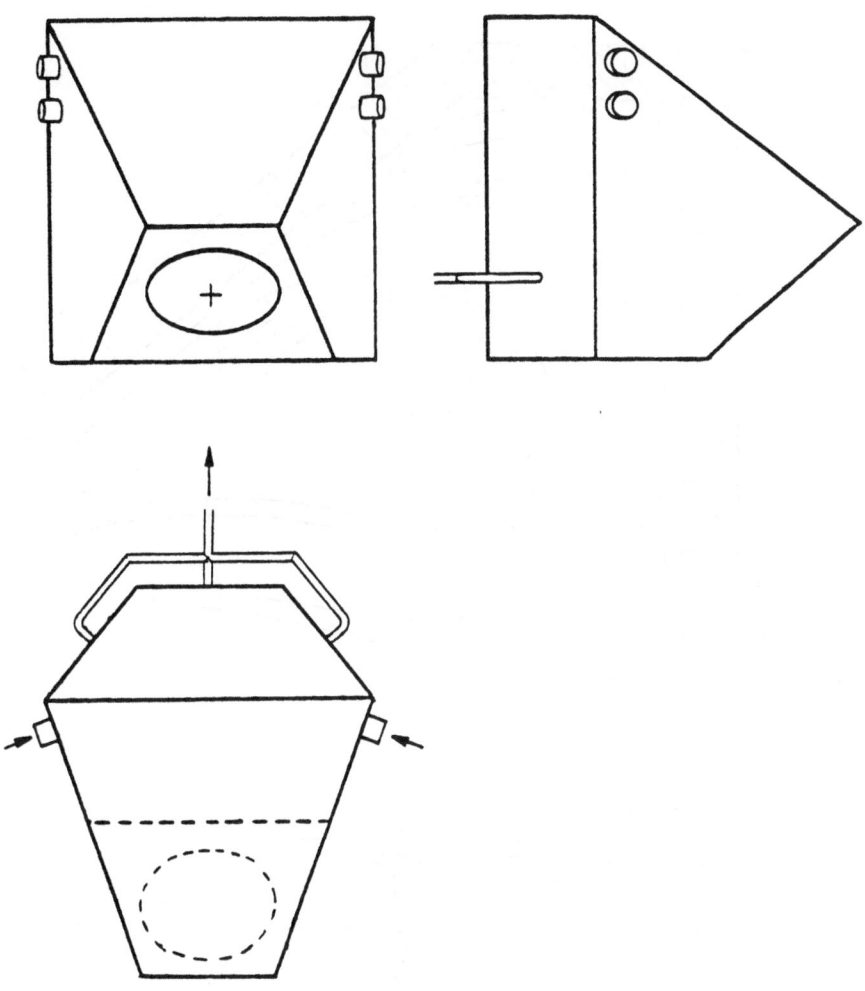

Fig. 3.1 - 9: STAP - Cavity

Example for arrangement of burners and exhaust pipes
for natural gas fired cavity heating equipment
(sketch on principle)

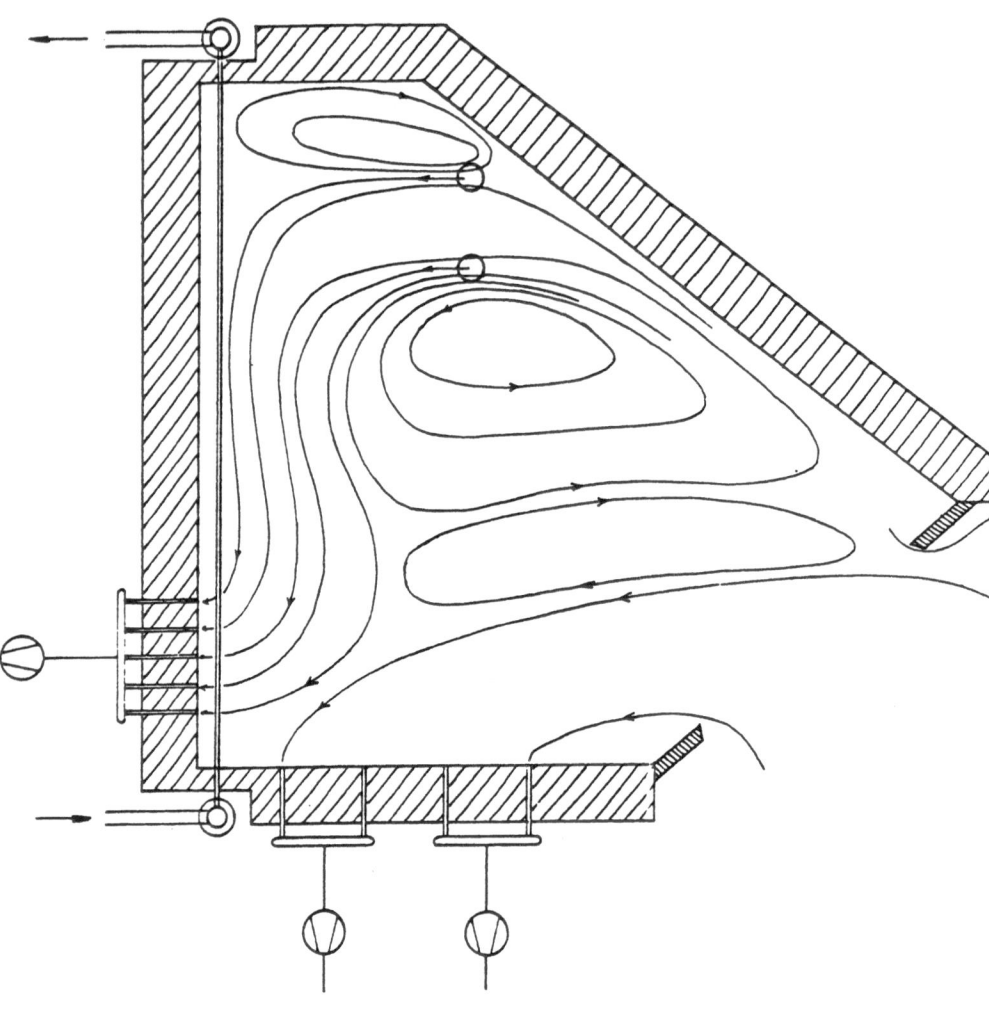

Fig. 3.1 - 10:Hot gas flow in the cavity.
Stream lines sketched in the cavity symmetric plane,
burner and exhaust are separate located.

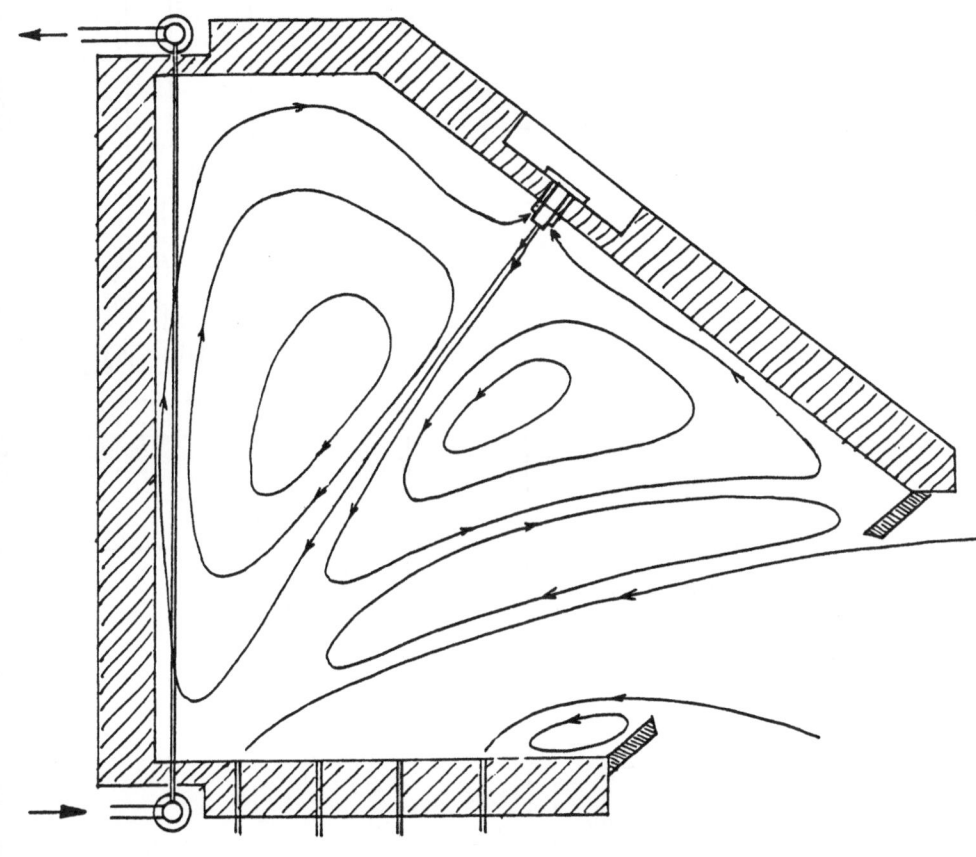

Fig. 3.1 - 11:   Hot gas flow in the cavity.
Stream lines sketched in the cavity symmetric plane,
exhaust opening integrated in the burner with
recuperator.

- 155 -

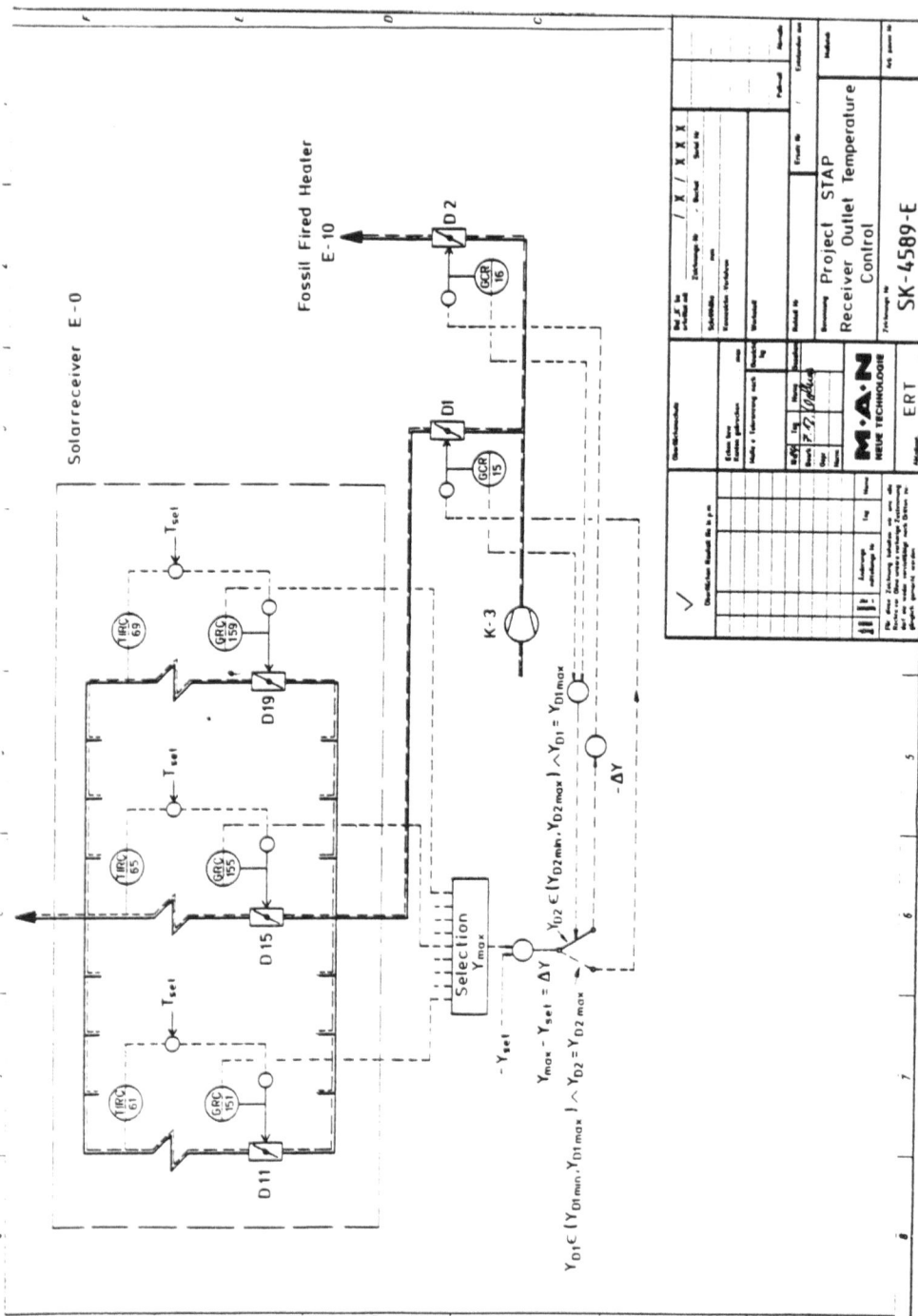

$Y_{D1} \in (Y_{D1\,min}, Y_{D1\,max}) \wedge Y_{D1} = Y_{D1\,max}$

$Y_{D2} \in (Y_{D2\,min}, Y_{D2\,max}) \wedge Y_{D2} = Y_{D2\,max}$

$Y_{max} - Y_{set} = \Delta Y$

Selection Y_{max}

Solarreceiver E-0

Fossil Fired Heater E-10

MAN NEUE TECHNOLOGIE   ERT

Project STAP
Receiver Outlet Temperature
Control

SK-4589-E

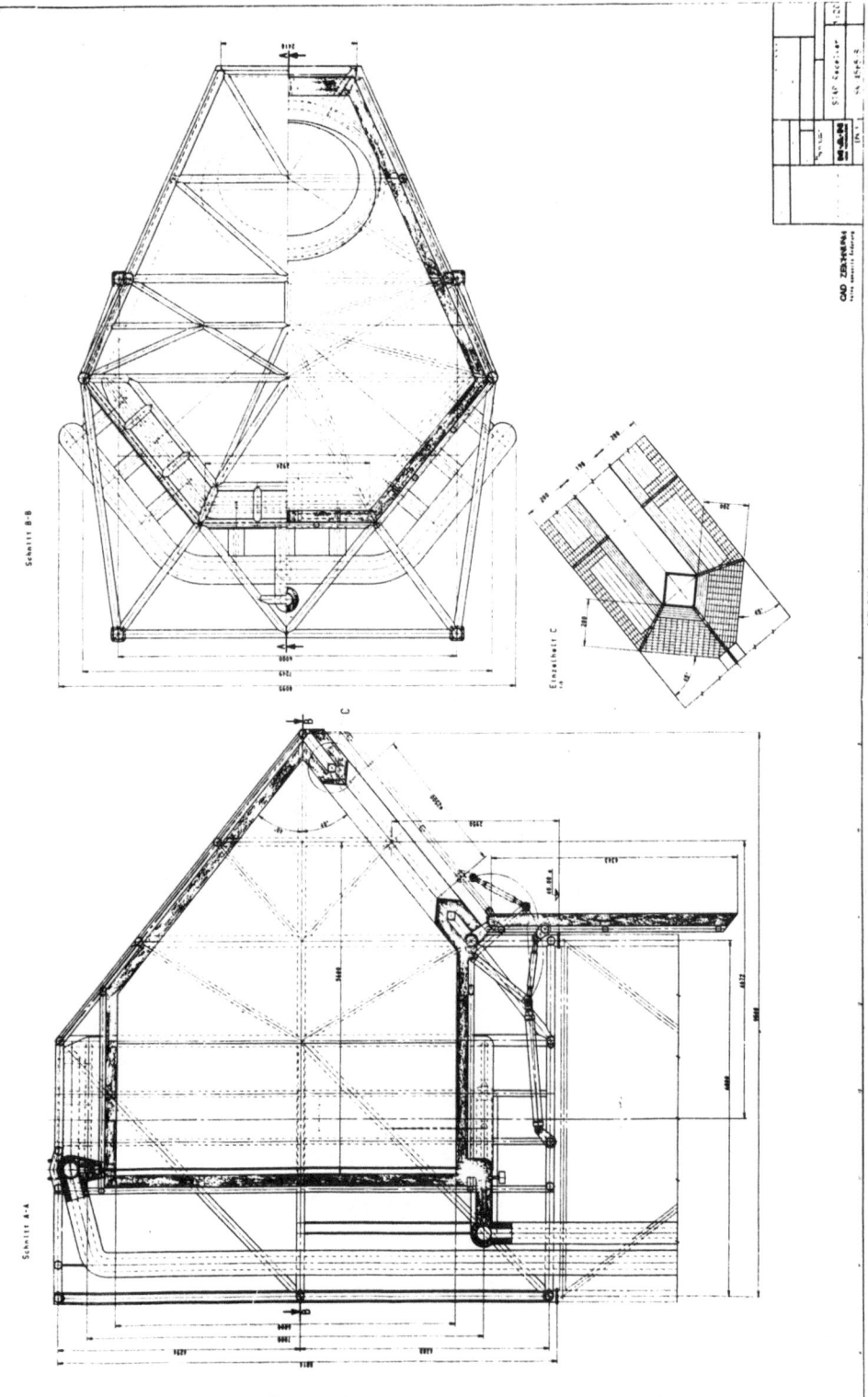

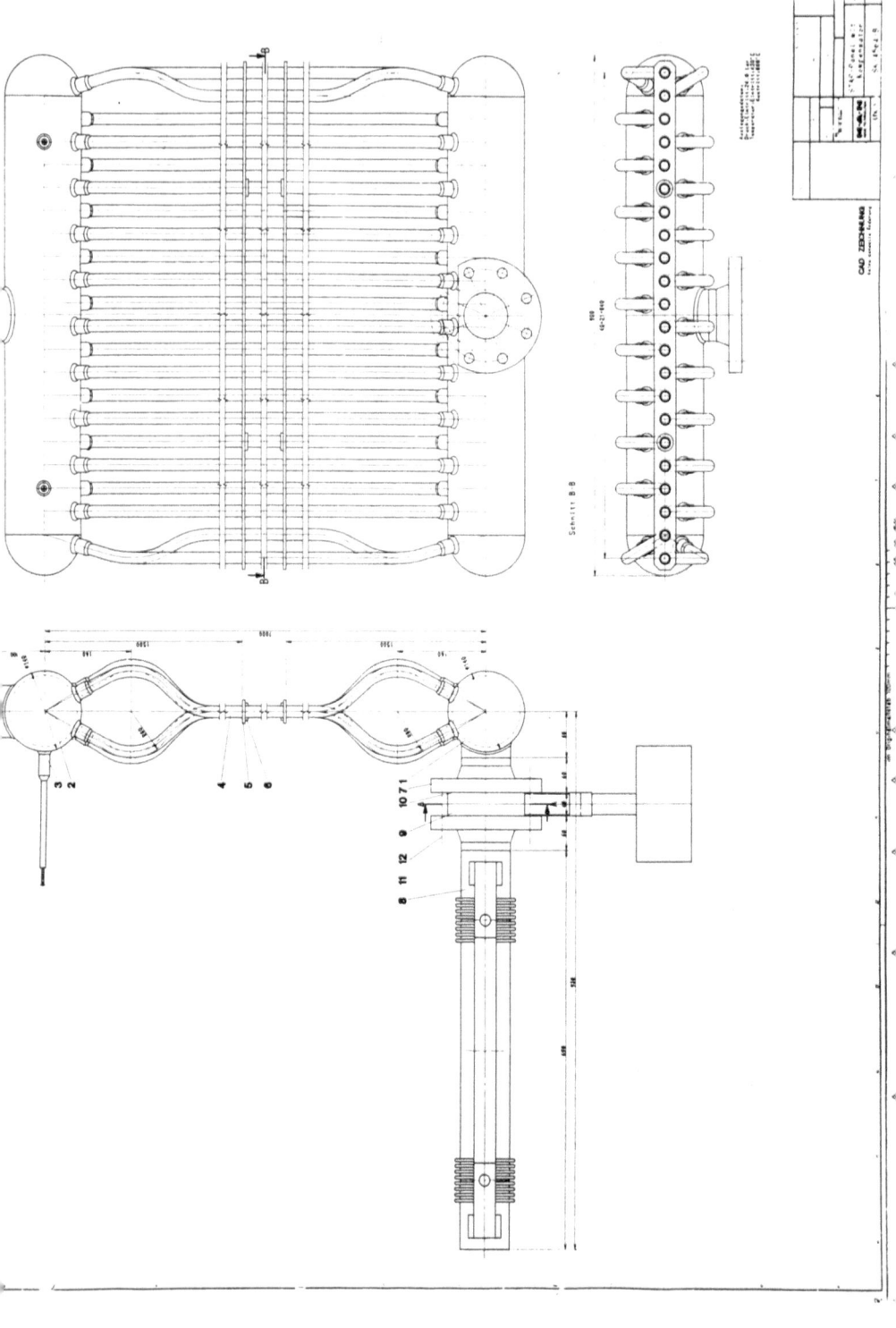

3.2 Air Heater E 10

See drawing D 6219200-0L00008 (also for E 11 and E 12)

The outlet temperatures of this exchanger shall be 800°C
The expected load variations are larger than 1 : 7. To
maintain tube wall temperatures below 900 $^{\circ}$C similar to
reformer technology of LURGI, a parallel flow exchanger
has been designed.

This of course will enlarge the total area to be installed
but satisfies the other conditions.

The flue gas temperature at the outlet of this exchanger
will be nearly identic with the air outlet temperature at
partial loads, namely 800 $^{\circ}$C, and will reach 900 $^{\circ}$C and
more with full load.

The main data of this heat exchanger will be:

surface	26,4 m^2
No. of tubes (passes)	56
tubes	25 mm x 2,5
material	x 10 NiCr AlTi 3220 H
	(Incoloy 800 H)
design lifetime	100 000 hrs.

3.3 Steam Superheater E 11

As this exchanger also is situated at the outlet of a
(secondary) combustion chamber the same principles apply
as for E 10. The situation is even more complex as flue
gas quantities, inlet temperatures and steam quantities
change independently from each other and therefore no
single design case can be fixed. Correspondingly the outlet temperature of the superheated steam may vary from
about 435 $^{\circ}$C to 680 $^{\circ}$C. Special investigations have been
made to better identify the heat transfer situation around

this heat exchanger. The "overheated" superheated steam will be quenched with boiler feed water afterwards to gain the steam with the required temperature level of about 350 °C.

The main data of the steam superheater will be:

surface   10,2 m^2
No. of tubes (passes)   27
tubes   20 mm x 2,0
material: x 10 CrNi AlTi 3220 H
                                (Incoloy 800 H)
design lifetime: 100 000 hrs.

3.4   Waste Heat Boiler E 12

This will be a conventional water tube boiler. It will be designed to maintain flue gas temperatures at maximum 300 °C at the outlet. It will provide steam for normal operation and start-up.

3.5   Prereactor R 1

See drawing No. D 6219200-0L00006.

With regard to the high temperatures, which are involved in the energy exchange of this reactor, high alloy material had to be selected for the active heat exchange surfaces. These parts will be dimensioned in such a manner that a lifetime of 100 000 hrs may be expected under normal operating conditions, which includes regular load changes and sudden, abrupt shut down procedures once in 2000 hrs.

The outer pressure bearing vessel parts can be protected from high temperatures by insulation brick lining, as then cheaper material can be selected for the pressure vessel, thereby causing a lower price in total. The details of such a construction are shown on the drawing as in reference above.

The main data of this reactor are as follows:

No. of catalyst tubes:	190
Tube size:	38 mm x 2,6
Catalyst volume:	0,48 m^3
Heat exchanger area:	68 m^2
Total volume of reactor:	about 8 m^3
Average heat load	24 kW/m^2
Average heat transfer coefficient, required	about 350 W/m^2K
Weight:	about 15 t.

3.6   Equipment List

The equipment for the modified solar plant at Almeria as
enumerated by the following list was being designed in
some detail (as for instance solar receiver, prereactor
and fossil fired air heater) or was dimensioned with adequate accuracy and fixed by datasheets to allow the price
calculations as being shown in chapter 5.1. Other material
as piping and valves, instrumentation, electrical equipment, steel structure etc. was evaluated partly on an item
by item calculation, partly by known percentages of overall costs for similar plants.

R-1   Prereactor
R-2 First Methanation Reactor
R-3/E-9 Final Methanation Reactor (internally cooled)

D-1   Ion Echanger
D-2   Ion Exchanger

E-0   Central Solar Receiver
E-1   Methanated Gas Preheater
E-2   Waste Heat Boiler I
E-3   Boiler Feed Water Preheater
E-4   Reformed Gas Air Cooler
E-5   Reformed Gas Reheater
E-6   Waste Heat Boiler II
E-7   Methanation Intermediate Air Cooler
E-8   Methanated Gas Air Cooler
E-9   see at R-3, Catalyst Heat Exchanger
E-10   Air Heater
E-11   Steam Superheater
E-12   Waste Heat Boiler III
E-13   Dissolved Gas Cooler
E-14   Air Cooled Steam Condenser
E-15   Methanated Gas Compressor Cooler
E-16   Air Compressor Recycle Cooler
E-17   Methane Pressurization Evaporator

E-18	Methane Vaporizer
E-19	Nitrogen Pressurization Evaporator
E-20	Nitrogen Evaporator
E-21	Air Cooled Water Cooler
E-22	Coolant Evaporator
E-23	Water/Water Exchanger
E-24	Instrument Air Water Cooler
F-1	Reformed Gas Condensate Separator I
F-2	Methanation Condensate Separator I
F-3	Methanation Condensate Separator II
F-4	Steam Drum
F-5	Freshwater Tank
F-6	Predegaser
F-7	Main Degaser and Boiler Feed Water Drum
F-8	Alkalinization Vessel
F-9	Water Filter
F-10	Expansion Drum
F-11	Water Treatment Condensate Separator
F-12	Fuel Oil Storage Tank
F-13	Reformed Gas Condensate Separator II
F-14	Liquid Seal
F-15	Knock-out Drum
F-16	Condensate Knock-out Drum
F-17	Liquid Natural Gas Storage Tank
F-18	Low Pressure Natural Gas Receiver
F-19	High Pressure Natural Gas Receiver
F-20	Liquid Nitrogen Storage Tank
F-21	Carbondioxide Storage Tank
F-22	Expansion Vessel I
F-23	Expansion Vessel II
F-24	Air Filter
F-25	Water Separator
F-26	Instrument Air Receiver
F-27 A/B	Instrument Air Dryers
F-28	Guard Filter

G-1	Freshwater Tank Pump
G-2	Freshwater Feed Pump
G-3	Predegaser Bottom Pump
G-4	Boiler Feed Water Pump
G-5	Dosing Pump
G-6	Expansion Drum Condensate Pump
G-7	Fuel Oil Feed Pump
G-8	Water Circulation Pump
G-9	Cooling Water Circulation Pump
	(All Pumps will have spare, except G-5)

K-1	Methanated Gas Compressor
K-2	Methanation Recycle Compressor
K-3	Hot Air Recycle Compressor
K-4	Flue Gas Blower
K-5	Predegaser Airblower
K-6	Dissolved Gas Compressor
K-7	Air Make Up Compressor
K-8	Combustion Air Blower
K-9	Cooling Air Blower
K-10	Methane Make Up Compressor
K-11	Coolant Compressor
K-12	Instrument Air Compressor

3.7   Arrangement Drawing

Due to the high temperatures of the heat carrier medium
(air) it should be avoided to have long gas piping between
the central solar receiver and the energy conversion
system, i.e. prereactor R 1 to avoid energy losses and additional costs for a specifically expensive piping. Therefore it was to be investigated whether it would be possible to arrange the main equipment, which should be linked
to the primary energy loop, near to the solar tower though
the space is limited at the SSPS-Plant at Almeria.

A solutioin of this problem is shown by the attached
arrangement drawing No. D 6219200-0L00007. Though this
drawing shoud be a first approach only it indicates that
this main aim, as outlined above, can be achieved without
special difficulties. This solution allows even to set up
the methanation plant within the steel structure at the
bottom of the solar tower. Only auxiliary equipment, as
liquid nitrogen storage, natural gas storage and the like
would have to be placed off the main process plants (and
therefore are not shown on the arrangement drawing),but as
there does not exist any restriction for the distance of
such auxiliary plants with respect to the distance from
the main plants, there would be ample space also at Almeria
and no difficulties are envisaged.

4.0 OPERATION OF THE TESTING FACILITIES

4.1 Proposals for a Test Program

The purpose of this testing facility should be the demonstration of the realizability of such a complex system,
which consists of several independent circuits of new technologies and to link those in such a manner that a stable
chemical reaction will be achieved, which is the basis for
any continuing chemical process. As the reaction step of
methane reforming is a stratified function of temperature,
pressure, throughput, time, energy flows and their changes
to each other, there is hardly any chance for predictions
of high accuracy of such a behaviour and an experiment of
the proposed nature is thought to be inevitable as a proof
of reliability, before the realization of any commercial
plant of this or similar nature can be approached.

In this context the main accents of the experiment should
be the operation of the methane reforming process under
different conditions like
. fluctuating energy input
. fluctuating temperature of the primary gas
. decreased temperature of the primary gas
and the investigation of the analysis of the product gas
under those varying conditions. Finally it should be investigated how much of the collected solar energy can be
found in the product under realistic solar conditions and
it should be tried to answer how much solar energy specifically could be gathered by a commercial plant. More general experiences will be gained also, in which way a solar
driven chemical plant has to be operated.

To evaluate the tests and to have a feedback of test experiences it is proposed to perform the test in measurement campaigns. First ideas of possible test tasks for
three measurement campaigns are described in the following:

First Measurement Campaign:
. Evaluation of conversion rates under operation of the
  fossil fired heater only
. Investigation of reactor inlet and outlet conditions
  (tube and shell side) during receiver start-up and shut-
  down procedures at different fossil load levels
. Increase of solar energy portion, verification of control system service at fluctuating solar irradiation
. Operation of the auxiliary cavity burners und different
  solar loads and investigation of energy balance with
  pure fossil and mixed operation
. Development of an operational strategy
Duration: approx. 6 months

Second Mearurement Campaign:
. Investigation of optimum arrangement of burners in the
  cavity and of the relevant exhaust gas outlet
. Investigation of heat distribution in the cavity and
  changes relevant to the first operating period
. Investigation of equalizing disturbances of the solar
  investigation by the auxiliary cavity burners
. Continuous operation with the solution, which has been
  found most suitable, for a period of threes months at
  least.
Total duration of this campaign: 8 months.

After the second mearuement campaign some parts of the
plant (of those which have been operated at the highest
temperatures) should be taken off and investigated metal-
lurgically for further prediction of lifetime of such
plants. With new exchanged parts the third measurement
campaign with increased risk of failures should be started

Third Measurement Campaign:

. Experiments of overload to determine reliable limits of
  future design
  (increased heat fluxes at design outlet temperatures,
  for instance with increased air mass flow, with all additional mirrors on duty, increased outlet temperatures
  of single panels of the solar receiver E 0, increased
  temperatures of the fossil fired heater E 10, increased
  temperatures of outlet of prereactor R 1, overload of
  cavity burners, change of molar steam/methane ratio at
  prereactor R 1)

. Continuous operation with overload operating mode, which
  indicates highest economic benefit, for at least three
  months

Duration of third campaign: approx. 6 months.

. Investigation of materials of construction.

4.2   Consumption Figures for Test Programs

To allow the calculation of the costs of an experiment the
consumption figures have been evaluated for different loads
of the proposed plant.

Four cases have been investigated:

Case A: Design load of the plant. The energy from the sun
         is at its design point, the fossil fired heater
         E 10 provides its minimum load of approx. 15% of
         its design load.

Case B: No energy from the sun available, total energy at
         design condition for the prereactor R 1 to be
         supplied by the fossil fired gas heater E 10;

Case C: No energy from the sun available, the fossil fired
         gas heater E 10 to be run at its minimum load of
         approx. 15 % to maintain reactions and temperatures
         within the plant.

Case D: The fossil fired heater E 10 is run constantly at
minimum load of approx. 15 %, while the sun is adding its energy during daytime as much as available.

Case D will be an operating mode to save energy and costs
in the experimental plant. This changing gas load will
not be preferable for commercial plants but it is expected
that in the proposed pilot plant this will be the operating
mode, which will be adhered to during long testing periods
and which therefore is thought to be typical for the prediction of consumption.

The daily consumption of feedstock and utilities for the
above reference load cases will amount to roughly the
following values.

	Case (A)[*]	Case (B)	Case (C)	Case(D)
Natural Gas Feed Stock m^3N/d	3 [*]	70	40	50
Fuel Oil kg/d	95 [*]	7000	1200	160
Diesel Oil kg/d	60 [*]	1450	300	750
Solar Thermal Input MWh/d	1,40 [*]	0	0	9,68
SNG Throughput m^3N/d	1028,9 [*]	24700	3500	9600
Electric Energy kWh/d	245 [*]	5900	4600	5100
Demineralized Water kg/d	450 [*]	10800	6500	7800
Cooling water m^3/d	no external utility; selfgenerated			
Nitrogen m^3N/d		20		

Hydrogen m^3N/d negligible, temporarily only

Carbondioxide m^3N/d max. 20

* For case (A) the values given are design rates on instant
basis as "quantity per hour", as this design case cannot
be maintained throughout the total day as a consequence
of the changing energy flux of the sun.

4.3   Manpower Requirements

For the continuous operation of the plant the following
personnel will be required as a minimum:

            1 Plant Manager
            1 Chemist
            1 Shift Supervisor   )
            1 Panel Operator   )   per shift
            1 Instrument Engineer)
            2 Operators   )

so that the staff will totally include
            27 persons
if five shifts will be nominated as presently common guiding rule.

This number of persons per shift is based on the assumption, that electricians, pipefitters, machinists and the
like are occasionally available during daytime. The indicated number of persons does not include any members of an
evaluation team.

5.0 COST ESTIMATES AND TIME SCHEDULE

5.1 Cost Estimates

On the basis of the technical data, which have been presented in this study herebefore, calculations of the pertaining work and deliveries has been performed by both partners, MAN and LURGI, to achieve a better judgement on the total costs of such an experiment.

The indicated prices are approximative with an estimated accuracy of about 20 %. They are based on prevailing prices of equipment as per cost index of November 1985. The indicated prices for deliveries are given under the assumption of German fabrication, delivered to the site of Almeria, but without customs or other duties outside Germany. The engineering portion has been evaluated on the assumption of undisturbed progress of this work of a start in 1985, in accordance with the attached time schedule and related only to the installation of the equipment and material as principally outlined herebefore.

The results of such cost estimates in million Deutsche Mark are as follows:

	LURGI	MAN	Total
1) Basic and detailed engineering	5,3	1,7	7,0
2) Delivery of equipment and material to the site and procurement of such deliveries	6,0	5,5	11,5
3) Civil work and erection of equipment, including pertaining supervision of the same and start-up guidance by Lurgi - MAN	4,5	1,4	5,9
Total	15,8	8,6	24,4

These figures do not include contingencies and are related to terms of payment in accordance with the schedule of labour and deliveries, i.e. in accordance with the relevant progress of performances.

5.2   Time Schedule

The attached time schedules indicate a total duration of 30 months for this project up to the start-up, which is thought adequate with respect to new construction elements and special material requirements with long delivery times. Details are to be taken from the graphs.

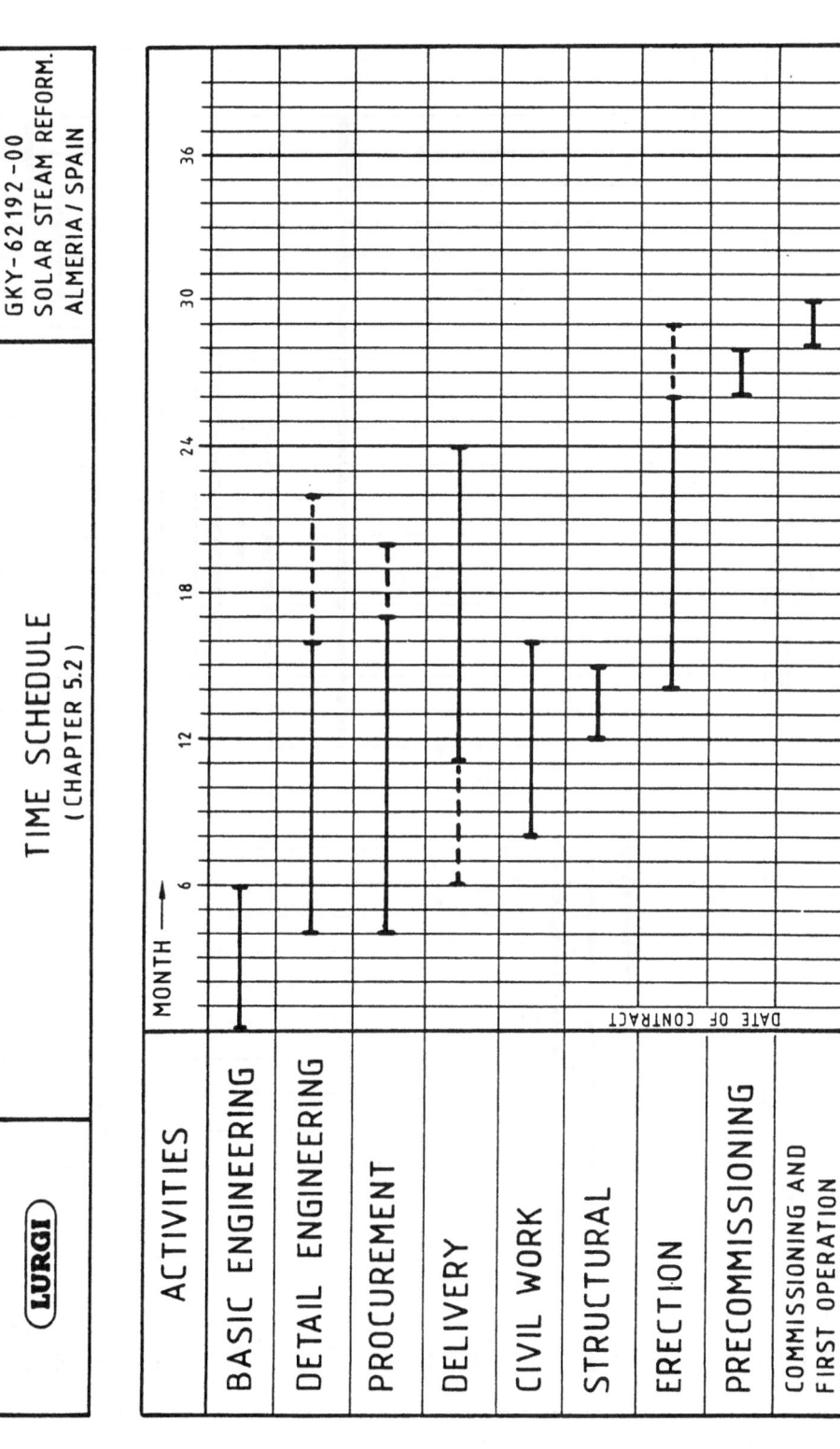

LURGI

TIME SCHEDULE
(CHAPTER 5.2)

GKY-62192-00
SOLAR STEAM REFORM.
ALMERIA / SPAIN

ACTIVITIES	MONTH →
BASIC ENGINEERING	
DETAIL ENGINEERING	
PROCUREMENT	
DELIVERY	
CIVIL WORK	
STRUCTURAL	
ERECTION	
PRECOMMISSIONING	
COMMISSIONING AND FIRST OPERATION	

6   12   18   24   30   36

DATE OF CONTRACT

PROJEKT: STAP

PLANART: General Time Schedule

PROJEKT-NR.
ANGEBOTS-NR.
A 630 341

PLAN-NR.

AUSGABE | PLANUNGS ZYKLUS | NAME

PLANUNG

PROJEKT LEITUNG | Dr. Melchior

Name | Unterschrift

1. Detailed Engineering
1.1 Management, Techn. Coordination, Quality Assurance
1.2 Thermal Design
1.3 Mechanical Design and Stress Analysis
1.4 Control System Design
1.5 Specification of Instrumentation and E-equipment
1.6 Construction-, Supply- and Quality-Control Documents

2. Follow-up Engineering Construction and Transport
2.1 Management, Technical Coordination
2.2 Receiver
2.2.1 Receiver Material Procurement
2.2.2 Structure Construction
2.2.3 Panel Construction
2.3 Primary Loop
2.4 I+C-equipment, E-equipment

3. Assembly and Commissioning
3.1 Cavity Assembly and Build-in of Panels
3.2 Primary Loop
3.3 Commissioning

6.0   SUMMARY

In accordance with the given order a system has been technically elaborated, which should allow the demonstration
of a workable solution in the larger context of "steam reforming of methane with solar energy". The proposed system
may be classified as conventional and conservative, but
it should be recognized that nevertheless the elements of
this system have not yet been tested under real solar conditions but have only been developed theoretically. Also
in the classification of conventional they have to be
looked at as elements of the latest available technology
at the limit of tolerable design. As endothermical cata-
lytical processes have to be classified as relatively slow,
but solar irradiation is known to change very rapidly it
seems inevitable that a demonstration is being performed
that will naturally combine those elements and will give
answer to the transient behaviour of such plants, which
hardly can be predicted by computer modelling due to lac
of reliable information. Only on the basis of such demonstration which also would show possible risks, solar
aided plants of large size will be realized commercially
in the future.

The outfit of the plant wiht a primary heat carrier loop
with a maximum temperature of 800 $^{\circ}$C will allow to also
link other processes than steam reforming to the proposed
demonstration plant with probably little additional costs,
if such processes will run at temperatures between 350 $^{\circ}$C
and 750 $^{\circ}$C. Thus similar answers on the transient behaviour of other (non-catalytic) processes could be gathered in future on a comparable scale, if only indirect heat
exchange would be acceptable for such a process. Ultimately
the installed equipment could also allow to make investigations with gases that would be heat carrier and reactant
in one fluid.

Though the costs for the proposed experiment have to be considered as rather high, one has to recognize that the costs of deliveries and engineering would be in much better relation to any product, if the plant would be built on a commercial scale. The size of the proposed testing facility comparably would allow the production of about 37 t/d methannol and an additional quantity of $24\ 000\ m^3N/d$ hydrogen only, which quantities are not being thought to be representative or economic for any commercial plant.

In this study the costs of elements, which would be linked directly to the solar needs of this plant, consequently have not been evaluated separately, as any costs of an ADAM-EVA-plant on a partially fossil supported plant would be speculative, too. It therefore may be advisable to continue at a later stage with additional cost studies, where other basic technical solutions of steam reforming plants have been investigated on the same accompanying conditions with regard to the geographical location and the available power. When technical solutions can be evaluated on a comparable basis this would allow at the same time to associate costs with the relatd solar systems and to fix specific increases of prices of a chemical plant by the needs of additional equipment for the use of the costfree solar energy. The cost investigations should then be carried out for such sizes as they are commercially used.

The differences caused by alternative systems should then clearly show up and reveal the attractiveness of any solar aided plant generally and for aspecial system specifically.

It is assumed that the present study is one of the necessary elements of this overall procedure.

SOLAR STEAM REFORMING OF METHANE
(SSRM) PROGRAM PROPOSALS

A. KALT

DFVLR, KÖLN

Contents

Solar Steam Reforming of Methane (SSRM)

Program Proposals

1. Program Scope

Within the intended development work to supply solar HT process heat
to industrial processes, especially chemical processes, the steam
reforming process is considered suitable in particular.
This is because that process

o is an endothermic process, is relatively flexible and can be
  considered as an example process
o leads to a product which can be used to produce fuels as well as
  other chemical products
o is a part of the 'Adam/Eve Process' which allows energy transport
  via a chemical heat pipe
o is well-known
o has a broad application, and
o because methane as a raw material is available in several solar
  areas.

This program shall advance the development of a solar steam reforming process and make available the required hardware as well as gain
system experience.
Subsystems and components shall be developed and promising systems
studied theoretically and experimentally. Hereby, the percentage
of supplied solar energy shall be as high as meaningful.

2. Objectives
2.1. Technical Objectives

Currently available hardware - especially the reformer itself - as
well as hardware under development shall be investigated in respect
to its suitability for solar methane reforming.

Necessary development efforts on component and subsystem level shall be defined.

New developed hardware shall then be experimentally investigated, separately and in combination with other required hardware.

In order to cover system aspects complete solar steam reforming systems shall be investigated.
Operational experience shall be gained during extended operation time.

2.2. Economical Objectives

Future solar methane reforming plants as well as other systems possibly generated by this program have finally to be economically competitive to other -conventional- solutions. Therefore, basic economic aspects have to be considered from the very beginning within this development program.
The study of economical aspects like basic costs, operational needs, etc. shall be included in the program.

2.3. International Cooperation

Solar methane reforming is of interest to several participating countries. Therefore, an international cooperation and a task sharing are aspired.

3. Approach
3.1. Phasing

Hardware for solar methane steam reforming (SSRM) is not available from the shelf. Some subsystems are under development like a HT solar receiver (800° C, gas-cooled). Other subsystem designs could be modified like the Adam and Eve reformer.
Generally, subsystem development is necessary before a complete solar steam reforming system can be built.

Therefore, two phases are proposed for this program.

- Phase 1:
  Hardware modification and component/subsystem development, including single and combined tests on a test-site (Almeria)
- Phase 2:
  Building, test and investigation of a complete solar steam reforming system to cover system aspects sufficiently.

Both phases shall be accompanied by studies, engineering work, theoretical investigations, etc. This work is already in progressl since 1984.

3.2. Proposed Development Work
3.2.1. Status

Steam reforming technology using fossil firing to supply process heat is well-known and used in the chemical industry.
Reforming temperatures are -depending on the required synthesis gas- between 740° C and 860° C.
Process heat supply by a heat transfer fluid (H_e) via a heat exchanger is realized in the Adam and Eve Project.
In furnaces flame temperatures beyond 1000° C are needed and can be supplied without problems. In the Adam and Eve heat exchanger fluid temperatures of 950° C are required.

Solar process heat supply with a temperature of about 1000° C is not possible at present. Solar receivers using ceramic technology are in an early state of development.
Reformers, operating at a lower level of process heat supply temperature and therefore allowing the use of metal solar receivers, are not yet developed.

3.2.2. Phase 1: Component and Subsystem Development
3.2.2.1. Continuation of Studies and Theoretical Work

Study activities carried out in the past and at present shall be continued. Besides this, design-, planning-, evaluation work, etc. will be required during the whole duration of the progam.

3.2.2.2. Step 1 of the Development and Experimental Work, Low Cost Basic Investigations

Using hardware available or presently under development as well as hardware adaptable or modifiable in short time, a steam reformer and a solar process heat supply system shall be built. (See Figure 1).

An air-cooled metal solar receiver and a reformer similar to that one used for the A/E Project but probably slightly modified for a smaller temperature difference between heat supply gas and reforming process shall be used.

An auxiliary heater shall allow an increase of the process heat supply temperature and enable in this way reformer temperatures up to 860° C.

This experimental setting-up shall be used for basic process investigations and it shall be as simple and low-costing as possible.

Experimental investigations intended on the test-site are among others

o design verification of the subsystems receiver, auxiliary heater and reformer;
o variation of reforming process parameter, like temperature and pressure, to verify possibly a lower temperature reforming concept;
o working together of reformer and process heat supply system, especially under varying solar energy input conditions, controlling of auxiliary heat supply, of mass flows, etc.;
o gain of basic operational experience.

3.2.2.3. Step 2 of the Development and Experimental Work Development of a Reformer for Lower Operating Temperatures

In order to increase the solar energy portion and to avoid the use of the auxiliary heater during direct solar operation periods, a reformer shall be developed which allows process heat supply at a lower temperature level. A concept for such a reformer is known. This reformer shall be investigated together with the metal solar receiver.

Furthermore, a thermal energy storage device to operate at temperatures of 800° C to 1000° C shall be developed.

Experimental investigations intended on the test-site are

o experimental verification of the design expectations of the new developed reformer, investigation of its properties and behaviour;

o system aspects: working together of the reformer and the solar metal receiver;

o design verification and investigation of the properties and the behaviour of the storage device;

o system aspects: working together of the storage device with the solar receiver and the reformer;

o operation of the reformer together with the auxiliary heater during non-solar hours;

o gain of basic operational experience with this subsystem combination.

3.2.2.4. Step 3 of the Development and Experimental Work
Supply of HT Solar Process Heat at about 1000° C

The other way to increase the solar energy supply portion is to produce process heat from solar energy at a temperature of about 1000° C. This requires the development of a corresponding solar receiver using ceramic technology. The storage device has also to be tested to operate at this temperature level. (See also Figure 2).

Experimental investigations intended in the test-site are

o development tests

o design verification of the ceramic receiver and the storage device, investigation of their properties and behaviours o   system aspects: working together of receiver storage device, reformer and auxiliary heater

o   gain of basic operational experience with this subsystem combination.

3.2.2.5. Step 4 of the Development and Experimental Work, Combined Receiver-Reformer

The problem to enable high reforming process temperatures may also be solved by combining solar receiver and reformer to one subsystem (receiver-reformer).
Temperature losses may be reduced and possibly a simplification of the system can be reached.
Therefore, such a receiver-reformer shall be developed and intensely tested. (See Figure 3).

Experimental investigations intended on the test-site are

o   development tests

o   design verification of the receiver-reformer, investigation of its properties and behaviour

o   working together of the receiver-reformer with an auxiliary heater which can be located outside or inside the receiver-reformer

o   gain of basic operational experience.

3.2.3. Phase 2: Extended System Investigations

During component and subsystem investigations, system aspects can be considered only partially. This is because

o   test duration is limited since a simplified testing set-up is used (see Figures 1 to 3)

o   an open reforming cycle is used which means CH_4 and H_2O must be fully and continuously supplied.

Therefore, the test setting-up must be supplemented by additional hardware to allow system tests with extended duration. It is recommended to use a methanization system as a load to utilize the produced synthesis gas and to supply most of the needed CH_4 and H_2O. (See Figure 4).

Receiver, reformer, storage device, auxiliary heater and other hardware like heat exchanger, control system, etc. may probably be taken from Phase 1.

At least two of such systems will have to be investigated experimentally:
- A separated receiver and reformer system, and
- an integrated system with a receiver-reformer.

Experimental investigations intended concerning system aspects are

o sufficient working together of all subsystems and components during extended operation periods and various operational modes
o investigation of operational behaviour, problems and needs during long time runs (reliability, failures, maintenance requirements, consumables, personnel, etc.)

3.3. Risk Areas

- Reformer for lower operating temperatures
  o new catalyst development is very difficult; existing catalysts should be used
  o then, development risk considered as moderate

- HT ceramic receiver
  o development is just starting
  o material investigations are still needed
  o estimated development time: approx. 15 years

- Receiver-reformer
  o development not really started
  o material, heat transfer and control problems

- HT thermal storage device
  o cowper storage devices in operation for lower temperatures; probably material problems at 1000° C
  o storage time only one to several hours so far; possibility of extension not sure.

4. Schedule and Milestones

The main schedule is shown in **Figure 5.**

Studies and theoretical investigations are in progress since 1984 and earlier.

In the beginning of 1986, a review of study results so far should be made, including a check of the ranking of activities as included in the schedule Figure 5.

Expected in the second half of 1986, study status will allow the design of an experiment using hardware available at present or under development to carry out basic experimental investigations. Construction can then start in 1987. This will be the beginning of Phase 1.

The experiment setting-up will contain a metal receiver with air as a heat transfer medium, a tube steam reformer for maximum 860° C operation temperature and an auxiliary heater to reach this temperature. (Step 1).

Step 2, the development of a reformer which can be operated with a lower heat supply temperature, could be already started during Step 1. Anyhow, theoretical investigations of the proposed design should begin in 1986.

Step 3, the use of a ceramic receiver to supply directly process heat at 1000° C, is expected to begin not earlier than 1990, since considerable development work is still needed for this receiver.

Step 4, however, the development of a receiver-reactor could be started in parallel to Step 2.
A larger development period is expected.

Phase 2, extended system investigations, requires a decision which systems should be investigated in more detail.

From the present view, two systems have to be investigated at least:

- a system with separated receiver and reformer, and
- a system with an integrated receiver-reformer.

The following milestones will be reached in 1986:

- Review of study results
- Decision on additional study work and on the design of the Step 1 experiment for basic investigations
- Decision on the start of theoretical investigations on a lower temperature reformer
- Decision on and continuation of study work for the receiver reactor
- Decision on the final design and the hardware construction of the Step 1 experiment (end of 1986, beginning of 1987).

5. Required Funds

Activities planned for the year 1986 are:

- Review of study results by a review group

- Required continuation/start of work on
  o   detail problems on solar steam reforming of methane
  o   realization of a lower temperature reformer
  o   layout possibilities of a combined receiver-reactor

- Design of the Step 1 experiment for basic experimental investigations.

The required funds can be estimated to be 1.5 to 2.0 million Deutsche Marks.

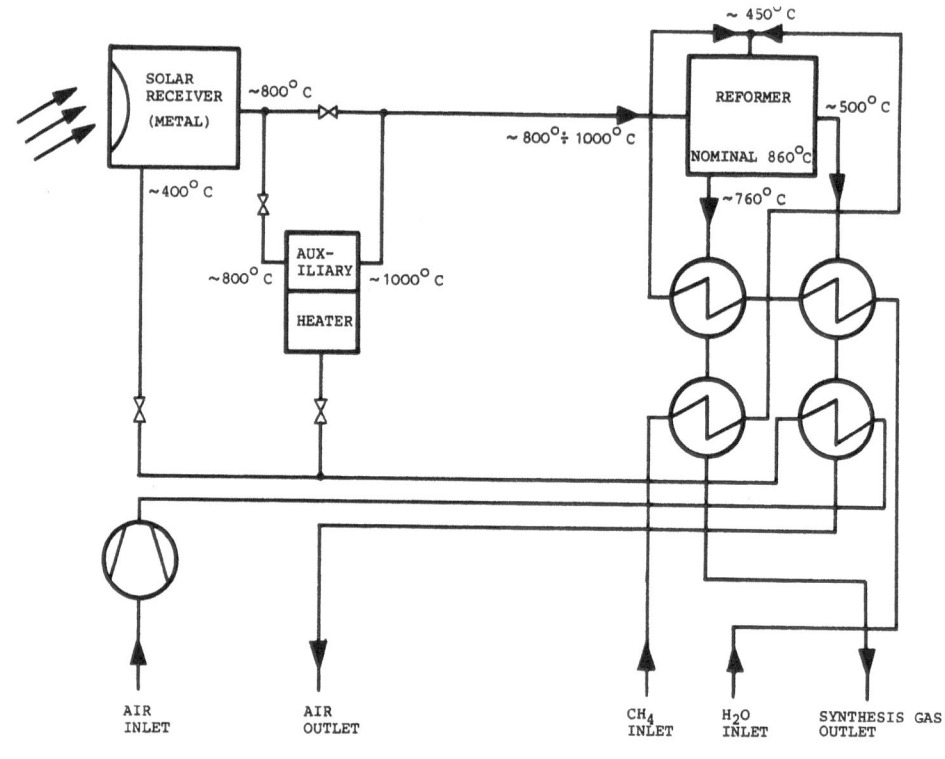

FIGURE 1: SOLAR STEAM REFORMING OF METHANE
PHASE 1   COMPONENT AND SUBSYSTEM DEVELOPMENT BASIC
INVESTIGATIONS USING SHORT-TERM AVAILABLE HARDWARE
INCREASE OF THE RECEIVER OUTPUT TEMPERATURE BY
FOSSIL HEATING

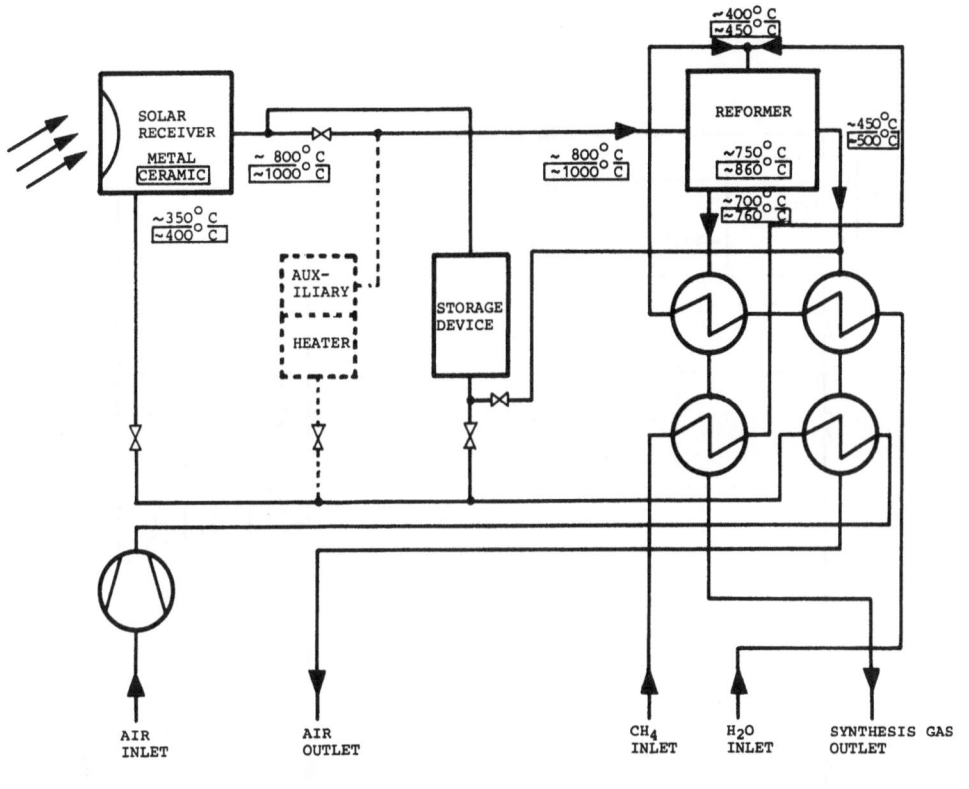

FIGURE 2: SOLAR STEAM REFORMING OF METHANE
PHASE 1 COMPONENT AND SUBSYSTEM DEVELOPMENT
MODIFICATION OF THE REFORMER TO 750° C OPERATION
TEMPERATURE OR OF THE RECEIVER TO 1000° C OUTPUT
TEMPERATURE

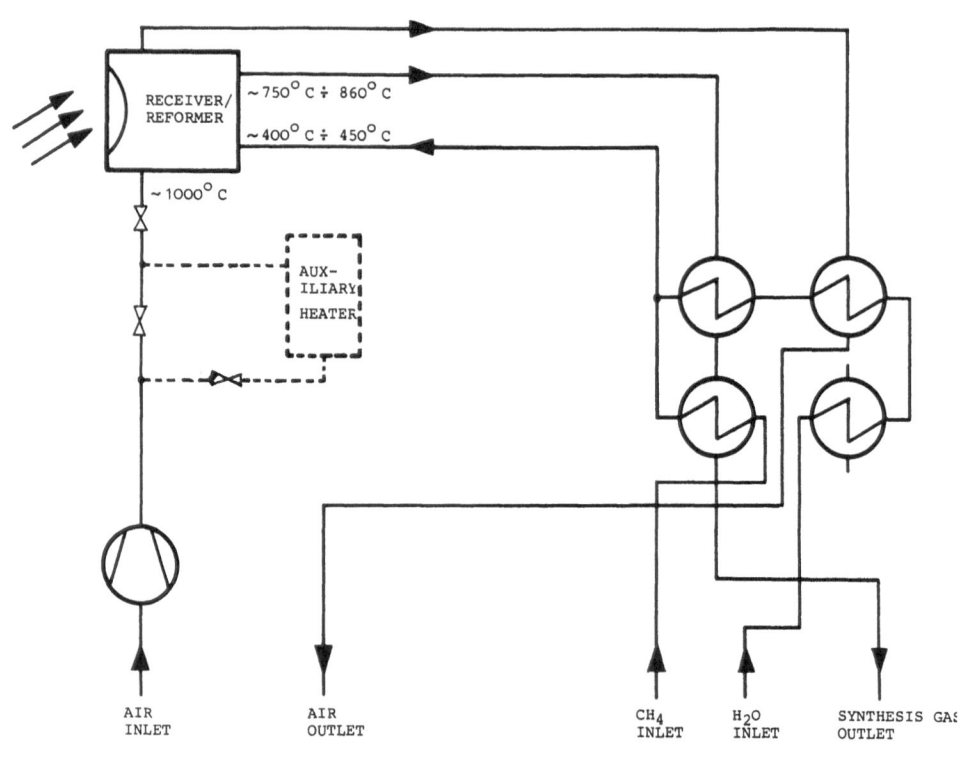

FIGURE 3: SOLAR STEAM REFORMING OF METHANE
PHASE 1   COMPONENT AND SUBSYSTEM DEVELOPMENT
INTEGRATED RECEIVER/REFORMER

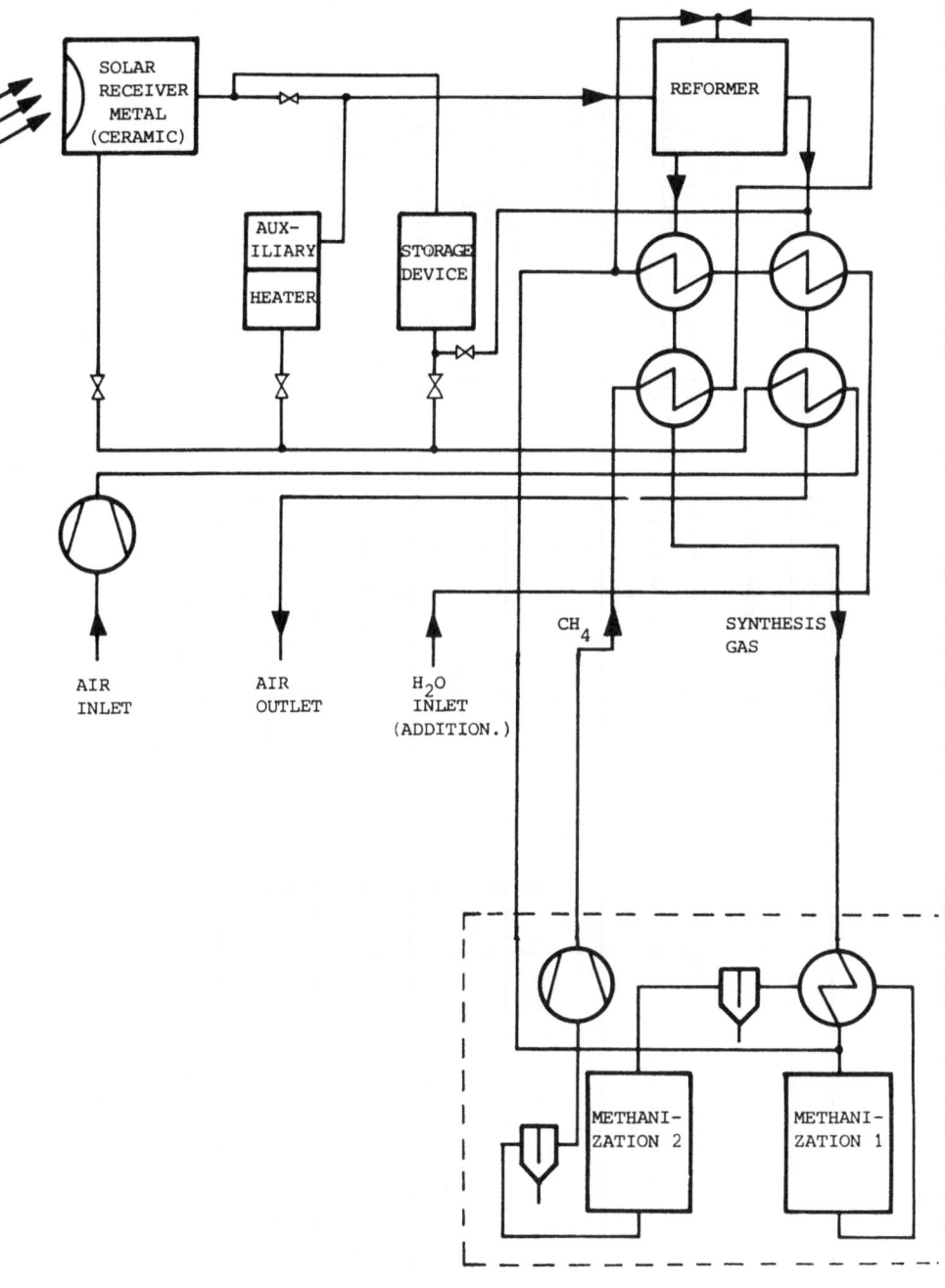

FIGURE 4:   SOLAR STEAM REFORMING OF METHANE
PHASE 2   EXTENDED SYSTEM INVESTIGATION
USING METHANIZATION (TO REDUCE CH_4/H_2O SUPPLY) AS A
LOAD

- 193 -

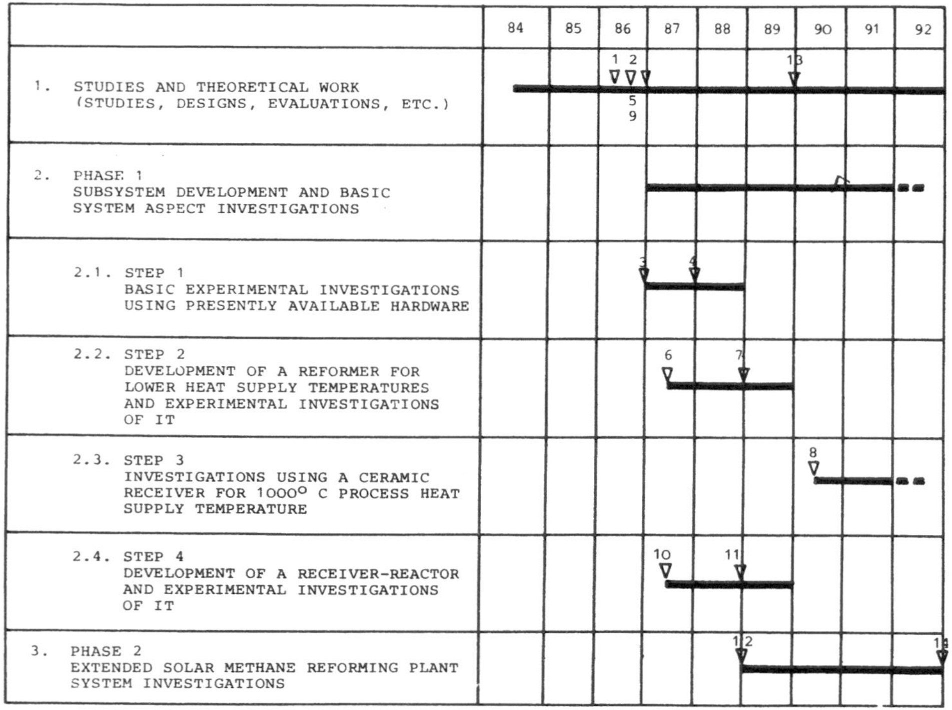

Milestones 1986

1 – Review of study results
   – Verification of the ranking
     of the development alternatives
   – Decision on additional
     study work

2 Decision of the start of
   the design phase for the
   Step 1 experiment

3 Decision on the start of
   the hardware phase of the
   Step 1 experiment

5 Start of studies for a
   lower temperature reformer

9 Continuation of study work
   for the receiver-reactor.

	84	85	86	87	88	89	90	91	92
1. STUDIES AND THEORETICAL WORK (STUDIES, DESIGNS, EVALUATIONS, ETC.)			1 2 ▽▽ 5 9			13 ▷			
2. PHASE 1 SUBSYSTEM DEVELOPMENT AND BASIC SYSTEM ASPECT INVESTIGATIONS									
2.1. STEP 1 BASIC EXPERIMENTAL INVESTIGATIONS USING PRESENTLY AVAILABLE HARDWARE				4 ▷					
2.2. STEP 2 DEVELOPMENT OF A REFORMER FOR LOWER HEAT SUPPLY TEMPERATURES AND EXPERIMENTAL INVESTIGATIONS OF IT				3	7 ▷				
2.3. STEP 3 INVESTIGATIONS USING A CERAMIC RECEIVER FOR 1000° C PROCESS HEAT SUPPLY TEMPERATURE			6 ▷			8 ▷			
2.4. STEP 4 DEVELOPMENT OF A RECEIVER-REACTOR AND EXPERIMENTAL INVESTIGATIONS OF IT			10 ▷		11 ▷				
3. PHASE 2 EXTENDED SOLAR METHANE REFORMING PLANT SYSTEM INVESTIGATIONS					12 ▷				14 ▷

FIGURE 5: MAIN SCHEDULE SOLAR METHANE REFORMING PROGRAM
           PHASES 1 AND 2

INTERATOM

SOLAR STEAM REFORMING OF METHANE
PROGRAM PROPOSALS

U. LEUCHS

INTERATOM, BERGISCH-GLADBACH

Contents

1   Objectives

Concentration of solar radiation to provide a high
temperature heat source is - principally - relatively simple. This can be seen in the fact that the
costs of highly concentrated solar heat are not
higher than the costs for a corresponding amount of
fossil fuels (table 1).

However, the consecutive energy conversion is comparatively complicated due to specific properties of
solar energy. Experience with existing solar thermal
power plants showed that surprisingly many troubles
were caused by conventional hardware and the solar
specific parts had good operating records. To find
the most suitable use is most important for economy
and utilization of solar high temperature heat. Until
now various types of central receiver systems have
been built and operated successfully, most of them
producing electric power.

This situation suggests the idea that a direct application of high temperature solar energy - e. g. in
chemical processes - could be advantageous. A highly
attractive chemical reaction for solar energy application seems to be the steam reforming of methane
because this process

-     requires a large input of process heat

- has a substantial and increasing market for its end products

- is comparatively uncomplicated, flexible to process conditions, easy to control and well proven for many decades.

The required heat for this process is conventionally provided by burning a part of the methane. By replacing this heat of combustion with solar heat about 20 % of the natural gas consumption could be saved. This would allow the use of all the natural gas as a chemical feedstock.

To reach economical competitiveness the solar specific hardware cost of a commercial plant has finally to be paid by the saved fuel and environmental advantages.

Therefore the development of solar steam reforming has to strive for a final process with

- a high input of solar energy

- low product costs.

The latter requirement includes low plant costs, low operating costs, a high plant efficiency (full- and partload) and a good product quality.

One should withstand the temptation of a high rate of fossil firing for economic reasons. This would not harmonize with the basic aim of solar energy utilisation, to save fossil fuels. And the economic advantage will vanish with raising prices of natural gas anyway.

The development effort for solar steam reforming should be low and yield a high benefit for other chemical processes which could be powered by solar process heat. The technologies developed here and the experience gained during operation of components or complete systems shall support the development of other solar or non-solar systems.

The program proposed in this paper shall advance the development of a solar steam reforming process with regard to these requirements. And - as shown in table 1 - if production of high temperature solar heat costs as much as fossil heat, the utilization of solar heat should not prove to be much more expensive.

Solar steam reforming is of interest to several members of the IEA. An international cooperation and task sharing is hoped for.

2   Present Status

2.1   Steam Reforming Technology

Steam reforming of methane is the prevailing process
for the production of synthesis gas and well proven for
many decades. A major fraction of the methanol-,
ammonia- and hydrogen-production in the world is based
on the steam reforming of natural gas.

Conventional steam reforming is carried out in fossil
fired (usually by natural gas) furnace chambers in
which the reforming tubes are arranged vertically.
The tubes are filled with a nickel containing catalyst
to promote the reforming reaction. Standard dimensions
of such reforming tubes are

-   heated length   8 - 12 m
-   inner diameter   80 - 130 mm
-   wall thickness   10 - 18 mm

A mixture of desulfurized natural gas and steam is
fed to the tubes at a pressure of about 15 - 20 bar
reaching a reaction temperature about 850 °C.

The composition of the yielded synthesis gas corresponds to a thermodynamic equilibrium and is mainly
determined by the following parameters:

-   temperature at the end of the tubes
-   pressure of process gas
-   initial ratio of steam/methane

A high conversion rate of methane is attained at a
high temperature, a low pressure and a high steam/
methane ratio (see fig. 1). A remaining methane fraction of 1.5 - 2.5 % by volume related to dry synthesis
gas is obtained typically. Only synthesis gas for
ammonia production contains 8 - 10 % remaining
methane at the outlet of the primary reformer. By
admixture of a corresponding amount of air in a
secondary reformer stage the needed nitrogen is
added and the methane content is lowered to final
values of about 0,2 - 0,4 % by volume.

As the steam reforming reaction is highly endothermic
a lot of thermal energy has to be transferred through
the tube walls. In conventional plants this heat is
provided mainly by heat radiation of the gas-burners.
But reformer tubes can also be heated convectively by
a heating fluid. The reformer plant EVA I und EVA II
at KFA in Jülich use hot helium of 950 °C as heating
gas. The design and testing of this plant has yielded
valuable experience with convectively heated reformer
tubes.

2.2   Solar Technology

A solar technology to provide high temperature process
heat has not yet been realized. However, the generation
of high temperature heat was subject of several projects
and has been investigated more or less intensively:

-       Power plant EPRI/BOEING
        1 MW bench scale metallic receiver, successfully
        tested with 816 °C air

- Power plant EPRI/BLACK & VEACH
  0,6 MW bench scale ceramic receiver, design
  temperature 1065 °C, manufactured

- GAST Technology Program
  0,5 MW metallic receiver panel is being tested,
  air 800 °C; test of a ceramic panel for 1000 °C is
  in preparation

3   Approach

3.1   Problems

When trying to combine the present technologies of
steam reforming with solar technology one has to meet
two main problems.

One problem is the temperature of the solar generated
process heat. Today solar technology can provide process
heat (hot air as heating medium) with a temperature of
about 800 °C or is expected to be able to within the imme-
diate future. Up to this temperature metallic materials
(austenitic alloys) are sufficient for the solar receiver.
For higher temperatures up to 1000 °C or more ceramic components are required which are not yet available. The
development of ceramic receiver technology has just been
started in the GAST program.

Conventional steam reforming technology however requires
temperatures of more than 800 °C in the process gas in order to achieve a satisfactory methane conversion rate.
This means - if heated convectively - the reformer requires a heating gas temperature of 950 °C - 1000 °C
due to temperature losses of 110 °C - 140 °C at the
tube walls. This is much more than solar technology
can provide today.

The second main problem arises from the discontinuity of solar radiation and is well known in the solar technology. This problem can be met by

- using a thermal storage
- additional fossil firing
- adapting the reforming process to a discontinous operation.

3.1.1   Temperature Problem

There are four different paths that could be followed to meet the temperature problem:
(see fig. 2)

1. Bridging the temperature gap by additional fossil heating.
2. Adapting the reformer to a heating temperature of 800 °C.
3. Adapting the solar receiver to a temperature of 1000 °C.
4. Adapting receiver and reformer by integration.

The different steam reforming concepts according to these paths are shown in fig. 3 and will be described in the next pages.

It can not yet be decided finally which of these concepts would be the most suitable and at first they all are taken into consideration. Some basic properties of these steam reforming versions are arranged in table 2.

Path 1 (Fuel Saver Version):

Only existing or shortterm available components are
required. Necessary modifications can be obtained with
low development effort. As receiver and reformer are
not optimally adapted to each other a low methane conversion rate is obtained and the quality of the generated synthesis gas has to be improved by additional
fossil heating. This can be done either in an air heater raising the air temperature from 800 °C to 1000 °C
before reaching the reformer (path 1a), or in a second
reformer which is heated conventionally (path 1b, see
fig. 3). A fossil heater (or thermal storage) to smooth
temperature transients of the air cycle is required in
any case.

Following this path solar steam reforming could be realized with low technical risks, but a high fossil and a
low solar energy portion in the product would be the
consequence.

The air cycle can be designed as an open or a closed cycle.
The properties of both cycle types differ significantly.
The main differences are:

-   The fossil fired air heater of an open cycle can
    be designed as a combustion chamber with direct
    firing into the cycle. The closed cycle however
    requires heat exchanging walls between air and
    flue gases. Such heat exchangers for air tempe-
    ratures up to 1000 °C are not yet developed and
    those of 800 °C require eventually little modifi-
    cations due to high temperature transients under
    solar conditions.

-   The open cycle needs a costly compressor driven
    by a turbine to obtain a suitable air pressure.
    The closed cycle can do with a comparatively
    simple air blower.

-   A closed cycle is much more flexible to choose
    and to change operating parameters such as air
    pressure and mass flow to meet the solar conditions.

Because of the last two points usually a closed cycle
seems to be preferable for process heat generation.

problems arising from path 1:

-   low solar energy portion in product
-   two reformers are needed (path 1b)

technical risk areas:

-   receiver for 800 °C outlet temperature not yet
    available but panels successfully tested
-   convectively heated steam reformers EVA I and
    EVA II have to be modified to be heated by air
    of 1000 °C (path 1a) or 800 °C (path 1b) instead
    of helium

-   piping (and eventually valves) for 1000 °C air between
    fossil heater and reformer (path 1a) not yet developed.

Path 2 (Low Temperature Version)

The steam reforming process is adapted to a heating gas
temperature of 800 °C, which is provided by solar technology. The adaptation can be performed by two alte-
rations of the process:

1. lowering the process gas pressure down to about
    five bar.

2. improving the heat transfer from heating gas to
    process gas.

Lowering the pressure of process gas leads to a signi-
ficant lower process gas temperature required to yield
the same rate of methane conversion (fig. 1). A pressure
reduction from 20 to 5 bar causes a reduction of the
required temperature from 860 °C to about 740 °C.

Heat transfer from heating gas to process gas can be
improved by a new reformer design.

Conventional steam reformer tubes are designed to be
heated mainly by heat radiation of hot burner flames.
If heated convectively by a heating gas of comparatively
low temperature the heat transfer through the tubewalls
is problematic.

Path 3 (High Temperature Version)

The receiver and air cycle is adapted to an air temperature of about 1000 °C required by state-of-the-art steam reforming technology. For air temperatures higher than about 800 °C metallic materials can not be used in the receiver. Development of ceramic technology has just started in the GAST technology program.

problems arising from path 3:

-   higher energy losses of the receiver and of hot gas piping due to raised temperatures

technical risk areas:

-   development of ceramic materials
-   development of ceramic manufacturing and connection technology
-   development of 1000 °C receiver
-   development of 1000 °C fossil heater (if closed cycle is choosen)
-   development of hot air piping and controlling.

Path 4 (Integrated Version)

The reformer tubes are integrated into the receiver and
heated by solar radiation. This concept allows omission of
the air cycle with its temperature- and energy-losses.
A good utilization of solar energy is expected and the
concept is comparatively simple. The reformer can be
heated directly by solar radiation or indirectly by ra-
diant flux from high temperature refractory walls.

problems arising from path 4:

- A thermal storage can not be connected because the reformer tubes are heated by radiation and radiation can not be stored.
- Fossil firing (either in the receiver/reformer or in a subsequent conventional reformer) is necessary for continuous operation and reduces the solar energy portion.
- Comparatively high reradiation losses at the receiver aperture due to high back wall temperature (in case of indirectly heated tubes).

technical risk areas:

- concept not yet developed
- high thermal stresses in the reformer tubes due to nonuniform and discontinuous heating causes material and design problems
- high thermal transients may destroy the catalyst

Of course combinations of these paths are possible. Especially a combination of path 2 and path 3 is interesting using a 1000 °C ceramic receiver and an adapted reformer which can be operated on a lower temperature level. This concept would allow to load a thermal storage by heat of 1000 °C and unload it down to temperatures of 900 °C or 800 °C. The utilization factor of the thermal storage would be increased significantly compared to concepts with almost the same temperature for loading and unloading.

3.1.2   Discontinuity Problem

The second main problem - the discontinuous energy
supply of the reforming process - can be solved by
three different conceptions. They are shown in fig. 4.
The problem should be met by usage of a thermal heat
storage in order to reduce fossil firing as far as
possible.

The integrated receiver/reformer (path 4) is the only
reformer concept considered here where no possibility
can be seen to connect a thermal storage in a convenient
manner. Fossil firing (in order to smooth the energy
supply) seems to be unavoidable. If following path 1
fossil firing is necessary too in order to raise the
air temperature. Only the paths 2 and 3 have the
potential to supply the reforming process to almost
100 % by solar energy. So these paths should be
prefered if a high part of solar energy in the
product is desired. A thermal heat storage should be
developed. A fossil back up system is required by every
path anyway.

The chances to adapt the reforming process to a discontinuous energy supply seem to be low because of
several technical reasons:

-    Plant efficiency drops rapidly in partload operation
-    Temperature transients in the reformer should not
     exceed about 1 °C/min. Otherwise catalyst may be
     destroyed (mainly a problem of the integrated re-
     ceiver/reformer).
-    A disturbed quality and quantity of process gas
     threatens the sensible catalyst and impedes the
     operation of the subsequent synthesis plant.

So the problem of discontinuous energy supply has to be
solved by a thermal heat storage and - if not avoidable -
by fossil firing.

3.2   Experiments

All considered versions of solar steam reforming plants
require a more or less novel design or at least modifications of the main plant components. In order to
demonstrate the practicability of solar steam reforming and to find out the most advantageous operating
procedures experimental work is necessary.

The experiments should be not too big in size in order to
save money, but they have to be big enough for drawing
conclusions for commercial scale plants.

The experimental device should be constructed as simple
and low-costing as possible. Therefore the experiments are planned to be carried out at the Plataforma
Solar in Almeria. A lot of already existing hardware
(heliostats, tower ...) can be utilized there. The experiments are restricted to the field of solar steam
reforming technology. Synthesis processes as methanol
or ammonia production are conventional technology and
need not to be tested here. The fossil powered main
reformer of paths 1b and 4b too, has not to be constructed and tested within this program for the same
reason.

3.2.1   Component test

The first experiment will be the 'component test',
which means the short term testing of the critical
components receiver, reformer and eventually the
thermal heat storage. The following results are
sought:

-      design verification of the critical components

- behaviour of the components under varying operating conditions. Static and dynamic variations
  of temperature, pressure and mass flow of heating gas and process gas. Changing of the initial
  H_2O/CH_4-ratio of process gas

- gain experience with start up and shut down
  procedures.

The maximum size of the components is limited by that
of the heliostat field (about 5MW maximum thermal
power) which is sufficient for the required experiments. The thermal input into the chemical process
is planned to be about 2 MW. So a continuous operation
during a large part of the day is possible and a solar
multiple >1 can be reached in phase 3.

According to the four paths four different types of
component test facilities are to be considered (fig.
5 - 8). In each case methane and water are preheated,
mixed and fed to the reformer which is directly or
indirectly heated by solar radiation.

The exit gas is partly used in an auxiliary heater
and the rest is disposed by a flare system. Flaring
the produced synthesis gas is the simplest and
cheapest way to carry out short term component tests
providing the basic informations required.

As methane and desalinated water are not sufficiently
available at the remote site these materials have to be
trucked in. Operation of the test experiment consumes
approx. 200 - 400 kg of liquefied natural gas per hour and
the fourfold amount of water if not recycled. One
truckload of methane lasts for several hours of operation.

3.2.2 System test

After finishing component tests the testing device is reconstructed to carry out long term system tests (fig. 9). The aim of the system tests is to

- investigate the interactions of the components with each other under various operating conditions

- reach a maximum of solar powering during operation

- gain basic operating experience of long term operation under solar conditions.

Flaring of the produced synthesis gas is too expensive during long term system tests. So it is recycled by a methanation unit which has to be added to the test facility.

3.3   <u>Phases</u>

The solar steam reforming program proposed here is
divided into three phases:

<u>Phase 1</u>

Different systems are investigated and a decision has
to be taken which solar steam reforming concept should
be followed.

<u>Phase 2</u>

Main components as receiver, reformer and eventually
heat storage are designed, constructed and tested at
a test facility on the Almeria site.

<u>Phase 3</u>

The test facility is extended to a complete solar
steam reforming system. Long term system tests are
carried out.

3.3.1   <u>Phase 1</u>

It cannot yet be decided finally which of the paths is
the most convenient one. The status of development differs
too much and especially path 3 and 4 are too less developped

Before the experimental phase is started it has to be
decided which path should be followed and which kind of
experiments should be performed at the Almeria site.
(If enough money is available more than one experiment
can be accomplished - successively in Almeria or in parallel on different sites. But within this program the
selection of only one path is planned.)

So the aim of phase 1 is the investigation and evaluation
of the different concepts of solar steam reforming, which
can serve as a basis for this decision. The evaluation
of the different concepts is made due to the following
criteria:

-    plant overall economy (including plant/operating
     costs, efficiency, reliability...)

-    solar energy fraction in the product

-    development effort required and technical risks.

For this purpose commercial scaled plants of about
40 MW thermal input have to be considered. Basic data
such as plant size, solar multiple, size of thermal storage
and kind of product (e. g. methanol) should be the same
in all cases because of comparability reasons.

Production of methanol is proposed because this product
has good transportation properties and the synthesis
process is relatively simple.

To reach this aim the work descibed below must be done
simultaneously:

- Management

An important task of phase 1 is the establishment
of an organization to perform the solar steam reforming program. This should hapen as early as possible
A responsible committee has to be founded together
with all national and international partners which
are interested to participate in this program.

On the other hand experienced industrial partners
have to be found which are qualified and interested
to carry out the design work and R + D including management tasks as coordination, reviews and documen-
tation. Together with these partners the program
plan has to be revised. The decision about the type
of experiment to be performed during next phase is
taken and a complete management plan has to be worked
out. Questions of financing, interfaces, work distribution have to be regulated.

- Research and Development

To support design work critical technical problems
have to be investigated which are essential for plant
design and which impede the estimation of development
effort and the technical risks of the different conceptions. Research and development work comprises
studies and small scaled tests as well. The following
problems are planned to be investigated:

Studies

- thermal storage (cowper type) 800 °C and 1000 °C
  loading and unloading procedure, possibilities
  of raising utilization factor, computer models
  (path 1, 2, 3)

- tube reformer heated by air 1000 °C
  comparison of different existing concepts
  (path 1a, 3)

- fossil heater for air temperatures up to 1000 °C
  practicability, concept proposals (path 1a, 3)

- air cycle
  possible improvements with replacing air by
  helium (path 1, 2, 3)

- integrated receiver/reformer
  basic design study, directly or indirectly
  heated reformer tubes, stress analyses of tube
  material and catalyst, rough thermodynamic comparison of different concepts, practicability of
  direct fossil heating, start up and shut down
  procedures (path 4)

- evaluation of results of GAST Technology Program
  due to receivers and hot gas piping

Tests

- honey combed catalyst continuation of development; laboratory tests of efficiency and degra-
  dation in already existing test facilities (path 2)

- small reformer tubes for integrated receiver/re-
  former; manufacturing and test in a test plant;
  test of tube material and catalyst behaviour during operation with high temperature transients
  (path 4)

Other problems to be investigated may occur during design work in phase 1.

Problems concerning receivers, ceramic technology and hot gas piping are treated in the GAST technology program and results are expected to be available in tim

- Concept Evaluation

A preliminary design of several commercial scaled plants - at least one modification of the plant types described before as path 1 to 4 - is carried out. Basic data such as plant size, solar multiple, size of heat storage and kind of product (e. g. methanol) have to be definded and should be the same in all cases because of comparability reasons.

Plant costs, plant efficiency and other data are calculated and estimated so that the product costs averaged over the year can be roughly determined. The development effort required for the novel design and novel components and the technical risks are estimated.

Then the required evaluation of the different steam reforming concepts is made.

3.3.2   Phase 2

At the end of phase 1 the decision is taken which type of experiment (described in chapter 3.2) should be performed on the Almeria site. If enough money is available more than one experiment can be constructed.

The goal of phase 2 is the design, manufacturing and performance of the component test experiment. Phase 2 is devided into two parts.

Phase 2a

During phase 2a all design work is accomplished. In contrary to phase 1 here only the selected experiment on the Almeria site is considered.

It begins with acquisition of site data (available solar power, available hardware, space on top of the tower, maximum weight of receiver...) and definition of a set of basic data (design point, size of heat storage, solar multiple). This data acquisition and definition could eventually be started during phase 1 already.

Based on these data a preliminary design of the test experiment is carried out. This comprises several technical modifications which have to be investigated and evaluated. At the end of preliminary design a review is made and a reference modification is defined. The set of basic data and interfaces are redefined and frozen. All results are documented and a specification is issued to control the design and interfaces of the subsystems.

A test and measuring program for the component test experiment is worked out and revised during detailed design work.

Technical problems which are not solved finally during phase 1 or which occured during preliminary design of the component test experiment are investigated. The scope of research and development work still required in phase 2a depends largely on the experiment selected, on the progress made during phase 1 and on the progress made in the GAST technology program.

No or low remaining demand is expected for path 1 or 2
and a relatively high demand seems still to be necessary
for path 3 and 4. The essential technical problems have
to be solved by the end of preliminary design.

Now the detailed design work can be started. After certain
periods of time (about 4 to 6 months) design reviews are
planned in order to control that all participants are working with the same sub-system configurations and the same
interfaces. At these occasions the actual status of work is
updated and documented according to work progress and the
detail of design.

At the end of detailed design work the complete design is
frozen and it has to be decided whether manufacturing can
start.

During phase 2a also the system experiment has to be
considered which is planned to be accomplished in phase 3.
The system experiment will probably be erected by recon-
structing the component test set-up of phase 2. This has to
be taken into account already during phase 2a and a preliminary design of the system test is worked out, too.

Phase 2b

During phase 2b the component test experiment is constructed and operated. Based on the detailed design
of phase 2a all required components and devices are manufactured and procured and shipped to Almeria. Mean-
while the existing and utilizable hardware on site is
checked and prepared and the operating crew is trained.
After construction, functional tests and commissioning
the test program is started which has been worked out during phase 2a.

Research and development work is continued on a lower
level to meet unexpected problems occuring during manufacturing and construction and for problems which arise
from design of the system experiment. An elaboration of
an operating program of the system test is started with
beginning of phase 2b. A rough evaluation of the component test results has to be finished before the final
design of system experiment is frozen. The end of phase 2
would be a decision to construct the system experiment in
phase 3.

3.3.3   Phase 3

During phase 3 system aspects are investigated. A complete solar steam reforming plant is operated and operating conditions are varied and optimized due to overall
plant efficency and solar energy input.

Processing of the yielded synthesis gas is not necessary
and not planned, but its quality should allow a methanol-
synthesis.

The easiest and cheapest way of erecting such a steam
reforming plant would be the reconstruction of the component test set-up of phase 2.

In long term operation flaring of synthesis gas is too
expensive. Therefore the raw materials methane and desalinated water are recycled by a methanization unit which
is supplemented. A thermal heat storage is added or extended in size to obtain a fairly continuous supply of
solar energy to the reforming process during the night.

The gained operating experience will allow a significantly
better estimation of the economical situation of solar
process heat and of the conditions on which commercial
competition with fossil fuels can be reached.

4   Steps

4.1   Phase 1

4.1.1   Management

1.   foundation of a responsible committee

2.   finding of apropriate industrial partners

3.   evaluation of plant concepts and selection
     of the type of experiment to be performed
     in Almeria

4.1.2   Research and Development

1.   studies

2.   manufacturing of test samples

3.   test performance and evaluation

4.1.3   Design

1.   definition of basic data for commercial scale
     plants to be compared

2.   preliminary design of the most promising modifi-
     cations of the four plant types for methanol pro-
     duction

3.   evaluation of product costs, solar energy input
     and development effort still required for the dif-
     ferent reforming concepts

4.2　　　Phase 2

4.2.1　　Component Test Experiment

1.　acquisition of site data

2.　definition of basic data

3.　preliminary design of selected component test experiment

4.　elaboration of test- and measuring program

5.　detailed design of component test experiment

6.　procurement and manufacturing of components and shipping to the site

7.　erection of component test experiment

8.　performance of component tests

9.　evaluation of test results

4.2.2　　Research and Development

All technical problems concerning the selected component- or system test experiment which have not yet been solved during phase 1 or which may occur in this phase.

4.2.3   System Test

1.   preliminary design of system test

2.   elaboration of test and measuring program

3.   detailed design of system test experiment

4.   procurement and manufacturing of additionally
     required components and shiping to site

4.3   Phase 3

System Test

1.   reconstruction of component test facility

2.   commissioning of system test experiment

3.   operation

4.   evaluation

5   Management Plan

The management plan describes the management methodolo-
gies, control mechanism and procedures that are used in
performance of this program. This plan depends largely
on the interests of the partners and institutions which
will perform this program. Therefore only a rough management plan can be given.

5.1   Organization

As the technology of solar steam reforming is a new
technical field the performance of the program shall
be entrusted to a group of industrial companies speci-
alized for the different technologies required.

This group should be small to reduce coordination losses and bureaucracy. But at least two industrial companies are needed; one for the solar specific part (mainly receiver) and the other for the reforming process. A third institution (may be an industrial company too) should be entrusted with management tasks as controll and coordination of work and the overall engineering as well. A group of three companies seems to be sufficient and suitable.

Each company may employ more or less subcontractors and is responsible for its special tasks. The responsibilities could be distributed e. g. as follows:

Company 1:   System-related activities

-    overall project responsibility and coordination

-    definition and controll of supply and scope of work and interfaces

-    overall systems engineering

-    site related activities

Company 2:   Chemical process

-    reformer design and construction

-    piping and instrumentation

-    methanation unit

-    subsystems

Company 3:   Air cycle

- receiver design and construction

- piping and instrumentation

- heat storage

- fossil heater

- subsystems

A rough organization structure is shown in fig. 10.

5.2   Work Breakdown Structure

The above described work distribution is closely related
the work breakdown structure (WBS). The WBS devides the
work to be done into major subunits and work elements. A
proposal of a WBS is given in fig. 11 - 13.

This WBS is valid for a steam reforming system with
separated receiver and reformer. If an integrated version is chosen the WBS has to be changed adequately.

5.3   Schedules

The component test is planned to be finished at the
end of 1990 and the system test will last until 1993.
Fig. 14 and 15 give a survey on the time schedule
proposed here.

5.4   Cost Estimate

The costs for the proposed program could only be estimated preliminarily because an accurate cost estimate
is not yet possible in this early moment.

Here only the costs are estimated of an experiment with an adapted reformer (low temperature version, path 2), because the technology is relatively well known. The scheme is shown in fig. 6.

Before a final cost estimate can be made comprehensive detailed work must be carried out. Therefore a detailed cost estimate must be carried out during phase 1 and 2.

Design and study work of phase 1 is in parts already being treated in a study which will be finished within the next months. It is assumed that the results will cover a large part of the work to be done in phase 1. Here only those costs are roughly estimated which are caused by the still remaining work. The cost estimates are based on experience gained at other programs.

The costs for the construction of the experiment are based partly on information of manufacturers given for the "GAST Pilot Experiment" proposal of April 1985. These costs were adapted to the special conditions of this program. The other part is based on rough design calculations of an experiment of type 2 made by Interatom.

Additionally it is assumed that the reformer can be situated close to the receiver on top of the tower in order to avoid long hot gas piping.

The estimated costs are listed in table 3 and the summarized fund flow is shown in fig. 16.

The following assumptions are made:

–	thermal power of receiver:	2.0 MW
–	air pressure:	20 bar
–	air temperature at receiver outlet:	800 °C

6   <u>Summary</u>

If present solar technology is combined with present
steam reforming technology one has to meet two main
problems.

One problem is the discontinuity of solar radiation
and is well known in solar technology. It can be solved
by usage of a thermal storage or additional fossil firing.

The second problem is that today solar technology can
provide only process heat of lower temperature than required by the reforming process. There are four principally different possibilities to solve this problem:

-     to bridge the temperature gap by fossil firing

-     to adapt the reforming process to lower temperatures

-     to adapt solar technology to higher temperatures

-     to develop another concept

These possibilities are described and their properties and
technical risks are discussed.

A program proposal for the development of solar steam reforming technology and the erection of a small scale test
plant on the Almeria site is presented.

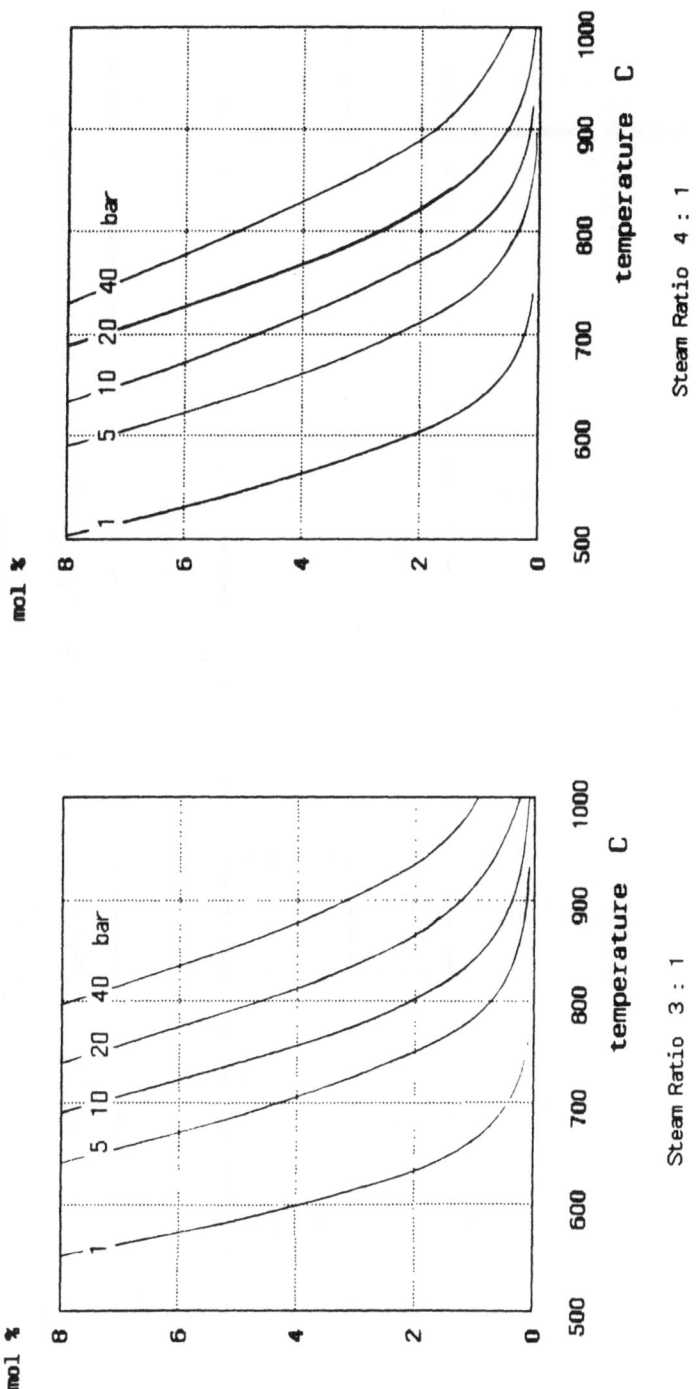

Fig. 1   Remaining methane in synthesis gas (wet) depending on final temperature and pressure of process gas in equilibrium

Source : KFA. Juelich

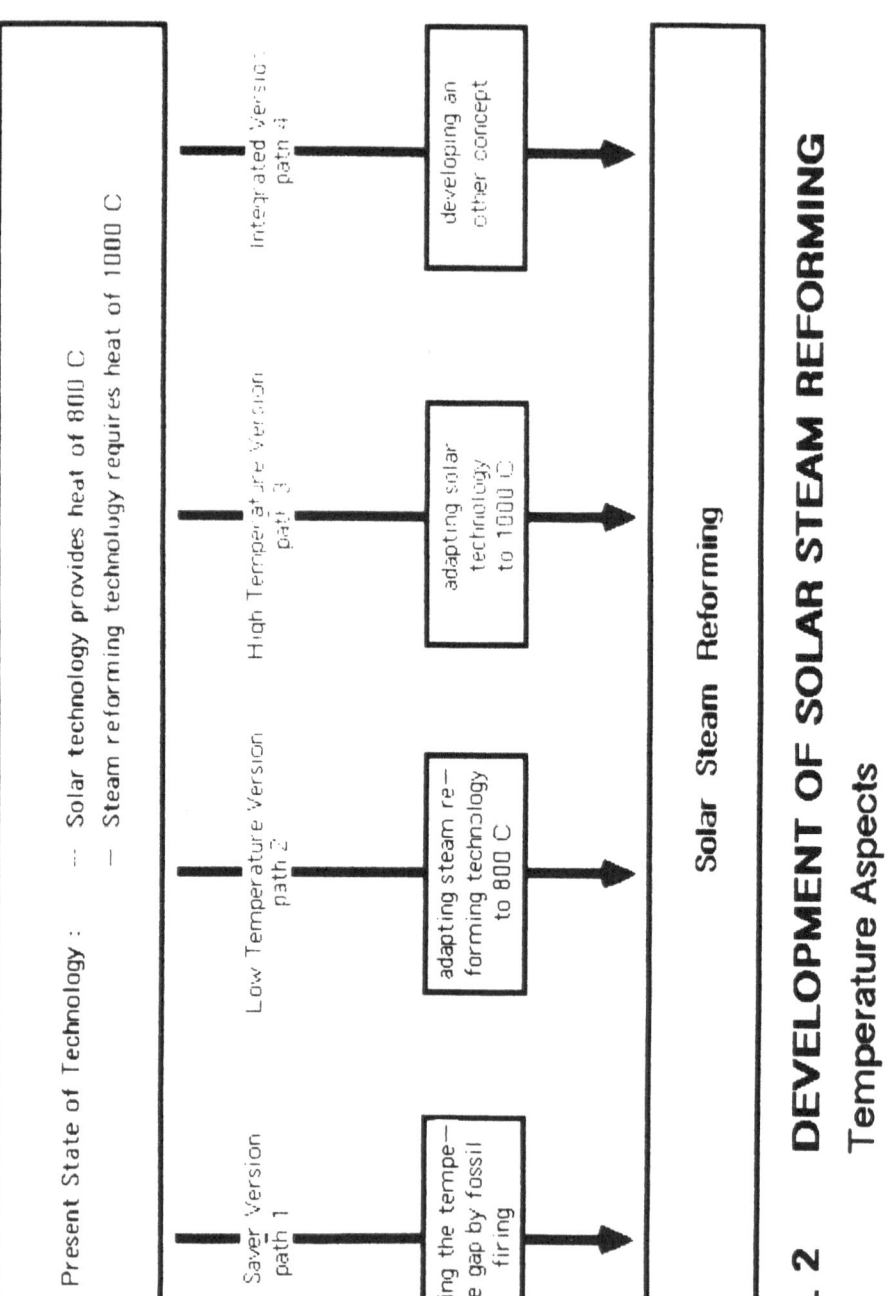

Present State of Technology : -- Solar technology provides heat of 800 C
                               — Steam reforming technology requires heat of 1000 C

Fuel Saver Version
path 1

bridging the tempe—
rature gap by fossil
firing

Low Temperature Version
path 2

adapting steam re—
forming technology
to 800 C

High Temperature Version
path 3

adapting solar
technology
to 1000 C

Integrated Version
path 4

developing an
other concept

Solar Steam Reforming

Fig. 2   DEVELOPMENT OF SOLAR STEAM REFORMING

Temperature Aspects

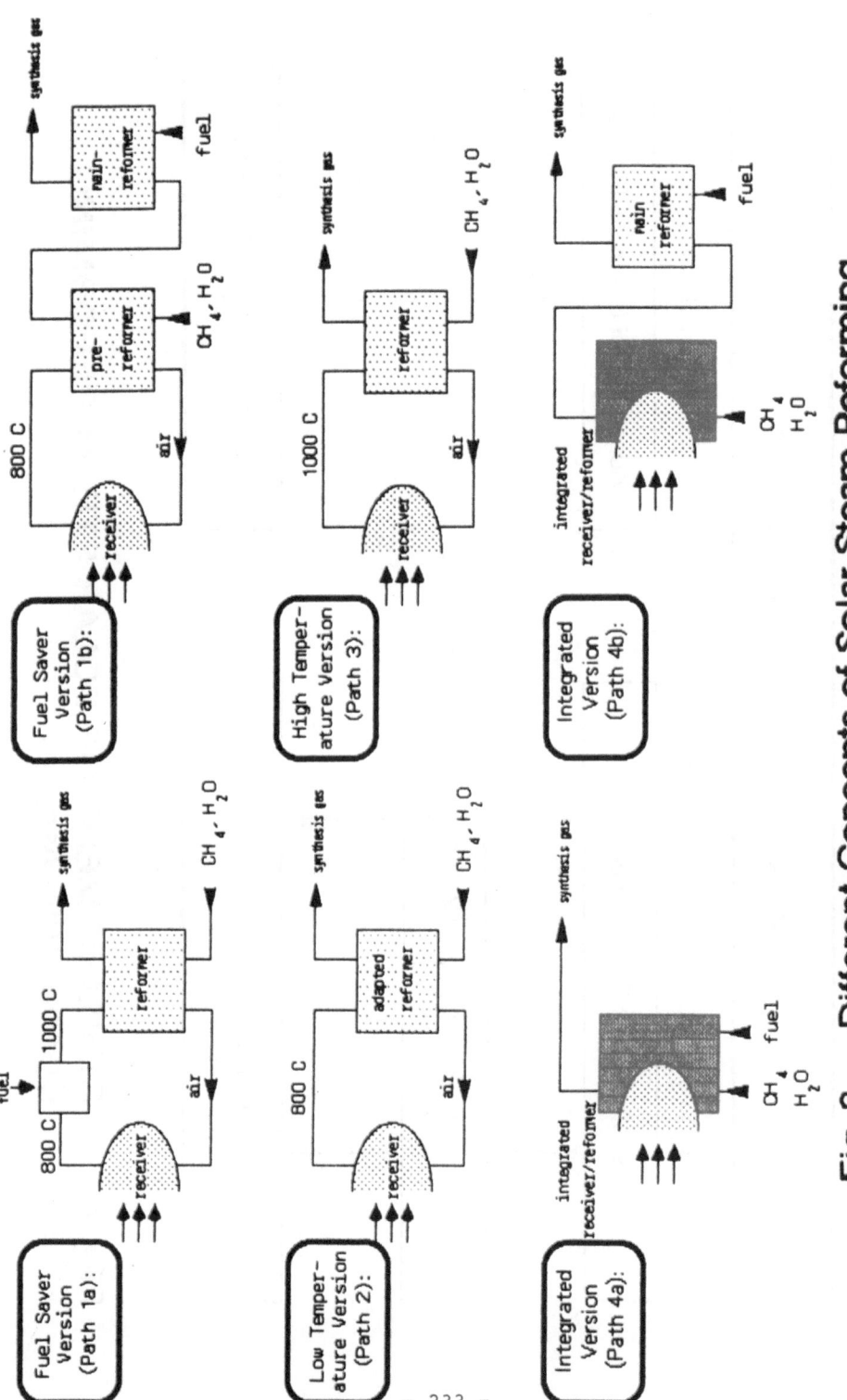

Fuel Saver Version (Path 1a):

1000 C
800 C
fuel
receiver
air
reformer
CH_4, H_2O
synthesis gas

Fuel Saver Version (Path 1b):

800 C
receiver
air
pre-reformer
CH_4, H_2O
main-reformer
fuel
synthesis gas

Low Temperature Version (Path 2):

800 C
receiver
air
adapted reformer
CH_4, H_2O
synthesis gas

High Temperature Version (Path 3):

1000 C
receiver
air
reformer
CH_4, H_2O
synthesis gas

Integrated Version (Path 4a):

integrated receiver/reformer
CH_4 H_2O
fuel
synthesis gas

Integrated Version (Path 4b):

integrated receiver/reformer
CH_4 H_2O
main reformer
fuel
synthesis gas

Fig. 3   Different Concepts of Solar Steam Reforming

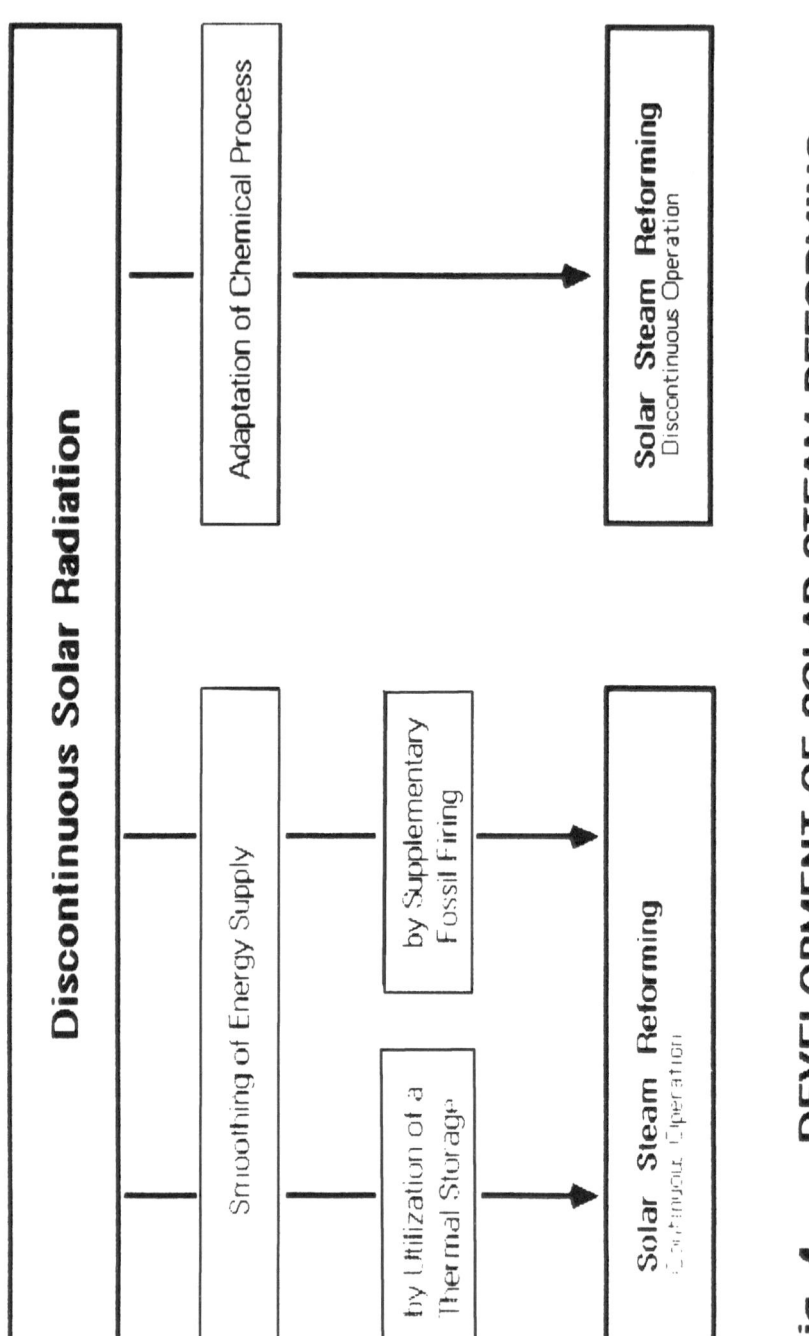

Fig. 4   DEVELOPMENT OF SOLAR STEAM REFORMING
Discontinuity Aspects

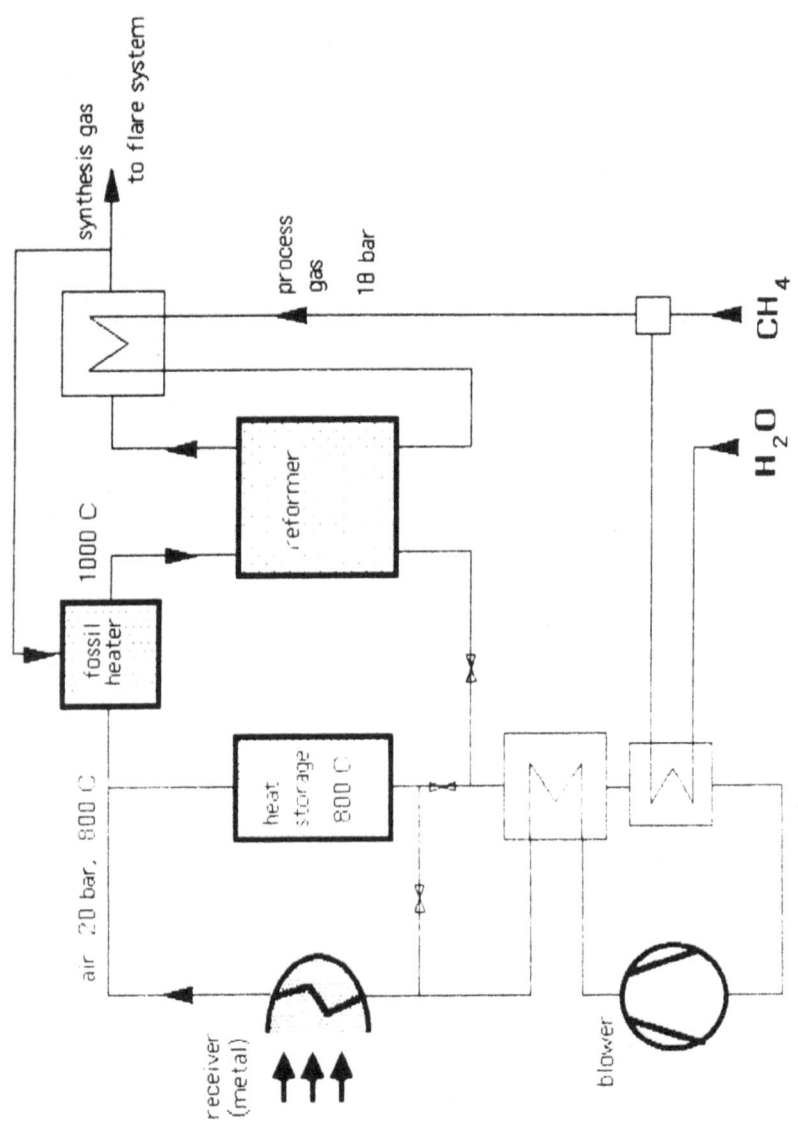

Fig. 5   Solar Steam Reforming of Methane, Component Test
Fuel Saver Version, Path 1

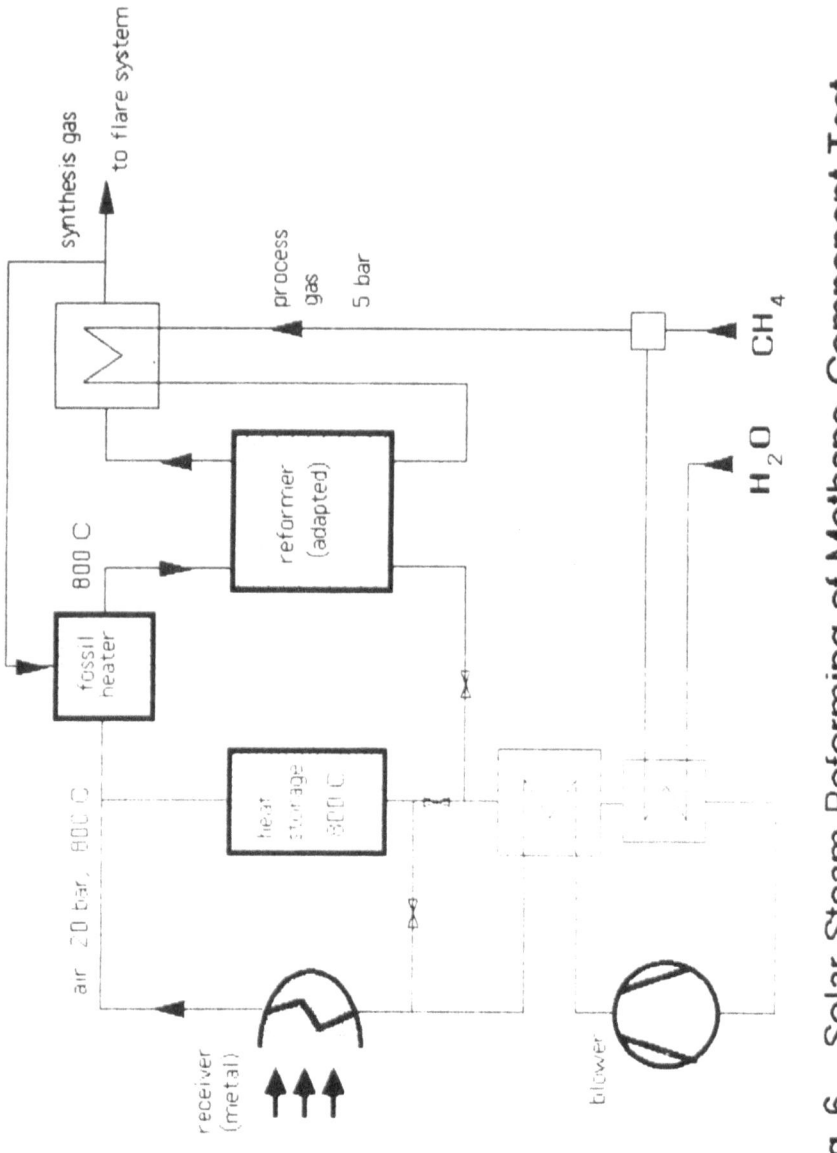

Fig. 6 Solar Steam Reforming of Methane, Component Test

Low Temperature Version, Path 2

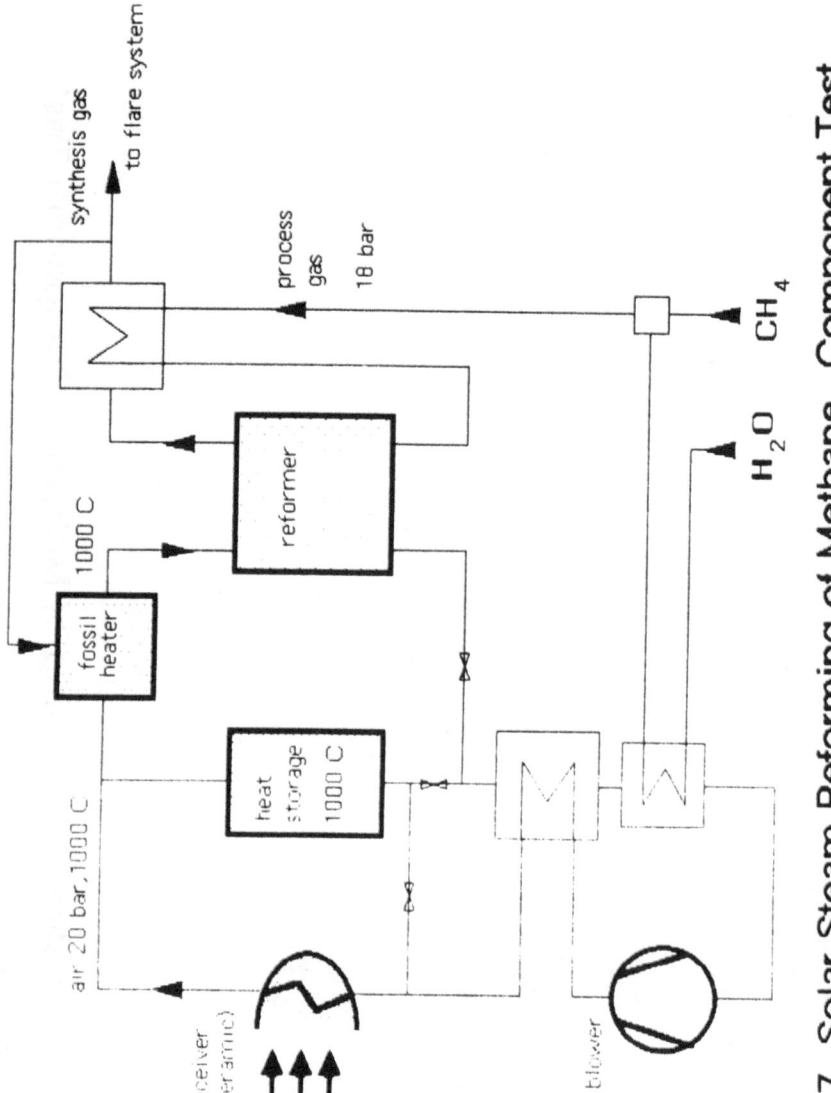

Fig. 7 Solar Steam Reforming of Methane, Component Test

High Temperature Version, Path 3

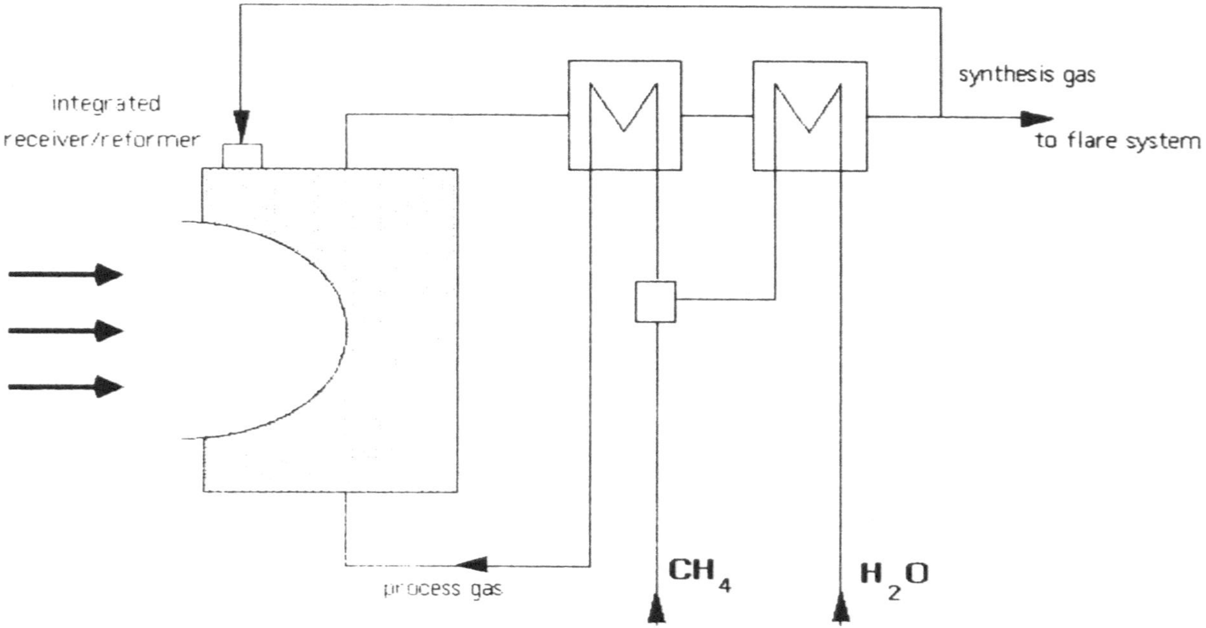

synthesis gas

to flare system

H_2O

CH_4

process gas

integrated
receiver/reformer

Fig. 8   Solar Steam Reforming of Methane, Component Test

Integrated Version, Path 4

- 238 -

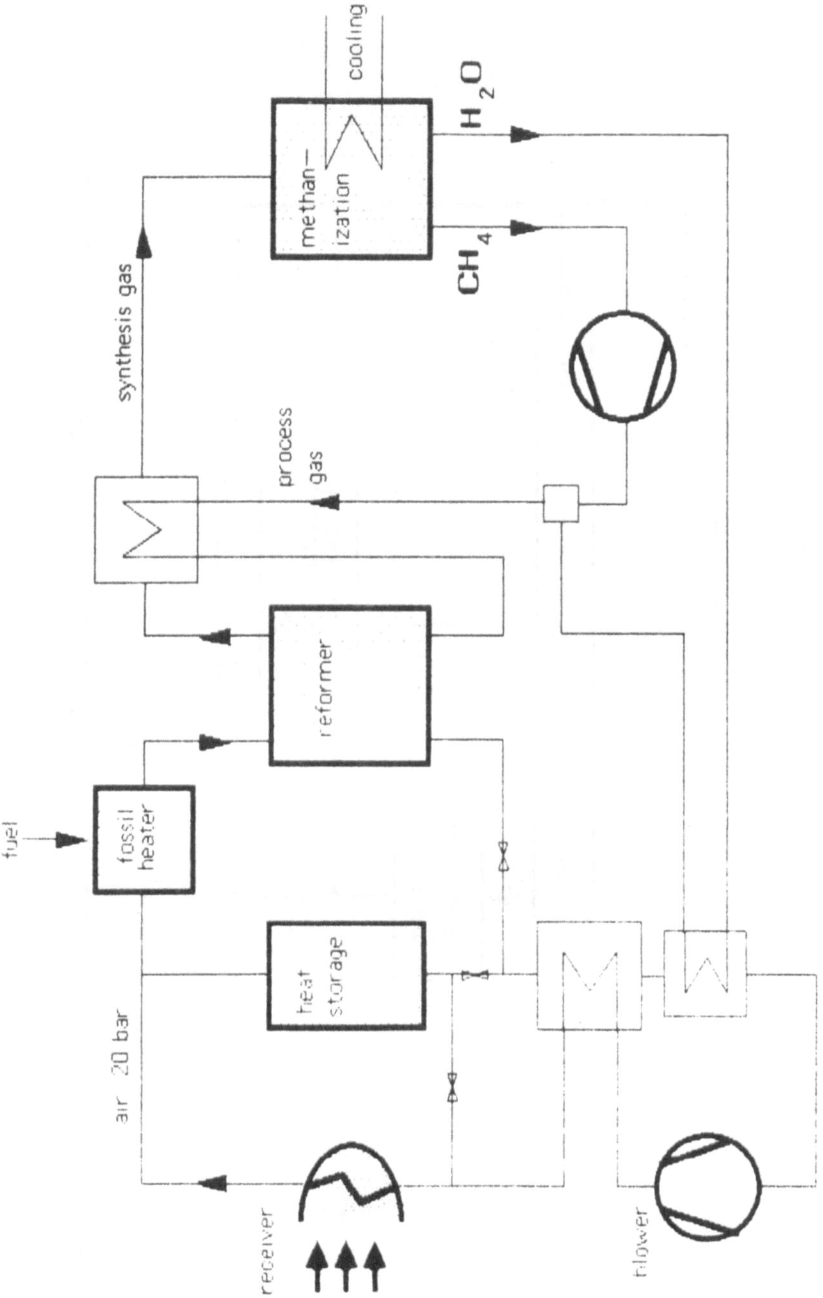

Solar Steam Reforming of Methane, System Test

Fig. 9

- 239 -

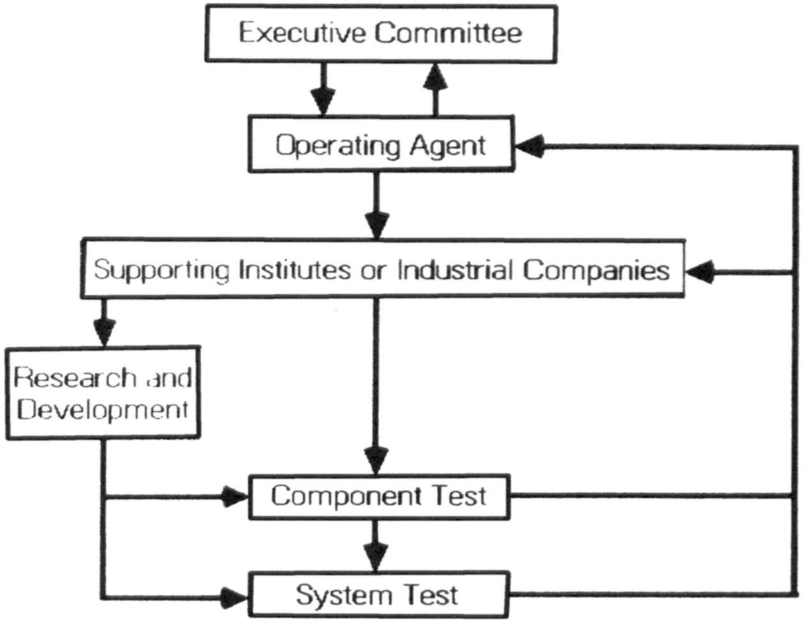
Executive Committee
Operating Agent
Supporting Institutes or Industrial Companies
Research and Development
Component Test
System Test

Fig.10 Organization Structure of the Solar Steam Reforming Program

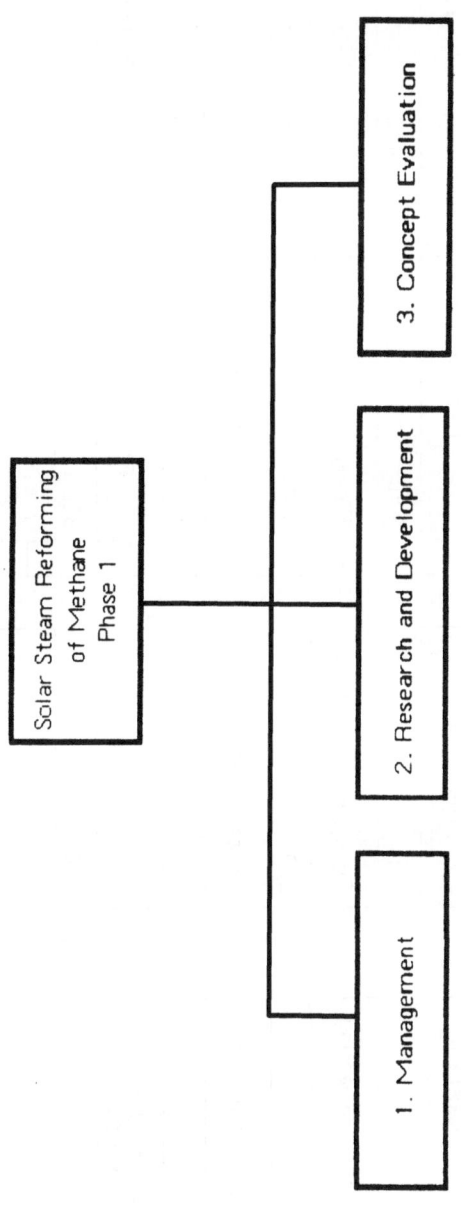

Fig. 11   Working Packages during Phase 1

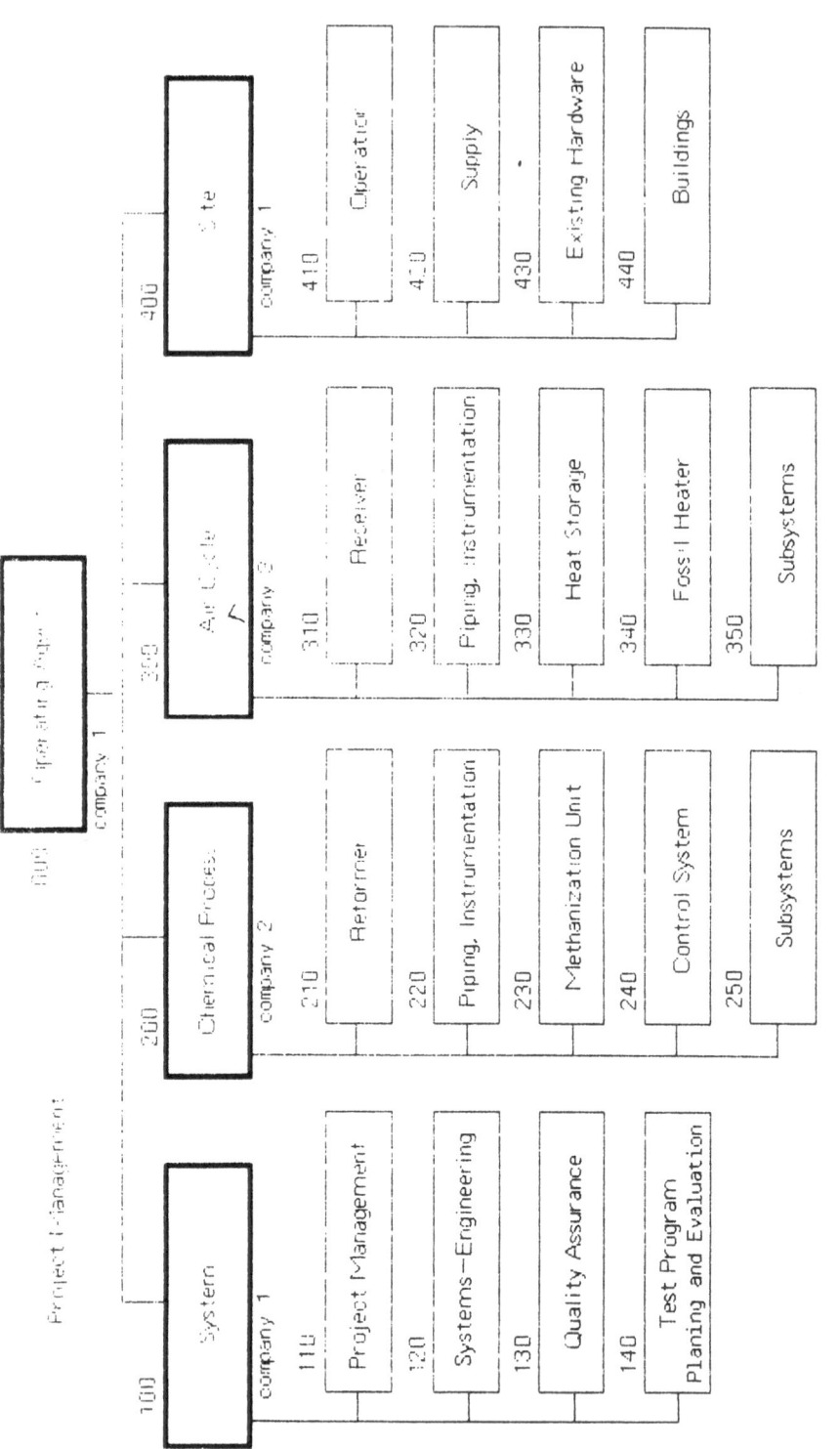

Fig. 12 Work Breakdown Structure, Phase 2 and 3, Overview

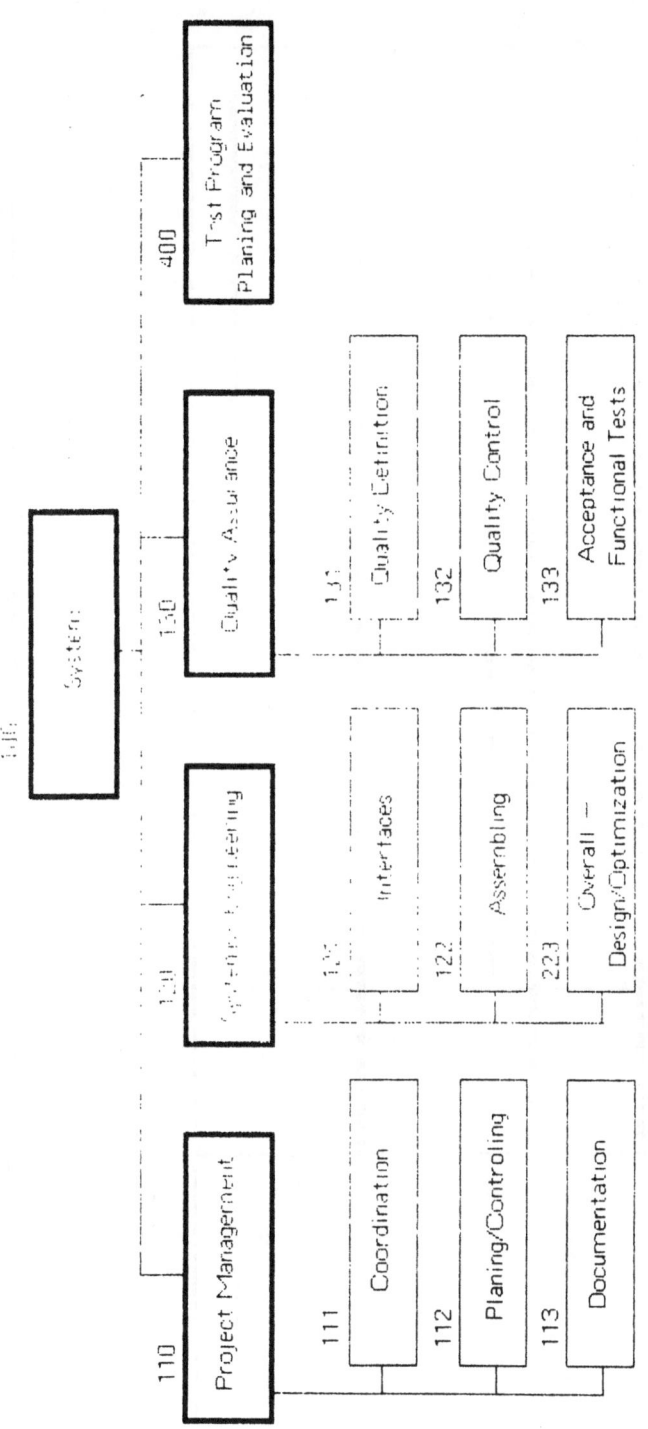

Fig. 13   Work Breakdown Structure, Phase 2 and 3

Work Package 100 — System

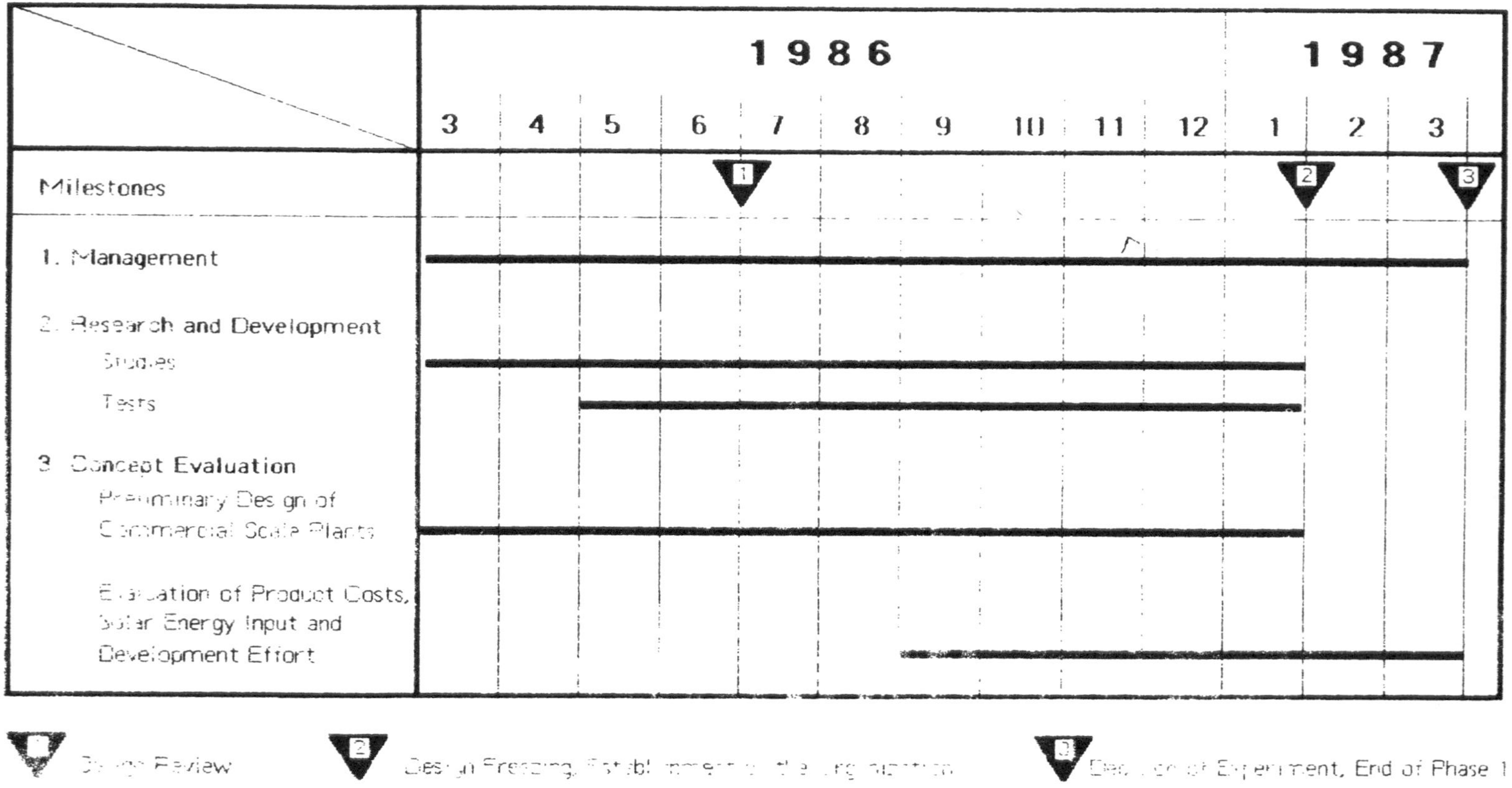

Milestones	1986										1987		
	3	4	5	6	7	8	9	10	11	12	1	2	3

1. Management

2. Research and Development

  Studies

  Tests

3. Concept Evaluation

  Preliminary Design of
  Commercial Scale Plants

  Evaluation of Product Costs,
  Solar Energy Input and
  Development Effort

▽ 1 Design Review   ▽ 2 Design Freezing, Establishment of the Test Program   ▽ 3 Design of Experiment, End of Phase 1

Fig. 14   Time Schedule Phase 1, Solar Steam Reforming Program

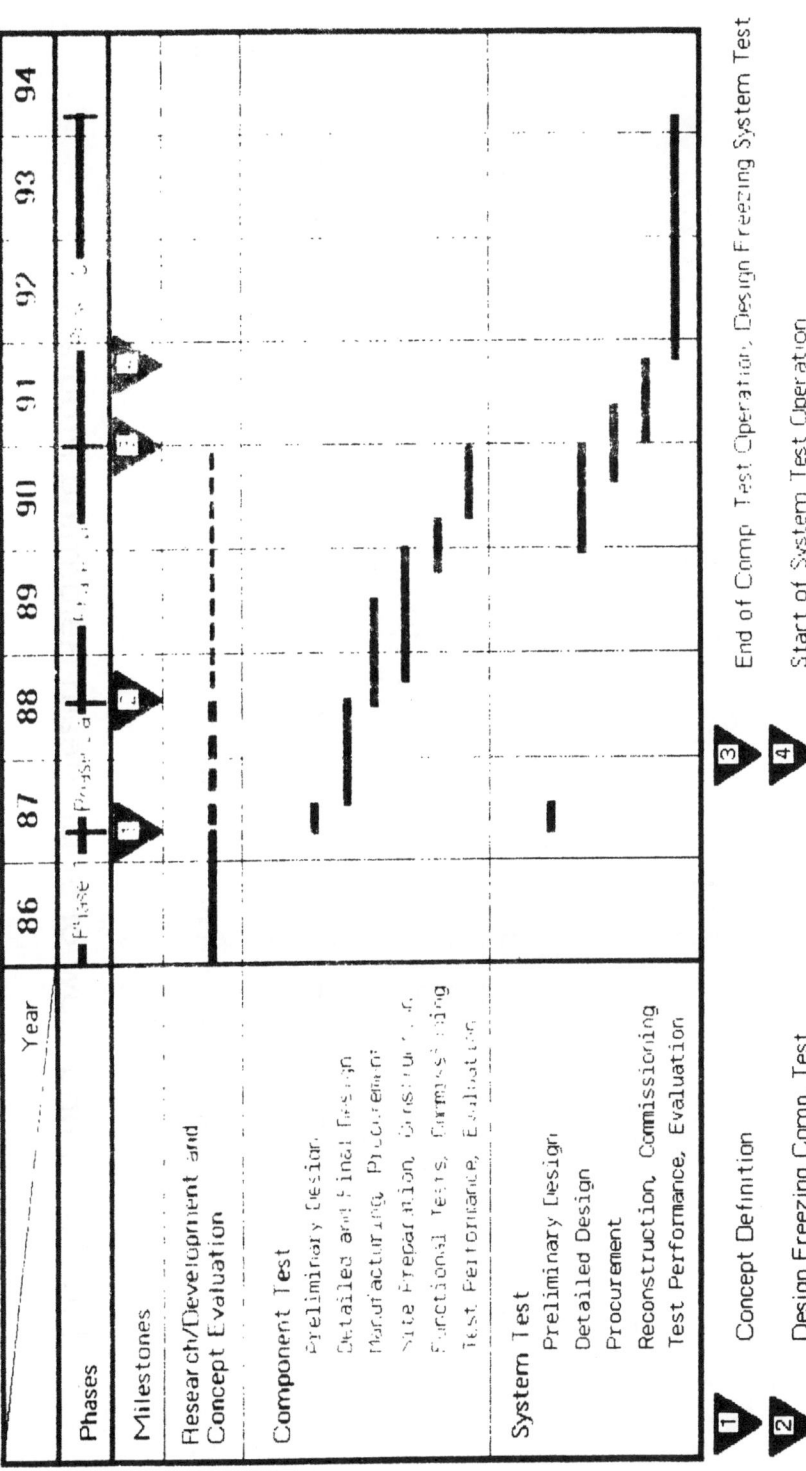

Fig. 15   Time Schedule Overview, Solar Steam Reforming Program
(Low Temperature Version)

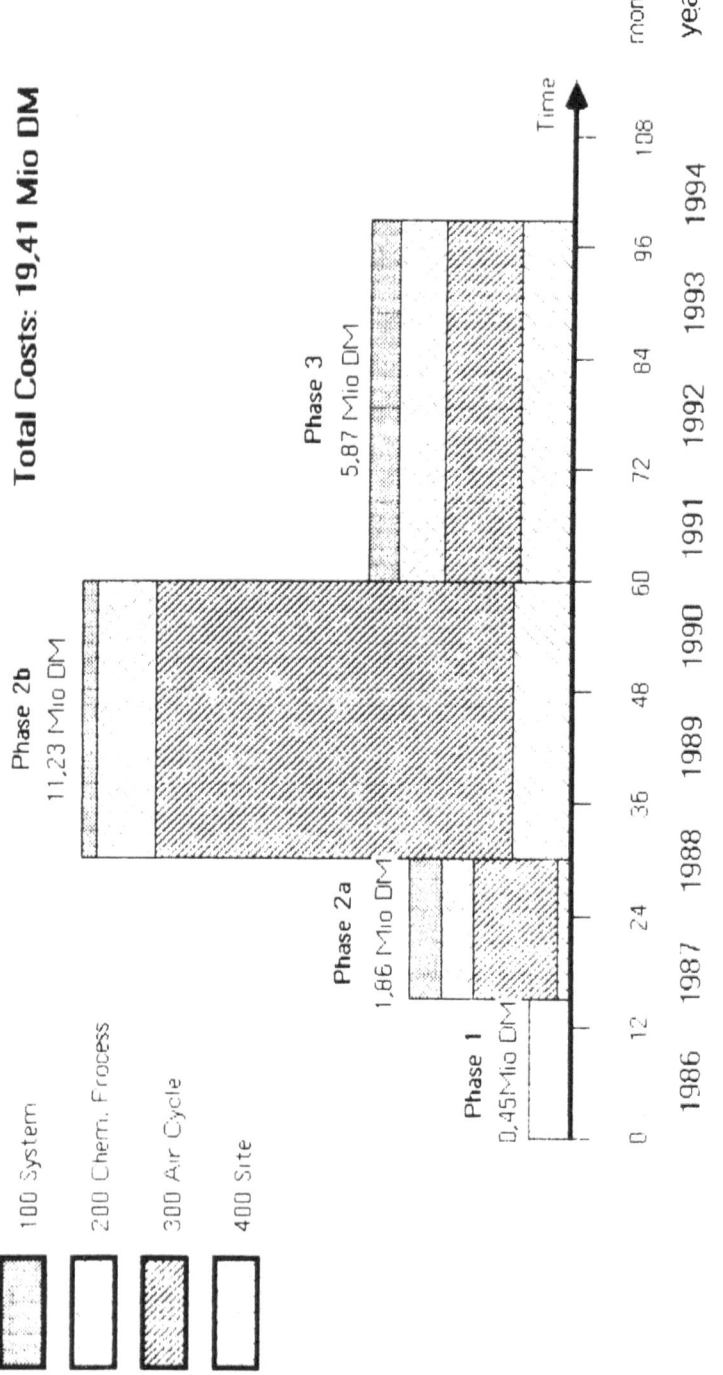

Fig. 16   Fund Flow of Solar Steam Reforming Program

Fuel	Fuel Costs 1985 DM / MWh $_{th}$
Oil, heavy	55
Oil, light	64
Natural Gas	50
Coal	25 ... 35
Concentrated Solar Radiation	36 ... 52 [*]

Assumptions:

Price of Heliostats	600 DM / m^2
Insolation	1,5 ... 2,0 MWh / a m^2
Annuity, Operating Costs (Related to Investment Costs)	13 % / a

Table 1   Comparison of Different Fuel Costs

	Fuel Saver Version (Path 1)	Low Temperature Version (Path 2)	High Temperature Version (Path 3)	Integrated Version (Path 4)
Receiver/Reformer	separated	separated	separated	integrated
Temperature gap bridged by	fossil firing	adapted reforming technology	adapted solar technology	conceptual changes
Reformer heated by	convective heat, storable	convective heat, storable	convective heat, storable	radiation, not storable
fossil heating principally required for	raising the process temperature	not reqired	not required	smoothing of energy supply

Table 2   Basic Properties of the Four Versions of Solar Steam Reforming

(in 1000 DM)	Phase 1	Phase 2a	Phase 2b	Phase 3	Sum
1. Management	50				50
2. Research and Development	300	80	30		410
3. Concept Evaluation	100				100
100 System		300	400	700	1.400
200 Chemical Process		400	1.300	1.470	3.170
300 Air Cycle		960	8.200	2.200	11.360
400 Site		120	1.300	1.500	2.920
Sum (in 1000 DM)	450	1.860	11.230	5.870	19.410

Total Costs : 19,41 Mio DM

Table 3   Cost Estimation of the Solar Steam Reforming Program

Costs in 1000 DM , Value 1985

COMPARATIV INVESTIGATIONS
AND RATINGS OF
DIFFERENT SOLAR SYSTEM USING
TUBULAR STEAM REFORMERS

W.D. MÜLLER
LURGI, FRANKFURT
H. FUHRMANN
MAN-TECHNOLOGIE, MÜNCHEN

Contents

1. Introduction - Tasks of the Study

This study has been prepared by

LURGI GmbH, Frankfurt/Main

and

MAN Technologie GmbH, München.

LURGI as process oriented engineering company was engaged for the overall aspects and the layout of steamreforming plant(s) as well as the appertaining synthesis and utility plants, while MAN's part was bound to the collecting and concentrating of solar energy and all related equipment including thermal storage and distribution of such energy.

Similar investigations with analogous share of tasks had been prepared already earlier buth with different aspects. While a first screening had taken place with a small study in 1984 ("Steam Reforming of Methane Using Solar Heat" hereinafter generally referred to as "prestudy") a second approach had been tried with a continuing study of the same title, but which evaluated possibilities and costs of an experiment, if steam reforming had to be tested already at this time at Almeria.

This present study purposely is enlarging on the aspects of the very first investigations of the prestudy. It should at the same time deal with possible modifications to better adapt the chemical process to the solar-specific conditions and it should provide cost data to compare the different systems between themselves and with available technology. For such systems, which seem to require highest interest with regard to defined aspects (high solar rates in product, economics, risks), experiments should be proposed that would lead to the reliable applicability of the investigated overall-systems.

2.   Investigated Chemical Production Complexes

2.1   Presentation of Studied Chemicals

2.1.1   Generald Considerations

Steam reforming of natural gas for the generation of synthesis gas is being applied worldwide for the production
of the following basic chemicals
- ammonia
- methanol
- hydrogen
- oxo-alcohols

In these plants natural gas is used as feed for the chemical reaction. The hydrogcarbon is catalytically split up
with steam according to the following equations

$$CH_4 + H_2O \longrightarrow CO + 3H_2$$
$$CH_4 + 2H_2O \longrightarrow CO_2 + 4H_2$$

thus resulting in a reformed gas or synthesis gas being
composed of carbonmonoxide, carbondioxide, hydrogen and
undecomposed methane and surplus steam, which are reacted
only to such portions as are limited by thermodynamic
equilibrium. The reactions will be more complete with
higher temperatures and lower pressures. Nitrogen and
argon will be present, if present in the feed.

If higher portions of hydrogen are wanted the socalled
catalytic shift conversion will be integrated. It is
working according to the formula

$$CO + H_2O \longrightarrow CO_2 + H_2$$

The output of hydrogen will be increased with lower temperatures, the pressure is of secondary importance for
this reaction.

Natural gas also is taken as fuel to provide energy for
the reforming of the methane. It is this portion of the
overall natural gas consumption, which can be diminished
or brought to zero by a solar operated steam reforming
plant. The natural gas for the chemical reaction cannot
be substituted.

On the other side not all gases in the synthesis gas can
be used by the specific synthesis or reacting gases can
be bound only to a certain efficiency. Consequently there
will be a surplus of unreacted gases - the so called
"purge gases" - that represent a certain value at least
with respect to their heating values. The purgegases
will vary with every process and specific conditions of
the process sequence, but there definitely has to be found
use of this purgegas to keep the overall economics of a
process as good as possible.

In the prestudy it already had been pointed out that the
production of hydrogen by solar aided steam reforming will
not be advisable as with modern technology (PSA-hydrogen
separation) there is a surplus of energy in form of steam
export from such reforming plants and with conventional
techniques the energy scheme nearly is balanced. When
adding solar energy to such a system, this will only increase the steam export. This would be a most expensive
generation with no use within the plant. Besides this,
hydrogen as a product will not be usable by its own and
will make a problem for transport with present techniques.
Especially because of the energy surplus hydrogen will not
range under the products to be found with solar supported
steam reforming plants.

Similar problems arise with methanol, as the purgegas quantity and quality normally are adequate to serve as fuel
for the steam reformer leaving only a small portion of
natural gas as fuel (balancing for control purposes).
Therefore, solar supported steam reformers in methanol pro-

duction plants will have hydrogen as a byproduct, which will give similar problems for use or transport as outlined before, but energywise such plants can be egalized.

It has been assumed within this study that pure hydrogen can be sold from such complex. If this was not the case, this hydrogen product also could be used as a fuel to generate electricity for local export.

2.1.2 Production of Chemicals, based on Solar-Supported Steam Reforming of Natural Gas

Due to the general criteria mentioned above,the production of the following chemicals has been envisaged within this study.
- Methanol plus hydrogen

- Ammonia via secondary reforming with air

- Ammonia via pure hydrogen and admixture of nitrogen from an air separation plant

- Octamix = Fuel Methanol
          = Methanol plus 45 % higher alcohols as mixture for blending of gasoline

The production of town gas and the so-called "ADAM-EVE" systems for energy transport have been excluded from this study as their applicability would require the use of the "product" within short distances from the solar plant, which would be a restriction with respect to general acceptance. The production of town gas should show similar results as with the studied chemicals, while the energy transport anyhow would require differentiated methods of consideration for comparison.

In all cases it was the basic condition that there should
be available only

- naturals gas

- sunshine

- raw water

and that it should be possible to run a grass-root plant
on these basic materials, even if they should not be taken
into account in all detail. For instance there have been
made provisions for the generation of electric energy in
the consumption figures for natural gas in all cases, but
the relevant investment, being equal and additive in all
production complexes for the same chemical, has not been
evaluated in detail. Similar statements could be made for
the generation of inert gases or the supply of cooling
water.

Emphasis was put on the comparability of technical conditions for the different plant concepts. This especially
is relevant for the main data as
- total absorbed energy of the steam reforming
  process
- pressure of the steam network

These two parameters were fixed for the purpose to allow
an exchange of moduls within studied systems. The first
request allowed a standardisation of solar related equipment, while the second kept savings by solar energy equal
in the process sequences for the same solar input.

By such "standardisation" the optimum requirements for a
specific process naturally got lost with the effect that
specific figures may not favourably compare with nowadays
standard conditions for such processes, but the simplifications allowed the correct comparison of different reforming conditions, when varying the chemical products.

It should be kept in mind that
- even large solar plants with storage systems, serving
  steam reformers (using two heliostat fields of approx.
  840 m x 620 m each), render relatively small chemical
  plANTS (about 400 t/d of ammonia or methanol only).
  Therefore the specific investment and hence the product
  costs of nowadays chemical plants, fossil fired, will
  be more favourable than used for the present study.
- the steam generation has been set for 40 bar in the
  steam drum. This pressure cannot be changed with
  Lurgi's low pressure methanol synthesis process and was
  then adopted also for the other processes for reasons
  as outlined above. The specific consumption figures
  could be better for ammonia plants, when using a
  sophisticated steam network with a steam pressure of
  about 100 bar starting and using counterpressure turbines, bleeds etc.

2.1.3 Codes and Used Terminology

Throughout this study a classification has been used,
which differentiates between

  - the type of chemical to be produced wihtin the
    comples and

  - the type of solar plant which shall be integrated
    for the task of steam reforming

Therefore tha code prefixes A, B, C, D and A' stand for
the chemical type, namely
  A - Methanol production with co-production of hydro-
      gen
      The steam reformer will be operated at about
      20 bar.

  B - Ammonia production by means of a secondary refor-
      mer using compressed air as oxidizing medium.

  C - Ammonia production from pure hydrogen by addition
      of compressed pure nitrogen, gained from an air
      separation plant.

D - Production of Fuel Methanol by the "Octamix"-process, designed for about 45 % by weight of higher alcohols.

A'- Methanol production with co-production of hydrogen.
The steam reformer will be operated at about 8 bar.
This plant type, though being very similar to "A" was introduced for study purposes,to show effects of varied reforming conditions depending from solar technically handable temperature levels.

The identification numbers (1) to (6) correlate with the type of solar cavity or solar operated reactor which shall be applied to perform the relevant duty of steam reforming. Though the details for the solar systems will be given later, for complete reference the identification numbers are listed herebelow. As a principle differentiation it must be known that

"Separated Systems" are entitled those that use a heat carrier to transport the energy from the solar receiver to the reactor. The heat carrier may be air or helium. Other carrier media as hydrogen or liquid salts have purposely been excluded from these investigations (refer also to chapter   ) . According to available technical knowledge and ideas only separated systems can be equipped with storages that will allow "solar" operation at night or with cloudy weather. The size of the storage and the "solar multiple" will fix the possible gain of time for "solar operation", when the sun actually is not visible.

"Integrated Systems" are entitled those, where the reformer
tubes for the catalytic steam refor-
ming reaction are directly reached
by the solar energy. The reformer
tubes, the integrated reactor, may
be irradiated directly from the heliostats in case of admissible low temperatures, but with higher pretended
temperature levels they also may be
arranged in the "shadow" of the cavities and may be irradiated by solar
light, which then will diffusly be
reflected from the walls of the cavity.

"Prereactor System" means systems with maximum steam reforming process temperatures of about
700 $^{\circ}$C. Such process temperatures
may be reached with heat carriers of
a maximum temperature of about 800°C,
which on their behalf will need tube
material temperatures of about 850°C
in the cavities. Such tube material
temperatures are thought to be the
maximum allowable for metallic tubes
under the specific conditions of solar irradiation.
Prereactor systems will need a suc-
cesssive process step with conventional, fossil fired steam reformers to
complete the required transformation
of methane into synthesis gas.

"Main Reactor Systems" will be called such arrangements,
that complete the chemical reaction
with solar energy in one step.
As maximum process temperatures of
more than 850°C are required, the use
of metallic materials in the cavities

is doubtful and needs special precautions. If "separated main reactor
systems" have to be considered, the
tube wall temperatures will be as
high as about 1050°C, which will require ceramic materials, still to be
developped.

With the terminology of this kind being available the
identification numbers (1) to (6) refer to the following
system:

1 - separated prereactor system without storage

2 - separated prereactor system with storage

3 - separated main reactor system without storage

4 - separated main reactor system with storage

5 - integrated prereactor system

6 - integrated main reactor system.

For id. numbers 2 and 4 additional qualifications have
been introduced and refer to the size of the storage and
the solar multiple.

2.1 and 4.1 refer to small storages, low solar multiple

2.2 and 4.2 refer to medium sized storages and moderate
solar multiples

2.3 and 4.3 refer to high storage capacity and high
solar multiples as they will be defined as
design principles later. The lower storage
capacities are fixed by available sizes of
storage modules.

The identification numbers (7) and (8) have been used only
in conjunction with the prefix A', i.e. with reforming
conditions at low temperature and low pressure. The id.
numbers then signalize:

7 - specially designed separated main reactor with
large storage system

8 - integrated main reactor

The reference cases for comparison purposes always have
been identified as - fossil. They refer to standard conditions as they would prevail with conventional, fossil
fired steam reforming plants for the production of the
different chemicals.

2.1.4   Block_Flow_Diagrams

As the title of the present study already is emphasizing,
great importance was attached to the comparison of different solar plant concepts and their influence on the chemical products.

Therefore basic concepts were established for the chemicals, which should be considered. Besides the general
criteria, which should be ruling according chapter
2.1.2,   it was agreed that the comparability should be
reached by using the same energy input to the chemical reaction from natural gas (methane) to reacted gas (synthesis gas).

In consultancy between the concerned parties, taking into
account earlier results of relevant studies, it was agreed
that this net energy input should amount to

approx. 40 MW.

With this basic figure the concepts of the production
lines could be fixed and the results are shown in the
attached block flow diagrams, figures 2.1 to 2.5. The
following explanations go with them:

Fig. 2.1, Methanol Production, Code Prefix: A
(and general considerations)

40 MW energy supply will be sufficient to process
12 000 m^3N/h natural gas into a synthesis gas for use in
a methanol synthesis plant. The appertaining conditions
for such steam reforming would be:

```
natural gas analysis:  CH₄      99,5 % by vol.
                       N₂        0,5 % by vol.
```

natural gas analysis: CH_4 99,5 % by vol.

N_2 0,5 % by vol.

steam/carbon-ratio approx.   2,6

natural gas/steam mixture
    inlet temperature to reaction   350 $^{\circ}$C

reformed gas outlet conditions

temperature   860 $^{\circ}$C

pressure absolute   18 bar

analysis: CO_2 approx.   7,3 % by mol

CO 14,9 % by mol

H_2 73,9 % by mol

CH_4 3,8 % by mol

N_2 0,1 % by mol

H_2O 0,385 mol/mol dry gas

reformed gas quantity, dry
    about 45,900 m^3N/h

Cooling down the reformed gas to ambient temperature and thereby condensing the surplus steam, will yield about 31 MW thermal energy, which can be used to a great portion within the process. Less than 7 MW will be lost as waste heat on a non recoverable level.

In a conventional, fossil fired steam reforming plant flue gases will leave the reformer at a temperature about 100 to 150 K above the highest process temperature.

The flue gas will be used to generate steam and to also superheat it. The heat load of the flue gas channel will exceed 20 MW, especially if fuel and combustion air will have to be preheated in such a rate that a flue gas outlet-temperature of about 150°C can be maintained, which was the general condition within the present study to keep systems comparable.

From 45 900 m^{3}N/h synthesis gas 13,75 t/h of pure methanol
can be gained. Details of a Lurgi methanol synthesis
plant and the adjacent distillation may be found in relevant publications as well as for the pressure swing adsorption unit, which separates 9.420 m^{3}N/h of high purity
hydrogen from the synthesis plant's purge gas. This
hydrogen will be available at a pressure of 25 bar,
approximately. (As the purgegas is available with a
pressure of more than 50 bar, the hydrogen could be made
available also with such high pressure, but then the separation efficiency of the PSA-unit would decrease and the
surplus residual gas would make a problem for use in solar
supported plants. This is another item, which does not
allow general optimizations, if different plant types
shall be compared).

The offgas from the PSA-Plant is used as fuel. Its amount
has been selected in such a way - or the separation efficiency of the PSA-plant has been fixed to such a value -
that with steam reforming plants of type A - 4.3 the offgas quantity (often referred to as "residual gas") will
be sufficient to supply all necessary heat in the flue gas
duct without any additional natural gas fuel. As will be
seen later there even is some surplus residual gas
available with A - 4.3, for which other users had to be
found.

As water is taking part in the reforming reactions (see
chapter 2.1.1) this is a "raw material", too. In the
blockflow diagrams only the quantity of demineralized
water, used up in the reforming plant, is shown to be provided from the water preparation plant. Internal reuse
of condensates and their thermal treatment has generally
not been shown as being technical standard and no special
task in connection with solar supported plants. The same
would apply for waste water treatment and further utility
plants. Their investment would equally increase (with

small margins) the total amount to be spent for such
plants, notwithstanding whether they are of type A-fossil,
A - 4.3 or any other A-type plant with the same capacity.

In chapter 4 natural gas equivalents for utility generation have also been taken into consideration as they are
of importance for the product prices, when depending from
total natural gas consumption within a chemical complex.
There the percentace of savings with solar plants may be
of interest, which was the reason to establish the total
consumption figures. As the clarity of the block-flow diagrams would suffer from too many details, such additional
natural gas consumers have been left off therein.

The blockflow diagrams have been worked out for fossil
fired conventional steam reformers. The indicated loads
of the flue gas ducts especially refer to this type of
plant. When supplementing or substituting the conventional reformer by solar reactors, the primarily available
flue gas quantity will be lowered or will be zero, needing
different air quantities for burning the fuel. Thus the
required load of the flue gas channel will vary at the
same time. It should therefore be recognized that indicated heat loads of the flue gas channel are valid for the
-fossil types only. As a general rule the duties will be
smaller with solar plants, but will not be much different
for the design case, when the sun is down. What comes out
from such reflections in that the investment for the flue
gas channel for a fossil operated and a solar operated
plant will not change significantly. The changes then
will be restricted to the reformer itself, what has been
marked in the special "block" of each block flow diagram.

For systems .....-5 and .....-6 a capacity, which is increased by 8,5 %, was adopted for reasons which will be
explained with the relevant detailed flow diagrams.

Fig. 2.2, Amminonia Production with Secondary Reformer
Code Prefix: B

40 MW energy supply will be sufficient to process 14.325 m^3/h natural gas of the same analysis as for Fig. 2.1. The reforming conditions would be:

steam/carbon-ratio　　　　approx. 3,25

natural gas/steam mixture

　　　inlet temperature to reaction　350 ^{o}C
outlet conditions from steam reformer
　　　temperature　　　　　　　　800 ^{o}C
　　　pressure absolute　　　　　　31 bar
　　　analysis: CO_2 approx.　10,4 % molar
　　　　　　　　CO　　　　　9,6 % molar
　　　　　　　　H_2　　　　70,1 % molar
　　　　　　　　CH_4　　　　9,8 % molar
　　　　　　　　N_2　　　　0,1 % molar
　　　　　　　　H_2O　　　0,668 mol/mol dry gas
　　　reformed gas quantity, dry
　　　　　　about　　47.940 m^3N/h

The outlet gas from the steam reformer is further processed by the addition of preheated air in the secondary
reformer and, after use of sensitive heat, the hydrogen
content of the gas is increased by the shift conversion
(high temperature and low temperature system) according
to the formula already presenten in chapter 2.1.1.
The appertaining outlet conditions from the process steps
will be as follows:

		secondary reformer	hightemp. shift	lowtemp. shift	methanation
temperature	°C	993	408	230	380
pressure absolute	bar	30	28	27	25
approx. analysis: % molar or mol/mol dry gas					
	CO_2	7,2	15,7	17,8	-
	CO	13,4	2,9	0,4	-
	H_2	57,2	61,2	62,1	73,9
	CH_4	0,2	0,2	0,2	1;5
	N_2+Ar	22,0	20,0	19,5	24,6
	H_2O	0,158	0,377	0,345	-
dry gas quantity m^3N/h		68.580	75.580	77.425	61.330

For the above figures it has been assumed that the CO_2-content of the shifted gas would be lowered to about 0,7 % molar, resulting in a CO-content of about 0,5 % molar. In methanation the reserve reactions than with reforming take place:

$$CO + 3H_2 \longrightarrow CH_4 + H_2O$$
$$CO_2 + 4H_2 \longrightarrow CH_4 + 2H_2O$$

The elimination of carbonoxides to a level of less than 10 ppm (molar) is necessary, not to poison the catalysts of the ammonia synthesis.

The synthesis gas after methanation will be compressed to about 150 bar for its use in the ammonia synthesis. The compressor was anticipated as being driven by a condensing turbine and the heat of reaction in ammonia synthesis was used to generate steam of about 40 bar for reasons as stated already earlier.

The energy system was balanced with the possibilities of the reforming plant. About 22,76 t NH_3/h will be produced when including a special hydrogen recovery in the purge gas stream of the synthesis. The ammonia can be made available at at pressure of about 20 bar or lower with temperatures corresponding to saturation. As pure CO_2 could be made available form the CO_2-washing system, the production of urea could follow, if desired.

The energy needs of the refrigeration unit, providing the necessary low temperatures for condensing the ammonia, have been integrated into the overall consumption figures of the complex. The purgegas will consist mainly of methane and some inerts and will be used as fuel in the reforming unit.

What can be seen immediately from the block flow diagram: The amount of heat to be handled in the flue gas heat recovery section is much bigger than with methanol production. Therefore the use of the residual gas as fuel is no problem, there is no need for residual gas storage in a gasholder, because the needed fuel quantity exceeds the offered one in every moment.

Fig. 2.3, Ammonia Production by Nitrogen-Admixture to Hydrogen
Code-Prefix: C

40 MW energy supply will allow to process 11.395 m^3N/h natural gas of the usual analysis as in Fig.2.1 (and 2.2). The reforming conditions would be:

steam/carbon-ratio at inlet   approx. 4,0

natural gas/steam mixture
            inlet temperature to reaction   350 ^oC

outlet conditions from steam reformer
            temperature   860 ^oC
            pressure absolute   18 bar
            analysis: CO_2 approx.   9,7 % molar
                      CO   12,5 % molar
                      H_2 76,2 % molar
                      CH_4 1,5 % molar
                      N_2 0,1 % molar
                      H_2O 0,630 mol/mol dry gas
dry gas quantity   approx.   47.815 m^3/h

As the purification of the synthesis gas shall be made by pressure swing adsorption techniques there is no need to significantly lower the CO-content. The installation of a high temperature shift conversion step will be suffi-cient to increase the hydrogen quantity by about 10 %.

The outlet conditions from the shift conversion will be as follows:

```
temperature                    415 °C
pressure absolute               16 bar
analysis: CO₂   approx.  17,5 % molar
          CO               2,8 % molar
          H₂              78,2 % molar
          CH₄              1,4 % molar
          N₂               1,0 % molar
          H₂O              0,490 mol/mol dry gas
     dry gas quantity  about  52.320 m³N/h.
```

The pressure swing adsorption plant will separate
34 780 m^3N/h hydrogen therefrom, leaving 17.540 m^3N/h
residual gas (dry) as fuel with a lower heating value of
about 1,762 kWh/m^3N. For systems C - 3 and C - 4 there
will be excess of energy, when operating with sun , making
the installation of a gasholder as a buffer necessary.

The required nitrogen quantity of 11.595 m^3N/h will be provided by an air separation unit of about 20.000 m^3N/h air
intake. The oxygen, which will be gained at the same
time, is supposed to be vented, as no direct user is found
within the investigated complex.

The ammonia production will be about 17,625 t/h or 420 t/d.
Design parameters and product availability will be as with
systems "B". As very pure gases will be used as make-up
gas for the synthesis unit, no purge gas will evolve from
the ammonia plant.

Fig. 2.4, Fuel-Methanol Production by the Lurgi-Octamix-
Process
Code-Prefix: D

40 MW energy supply will be used to steam reform 10.290
m^3N/h natural gas of the "standard analysis" and 14.350
m^3N/h carbondioxide, which will be gained within the process sequence and which will be recycled. The purpose of

such superimposed CO_2-recycle is to lower the hydrogen/CO-
ratio at the outlet of the reformer as far as possible.
Besides those reactions listed in chapter 2.1.1 also the
conversion to the equation

$$CH_4 + CO_2 \longrightarrow 2\ CO + 2H_2$$

will take place. But, because of the overall hydrogen
yield from the methane-water vapour reaction, there will
be an excess of hydrogen, which cannot be bound in the
alcohol-molecules. To maintain a proper hydrogen/carbon-
monoxide ratio in the synthesis gas, necessary for the reaction, a part of the hydrogen, being generated in the
steam reformer, has to be separated ahead of the synthesis
plant and will be available as secondary product (as with
the methanol production, too). As pure hydrogen is more
valuable than it would be, if taking its calorific value
into account only, it will be more economic, to sell such
hydrogen than to burn it as fuel on site and to lower the
amount of natural gas for fuel, what principally would be
possible, of course.

The CO_2-content in the feed gas to the synthesis reactor
has to be limited to prevent side reactions. As CO_2 on
the contrary also will be built during the catalytic reaction, CO_2 will also have to be removed from the synthesis
recycle.

The following main characteristiques have been adopted for
the shown process sequence.

steam/carbon-ratio at reformer inlet   approx. 2,6
(C' calculated without CO_2)

CO_2/CH_4-ratio at reformer inlet   approx. 1,4

feedgas mixture
   inlet temperature to reaction   350 $^{\circ}C$

outlet conditions from steam reformer
   temperature   860 $^{\circ}C$
   pressure absolute   18 bar
   analysis: CO_2 approx.   21,7 % molar
              CO 26,7 % molar
              H_2 49,7 % molar
              CH_4 1,8 % molar
              N_2 0,1 % molar
              H_2O 0,430 mol/mol dry gas

   dry gas quantity   about 49 040 m^3N/h

From this reformed gas after cooling procedures the carbondioxide will be removed in a chemical scrubbing system to
a residual level of about 1 % by vol.

About 60 % of the purified gas will be sent to a pressure
swing adsorption unit, wherein about 11 280 m^3N/h pure
hydrogen will be separated. The offgas of this unit will
consist predominantly of carbonmonoxide and has to be recompressed to the purified gas stream after the carbondioxide removal unit. The mixture has the following properties:

pressure absolute   about 15 bar
temperature   about 40 $^{\circ}C$
dry gas quantity   about 27530 m^3N/h
analysis: CO_2 approx 1,4 % molar
          CO 47,6 % molar
          H_2 47,6 % molar
          CH_4 3,2 % molar
          N_2 0,2 % molar

- 276 -

The gas mixture will be compressed to values as is usual for low pressure methanol plants, i.e. 50 to 80 bar approx. The recycle of the synthesis will include a secondary CO_2-removal system, which will keep the CO_2-level at the same height as with the synthesis gas. By the recycle loop the water content of the reactor feed gas also will be kept at a low value, thereby finally gaining a product of less than 0,5 % by weight of water.

About 1800 m^3N/h of dissolved gases and low boiling byproducts are being separated from the raw product stream in a stabilizing column and then the product quantity will amount to 8,625 t/h of fuel methanol with about 45 % by weight of higher alcohols.

The lower heating value of the product will amount to about 7,05 kWh/kg at a specific weight of about 0,80 kg/l.

Due to high energy consumption within the overall process even with high solar energy rates for the reforming, the heating value of the residual gases is not sufficient to provide all necessary energy in the flue gas channel and hence there is no need for the installation of a gas storage.

Fig. 2.5, Methanol Production with Reforming at Low Pressure
Code-Prefix: A'

The process sequence is identic with that one of Fig. 2.1. Yet, to be able to operate a solar supported "main reactor steam reformer" with nowadays available techniques, the pressure of reforming and the appertaining temperature were lowered. As a consequence the synthesis gas will be available at lower pressure only and its compression to an equivalent pressure, as it is normally available with the A-system, needs so much additional energy that only little specific overall energy savings are to be expected, thus eliminating the interest for further development of such system. Details will be given later.

Deviating from usual design principles the energy input
could not be fixed as identical for the two competing concepts. For the separated concept it will be technically
possible to install elements in the steam reformer tubes
that will exchange energy to the gas, which shall be reformed. Thus only 40 MW solar energy input will be needed
to reform 15.125 m^3N/h of natural gas, while 46,129 MW
solar energy will be needed to reform the same amount of
natural gas without such special heat exchange elements.
Therefore in the A'-systems the natural gas throughput has
been fixed by one case and the appertaining solar energy
input of 40 MW, while the other case has been fixed
through the natural gas throughput and the solar energy
input as a following paramter.

The following mass balance has been the basis for Fig. 2.5:

natural gas intake (standard analysis) 15.125 m^3N/h
steam/carbon-ratio   about 3,0
reforming conditions (equilibrium)
    pressure absolute   8 bar
    temperature   760 ^oC
    analysis: CO_2 approx.   9,7 % molar
          CO 11,7 % molar
          H_2 73,9 % molar
          CH_4 4,6 % molar
          N_2 0,1 % molar
          H_2O 0,473 mol/mol dry gas
    dry gas quantity about 57.900 m^3N/h
  absolute inlet pressure to synthesis gas compression
about 4,8 bar

The product quantity will amount to about 16,35 t/h of
pure methanol. From the purge gas a specific hydrogen
quantity as in the A-system should be withdrawn by a pressure swing adsorption plant, thus resulting in a pure
hydrogen quantity of 11.195 m^3N/h. Due to the higher
energy demand of the A'-systems, the residual gas will be
used up totally in the reforming unit even with high solar

energy input and therefore the installation of a gasholder
for the PSA-offgas will not be necessary.

2.2   Steam Reforming Plants

2.2.1   General Considerations

From the explanations going with the block flow diagrams
it can be derived that for the chosen process sequences
the heat and mass balances as well as the greatest part
of the utility consumption will be the same, if the same
amount of energy will be spent for the endothermic reaction in the reforming unit.

The present study therefore is made up in such a way that
the energy input for the reforming of the natural gas will
always be the same for a given process sequence. This
will be found, whether this reaction energy is brought in
by fossil firing only, by solar heating only or by a mixture of both energy sources.

This concept fixes also that the recuperable heat from the
reformed gas will be the same and that the duty for any
flue gas channel as supporting device will not change
either.

This concept and procedure allows to exchange the fossil
fired tubular steam reformer against a main reactor system
or against a prereactor system with a consecutive fossil
fired steam reformer of reduced load, without affecting
the duty of the rest of the plant.

It will therefore be adequate to present the purely fossil
operated reforming plants first and then to discuss the
changes only which result from the adaption to solar conditions.

2.2.2   Schematic Flow Diagrams of Steam Reforming Plants

2.2.2.1   Schematic Flow Diagrams of Fossil Fired Steam Reforming
Plants

2.2.2.1.1
Flow Diagram "A-fossil", Fig. 2.6

12.000 m^3N/h natural gas will be reformed to generate
methanol synthesis gas. The main process parameters have
been given in chapter 2.1.4.

The natural gas will be preheated to about 350 ^{o}C. A de-
sulphurization step (with recycled hydrogen) could be included but has generally been omitted in the flow sheets,
assuring a sulphur free feed gas. The preheated natural
gas will be mixed with about 25.070 kg/h steam, which shall
be provided with such conditions that also the mixture of
steam and natural gas has a temperature of 350 ^{o}C to enter
the steam reformer. This low temperature has been chosen
despite of the fact that usual inlet temperatures will be
higher,not to form carbon black and to achieve reasonable
conversion rates already in the inlet zone of the reformer. The low inlet temperature has been chosen in the
present study to achieve comparable data with the "prereactor systems", which will use a different, more active
catalyst. It may be necessary to also install such highly
active catalyst in a fossil fired steam reformer in the
inletzone, if such low inlet temperatures shall be used,
but tests for such an outfit would be necessary beforehand.

The methane/steam mixture will pass through the tubular
reformer and the catalytic conversion will take place.
By firing of preheated fuel with preheated air the necessary reaction heat of 40 MW will be provided. The reformed
gas will have a temperature of 860 ^{o}C at the outlet of the
reformer and will be cooled down in a waste heat boiler E 8
        in the feed preheater E 9 and the boiler feed water
preheater E 10. The gas will be cooled down to 130 ^{o}C

approximately, leaving energy only for (not shown) thermal boiler feed water pretreatment and air cooling. Any surplus water vapour, which had not taken part in the reaction, will be condensed and knocked out from the gas ahead of the synthesis gas compression.

The flue gas, generated in the fire box of the reformer, will have a temperature of about 950 to 1000 °C. Its heat capacity will be used to generate saturated steam of the same pressure as in the methanol synthesis plant, i.e. 40 bar absolute, and to superheat it. Any deficiency of energy will be made up by firing of additional fuel in additional fire chambers. Those have generally been shown ahead of the waste heat boiler and ahead of the steam superheater. The effective sequence and arrangement of such equipment and the number of additional fire chambers will be determined by detailed engineering work only. The arrangement shown was adapted only for the purpose of easier heat balances and appertaining distribution of fuel. Preheating of fuel and combustion air has been chosen to such an extent that a flue gas temperature of 150 to 170 °C was reached at the outlet of the duct, which meant that the flue gas temperature at the outlet of the steam superheater E 5 should be about 100 to 120 K above the preheating temperatures of fuel and combustion air. In the attached flow diagrams connecting lines for fuel and combustion air have been left off for clearness of the picture. Points masked "A" and "B" will be connected by relevant piping.

2.2.2.1.2

Flow Diagram "B-fossil", Fig. 2.7

14.325 m^3N/h natural gas will be reformed with an energy input of 40,0 MW to generate synthesis gas for ammonia production.

With regard to the lower exit temperatures of the steam reformer D 1, which can be achieved with a subsequent, air-

operated secondary reformer, also the pressure level has
been elevated. Detailed gas analysis and reaction conditions have been given in chapter 2.1.4 with Fig. 2.2.
The big amount of steam, necessary to drive the steam turbines of the synthesis gas compressor and the process air
compressor, needs high energy input to the flue gas duct.
Even with high solar substitution of the energy for reaction, the residual gas (purge gas of synthesis) will not
be sufficient. Therefore no gasholder as buffer is
necessary for fuel storage. As there is a big demand for
low pressure steam in the CO_2-removal section, there has
been provided an installation to reduce and quench saturated 40 bar steam to gain low pressure steam. An optimization of the total steam system could lead to other
solutions (as outlined in earlier paragraphs), but this
concept has been adopted for the reason of comparability.
For other general considerations also the explanation of
the preceeding chapter will be valid.

2.2.2,1.3

Flow Diagram "C-fossil", Fig. 2 - 8

11.395 m^3N/h natural gas of the standard analysis will be
reformed with an energy input of 40,0 MW to generate hydrogen for ammonia production.

The flow scheme is very similar to Fig. 2 - 6, but differs
in the steam/carbon-ratio for the reformer feed gas and
by the introduction of the catalytic high temperature
shift conversion step (reactor D - 3) to increase the
amount of recoverable hydrogen from a given quantity of
natural gas. As more energy is necessary in this kind of
ammonia plant than with the methanol synthesis of Fig.
2 - 6, also the duties of the flue gas channel are increased. Nevertheless with high instantaneous solar energy rates less fossil energy will be required than will be
available from the pressure swing adsorption unit. Gasholders therefore will be necessary for systems C-3, C-4.
and C-6. Further details on analyses of the reforming
unit have already been given in chapter 2.1.4 with Fig. 2 - 3

2.2.2.1.4

<u>Flow Diagram "D-fossil", Fig. 2 - 9</u>

10.290 m^3N/h natural gas of the standard analysis and
14.350 m^3N/h within the process steps regained and recycled carbondioxide will be reformed together by an input
of 40,0 MW to gain a reformed gas, which will be adequate
to produce fuel methanol after further upgrading.

The steam reforming process will be fitted out with very
similar equipment as for methanol production, but, due to
higher energy requirements for compression and for low
pressure steam, the energy requirements of the fluegas
duct are rather high so that residual gases can be used
up at any time and need no intermediate storage in gasholders.

As a <u>speciality</u> of this process arrangement the tubular
reformer has been furnished with <u>internal heat exchange
devices within the reformer tubes.</u> While the maximum reforming temperature for equilibrium is maintained with
860 ^{o}C at the lower end of the tubes, <u>helical coils within
the reformer tubes,</u> also bedded within the catalyst, release sensible heat to the gas, which is passing through
the catalyst. Thus the amount of energy, which has to be
introduced from outside, will be reduced or with the same
amount of energy (here 40,0 MW) a higher quantity of feedgas can be handled for the unchanged outlet analysis. The
helical coils have to be of special material, not to influence the reformed gas analysis.

This arrangement would generally be applicable as in total
it will not be more expensive than the "once-through" tubular reformer with its waste heat system. In fossil operated plant systems the introduction of such helical coils
always will be interesting, when the energy balance shows
an excess of heat, causing export steam for instance. Then
the helical coil system will reduce or eliminate such excess energy and it should be applied.

Thus in the context of this study its application for
methanol production could be advisable.

On the other side the construction of the reformer is
slightly more complicated. While this would not embarras
for normal, fossil operated reformers, there are cases
with solar reformers that need a split up of tubes. Instead of installing one tube of 12 m length there will be
used panels of 4 m length only, making it necessary to
arrange tubes of this length in series. With the connections of the internal coils this would make rather complicated structures, which were not thought to be advisable
in the first approach though principally feasible. Though
originally helical coil tubular reformers were anticipated
within this study by considerations of energy balances,
such preconditions were given up later on. (some effects
of such "historical" procedure will be explained with
system A - 5 and A - 6).

What would be possible as variations shall be shown by the
following table

	Helical Coil System	Energy Demand	Natural Gas Demand	Product Quantity
a)	yes	40,0 MW	10.290 m^3N/h	8,625 t/h
b)	no	43,73 MW	10.290 m^3N/h	8,625 t/h
c)	no	40,0	9,410 m^3N/h	7,900 t/h.

The load capacity variation amounts to about 9,3 %. It
could be influenced by the choice of the helical coil outlet temperature and it will vary also with specific process conditions as steam/carbon-ratio for instance.

What can be seen from this comparison: A higher portion
of the supplied energy will be found in the product with
h.c., because in case (b) the additional energy of 3,73
MW will only be used to change the energy balance of steam
production: More steam will be won from the process

stream, less will be left to the flue gas side. Thus, in
case these additional 3,73 MW were provided from the sun,
this energy would have been used for steam production but
not for reforming purposes. Case (c) specifically is
identic with (b) but adopts throughput to given energy
values.

As has been explained before, the helical coil steam reformer system has been excluded from the main investigations of this study because of constructional difficulties, when going to solar supported cavities/reactors and
when to keep common basis for comparisons throughout. But
to also show eventual effects of such helical coil reformers, the D-Systems have theoretically be investigated,
as if such helical coils could be made available also in
solar service. As the natural gas quantity was expected
to be the smallest with the D-Systems,these were chosen
for such specific investigation, because it was expected
that any effects from the special coils could be found
therein the easiest. By forestalling later results it has
to be admitted, that such effects could not be found and
that therefore helical coil systems for solar service do
not seem to have high priority in development aims.

2.2.2.2 Schematic Flow Diagrams of Solar Supported Steam Reforming
         Plants for Methanol Production

2.2.2.2.1
         General Considerations

In the chapters dealing with block flow diagrams and reforming units it was explained, which principles of classification have been adopted within this study. Such principles allow to exchange in a sort of modules the classic
fossil fired steam reformer of the chapter 2.2.2 by solar
supported steam reformer combinations as shown by the figures 2 - 10 to 2 - 13. Figures 2 - 14 and 2 - 15 are

special cases as "main reactors" operating at low pressure
and low temperature. They have no comparative fossil
scheme by their own, but the same principles apply.

There have been included in this report only the solar
schemes for methanol production, code-prefix A. In course
of the study similar detailed schemes have been developed
also for systems B, C and D, but they are not shown for
reasons of simplification. All necessary details can exem-
plarely be shown by the solar supported steam reforming
plant for methanol production.
Systems A - 1 and A - 2 only differ by the inclusion of
solar thermal energy storages. Leaving off storage S 1
would mean that one gets system A -1 instead of any A -
2 - systems. The same relation is valid for systems A -
3 and A - 4

Therefore
Fig. 2 - 10 is valid for systems A - 1 and A - 2 (though specifically marked with storage data of A - 2.3) and

Fig. 2 11 is valid for systems A - 3 and A 4 (though specifically marked with storage data of A - 4.3)
Fig. 2 –12, 2 - 13, 2 - 14 and 2 - 15 are for such cases
as specifically entitled.

Fig 2 - 10 and Fig. 2 - 11 are related to capacities as shown in
the block flow diagram, Fig. 2 - 11, i.e. 13,75 t/h metha--
nol and 9 420 m^3N/h pure hydrogen.

Fig. 2 - 12 and Fig. 2 - 13 are related to capacities of 14,920
t/h methanol and 10.220 m^3NH_2/h. These increased capacities were caused by leaving out helical coils at an advanced state of work with the solar related part of the
plant. With at this time agreed solar design energy of
43,4 MW for the integrated solar main reactor, the chemical capacity was fixed (as given now) and the integrated
solar prereactor system was adopted accordingly. (See also
chapter 2.2.2.1.4 for effects of helical coil systems).

2.2.2.2.2

Specific Comments to Flow Diagrams

Fig. 2 - 10, Reforming Unit with Solar Prereactor

Shown is a system A - 2.3 with separated solar circuit and storage. If there was no storage, the system would correspond to code A - 1 and with smaller storage (and less solar multiple of the reflectorfield) it would correspond to codes A - 2.1 and A - 2.2. It is anticipated that the flow conditions at the solar prereactor D 1, i.e. 164.000 m^3N/h air of 780 $^{\circ}C$, will be maintained by all means. 21,5 MW (net) will be extracted from this air, which then will have a temperature of 450 $^{\circ}C$ at the outlet of this prereactor. With a storage (S 1) and less solar energy supply from the cavity, for instance 10 MW at a clear day between sunrise and noon, the missing energy will have to be supplied from the fossil heater C 2. For the chosen split 76.280 m^3N/h air would pass through cavity receiver K 1 and 87 720 m^3N/h air would pass through heater C 2, both generating 780 $^{\circ}C$. During night all air would pass through the fossil heater C 2.

For a design with solar multiple and storage (A - 2) the energy offer from the cavity will exceed 21,5 MW at a clear day before noon. For an energy offer exceeding 21,5 MW the flow rate of the recycle blower G 1 will have to be increased and the excess air will have to be lead through the storage vessel S 1. When the solar energy offer from the cavity will fall below 21,5 MW a part of the air will have to be routed backwards through the storage S 1. Thereby it will be possible to maintain the necessary air quantity and temperature at the inlet of the prereactor. When the storage should be empty before full capacity (21,5 MW) can be supplied again from the cavity, the missing part - up to 21,5 MW - will have to be supplied from the fossil heater C 2.

21,5 MW energy will change the natural gas/steam mixture
of 350 $^{\circ}$C inlet temperature to a reformed gas of about
700 $^{\circ}$C at 20 bar (absolute) having a composition of about

$$CO_2 \quad 11,1 \text{ \% molar}$$
$$CO \quad 5,7 \quad "$$
$$H_2 \quad 61,3 \quad "$$
$$CH_4 \quad 21,7 \quad "$$
$$N_2 \quad 0,2 \quad "$$

$$H_2O \quad 0,727 \text{ mol/mol dry gas}$$

dry gas quantity   31 040 m^3N/h

As the outlet temperature of this process stream is around
700 $^{\circ}$C it will be necessary to have steam reforming catalyst in the final reaction zone of the prereactor D 1.
At the inlet zone a special form of a rich gas catalyst,
thermally stable up to temperatures of 650 $^{\circ}$C, will be provided to achieve chemical conversion already at the low
inlet temperature of 350 $^{\circ}$C.

Further detials on gas compositions and duties have already been given with Fig. 2 - 1 and Fig. 2 - 6.

With 21,5 MW reaction energy being provided from the solar plant (from cavity or storage) there will be still a
demand of 1262 m^3N/h natural gas for firing in excess of
the residual gas quantity of 3840 m^3N/h, to provide the
energy for the tubular reformer C1/D2 and of the fluegas
channel (E 4, E 5, E 6, E 7). If the solar plant (cavity
or storage) cannot supply any energy, another natural gas
quantity of 2.320 m^3N/h, then totalling 3.582 m^3N/h
natural gas will be required to provide the missing solar
energy via fossil fired heater C 2.

Details on consumption figures will be given later.

Fig. 2 -11, Reforming Unit with Solar Main Reactor

Shown is a system A - 4.3 with separated solar circuit and
storage. If there was no storage, the system would correspond to code A - 3 and with smaller storage (and less solar multiple of the reflector field) it would correspond
to codes A - 4.1 and A - 4.2. It is anticipated that the
flow conditions at the solar main reactor D 1, i.e.
190 600 m^3N/h air of 1000 ^oC, will be maintained by all
means. 40,0 MW (net) will be extracted from this air,
which then will have a temperature of 480 ^oC at the outlet
of the reactor D 1. With no storage (S 1) and less solar
energy supply from the cavity K 1, for instance 25 MW at
a clear day between sunrise and noon, the missing energy
will have to be supplied from the fossil heater C 2. For
the chosen split 119.125 m^3N/h air would pass through
cavity receiver K 1 and 71.475 m^3N/h air would pass
through heater C 2, both items generating 1000 ^oC. During
night all air would pass through the fossil heater C 2.

For a design with solar multiple and storage (A - 4) the
energy offer from the cavity will exceed 40,0 MW at a
clear day before noon. For an energy offer exceeding
40,0 MW the flow rate of the recycle blower G 1 will have
to be lead through the storage vessel S 1. When the solar
energy offer from the cavity will fall below 40,0 MW a
part of the air will have to be routed backwards through
the storage S 1. Thereby it will be possible to maintain
the necessary air quantity and temperature at the inlet
of the reactor. When the storage should be empty before
full capacity (40,0 MW) can be supplied again from the cavity, the missing part - up to 40,0 MW - will have to be
supplied from the fossil heater C 2.

Details for the process gas composition and the quantities
are given already with Fig. 2 - 1. As with the prereactor
system special highly active catalyst will be provided in
the first part of the reforming tubes, while the portion
with higher temperatures will be made up from normal reforming catalyst.

With 40,0 MW reaction energy being provided from the solar
plant (from cavity or storage) there will be an excess of
residual gas, which will have to be stored in a gasholder
and will have to be used during times, when no solar energy will be available. If no gaseous storage will be installed for such duty, the maximum amount of natural gas
to be fired additionally to the residual gas would amount
to 3578 m^3N/h natural gas when no solar energy is available, just as with the normal A-fossil case. The average
consumption for natural gas for fossil firing therefore
will be between 0 and 3578 m^3N/h, depending on all storage
capacities, installed solar multiple etc. With a gasholder
being installed for settling the demand fluctuation of one
day, the maximum natural gas flow rate for fossil firing
with a system A - 3 already will go down to 3.270 m^3N/h
at night and with a system A - 4.3 the relevant figure
will be 0, leaving also some residual gas for purposes outside the reforming unit. Details can be seen from the
tables for consumption figures.

Fig. 2 - 12   Reforming Unit with Integrated Solar Prereactor Code
               A - 5

The principal layout for the chemical reaction scheme
corresponds to Fig. 2 - 10, but the prereactor is "integrated", which implies that no energy storage for the solar system can be provided and that therefore also the solar multiple will be about one only. Due to the increased
natural gas throughput through the system, the achievable
maximum temperatures of the prereactor will be lower by
about 20 K than with the A - 2 system. The outlet analysis from this reactor will be

CO_2	about	11,2 % molar
Co		4,9
H_2		59,6
Ch_4		24,1
N_2		0,2

H_2O 0,778 mol/mol dry gas

dry gas volume about 32 200 m^3N/h

The outlet conditions from the fossil fired tubular steam reformer are found with the corresponding block flow diagrams. The increased capacities are valid herewith. Natural gas will be necessary as auxiliary energy source under all circumstances, which is the reason that no gasholders have to be provided for the storage of excess residual gas.

When no sunshine is available or when the energy is not sufficient in quantity, the missing amount has to be supplied by fossil means. It has been anticipated that firing directly into the cavity will be possible and that the resulting flue gases can be kept from escaping by the opening for the solar irradiation. Reference is made to the earlier study - GKY-62192-40 - where the feasibility of such a system was investigated and estimated already in more detail.

Fig. 2 - 13 Reforming Unit with Integrated Solar Main Reactor
Code A - 6
The principle layout for the chemical reaction scheme corresponds to Fig. 2 - 11, but the main reactor is "integrated", what implies that no energy storage for the solar system can be provided and that therefore also the solar multiple will be about one only. The outlet composition of the reactor will be identic with other reformed gas qualities of any A-system and have already been given with the block flow diagram. The increased capacities (+ 8,5% are valid herewith.

Only around noon there will be an excess of residual gas, which will have to be stored in a comparatively small gas holder for subsequent use in the reforming plant. As with the system A - 5 a cavity firing installation will be necessary for the system A - 6, but due to the higher reaction temperatures also the flue gas temperature from the cavity will be increased.

Details on the cavity structure, the materials, which shall be used, and solar preconditions will be explained in more detail in chapter 3. For consumption figures see the following chapter 2.2.3.

Fig. 2 - 14 Reforming Unit with Separated Main Reactor at Low Pressure
Code: A' - 7.

The principle layout is similar to Fig. 2 - 11 but has assumed the installation of a helical coil system in the reactor to increase the solar specific rate of energy in the product as far as possible. Restrictive considerations of construction have not been deemed essential for the basic evaluations as they have been performed for such systems at low pressure.

The solar plant has not been worked out in detail but has been estimated as a mixture of system - 2 for temperatures but of system - 4 for sizes. It has been assumed that the size-factor will be predominant with such considerations.

As a prefixed condition the upper temperature of the solar circuit had been set to 800 ^{o}C. An approach of 40 K is thought to be the lowest technically still admissible one. At a resulting process temperature of 760 ^{o}C and at an absolute pressure of 8,0 bar the following reformed gas conditions can be achieved.

CO_2	9,7 % molar
CO	11,7
H_2	73,9
CH_4	4,6
N_2	0,1
H_2O	0,473 mol/mol dry gas

dry gas quantity 57.900 m^3N/h.

By the helical coil system an energy equivalent of 6,13 MW is transferred to the reacting gas, when the reformed gas is cooled down to 600 ^{o}C. Due to process conditions,

as will be further explained in chapter 4, the outlet
temperature of the circulating air from reactor D 1 cannot
be lower than 500 $^{\circ}$C. 333 800 m^{3}N/h of air have to be recycled for a duty of 40,0 MW with a temperature difference
of 300 K. This is 1,75 times the quantity used in other
separated main reactor systems because of such limitation
in temperature difference. The energy requirement for the
recycle is further enlarged beyond this factor, as the inlet temperature to the recycle compressor is increased.
Not to further worsen the energy balance, the pressure
level of the solar air recycle has been left at 20 bar,
as the energy requirements for system A'-7 would further
increase with lower overall pressure, though such lower
pressure would be desirable for cheaper construction of
the reactor, where the differential pressure between energy transport medium and process medium is of great importance. Due to the low inlet temperature of the natural
gas-steam-mixture the first portion of catalyst will have
to be high active, while the second portion will be normal
steam reforming catalyst.

The energy requirements of the flue gas channel have been
adjusted, mainly to compensate the specific higher steam
demand of the synthesis gas compressor driver. Because
of the lower dew point of the reformed gas the boiler feed
water preheater cannot regain specifically as much energy
as with higher pressure, which causes higher heat losses
at the air cooled cooler E 11 or the relevant cooling
equipment. Due to the high energy demand natural gas for
firing will be necessary also at full energy supply from
the solar system. A gasholder for residual gas therefore
will not be required.

Fig. 2 - 15   Reforming Unit with Integrated Main Reactor of Low
Pressure
Code: A' - 8

The principle layout is similar to Fig. 2 - 13. The energy input has been increased to 46,13 MW to maintain gas
throughput and analyses as with Fig. 2 - 14. The heat
load of waste heat boiler E - 8 is increased because of
the higher exit temperature from the reactor (helical
coils have been left off as their application seemed doubtful in direct solar service), thereby changing also the
overall load of the flue gas channel. The energy balance
also is influenced by the fact that the pressure upstream
of the synthesis compressor will be higher by about 1,5
bar as the pressure drop of the helical coil system has
been eliminated. The overall energy balance also is improved, as no recycle energy for the solar energy transport is needed, yet on the effect of absence of a storage
system. The specific energy balance is improved by these
means versus system A' - 7, but it is still poor in comparison with the normal A-fossil system, as will be shown
by utility consumption and economics in later chapters.
Due to the high energy demand of system A' - 8, the residual gas can be used up for firing at all times, which is
the reason that no gaseous storage will be necessary
within this overall scheme.

2.2.3   Consumption Figures for Investigated Systems

For each system, as they are listed in tables 2 - 1 to 2
2 - 4, two working flowsheets of similar kind as having
been shown in the preceeding chapter have been established,
one for operation at full solar energy supply and one for
operation of zero energy supply from the sun. These
working flowsheets have been the basis for all consumption
figures as they are given in this chapter. These numbers
also are fundamental for all derived specific data of later evaluations. The abbreviations, used within these
tables, will be explained in detail here below.

2.2.3.1 Data of Table 2 - 1

System   in accordance with chapter 2.1.3

$V_{E\ Proc.}$ represents the quantity of natural gas (99,5%
              molar CH_4, 0,5% molar N_2) being used as pro-
              cess feed for catalytic reforming with steam
$G_{Prod.}$ indicates the quantity of pure methanol being
              produced and released from the methanol distil-
              lation plant. Methanol of specification, Grade
              A, has been assumed.

$V_{H2byprod.}$ represents the quantity of pure hydrogen,which
              will be produced as a byproduct for reasons
              as generally explained in 2.2.1. The hydrogen
              quality will be higher than 99,9 %.

Q_{solar} is the annual solar energy, which will be
              available as net output from the receiver or
              at the nozzles of the solar operated reactors.
              The figures were calculated for ideal weather
              conditions all over the year and with 365 d/a.
              (In economic evaluations only 8160 h/a have
              been adopted as possible annual operating time.
              Then the downrating was made linear, i.e. the
              figures of tab. 2 - 1 were mulitplied by 8160/
              8760)
              The instantaneous solar energy, which can be
              used by the reaction at best, has been given
              at many places but shall be tabulated for com-
              parisons here again

System	$Q_{design\ solar}$
...- 1, ...- 2	21,5 MW
...- 3, ...- 4	40,0 MW
...- 5	21,5 MW
...- 6	43,4 MW
A' - 7	40,0 MW
A' - 8	46,13 MW

$Q_{sol.int.}$ indicates the annual energy consumption of the secondary devices of the solar plant. This figure mainly is made up from the consumption of the recycle blower G 1 for separated systems. The figure is given on a thermal basis, assuming an average efficiency of 35% The figure corresponds to MAN-Table 5.1 (Cost and Performance Compilation). The annual time realtions are equal as explained with Q_{solar}.

$V_{E\ fuel\ d}$ indicates the instantaneous quantity of natural gas, which will be needed as fuel for fossil firing in the specific system, when the solar design rate is reached.

$V_{E\ fuel\ n}$ indicates the instantaneous quantity of natural gas, which will be needed as fuel for fossil firing in the specific system at times, when no solar energy is available.

η_{solar} represents the number of annual possible full load hours in relation to the annual energy consumption of the relevant solar reaction system. If Q_{design} represents this latter value (21,5 MW for prereactor-systems, about 40 MW for main reactor system) the following equation is used

$$\eta_{solar} = \frac{Q\ solar}{8760 \cdot Q_{design}}$$

$\overline{V}_{E\ fuel}$ is the average natural gas fuel consumption of the specific system.

$$\overline{V}_{E\ fuel} = \eta_{solar}\ V_{Efuel\ d} + (1 - \eta_{solar}) V_{Efueln}$$

$V_{E\ utilit.}$ indicates the average consumption of natural gas to generate utilities. This figure is made up from electric consumption for smaller consumers (for instance: combustion air blower, flue gas blower, pumps, fans of air cooled coolers etc.) and from an equivalent

in value for raw water intake and cooling water
supply. As with $Q_{sol.int.}$ a thermal efficiency
of 35 % has been assumend for electricity
generation.
The specific figure of $V_{E\ util.}$ has been maintained for all chemical complexes of the same
nature (A, B, C, D-systems), independent from
known small deviations. $V_{E\ util.}$ was introduced into the tables to finally get the
approximate overall consumption of natural gas
and to thereby get a relative change of overall consumption in natural gas for any system.
Most other criteria are influenced only by the
absolute difference in such consumptions,where
$V_{E\ util.}$ could be let off.

2.2.3.2 Data of Table 2 - 2

The abbreviations correspond to table 2 - 1 with the exception:

G_{Prod} here represents the quantity of pure ammonia
                being produced within the complex (see also
                chapter 2.1.4 for possible product conditions)

$V_{H2\ byprod.}$ this line is missing as no pure hydrogen will
                be manufacturized with B-systems.

2.2.3.3 Data of Table 2 - 3
The abbreviations correspond to table 2 - 2, the same explanations are valid.

2.2.3.4 Data of Table 2 - 4
The abbreviations correspond to those of table 2 - 1, but

$G_{Prod.}$ indicates the quantity of Octamix-Fuel-Metha-
                nol, being produced within the complex. The
                product specification can be found in chapter
                2.1.4.

GKY-63141-00

AMMONIA PRODUCTION

WITH SECONDARY REFORMER

CODE PREFIX: B

Fig. 2 - 2

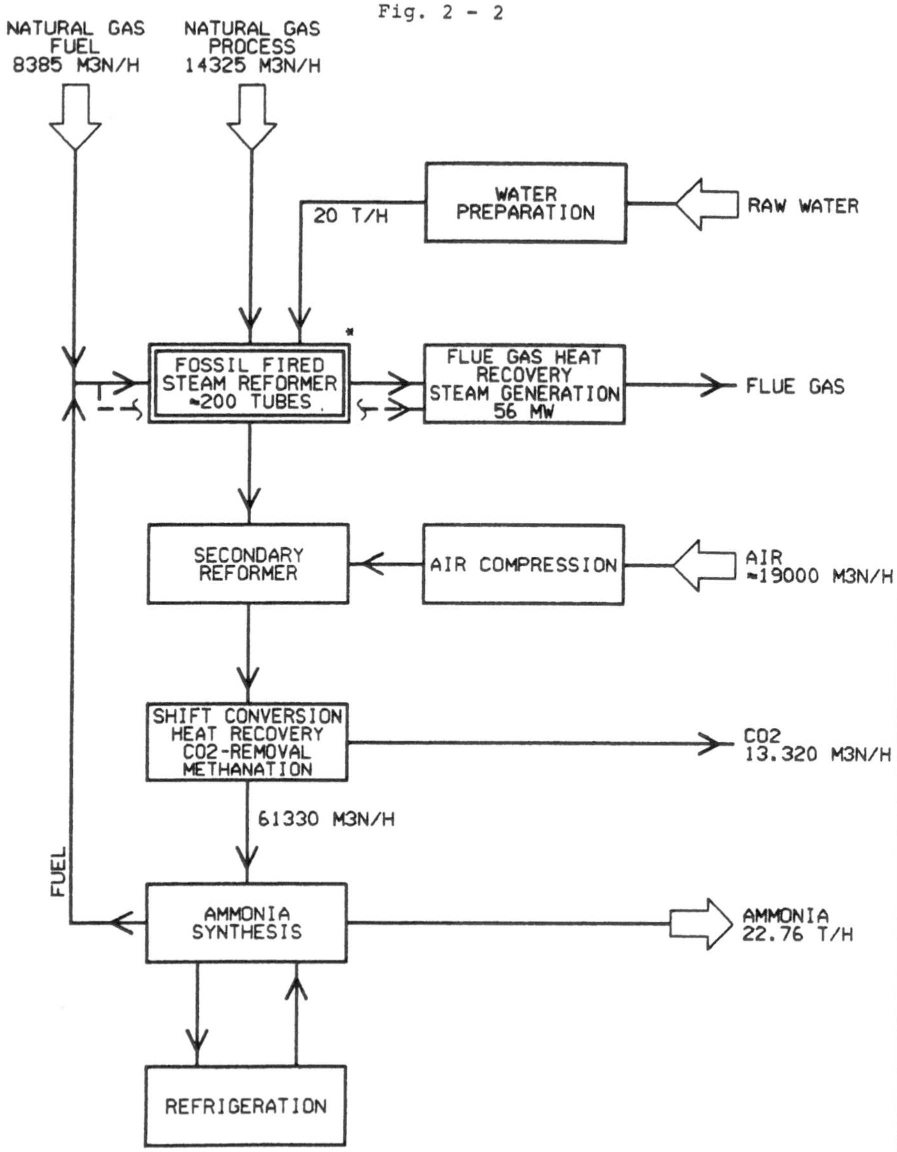

NATURAL GAS
FUEL
8385 M3N/H

NATURAL GAS
PROCESS
14325 M3N/H

WATER
PREPARATION

20 T/H

RAW WATER

FOSSIL FIRED
STEAM REFORMER
~200 TUBES

FLUE GAS HEAT
RECOVERY
STEAM GENERATION
56 MW

FLUE GAS

SECONDARY
REFORMER

AIR COMPRESSION

AIR
~19000 M3N/H

SHIFT CONVERSION
HEAT RECOVERY
CO2-REMOVAL
METHANATION

CO2
13.320 M3N/H

61330 M3N/H

FUEL

AMMONIA
SYNTHESIS

AMMONIA
22.76 T/H

REFRIGERATION

* FOR SOLAR OPERATED SYSTEMS B-1 TO B-6 THE ITEMS OF
THE SCHEMES OF MAN-COMPILATION WILL BE FOUND IN THIS
BLOCK,THE STEAM REFORMER WILL BE DIFFERENT OR MISSING.
FOR B-5 AND B-6 CAPACITY IS INCREASED BY 8.5%

001 10.J4.86 BUE   STATUS-
CAD I [141,4] A3141B002.DGN

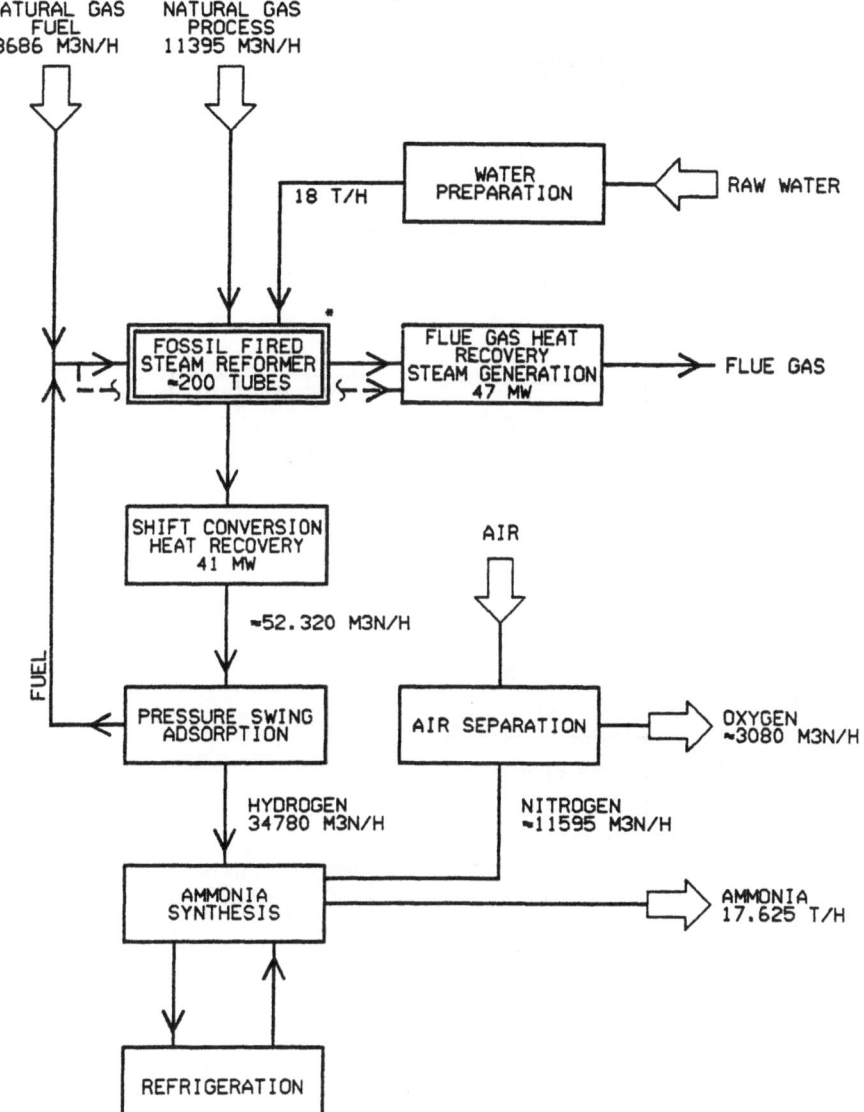

GKY-63141-00

AMMONIA PRODUCTION

HYDROGEN WITH NITROGEN-ADMIXTURE

CODE PREFIX: C

Fig. 2 - 3

NATURAL GAS FUEL 3686 M3N/H

NATURAL GAS PROCESS 11395 M3N/H

WATER PREPARATION ← RAW WATER

18 T/H

FOSSIL FIRED STEAM REFORMER ~200 TUBES

FLUE GAS HEAT RECOVERY STEAM GENERATION 47 MW → FLUE GAS

SHIFT CONVERSION HEAT RECOVERY 41 MW

~52.320 M3N/H

AIR

PRESSURE SWING ADSORPTION

AIR SEPARATION → OXYGEN ~3080 M3N/H

FUEL

HYDROGEN 34780 M3N/H

NITROGEN ~11595 M3N/H

AMMONIA SYNTHESIS → AMMONIA 17.625 T/H

REFRIGERATION

* FOR SOLAR OPERATED SYSTEMS C-1 TO C-6 THE ITEMS OF THE SCHEMES OF MAN-COMPILATION WILL BE FOUND IN THIS BLOCK, THE STEAM REFORMER WILL BE DIFFERENT OR MISSING. FOR C-5 AND C-6 CAPACITY IS INCREASED BY 8.5%

FIG. 2-3

002 ...06.86 HIH/AÜGG   STATUS-
CAD 1 [141,004] A3141B003.DGN

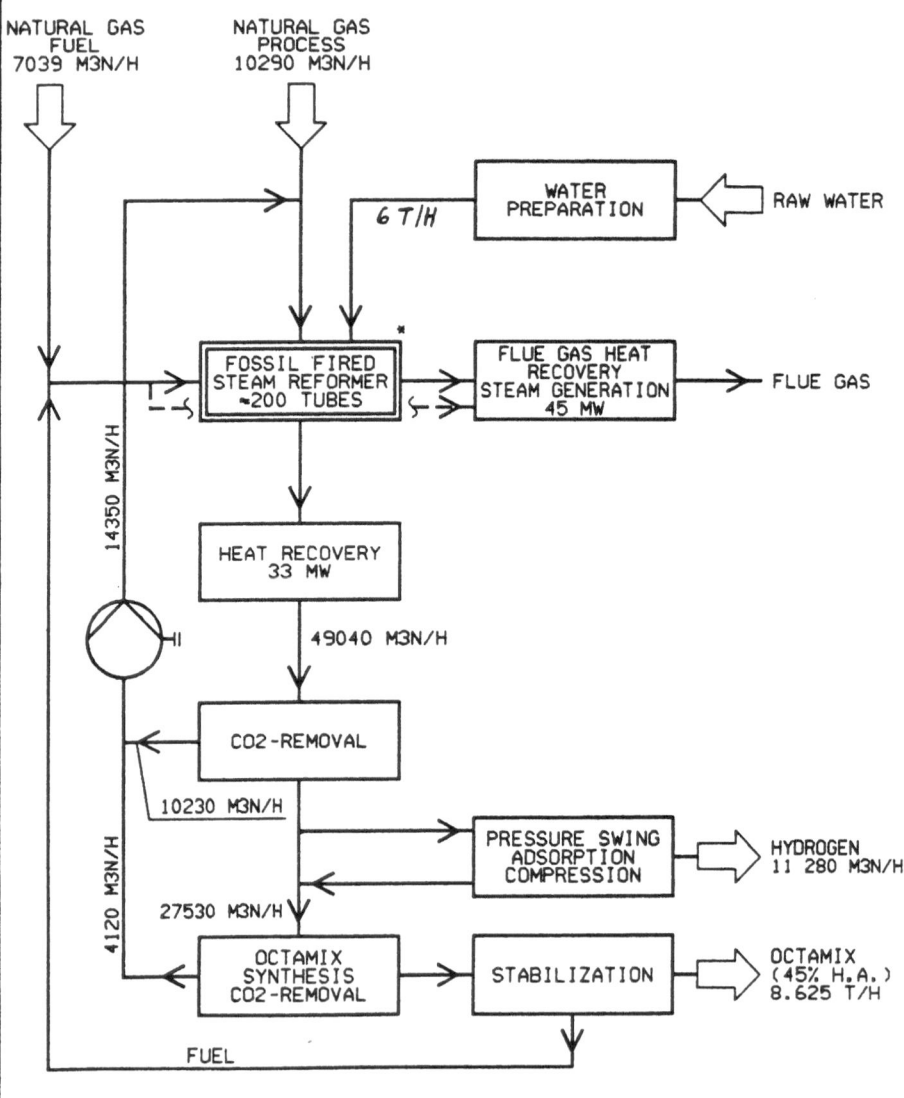

GKY-63141-00

OCTAMIX

FUEL METHANOL PRODUCTION

CODE PREFIX: D

Fig. 2 - 4

NATURAL GAS FUEL 7039 M3N/H

NATURAL GAS PROCESS 10290 M3N/H

WATER PREPARATION ⟵ RAW WATER

6 T/H

FOSSIL FIRED STEAM REFORMER ~200 TUBES *

FLUE GAS HEAT RECOVERY STEAM GENERATION 45 MW → FLUE GAS

14350 M3N/H

HEAT RECOVERY 33 MW

49040 M3N/H

CO2-REMOVAL

10230 M3N/H

PRESSURE SWING ADSORPTION COMPRESSION ⟹ HYDROGEN 11 280 M3N/H

4120 M3N/H

27530 M3N/H

OCTAMIX SYNTHESIS CO2-REMOVAL → STABILIZATION ⟹ OCTAMIX (45% H.A.) 8.625 T/H

FUEL

* FOR SOLAR OPERATED SYSTEMS D-1 TO D-6 THE ITEMS OF THE SCHEMES OF MAN-COMPILATION WILL BE FOUND IN THIS BLOCK, THE STEAM REFORMER WILL BE DIFFERENT OR MISSING. FOR D-5 AND D-6 CAPACITY IS INCREASED BY 8.5%

001 .~.04.86 BUE STATUS-
CAD] [141,4] A3141B004.DGN

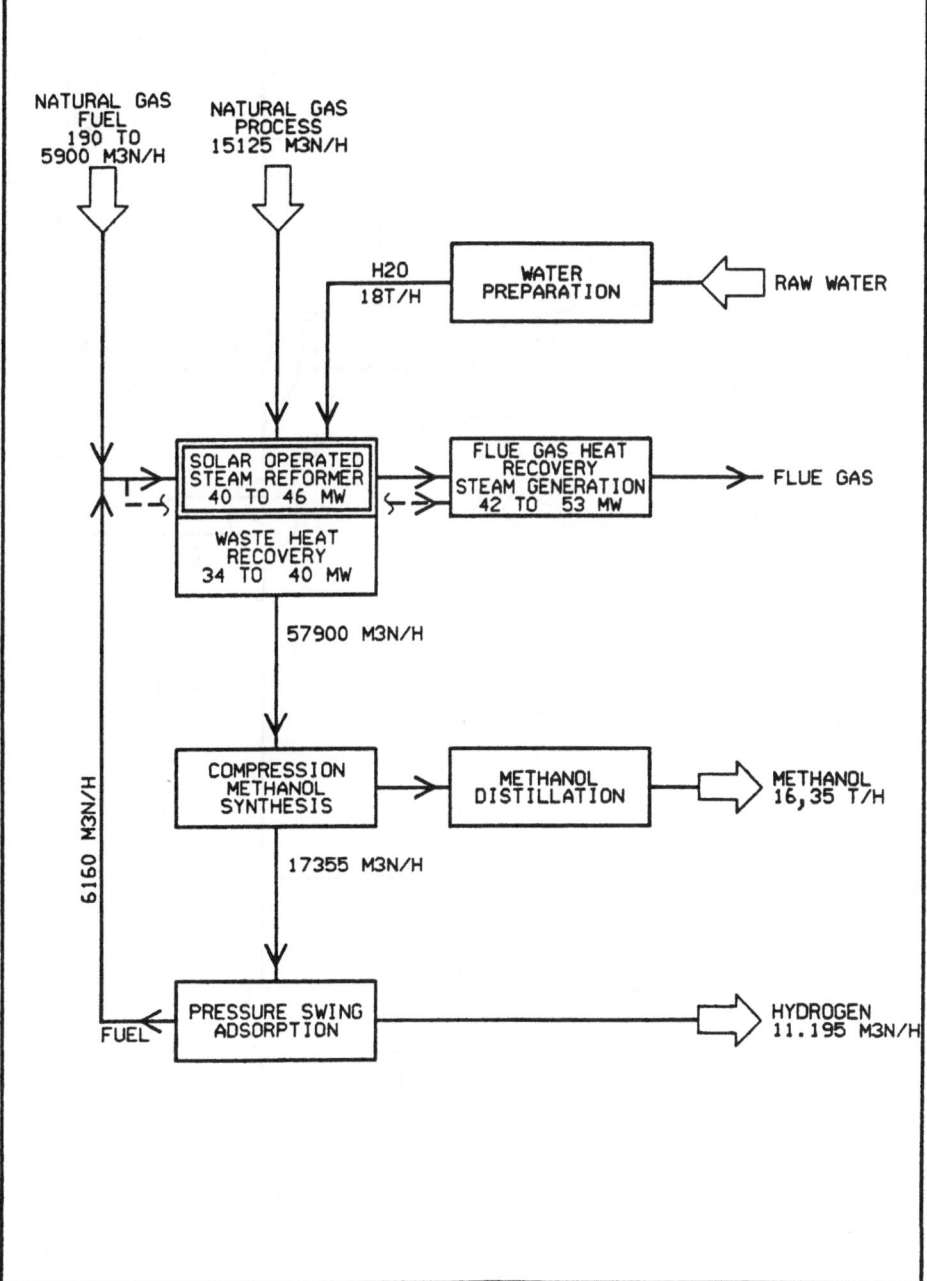

GKY-63141-00

METHANOL PRODUCTION
REFORMING AT LOW PRESSURE

CODE PREFIX: A

Fig. 2 - 5

NATURAL GAS FUEL 190 TO 5900 M3N/H

NATURAL GAS PROCESS 15125 M3N/H

H2O 18T/H

WATER PREPARATION — RAW WATER

SOLAR OPERATED STEAM REFORMER 40 TO 46 MW

WASTE HEAT RECOVERY 34 TO 40 MW

FLUE GAS HEAT RECOVERY STEAM GENERATION 42 TO 53 MW → FLUE GAS

57900 M3N/H

COMPRESSION METHANOL SYNTHESIS

METHANOL DISTILLATION → METHANOL 16,35 T/H

17355 M3N/H

6160 M3N/H

PRESSURE SWING ADSORPTION → HYDROGEN 11.195 M3N/H

FUEL

001 15.04.86 BUE STATUS-
CAD 1 [141,004] A3141B005.DGN

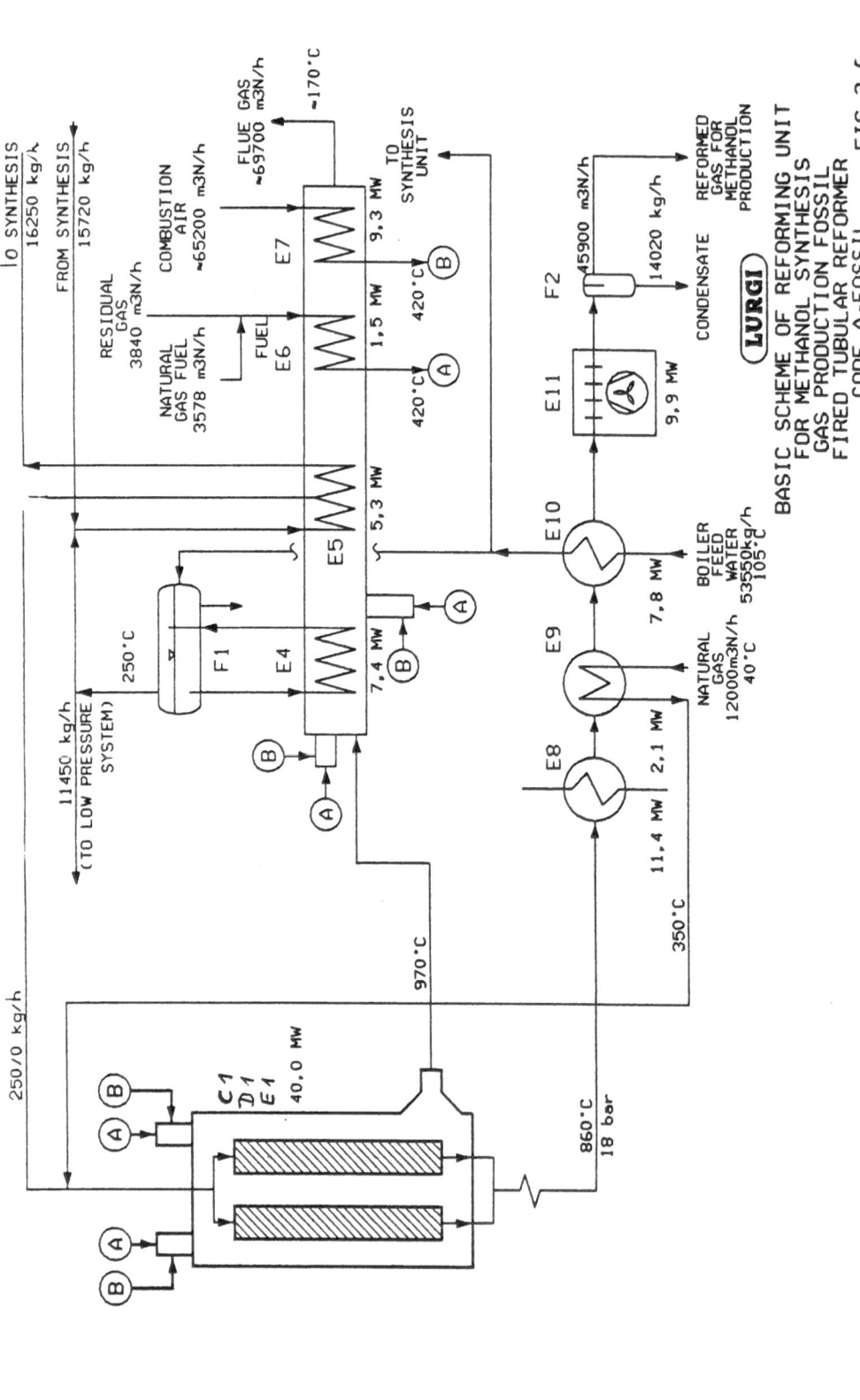

BASIC SCHEME OF REFORMING UNIT
FOR METHANOL SYNTHESIS
GAS PRODUCTION FOSSIL
FIRED TUBULAR REFORMER
CODE A-FOSSIL   FIG.2-6

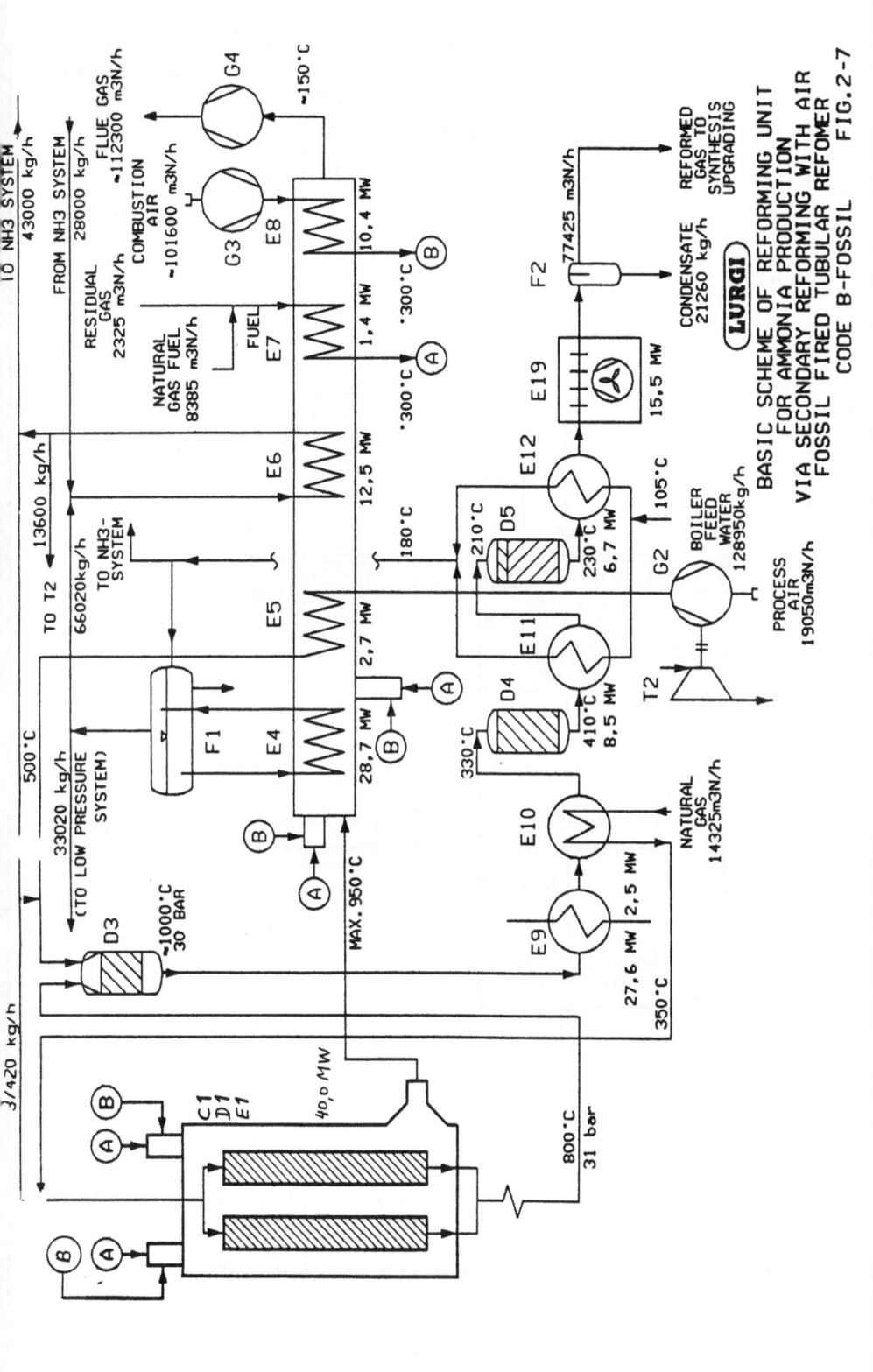

BASIC SCHEME OF REFORMING UNIT
FOR AMMONIA PRODUCTION
VIA SECONDARY REFORMING WITH AIR
FOSSIL FIRED TUBULAR REFORMER
CODE B-FOSSIL   FIG.2-7

LURGI

METHANOL PRODUCTION

CODE PREFIX: A

Fig. 2 - 1

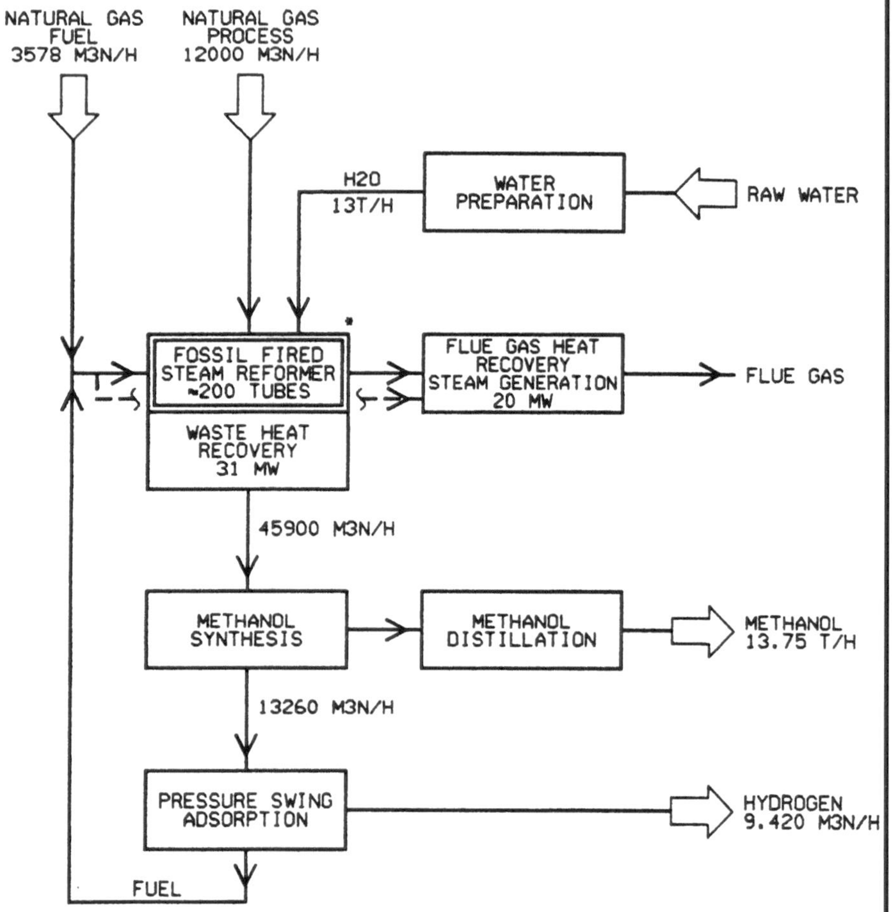

NATURAL GAS
FUEL
3578 M3N/H

NATURAL GAS
PROCESS
12000 M3N/H

H2O
13T/H

WATER
PREPARATION

RAW WATER

FOSSIL FIRED
STEAM REFORMER
~200 TUBES

WASTE HEAT
RECOVERY
31 MW

FLUE GAS HEAT
RECOVERY
STEAM GENERATION
20 MW

FLUE GAS

45900 M3N/H

METHANOL
SYNTHESIS

METHANOL
DISTILLATION

METHANOL
13.75 T/H

13260 M3N/H

PRESSURE SWING
ADSORPTION

HYDROGEN
9.420 M3N/H

FUEL

* FOR SOLAR OPERATED SYSTEMS A-1 TO A-6 THE ITEMS OF
THE SCHEMES OF MAN-COMPILATION WILL BE FOUND IN THIS
BLOCK, THE STEAM REFORMER WILL BE DIFFERENT OR MISSING.
FOR A-5 AND A-6 CAPACITY IS INCREASED BY 8.5%

001 09.04.86 BUE   STATUS-
CAD 1 [141,4 J A3141B001.DGN

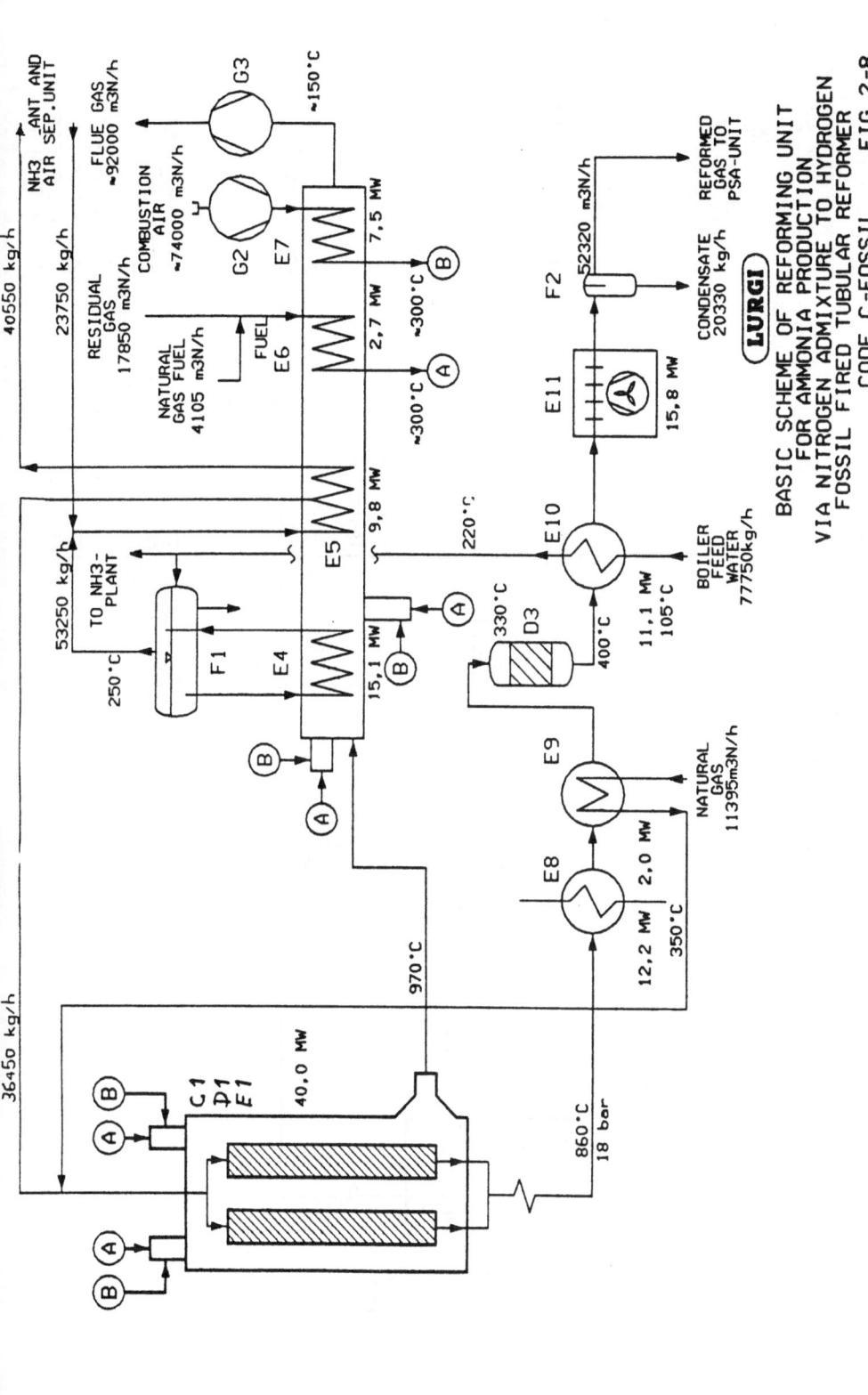

BASIC SCHEME OF REFORMING UNIT
FOR AMMONIA PRODUCTION
VIA NITROGEN ADMIXTURE TO HYDROGEN
FOSSIL FIRED TUBULAR REFORMER
CODE C-FOSSIL   FIG.2-8

LURGI

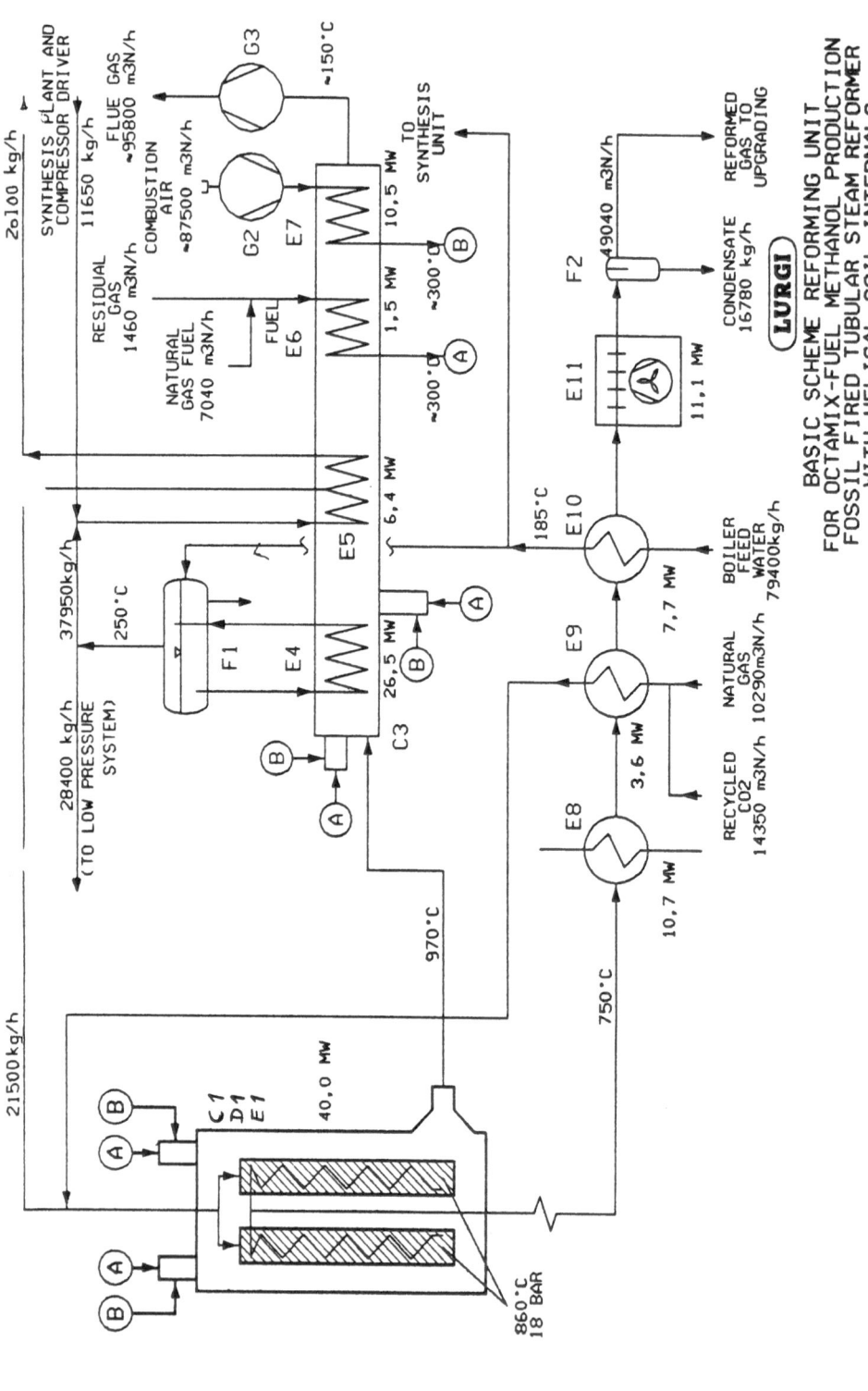

BASIC SCHEME REFORMING UNIT
FOR OCTAMIX-FUEL METHANOL PRODUCTION
FOSSIL FIRED TUBULAR STEAM REFORMER
WITH HELICAL COIL INTERNALS
CODE C-FOSSIL   FIG.2-9

LURGI

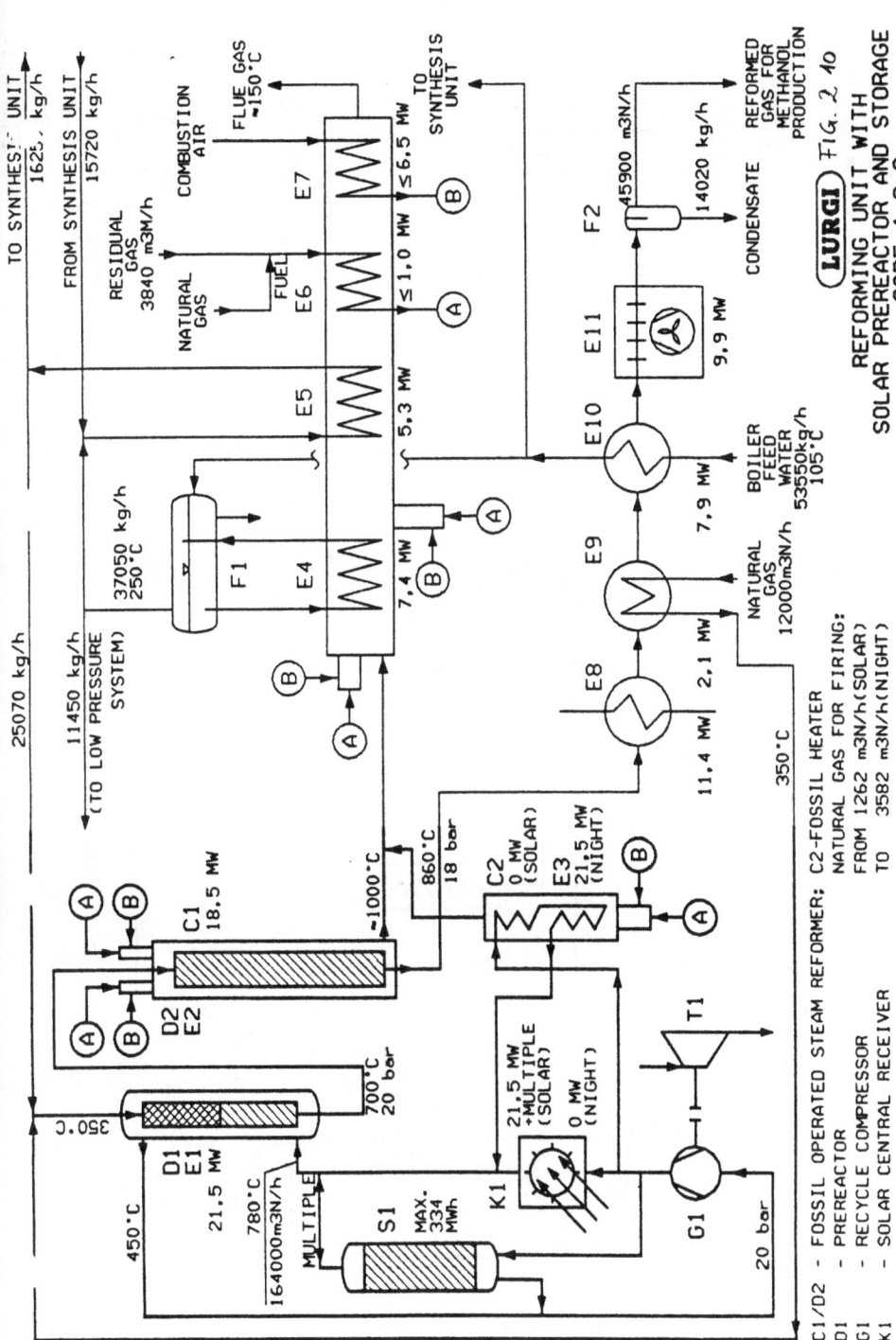

LURGI FIG. 2.10

REFORMING UNIT WITH
SOLAR PREREACTOR AND STORAGE
CODE A-2.3

C1/D2 - FOSSIL OPERATED STEAM REFORMER; C2-FOSSIL HEATER
D1 - PREREACTOR   NATURAL GAS FOR FIRING:
G1 - RECYCLE COMPRESSOR   FROM 1262 m3N/h(SOLAR)
K1 - SOLAR CENTRAL RECEIVER   TO 3582 m3N/h(NIGHT)
S1 - STORAGE

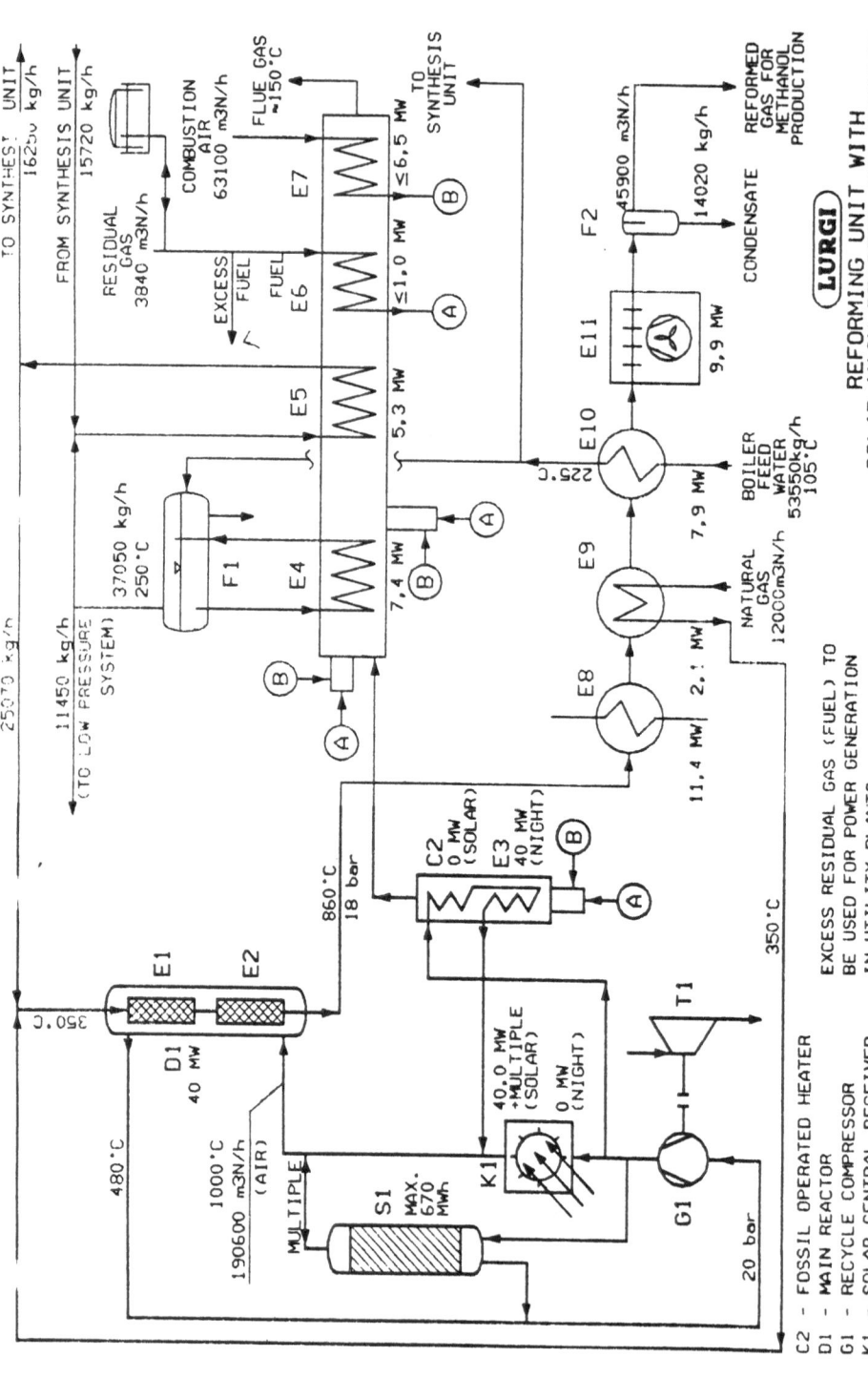

LURGI

REFORMING UNIT WITH
SOLAR MAIN REACTOR AND STORAGE
CODE A-4.3   FIG 2-11

C2 – FOSSIL OPERATED HEATER
D1 – MAIN REACTOR
G1 – RECYCLE COMPRESSOR
K1 – SOLAR CENTRAL RECEIVER
S1 – STORAGE

EXCESS RESIDUAL GAS (FUEL) TO
BE USED FOR POWER GENERATION
IN UTILITY PLANTS

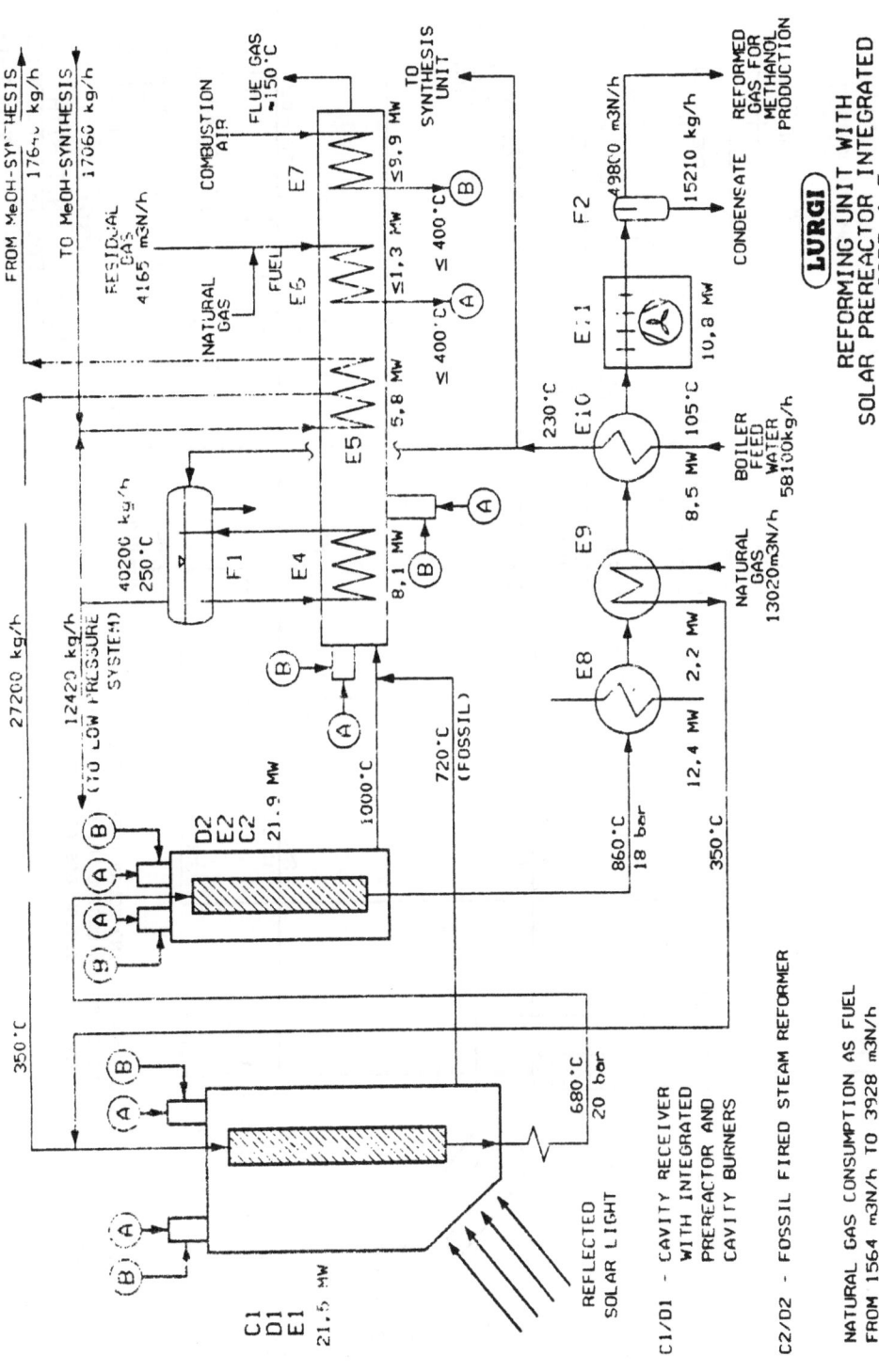

FROM MeOH-SYNTHESIS 1760 kg/h

TO MeOH-SYNTHESIS 17060 kg/h

COMBUSTION AIR

FLUE GAS ~150°C

E7 ≤9,9 MW

TO SYNTHESIS UNIT

RESIDUAL GAS 4165 m3N/h

FUEL E6 ≤1,3 MW ≤400°C

NATURAL GAS ≤400°C

Ⓐ Ⓑ

REFORMED GAS FOR METHANOL PRODUCTION

49800 m3N/h

F2

15210 kg/h

CONDENSATE

27200 kg/h

12420 kg/h (TO LOW PRESSURE SYSTEM)

40200 kg/h 250°C

F1   E4 8,1 MW

E5 ~5,8 MW

230°C

E10 E11

10,8 MW

8,5 MW 105°C

BOILER FEED WATER 58100 kg/h

Ⓐ Ⓑ

1000°C

720°C (FOSSIL)

E9

2,2 MW

NATURAL GAS 13020 m3N/h

Ⓑ

350°C

D2 E2 C2 21,9 MW

Ⓐ Ⓐ Ⓑ

860°C 18 bar

E8

12,4 MW

350°C

350°C

Ⓐ Ⓑ

680°C 20 bar

REFLECTED SOLAR LIGHT

C1 D1 E1 21,5 MW

Ⓑ Ⓐ

C1/D1 - CAVITY RECEIVER WITH INTEGRATED PREREACTOR AND CAVITY BURNERS

C2/D2 - FOSSIL FIRED STEAM REFORMER

NATURAL GAS CONSUMPTION AS FUEL FROM 1564 m3N/h TO 3928 m3N/h

LURGI

REFORMING UNIT WITH SOLAR PREREACTOR INTEGRATED CODE A-5

FIG 2-12

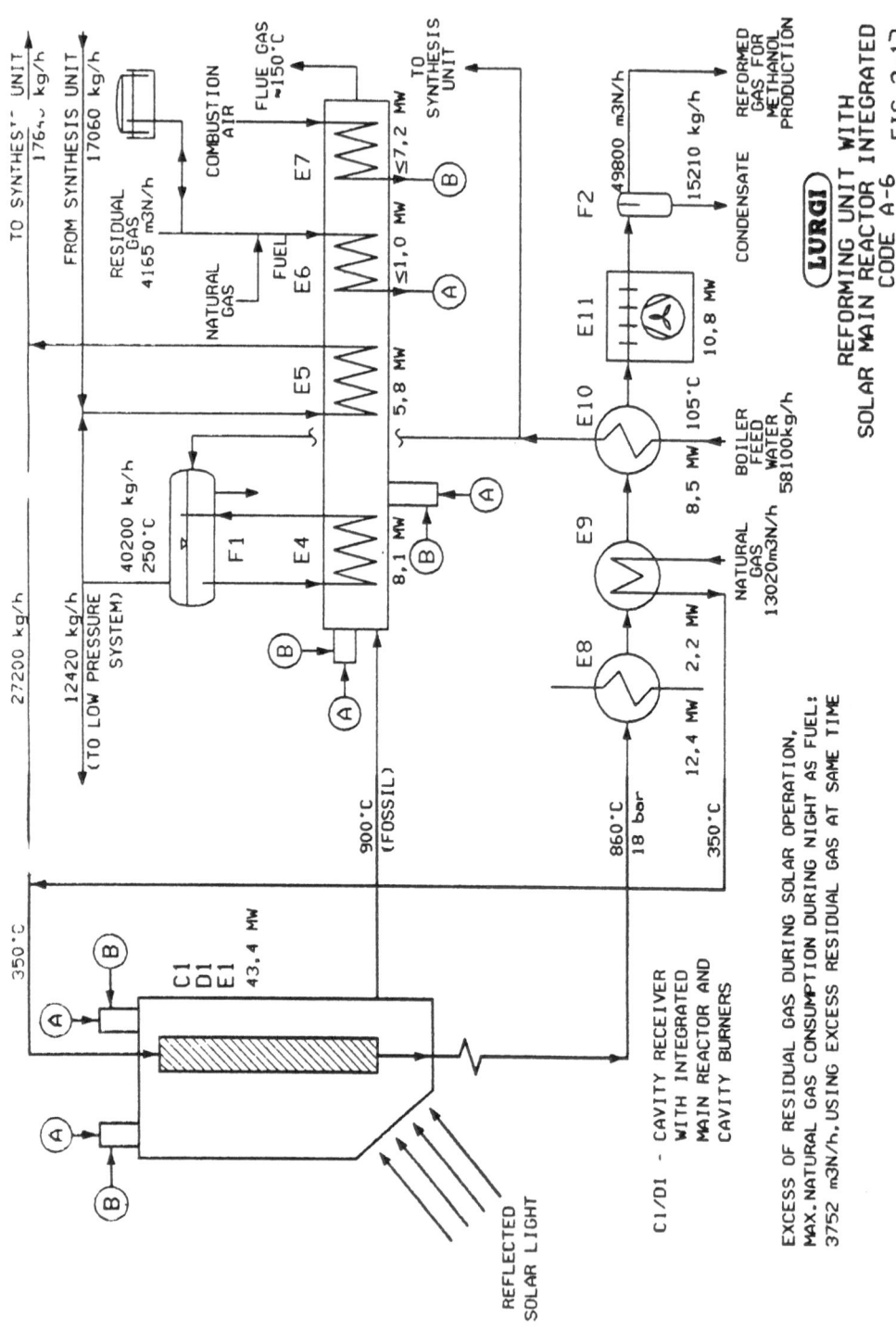

REFORMING UNIT WITH
SOLAR MAIN REACTOR INTEGRATED
CODE A-6   FIG 2-13

LURGI

EXCESS OF RESIDUAL GAS DURING SOLAR OPERATION,
MAX.NATURAL GAS CONSUMPTION DURING NIGHT AS FUEL:
3752 m3N/h,USING EXCESS RESIDUAL GAS AT SAME TIME

C1/D1 - CAVITY RECEIVER
         WITH INTEGRATED
         MAIN REACTOR AND
         CAVITY BURNERS

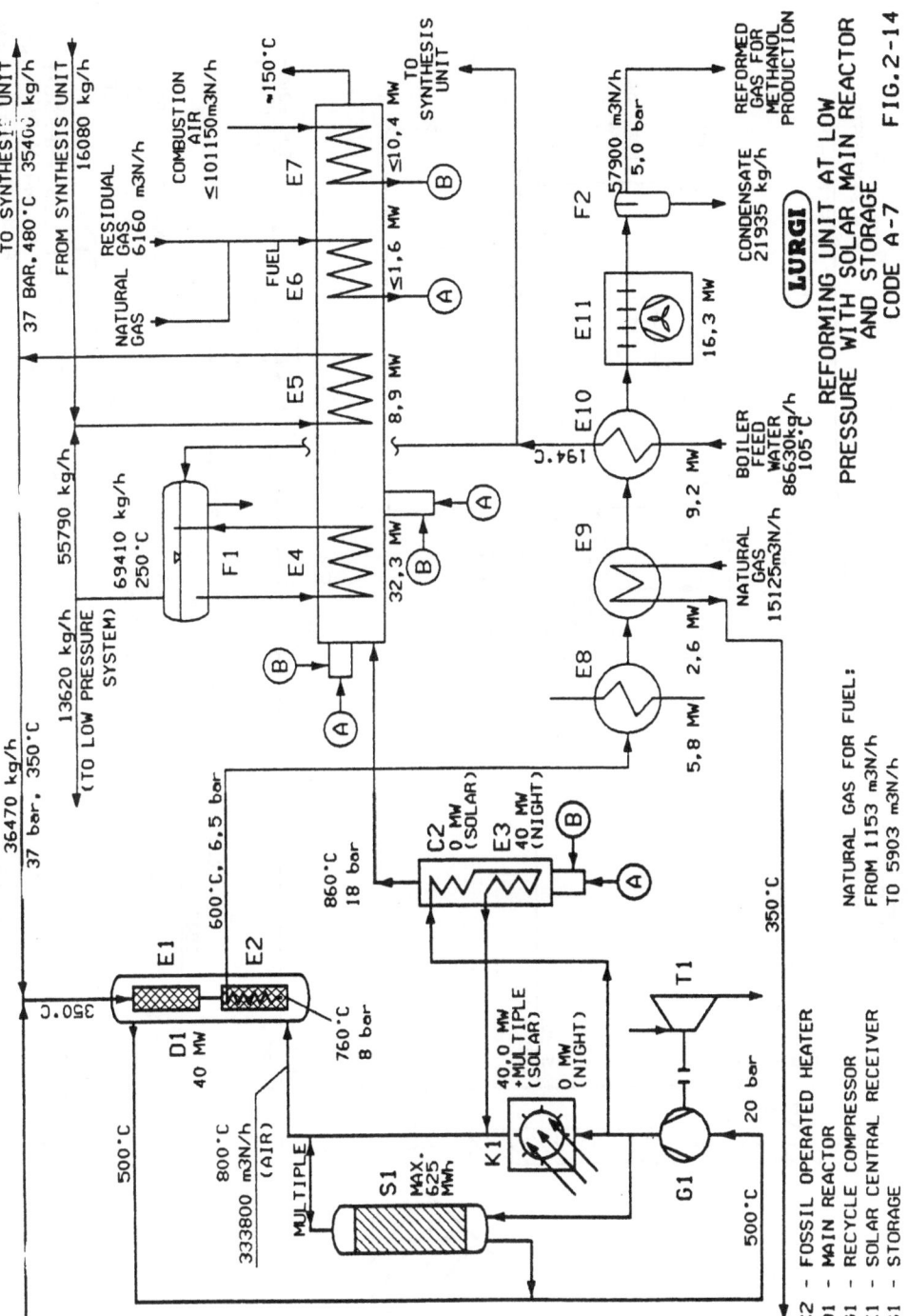

REFORMING UNIT AT LOW
PRESSURE WITH SOLAR MAIN REACTOR
AND STORAGE
CODE A-7

LURGI

FIG.2-14

C2 – RECYCLE COMPRESSOR
D1 – MAIN REACTOR
G1 – FOSSIL OPERATED HEATER
K1 – SOLAR CENTRAL RECEIVER
S1 – STORAGE

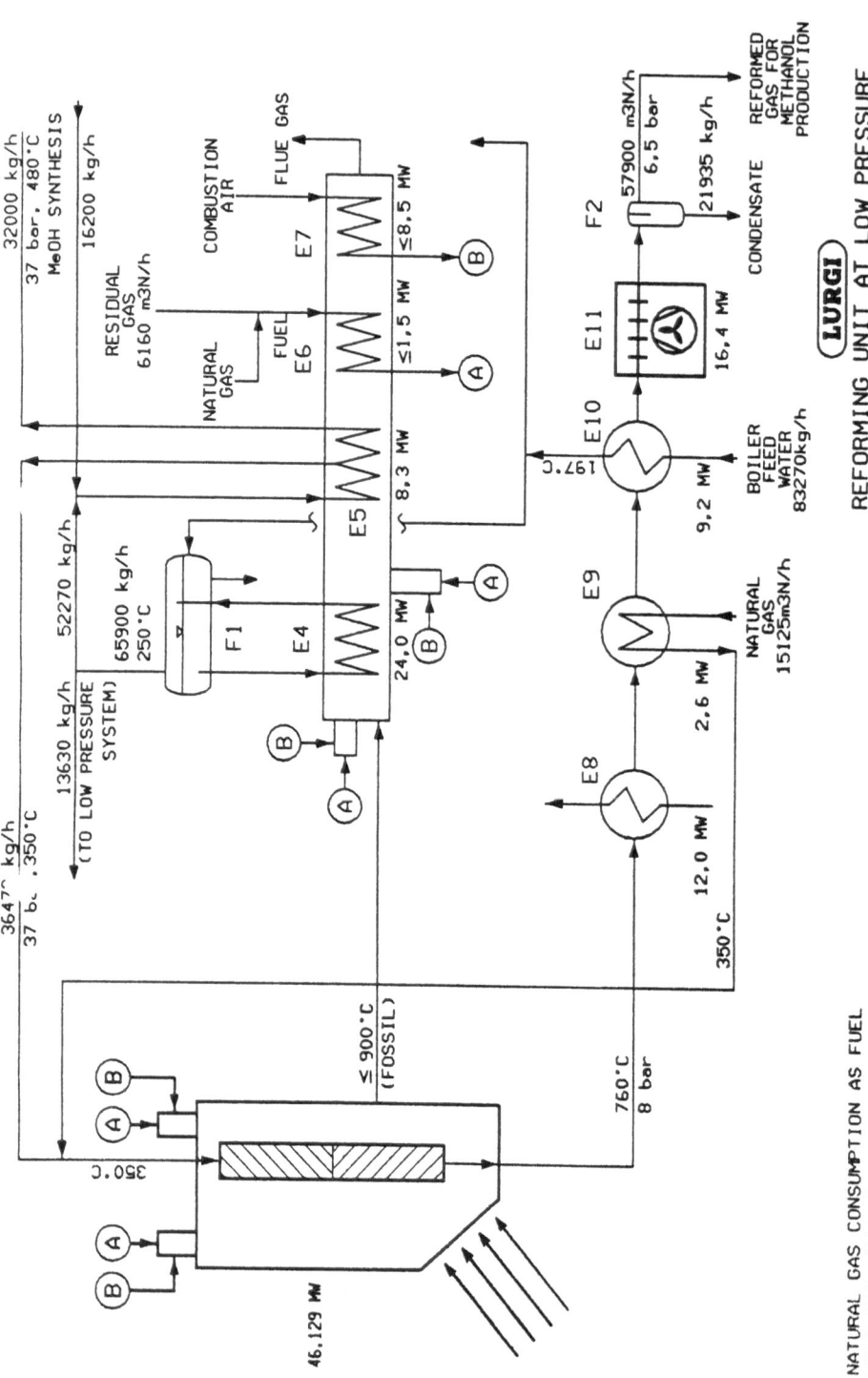

REFORMING UNIT AT LOW PRESSURE
WITH INTEGRATED SOLAR MAIN REACTOR
CODE A-8

FIG. 2-15

LURGI

NATURAL GAS CONSUMPTION AS FUEL
FROM 193 m3N/h TO 5171 m3N/h

GKY-63141-00

A-System, Listing of Consumption Figures   TAB. 2 - 1

System	$V_{E\ PROC.}$ M3N/H	$G_{PROD.}$ T/H	$V_{H2BYPROD.}$ M3N/H	Q_{SOLAR} GWH/A [1]	$Q_{SOL.INT.}$ GWH/A [1]	$V_{E\ FUEL\ D.}$ M3N/H	$V_{E\ FUEL\ N.}$ M3N/H	ζ_{SOLAR} -	$\bar{V}_{E\ FUEL}$ M3N/H	$V_{E\ UTILIT.}$ M3N/H
A-FOSS.	12.000	13.75	9.420	0	0	3578	3578	0	3578	550
A-1	12.000	13.75	9.420	56.0	15.2	1262	3582	0.2973	2892	550
A-2.1	12.000	13.75	9.420	78.4	17.3	1262	3582	0.4163	2616	550
A-2.2	12.000	13.75	9.420	109.4	19.5	1262	3582	0.5809	2234	550
A-2.3	12.000	13.75	9.420	172.1	29.3	1262	3582	0.9138	1462	550
A-3	12.000	13.75	9.420	102.9	20.6	$0^{3)}$	$3273^{4)}$	0.2957	2312	550
A-4.1	12.000	13.75	9.420	150.6	25.1	0^{3}	$3025^{4)}$	0.4298	1725	550
A-4.2	12.000	13.75	9.420	205.8	29.6	0^{3}	$2532^{4)}$	0.5873	1045	550
A-4.3	12.000	13.75	9.420	339.3	$(41.0)^{2)}$	$0^{3)}$	$0^{3)}$	0.9683	$0^{3)}$	$422^{5)}$
A-5	13.020	14.90	10.220	53.3	2.2	1564	3884	0.2830	3228	597
A-6	13.020	14.90	10.220	104.7	4.2	$0^{3)}$	3594	0.2754	2605	597
A'-7	15.125	16.35	11.195	346.7	78.5	1153	5903	0.9894	1205	790
A'-8	15.125	16.35	11.195	111.3	4.5	193	5171	0.2754	3800	790

1) 8760 H/A   2) THEORETICAL FIGURE ONLY. CONSUMPTION WILL BE MADE UP BY EXCESS FUEL (RESIDUAL GAS)
3) EXCESS FUEL (RESIDUAL GAS) AVAILABLE, TO BE STORED IN GASHOLDER.
4) FIGURE TAKES USE OF STORED EXCESS FUEL INTO ACCOUNT. FIGURE WILL CHANGE WITH DIFFERENT $\zeta_{SOL.}$
5) CONSUMPTION FIGURE TAKES EXCESS FUEL INTO ACCOUNT.

- 313 -

B-System, Listing of Consumption Figures   Tab. 2 - 2

System	$V_{EProc.}$ M3N/H	$G_{Prod.}$ T/H	Q_{Solar} GWH/A[1]	$Q_{Sol.int.}$ GWH/A[1]	$V_{E\ Fuel\ D.}$ M3N/H	$V_{E\ Fuel\ N.}$ M3N/H	ζ_{Solar} -	$\bar{V}_{E\ Fuel}$ M3N/H	$V_{E\ utilit.}$ M3N/H
B-Foss.	14,325	22.76	0	0	8385	8385	0	8385	915
B-1	14,325	22.76	56.0	15.2	6065	8385	0.2973	7695	915
B-2.1	14,325	22.76	78.4	17.3	6065	8385	0.4163	7420	915
B-2.2	14,325	22.76	109.4	19.5	6065	8385	0.5809	7038	915
B-2.3	14,325	22.76	172.1	29.3	6065	8385	0.9138	6265	915
B-3	14,325	22.76	102.9	20.6	4068	8385	0.2937	7117	915
B-4.1	14,325	22.76	150.6	25.1	4068	8385	0.4298	6530	915
B-4.2	14,325	22.76	205.8	29.6	4068	8385	0.5873	5850	915
B-4.3	14,325	22.76	339.3	41.0	4068	8385	0.9683	4205	915
B-5	15,543	24.695	53.3	2.2	6581	9098	0.2830	8386	993
B-6	15,543	24.695	104.7	4.2	4414	9098	0.2754	7808	993

1) 8760 H/A

C-System, Listing of Consumption Figures   Tab 2 - 3

System	$V_{E\,PROC.}$ M3N/H	$G_{PROD.}$ T/H	Q_{SOLAR} GWh/a[1]	$Q_{SOL.INT.}$ GWh/a[1]	$V_{E\,FUEL\,D.}$ M3N/H	$V_{E\,FUEL\,N.}$ M3N/H	Z_{SOLAR} -	$\overline{V}_{E\,FUEL}$ M3N/H	$V_{E\,UTILIT.}$ M3N/H
C-foss.	11.395	17.625	0	0	4105	4105	0	4105	640
C-1	11.395	17.625	56.0	15.2	1785	4105	0.2973	3416	640
C-2.1	11.395	17.625	78.4	17.3	1785	4105	0.4163	3140	640
C-2.2	11.395	17.625	109.4	19.5	1785	4105	0.5809	2758	640
C-2.3	11.395	17.625	172.1	29.3	1785	4105	0.9138	1985	640
C-3	11.395	17.625	102.9	20.6	0[2]	3932[3]	0.2937	2777	640
C-4.1	11.395	17.625	205.8	25.1	0[2]	3791[3]	0.4298	2162	640
C-4.2	11.395	17.625	205.8	29.6	0[2]	3512[3]	0.5873	1450	640
C-4.3	11.395	17.625	339.3	41.0	0[2]	0[2]	0.9683	0[2]	367[4]
C-5	12.365	19.125	53.3	2.2	1937	4454	0.2830	3742	695
C-6	12.365	19.125	104.7	4.2	0	4367*	0.2754	3165	695

1) 8760 H/A   2) Excess fuel has to be stored in gasholder

3) Figure takes use of stored excess fuel into account. Figure would change with different $Z_{SOL.}$

4) Consumption figure takes excess fuel into account.

- 315 -

D-SYSTEMS, LISTING OF CONSUMPTION FIGURES   TAB. 2 - 4

SYSTEM	$V_{EPROC.}$ M3N/H	$G_{PROD.}$ T/H	$V_{H2BYPROD.}$ M3N/H	Q_{SOLAR} GWH/A	$Q_{SOL.INT.}$ GWH/A	$V_{E\ FUEL\ D.}$ M3N/H	$V_{E\ FUEL\ N.}$ M3N/H	Q_{SOLAR} -	$\overline{V}_{E\ FUEL}$ M3N/H	$V_{E\ UTILIT.}$ M3N/H
D-FOSS.	10.290	8.625	11.280	0	0	7040	7040	0	7040	785
D-1	10.290	8.625	11.280	56.0	15.2	4719	7040	0.2973	6350	785
D-2.1	10.290	8.625	11.280	78.4	17.3	4719	7040	0.4163	6074	785
D-2.2	10.290	8.625	11.280	109.4	19.5	4719	7040	0.5809	5692	785
D-2.3	10.290	8.625	11.280	172.1	29.3	4719	7040	0.9138	4920	785
D-3	10.290	8.625	11.280	102.0	20.6	2722	7040	0.2930	5772	785
D-4.1	10.290	8.625	11.280	150.6	25.1	2722	7040	0.4298	5185	785
D-4.2	10.290	8.625	11.280	205.9	29.6	2722	7040	0.5873	4505	785
D-4.3	10.290	8.625	11.280	339.3	41.0	2722	7040	0.9683	2859	785
D-5	11.170	9.36	12.240	53.3	2.2	5120	7640	0.2830	6927	852
D-6	11.170	9.36	12.240	104.7	4.2	2954	7640	0.2754	6350	852

3.   Solar Section of Plant-Design and Design Results

3.1   Overview of the Examined Systems

The various concepts which have been examined for
coupling process heat produced by solar means in a
chemical plant for steam reforming of methane are
identified with the numbers 1 to 6 in the overview
table Tab. 3-1.

In the case of the concepts 1 to 4, the solar generated
heat in a primary circuit is transferred to a reactor
separated from the receiver. In versions 1 and 2, this
solar circuit operates a prereactor, in versions 3 and 4
a main reactor. Provisions are made for a fossil fuel
heater to act as a substitute for the lack of solar
energy during 24-hour operation as required. Compared to
the plants 1 and 3, the plants 2 and 4 feature the
possibility of storing solar energy. As a consequence of
energy storage operation, these circuits have been
extended to include a recuperator and cooler, as will be
discussed in detail at a later point. The storage systems
are in turn subdivided according to their different solar
multiples and are identified by an index, e.g. 2.2 or 4.1.

In addition to the concepts with indirectly heated
reactors, the selection of examined systems also includes
such systems with direct coupling of solar energy in the
reformer pipes integrated in the receiver.

Version 5 represents a system with integrated prereactor,
version 6 a system with integrated main reactor.

3.2   Design and Description of the Subsystems

In order to achieve a comparison between the various
systems, the thermal output to be transferred to the
prereactor and main reactor of the relevant chemical
plant is defined as approx. 40 MW. The connection data of
the solar circuit to the reactor or the process gas data

in the case of the integrated concepts are listed in
Table 3-1. The heat transfer medium of the systems 1 to
4, assumed for design purposes, is air. A permissible
pressure loss totalling 1.8 bar, divided into 50 % for
the solar section and 50 % for the reactor, must be
maintained.

The design time base for the solar components is 21.6.,
12 noon summer time with a defined insolation of 900 W/m².

The output range of the storage systems has been
selected, based on the defined solar multiples, such that
the influence of various solar coverage values can be
compared with each other.

In addition, it is also necessary in the case of the
concepts 2.3 and 4.3 that the total energy requirements
of the reactors are covered by the sun on at least one
day. The design time base for the necessary storage
capacity is 21.3. 12 noon - not 21.6. - in order to
ensure good storage utilization over a longer period of
time.

3.2.1   Heliostat Field

The two subsystems heliostat field and receiver are
closely linked. The best possible degree of conversion of
the solar energy results in the optimum of the product of
the field efficiency and receiver efficiency. This
relationship and the objective, based on the receiver
output required, of designing   the heliostat field,
renders an iterative method of calculation necessary.

However, neither repeated field and receiver calculations
nor optimization of all influencing parameters can be
carried out for the ten systems as part of this study.
For this reason, the complete calculation method will be

described only in the case of plant 1. The heliostat
fields of the other plants are defined with the aid of
the degree of receiver efficiency estimated in advance
and the interface aperture defined once. The field output
will be slightly corrected as part of subsequent receiver
design.

With regard to the definition of the aperture size and
inclination with respect to the field, the experience
gained in other projects can be well utilized here. A
field design is aimed for in which the product of the
aperture and receiver efficiencies lies within the area of
the weakly emphasised optimum, refer to /1/.

With regard to the degree of receiver efficiency, a small
opening and a shallow angle of incidence of the aperture
should be aimed at. These factors determine the spillage
losses of the field and therefore also the radiation load
at the edge of the aperture.

An average concentration relationship in the opening of
1150 and an aperture angle of incidence with respect to
the horizontal of 35° provide useful reference values for
dimensioning.

The field design calculations have been carried out with
the computer program HEFLD developed at M.A.N. /2/. Its
functional principle is briefly described in the
following.

The computer program determines in the computer mode used
the insolation in the aperture at a defined time
dependent on the number and arrangement of the heliostats. In addition, it provides the bearing distribution
of the intensity in the aperture. If necessary, the
program can calculate in an optimization stage a heliostat field for a preset output. For this purpose, an
adequate surface is subdivided into radial sectors and

zones, on which heliostats are arranged in accordance with a specified mounting schedule. At the required optimization time, the preset output is obtained from the field zones in such a way that zones with the better degree of efficiency receive priority. The input data required are the geographical location data, the optimization times, the position and size of the aperture, the data of the individual heliostat and the arrangement of the mirrors in the field zones. Field output and field efficiency are specified in detail divided into individual components.

The decision with regard to a 1 or 2-segment configuration is made by limiting the cavity size. The cavity dimensions in the case of GAST (28.5 MWth per module) should not be exceeded to any great extent since detailed design examinations for the receiver in the GAST project are being carried out up to this size.

With the exception of 1, 2.1 and 5, this results in heliostat fields with two segments for all plants. Due to the high output, plant 4.3 features two fields with two towers and a total of 4 cavities.

Selection of the tower heights is an optimization task.

However, optimization of the tower height for the 10 systems examined in this study was not possible. The tower heights defined were such that only a few heliostats could be saved if the height of the tower was further increased. This optimization will still be carried out during subsequent detail design of the plants.

The results of the heliostat field arrangement are listed in Tab. 3-2. The field efficiency takes into consideration the following loss components: Cosinus, shading, blocking, absorption, reflexion and spillage.

The heliostat fields consist of individual heliostats of GAST configuration (Fig. 3-1) and are calculated for a location 37° latitude north (Almeria, Southern Spain). The insolation data used were obtained from /3/ and corrected to the defined design value of 900 W/m^2 on 21.6.

The insolation progressions on the four representative days established for determining the annual energy are shown in Fig. 3-2.

Taking plant 1 as an example, Fig. 3-3 shows the progression of the field efficiency over time on these days. The 12 noon values show that the field is not optimized for 21.6. but rather 21.3. or 21.9., resulting in a higher annual energy yield. To limit calculation requirements, the heliostat fields of the remaining plants are designed based only on the time 12 noon on the four days for different tower heights. The efficiency progression over the course of the day is derived from plant I.

3.2.2   Receiver

The receivers envisaged for the various plants differ. Their design is particularly influenced by criteria such as whether the receiver is intended for a plant with separate heat transfer circuit or whether the reactor pipes are integrated in the receiver and which gas outlet temperatures must be reached.

The cavities of the plants with a separate reactor are suitable for directly exposed heat transfer pipes and are of the known GAST design.

The concepts with a chemical reactor within the cavity differ in design. In order to achieve a homogeneous radiation load and a more favourable transfer ratio during

- 321 -

transient procedures, the reactor pipes are not arranged in the directly exposed part of the cavity. The areas required in the "radiation shadow" are achieved by uprighting and enlarging the front cavity walls.

In addition to these differences in the design of the cavities and the heat absorber pipes (air pipes) and process gas pipes with catalytic filling) there is a further differentiating feature, the material of the heat absorber pipes. With the exception of concepts 3 and 4, whose pipes in the area exposed to radiation are made of ceramic material (SiSiC) corresponding to the high temperature level, it was assumed in the case of all other receivers that the heat transfer media will be made of Alloy 800 H.

The aim of the design calculations is to achieve the best possible receiver efficiency while taking marginal conditions into consideration: Limitation of the pressure loss in the primary circuit and fulfilling the required service life. The required service life of 20,000 hours of operation at nominal load with 10,000 full load interruptions has been allowed for in the temperature-dependent material data of IN 800 H and SiSiC.

The following assumptions relating to the surfaces have been made for calculating the energy flow in the cavity:

Solar spectrum

  Absorption coefficient of the heat exchanger pipes metal/ceramic: 0.92/0.8

  Absorption coefficient of the cavity inner walls: 0.15

IR spectrum

  Absorption coefficient of the heat exchanger pipes metal/cermaic: 0.94/0.8

Absorption coefficient of the cavity inner walls: 0.47

Thermal transmission coefficient through the cavity
walls: 0.4 W/m²K

Convection losses are taken into consideration in a
simplified model which contains a cavity zone in an
exchange arrangement with the surrounding area dependent
on the angle of incidence of the aperture.

The receiver calculations for the concepts 1 to 4 are
carried out with the computer programs CREAM and TUTEST
developed as part of the GAST project. CREAM determines
from the balance of energy flow the effective power
output, whereas TUTEST determines the heat transfer to
the circuit medium in detailed form. CREAM cannot be used
in the case of concepts 5 and 6 due to the deviating
cavity design. In this case, a simplified method of
calculation (computer program ETACAR) is used which
abandons a differentiated consideration of the local
flows of energy.

The receiver output ratings of the concepts 1 to 4 at the
design point are selected such that they exceed the
reactor output by 0.5 MW (see Tab. 3-1). This global
amount is intended to take into consideration the heat
losses of the lines and of the accumulators.

A survey of the results of receiver design relating to
the main dimensions of the cavities is provided in Tab.
3-3, the results regarding the main thermodynamic data
as well as the piping geometry are given in Tab. 3-4.

Tab. 3-4 requires explanation. The degrees of cavity efficiency given in this table are obtained with the aid of
specified circuit connection data. During the course of
this study, the cooling air inlet temperatures in the
receiver have changed in the case of plants 2 and 4.

The enthalpy remaining in the gas during the storage
charging procedure can only remain in the circuit by
means of recuperative thermal transmission to the section
of the circuit between the blower and receiver, resulting
in an intermittent increase in the receiver inlet
temperature. These change only have a
slight effect on the degree of efficiency since the
average temperature level in the cavity is not
significantly increased. It was therefore not considered
necessary to redesign the relevant receivers in order to
allow for this influence.

3.2.2.1 Receivers for Plant Concepts with Separate Reactor

The results of receiver design for concepts with a
separated reactor are illustrated by way of the example
of plant 1.

The receiver of this system consists of a cavity with
metallic heat exchanger units.
Dimensions and arrangement of the cavity are shown in
Fig. 3-4. The three rear walls of width B are intended to
act as a mount for the heat transfer pipes. Sets of 25
pipes at ‘a spacing ratio of 2.0 are combined to form a
panel and arranged on two levels on the cavity walls
(Fig. 3-5). The lower panels have a radiated length of
8 m, whereas the panel arranged above them has a radiated
pipe length of 7.5 m. Panels arranged one on top of the
other are connected in series in all versions with
isolated reactor with the flow being arranged from the
base of the cavity to the top for efficiency reasons.
Control is implemented after the outlet temperature from
the upper panel since this temperature must be maintained
at the value required by the reactor during solar
operation. At the design point, the receiver of plant 1
achieves an efficiency of 84.8 % (refer to Tab. 3-4)
which changes as shown in Fig. 3-6 at reduced insolation.

In conjunction with the insolation and the field
efficiency, the degrees of cavity efficiency at the
various times of day are obtained on the selected days
(Fig. 3-7), assuming cloudless skies and trouble-free
operation. This partial load characteristic of the
receiver is assumed for all other versions with separate
reactor since they are also operated in accordance with
the same control concept.

The receivers of the remaining plants with separate
prereactor are of identical design (see Fig. 3-8, 3-9
and Tab. 3-4). The two cavities of the plant 2.2 are
identical to the cavity of plant 1.

The cavities of the plant concept with separated main
reactor (3 and 4) have a piping system made of SiSiC. In
the design of these heat transfer pipes, it is assumed
that at the time of implementation of such a plant, the
production methods for manufacturing minimum pipe length
of 8 m will have been developed and tested. Considerably
shorter pipes would result in a more involved design
regarding the connection points. The shielding from
direct radiation necessary at these points would have an
adverse effect on the cavity efficiency. The ceramic
panels are arranged in two storeys in the cavity of plant
3, and in three storeys in the cavities of plant 4. The
flow through pannels arranged one on top of the other is
from the cavity base to the top. Fig. 3-10 to 3-12
contain the cavity dimensions of plant 3, 4.1 and 4.2. A
three-storey piping system in plant 4.2 is shown by way
of example in Fig. 3-13. In the case of concept 4.3, four
cavities (Fig. 3-14) from two solar plants supply one
reactor with the required thermal output.

A comparison of the degrees of cavity efficiency in Tab.
3-4 clearly shows a reduction in systems 3 and 4 of 3 to
5 percent compared with the systems 1 and 2. This is the
result of the increased average cavity temperature in the
case of the 1000 °C plants.

3.2.2.2 Receivers for Concepts with Integrated Reactor

As already mentioned, the reactor pipes are arranged in the receivers with integrated reactor in the area of the cavity not directly exposed to radiation. This requirement leads to a cavity design as shown in Fig. 3-15, selected for the plant 6 with an integrated main reactor. The front 12 m wide walls which are not directly exposed to radiation are provided for equipping with the reformer pipes. The two walls arranged opposite the aperture reflect and emit the solar radiation onto the pipes. Their arrangement can be seen in the development of the cavity walls (Fig. 3-16).

The dimensions of the cavity with integrated prereactor (plant 5) as well as the arrangement of the reactor pipes in front of the walls are shown in Fig. 3-17 and Fig. 3-18.

This concept of a prereactor integrated in the cavity has already been examined and described in a previous study /4/. A general overview of the integrated prereactor, Fig. 3-19, has been adopted from this study.

The arrangement of the pipes in one or several storeys is the result of the pipe length which can be realized in these concepts. These were stipulated by the process.

The maximum length of the selected 60.3 x 3.6 mm pipes at an average heat flow density of 30 kW/m² is 4.65 m in the case of the integrated prereactor; in the case of the integrated main reactor the length amounts to 9.7 m.

The main reactor pipes are equipped in the lower section 1 (up to 2.2 m) with a standard catalyst, and above this with a rich gas catalyst, resulting in different heat transfer rates in axial direction.

In addition to the pipe dimensions and pipe spacing ratio, the space requirements for filling and draining the pipes determine the size of the cavity (see Fig. 3-19).

To guarantee 24-hour operation as required, the concepts with integrated reactors require an auxiliary fossil fuel heating system arranged in the cavity. It is operated during solar operation to supplement inadequate insolation with the aperture open and during the night when the aperture is closed. Night operation corresponds to that of a fossil fuel heater, the operability of which will not be examined as part of this study. On the other hand, mixed solar and fossil fuel operation is a new, untested concept. The combustion in the cavity influences the thermal loss flow through the aperture, either through the radiation of the combustion gas or the altered permeability of the radiation from the inner walls. It is also assumed in this respect that it will be possible to maintain the convective losses as low as possible by drawing off the hot combustion gases. It is intended to couple the enthalpy in the flow of waste gas to the chemical plant at a different point.

For this purpose, the waste gas temperature and the temporal progression of the volume of waste gas to be re-fed into the process are estimated with an approximation calculation introduced in /4/ and specified in Section 3.3, Fig. 3-61 and Fig. 3-62.

Theoretically, combustion and flow procedures during mixed operation are, however, extremely difficult to determine so that a function verification can only be presented on an experimental basis. This also applies to the arrangement of the burners in the cavity. It was formerly /4/ proposed to arrange the burners at the top of the cavity.

The design calculations for additional firing in the concepts with an integrated reactor are therefore based on idealized assumptions:

- No change in the physical surface properties of pipes and walls as the result of combustion residues and oxidation

- The average temperature of the combustion gases is 40 K above the process gas outlet temperature

- The air ratio during the combustion of natural gas is X λ= 1.05.

Reference is made in the following to further problems relating to the integrated main reactor. This receiver, in which the process gas is to be heated to 860 °C in the reactor pipes arranged in the cavity, is also to be equipped with Alloy 800 H pipes. Taking the creep and alternating loads into consideration, only provisional statements can be made in this study relating to their function. For this purpose, the temperature and stress distribution of the reactor pipes of the integrated main reactor were calculated on the basis of the assumed axial profile of the heat flow density over the pipe length as shown in Fig. 3-20. Specification of this idealized distribution results in the temperatures and relative material loads shown in Fig. 3-20. The load or stress factor ß is a measure for the local material utilization, whereby values greater than 1.0 already signify a reduction in the target service life. The stress factor takes into consideration creep and material fatigue during 20,000 hours at 10,000 load changes. However, since the operating times of the receivers with an integrated reactor differ from those of the GAST receiver by a factor of 4, the creep load also increases correspondingly. This distinctly lower number of load changes - theoretically zero during undisturbed operation - results in a reduction of the alternating load, however, at material temperatures higher by approx. 60 K compared to the GAST receiver.

Attention is drawn to further particular features of the system concepts 5 and 6 as part of testing the receiver design in the material testing department. The involvement of toxic and explosive substances renders an increased survival probability necessary compared to the separated concepts. The static strength is reduced as the result of increased carburization of the material Alloy 800 H at temperatures up to approx. 900 °C. The simultaneous oxidation of the surfaces can scarcely form an effective protective barrier against the penetration of further carbon in the matrix since the oxide layer can be expected to flake off in the event of cyclic load. It is therefore recommended to include materials of higher strength in comprehensive examinations. To reduce stress, the lowest possible heat flow densities should be processed, a larger pitch ratio should be implemented and greater pipe wall thicknesses should be selected.

Tab. 3-3 and Tab. 3-4 provide a survey of the results of receiver design.

Fig. 3-21 shows the progression of receiver efficiency of plant 6 during mixed solar-fossil fuel operation on 21.6.

3.3   Energetic Observation of the Systems

3.3.1   Thermal Output of the Receivers

The effective solar thermal output transferred to the
receivers of the plants 1 to 6 on the 4 selected days can
be easily determined with the following relationship:

$$P_{Rec}(t) = \frac{\dot{S}(t)}{\dot{S}_o} \cdot \eta_F(t) \cdot \eta_C(t) \cdot P_{F,o}$$

with $\eta_F(t) = \eta_F (\dot{S}(t))$

$$\eta_C(t) = \eta_C (\dot{S}(t), \eta_F(t)).$$

In this formula $P_{Rec}(t)$ signifies the actual receiver
output, $\dot{S}$ the non-concentrated insolation, η_F the field
efficiency, η_C the cavity efficiency and P_F the
radiated power striking the heliostats vertically. The
index o describes the design point: 21.6., 12 noon summer
time.

The calculations apply to idealized solar operation
without insolation interruptions and without power losses
when starting and stopping the receiver.

The atmospheric conditions at Almeria and the heliostat
field design in accordance with the criterion of the most
favourable annual energy yield, result in a receiver
overload on 21.3. and 21.9. as shown in Fig. 3-22 with
the system 1 taken by way of example. The calculations
for the annual energy produced by solar means assume that
this additional output can be coupled into the process
without any risks for the receiver. The comparability of
the various concepts remains unchanged by this procedure.

The results of the receiver power output calculation for the concepts 1 to 6 are listed in the Figs. 3-23 to 3-60. They represent the solar input over the entire mirror area of the heliostat field, the solar output in the receiver and the power consumption of the reactor on the 21st of the months of March, June, September and December. The daily energy values added in brackets are obtained by means of integration of the curves over time. As a comparison of the individual concepts, these results are presented in combined form in Tab. 3-5.

3.3.2   Determining the Annual Power Output at the Reactor

The amount of solar energy transferred in the receiver annually is calculated, as agreed, from the daily values of the 4 days representative for a three months period.

i.e.:

$$P_{Rec}(a) = 91.25 \times (P_{Rec}(21.3.) + P_{Rec}(21.6.) + P_{Rec}(21.9.) + P_{Rec}(21.12.)$$

In the concepts under examination, the receiver output is transferred to the chemical plant by various means. In the case of the plants without storage facilities, it is transferred directly to the reactor, reduced only by the thermal output losses along the transport path. In the case of the plants with thermal storage, a part of the energy is fed to the reactor only after intermediate storage. The circuit simulation calculations necessary for these concepts are extremely extensive, rendering restrictions in calculation requirements necessary. For this reason, the annual solar energy for the storage systems and fed to the reactor is determined on the basis of only one calender day. The 21.9. is selected for this purpose, since the energy yield on this day deviates from the annual average by only 2.6 %.

3.3.2.1 Energy Balance of the Plants without Storage Facilities

The annual thermal energy transferred from the sun to the reactor of the plants 1 and 3 is derived from the annual receiver energy $P_{Rec}(a)$, reduced by the annual heat transport losses $P_L(a)$. Due to the requirement relating to comparability with the remaining systems, the increase in the enthalpy in the circuit by the blower (approx. 4 MWh) during receiver operation is included in the calculations. The annual solar energy delivered to the reactor is shown in Tab. 3-1 and Tab. 5.1-1.

Making this energy available also involves high energy requirements for operation of the blower, for the power supply of the heliostats and other auxiliary systems.

During operation of the primary circuits, the blower is responsible for approx. 90 % of the internal power consumption. The thermal energy equivalent of the required blower output is calculated by assuming a conversion efficiency into mechanical energy of 35 %. In the case of systems with a comparably low percentage of solar energy (plant 1), a particularly unattractive ratio is obtained between converted solar energy and internal power consumption (56 to 36.1 MWh) when air is used as the heat transfer medium. For this reason, the plant-own power consumption was determined also with helium as the heat transfer medium in the primary circuit, which provided a considerably more favourable result as shown in Tab. 5.1-1.

The plants with an integrated reactor (5 and 6) do not require a blower in the solar section and this is clearly reflected in the distinctly lower own power requirements (Tab. 5.1-1). However, the systems can compete in 24-hour operation only when the enthalpy of the hot combustion gases is used at a different point in the process. The mass flow distributions shown in Fig. 3-61 (system 5) and Fig. 3-62 (system 6) are obtained for the accumulation of

combustion gases. The annual solar energy fed to the
integrated reactors amounts to 53.4 GWh in the case of
concept 5 and 104.7 GWh for concept 6 (see Tabs. 3-1 and
5.1-1).

3.3.2.2 Energy Balance of the Plants with Storage Facilities

This section was compiled in cooperation with DFVLR,
Institut für Technische Thermodynamik, Stuttgart: refer
to /6/.

The plants 2 and 4 are equipped with storage systems, in
which the energy obtained from the heat transfer medium
in the receiver is stored as sensible heat and can be
fed to the reactor as required.

The design of the storage has been based on the work
carried out by GEYER /7/. Here, a rock storage is
proposed for gas-cooled solar power stations and this has
been adopted for the objectives of this study.

A thermal solid matter storage basically consists of
three storage elements:

the heat exchanger,   in which the energy is exchanged
                          between the storage mass and heat
                          transfer medium

the filler material,   which absorbs or gives off energy
                          as the result of a change in
                          temperature corresponding to its
                          material-dependent specific
                          thermal capacity and

the storage containment, which insulates the hot storage
material from the ambient
temperature and which also serves
as the container for the storage
material.

An energy exchange takes place between the filler
material and the heat transfer medium when there is a
driving temperature difference between both media: If the
temperature of the heat transfer medium is higher than
the local filler temperature, the medium
transfers at this point energy to the filler while
cooling - the storage is "charged". The reverse
process takes place during "discharge": Here the local
filler material temperature is higher than the
temperature of the heat transfer medium - the storage
gives off energy to the heat transfer medium while
cooling. The charging and discharging procedure of the
accumulator is designed for counterflow operation.

The heat exchanger is particularly uncomplicated when the
heat   transfer medium is a gas and the heat transfer
takes place with direct contact between gas and solid.
In this case, gas ducts are integrated directly in the
structure of the filler material; their surface then forms
the heat exchanger area. In the solid matter storage
the arrangement of the gas ducts through the filler
material defines the heat exchanger surface. On its
way through the storage, the gas absorbs thermal
energy during discharge or gives off thermal energy to
the storage mass during charging. Along the path, a
temperature profile is developed in the solid storage
mass, with its form depending on the material-specific
variables of the heat transfer and the geometry of the
filler material.

With regard to the storage capacity, a compromise must
therefore be found between

. the utilization of the capacity of the storage mass and
. the utilization of the energy content available in the
  gas.

Fig. 3-63 shows the storage configuration taken as
the basis of this study. With regard to its design, this
storage module is based on the regenerators commonly
found in the steel industry. The possibility of direct
firing of the storage module, as indicated in Fig. 3-63,
was not examined as part of this study. The main data of
the accumulator are listed in Tab. 3-6.

Three different plant and corresponding storage sizes
are examined for the plant concepts 2 and 4. The plants
2.1 and 4.1 have a solar multiple of 1.5, the plants 2.2
and 4.2 have a solar multiple of 2.0. The solar multiple
is not specified for the plants 2.3 and 4.3, but rather a
stipulation is made in that on 21.3. so much solar energy
must be collected as to ensure the systems can be
operated for 24 hours without additional firing with
fossil fuels.

Fig. 3-64 shows a flow scheme for the storage
systems 2 and 4. During operation, the following gas
status data are obtained, in each case at the gas inlet
of the storage:

	Plants 2	Plants 4
Charging:		
Temperature	800 °C	1020 °C
Pressure	20 bar	20 bar
Discharging:		
Temperature	470 °C	500 °C
Pressure	22 bar	22 bar

The permissible minimum temperature during discharge at
the gas outlet is 780 °C for plants 2 and 1000 °C for the
plants 4.

After several charging and discharging cycles, the
temperature distribution is obtained in the storage
material as shown in Fig. 3-65. Assuming the distribution
of the filler material temperature indicated by the upper
dotted line (Fig. 3-65) and if discharging is started at
this point, then air flows in the example shown in Fig.
3-65 at a temperature of 450 °C from the "right" to the
storage where it is heated and leaves the storage
at a maximum temperature of 790 °C. The storage is
discharged as soon as the temperature at the gas outlet
drops below 780 °C and the lower temperature profile
indicated with a dotted line is reached. The charging
cycle now begins. Air - in this calculation example at a
temperature of 790 °C - flows from the "left" to the
storage   and heats the storage material. As a result,
the temperature of the gas which leaves the storage
also gradually increases. However, this also means that
only a part of the energy fed to the gas in the receiver
is stored. Fig. 3-66 illustrates these conditions. The
energy remaining in the accumulator is not sufficient to
operate the plant 2.3 for 24 hours only with the sun as
required when the energy contained in the gas leaving the
storage   is not used. For this purpose, it is
necessary to modify the flow scheme shown in 3-64
since the permissible gas inlet temperature of blowers
available today is maximum 400°C - 450°C. For this
reason, a recuperator is integrated in the circuit which
recuperates a percentage of the effective heat still
contained in the gas after the storage behind the
blower (Fig. 3-67). Depending on the size of the
storage   and recuperator, a cooler is still necessary
to ensure that permisible blower inlet temperature is not
exceeded. The permissible blower temperatures are defined
for the study at 500°C for the systems 2 and 550°C for
the systems 4, based on the assumption that such blowers

will be available at the time the solar methane production plants are built.

Furthermore, the energy exchange between the receiver and reactor can now only be considered as a whole since the blower and the receiver inlet temperature and (since the receiver outlet temperature is to be maintained constant) also the receiver mass flow depend on the charge status of the storage. To calculate the energy balance of the storage systems in accordance with concept 2 and 4, the computer program RECIR was therefore used for a closed circuit in which the storage is, however, represented as a simplified model. With the aid of parameters, the model is adapted to the results of DFVLR (see /6/) (see Tab. 3-7 and Tab. 3-8). This program was used for dimensioning the circuit components storage, recuperator, cooler and blower and to calculate the energy transferred from the receiver via the storage to the reactor as well as the own consumption of the blower and is described in detail in /8/.

The four methods of operation which are to be simulated with the computer program are illustrated in Figs. 3-68 to 3-71, based on the fundamental circuit diagram. Basically, the circuit includes the seven components; process reactor, receiver, fossil fuel heater, heat storage, recuperator, cooler and blower. The abbreviations signify the designations for the circuit temperature.

Apart from the fossil fuel heater, all components are in operation during pure charging mode (Fig. 3-68). The air circulated by the blower is preheated in the recuperator and heated in the receiver to the upper circuit temperature.
This receiver mass flow is then subdivided

- into an initial proportion (reactor mass flow) which
  flows to the reactor, in which the required power

output is fed to the process at given inlet and outlet
temperatures and

- a   second proportion (accumulator mass flow), which
  firstly gives off heat in the heat storage
  corresponding to the relevant charge status, then in
  the recuperator corresponding to the relevant gas
  status on the secondary side and finally in the cooler.

The cooler prevents the maximum permissible blower inlet
temprature from being exceeded. The temperature at the
blower inlet is obtained by mixing the reactor mass flow
at constant temperature and the accumulator mass flow,
whose temperature - and therefore also that at the
recuperator outlet on the primary side - increases as the
charge level increases.

In this connection, it should also be mentioned that, at
adequate discharge of the storage, the direction of
the heat flow in the recuperator can reverse so that heat
is absorbed from the receiver mass flow which under
certain circumstances can be coupled out of the circuit
via the cooler.

This heat loss and the pressure loss which then un-
necessarily occurs in the recuperator can be avoided by
bypassing the recuperator.

In the case of mixed discharge (Fig. 3-69), the receiver
still gives off energy; in order to cover the
requirements of the reactor, however, the storage is
also discharged and the fossil fuel heater is operated
parallel to this as required. The cooler and recuperator
are no longer in operation.

Pure discharge operation (Fig. 3-70) occurs when the
receiver can no longer give off energy, so that the
reactor output is provided solely by the heat storage
and possibly by the fossil fuel heater as well.

Pure natural gas operation (Fig. 3-71) should be possible in order to be able to cover downtimes as the result of malfunctions, maintenance and repair on the components of the solar section as well as during periods of poor weather.

The temporal progression of several circuit temperatures and of the outputs of the circuit components is represented in Fig. 3-72 for the plant 2.3 on 21.3. During the night, the power requirements of the reactor are covered by the storage (pure discharge operation in accordance with Fig. 3-70), at approx. 7 a.m., the receiver begins to give off energy and the power drain from the storage is reduced (as shown in Fig. 3-69). At approx. 7.45 a.m. the receiver gives off more power than required by the reactor; the new charging cycle of the accumulator is started with the excess power (as shown in Fig. 3-68).

The mass flow in the circuit also increases and therefore the blower power output corresponding to the increasing receiver output. The power output of the blower does not have the same progression as the receiver output since the receiver inlet temperature additionally increases as the storage is charged. Since the receiver outlet temperature for the plant 2.3 is specified at 810 °C, the mass flow through the receiver must be increased correspondingly.

As the charge of the storage increases, the storage outlet temperature also increases and correspondingly the blower inlet and the blower outlet temperatures. The recuperator is bypassed (dashed line of the recuperator output) for as long as the storage outlet temperature is still too low to recuperate the heat. Heat is given off in the cooler as soon as the permissible blower inlet temperature of 500 °C has been reached.

The sum of the energy given off by the receiver directly
to the reactor and the stored energy, of which it is
assumed that it can be completely discharged once again,
results in the process heat fed to the reactor on 21.3.
and obtained from solar energy.

The annual energy transferred by the storage systems to
the reactor is discussed below. However, definition of
the size of the storage and recuperator should
firstly be mentioned.

The storages represent a considerable share of the
total investment costs of the plant. Great care is
therefore called for when defining the storage capacity
during the detailed design of a plant. Size optimization
for the storage was not possible as part of this
study. The size of the storage was determined based
on the example of the plant concept 2.3, the storage size
of the other plants was then estimated in relation to the
quantities of energy produced by the plants annually by
solar means.

The requirement which the plant concept 2.3 must meet is
that the total heat requirements of the prereactor can be
covered during 24-hour operation on 21.3. solely by
process heat generated by solar means.
With this requirement, the size of the heliostat field
and the receiver output are defined corresponding to the
estimated insolation conditions if it is assumed that
the total energy collected can be stored or transferred
to the reactor.

However, in the case of the storage system used,
allowance should be made for the fact that the gas outlet
temperature from the storage rises during charging as
the accumulator is increasingly filled. As described
above, it may then be necessary to eliminate a part of
the collected solar energy by means of cooling. This part
ε , with

$$\varepsilon = 1 - \frac{Q_{Cooler}}{Q_{Receiver}}$$

Q_{Cooler} - Energy taken off the
circuit via the cooler

$Q_{Receiver}$ - Energy fed to the circuit
by the receiver

is not neutral with regard to costs since - at fixed
field and receiver output - it depends on the size of
the storage, the size of the recuperator and the
cooler capacity. Fig. 3-73 shows for the plant 2.3
how the sum of the investment costs for storage,
recuperator and cooler is influenced by the selected
number of storage modules.

If, for example, storage sizes of approx. 2.1 - 2.3
modules are assumed, only $\varepsilon = 0.8$ can be achieved
even in the case of an extremely large recuperator
since the storage is charged almost to its theoretical capacity limit. It is not possible to increase
the field and receiver output in order to collect
the missing 20 % of energy since this energy can no
longer be accommodated in the storage filled to
almost its theoretical limit (filler material temperature = charge temperature). This is only possible when
approx. 3.5 storage modules and a recuperator with a
heat exchanger surface of 10,000 m^2 are used so that
$\varepsilon = 0.95$ can be achieved. The 5 % of the energy
given off by the receiver which must now be cooled
off must, (because this amount of energy is still
missing) in order to enable 24-hour operation on a pure
solar basis of the plant 2.3 as required be additionally

collected by increasing the field and receiver output.
Increasing the field and receiver output by 5 % results
in an increase in costs amounting to approx. 10 million
DM and is therefore still more favourable than attempt-
ing to achieve $\mathcal{E} = 1$ by increasing the size of the
accumulator and recuperator.
In the following, the plant 2.3 has been based on four
storage modules and a recuperator surface area of 5000 m^2.

The number of modules and the size of the recuperator
for the other plants have been defined on this basis
as shown in Tabl. 3 - 9.

Based on this table which shows the blower inlet
temperature $T_{Blow,in}$ and $\mathcal{E}$, the selection of the
accumulator size can be checked with regard to its
plausibility.

In the case of a plant with an inadequately sized
storage, $\mathcal{E}$ is less than 1. This is the case in the
plants 2.2, 2.3 and 4.3. As explained above, however, in
the plants 2.3 and 4.3 an increase in the storage
capacity results in higher specific energy costs. The
value $\mathcal{E} = 1$ is achieved in the other systems and a check
should be carried out as to whether the storage has
been selected too large. A measure for this purpose is
whether the maximum permissible blower intake temperature
of 500 °C for the plants 2 and 550 °C for the plants 4 is
utilized. This is not quite the case in the plants 2.1,
4.1 and 4.2, indicating that more precise design
optimation could result in slightly smaller storages
for these plants.

The storage size, recuperator area, cooling capacity
and blower capacity for the storage systems are itemized
in Tab. 3-10.

The energy fed to the primary circuit by the receiver and
the blower in the plants as well as the energy supplied
directly to the reactor or via the storage to the
reactor are listed in Tab. 3-11. The energy removed from
the circuit via the cooler is also specified. The annual
energy values specified in Tab. 5.1-1 were determined from
these data as described above.

Bibliography to Section 3

/1/ Steam Reforming of Methane Utilizing Solar Heat,
    Final Report of MAN-Neue Technologie and LURGI
    GmbH, December 1984

/2/ Reichle, R.,
    Feldauslegungsergebnisse mit HEFLD für DFVLR-
    Studie, pers. Mitteilungen im September und
    Oktober 1985

/3/ INTERATOM GmbH,
    Basisdaten für die Testanlage in Almeria,
    Dokument GASTP IAS RL 100200014 A vom 15.4.1982

/4/ Steam Reforming of Methane Utilizing Solar Heat,
    Abschlußbericht der Firmen LURGI und MAN-Neue
    Technologie, Februar 1984

/5/ Dr. Agatonovic, Bau eines Receiver für Methanre-
    formierungssystem, EGS Strukturnotiz vom 13.2.1986

/6/ Klaiß, H., G. Merkel, M. Geyer
    "Speicheranalyse im Rahmen der Studie
    'Vergleichende Untersuchung und Bewertung ver-
    schiedener solarer Methanreformierungs-Systeme
    mit Folgeanlage'"
    DFVLR EN-TT, IB 444 005/86, 1986

/7/ Geyer, M.,
    "Betriebsverhalten und Wirtschaftlichkeit eines
    modularen thermischen Gesteinsspeichers als
    Systemelement in Solarturmkraftwerken"
    (Veröffentlichung in Vorbereitung)

/8/ Fauvel, G.,
"Berechnung des Leistungsverhaltens eines Re-
ceiverkreislaufs incl. Rekuperator und Wärme-
speicher."
Aktenvermerk A 630 357-ENP4-006, 1986 der MAN-
Technologie GmbH

Current No.	1	2	3	4	5	6
Solar Coupling	Solar heated separated pre-reactor without storage	Solar heated separated pre-reactor with storage	Solar heated separated main reactor without storage	Solar heated separated main reactor with storage	Direct solar heated integrated receiver/pre-reactor	Direct solar heated integrated receiver/main reactor
Simplified Flow Scheme						

Common Plant Features

- Design Point: 21.6., 12:00 o'clock (solar time); Insolation: 900 W/m²
- The sum of the rated thermal power of the pre-reactor and the main reactor yields appr. 40 MW in each case.
- Only the receiver resp. the integrated receiver/reactor are located on the tower.
- Air is used as coolant for the primary loop.
- Hot gas lines designed for: 800 °C

Main Plant Data:	1-modular		2-mod.	2-mod.	2-modular		2-mod.	2-mod.	2-mod.	1-modular	2-modular
Receiver rated output (21.6., 12.00 h)	22 MW	33 MW	44 MW	70 MW	40,5 MW	61 MW	81 MW	2x65 MW	21,5 MW	43,4 MW
Solar multiple	1	1,5	2,0	3,2	1	1,5	2,0	2x1,6	1	1
Number of heliostats	765	1196	1478	2552	1450	2173	2948	2x2318	764	1576
Tower hight	130 m	160 m	170 m	220 m	170 m	210 m	250 m	2x220 m	130 m	180 m
Number of storage moduls	-	1	1,5	4	-	1,5	3	2x3	-	-
Annual solar energy	56,0 GWh	78,4 GWh	109,4 GWh	172,1 GWh	102,9 GWh	150,6 GWh	205,8 GWh	169,6 GWh	53,4 GWh	104,7 GWh
Solar energy efficiency	0,49	0,44	0,46	0,47	0,44	0,43	0,44	0,46	0,47	0,42

Hot gas lines designed for: 800 °C · Hot gas lines designed for: 1000 °C · 2 Towers · Process gas lines designed for: 700 °C / 900 °C

Integrated receiver/reactor
PR-Pre-reactor
MR-Main reactor

Receiver

Secondary energy source

Fig. 3-1: Summary of investigated plant concepts

Concept no.	1	2.1	2.2	2.3	3	4.1	4.2	4.3	5	6
Thermal receiver power (MW)	22	33	44	70	40.5	61	81	2x65	21.5	43.4
Number of field segments	1	1	2	2	2	2	2	2x2	1	2
Number of heliostats	765	1196	1478	2552	1450	2173	2948	2x2318	764	1576
Tower height (m)	130	160	170	220	170	210	250	2x220	130	180
Field efficiency	0.726	0.702	0.705	0.702	0.705	0.691	0.68	0.693	0.726	0.705

Tab. 3-2:   Main data of field design

Plant concept		1	2.1	2.2	2.3	3	4.1	4.2	4.3	5	6
Thermal receiver power	(MW)	22	33	44	70	40,5	61	81	130	21,5	43,4
Number of cavities		1	1	2	2	2	2	2	4	1	2
Cavity height	(m)	17	24	17	25	17	22	24	22	21	20
Cavity depht	(m)	16,5	23,1	16,5	23,8	12,5	16,0	19,2	16,5	16,5	18
Cavity width	(m)	20	26	20	27	16	20	23,2	21	23,2	25,4

Tab. 3-3:   Main cavity dimensions

Plant concept	1	2.1	2.2	2.3	3	4.1	4.2	4.3	5	6
Number of cavities	1	1	2	2	2	2	2	4	1	2
Inlet pressure p_E (bar)	22	22	22	22	22	22	22	22	22*	22*
Inlet temperature T_E (°C)	450	450	450	450	480	480	480	480	350*	350*
Outlet temperature T_A (°C)	790	810	810	810	1010	1030	1030	1030	680 *	860*
Massflow $\dot{m}$ (kg/s)	58,2	84,1	56,1	89,2	33,6	49,2	65,4	52,4	10.2*	10.2*
Av. heatflux density (kW/m²)	29,5	25,1	29,5	24,4	40,6	39,8	37,1	40,0	30	30
Thermal receiver power (MW)	22	33	22	35	20,25	30,5	40,5	32,5	21.5	43.4
Cavity efficiency	0.848	0.842	0.848	0.842	0.795	0.814	0.808	0.814	0.830	0.784
Panel informations										
Number of panels	36	36	36	.36	36	54	54	54	26	9
Panels in series	2	2	2	2	2	3	3	3	1	1
Tubes per panel	25	23	25	24	15	19	23	19	36	44
Tube length (m)	8/7,5	12/10	8/7,5	12/11	7/7	6/5/6	7/6/7	6/6/6	4	9.7
Outside tube diameter (mm)	34	46	34	46	42	42	42	42	60.3	60.3
Tube wall thickness	3	4	3	3	5	5	5	5	3.6	3.6
Tube spacing ratio	2	2	2	2	2	2	2	2	1.5	1.5

*process gas

Tab. 3-4:   Main receiver data for the different plant concepts

Date	Energy De-livery/Re-quirement		Plant Concept No.									
			1	2.1	2.2	2.3	3	4.1	4.2	4.3	5	6
21.3.	Heliostatf.		338	516	697	1094	685	1025	1397	2186	338	744
	Receiver		166	249	332	528	306	460	612	982	162	321
	Reactor		516	516	516	516	960	960	960	960	516	1042
21.6.	Heliostatf.		379	578	782	1228	768	1150	1567	2452	379	835
	Receiver		170	255	340	540	313	471	626	1005	165	324
	Reactor		516	516	516	516	960	960	960	960	516	1042
21.9.	Heliostatf.		310	473	640	1004	628	940	1281	2004	310	682
	Receiver		154	231	309	489	284	427	567	911	150	296
	Reactor		516	516	516	516	960	960	960	960	516	1042
21.12.	Heliostatf.		224	342	463	726	454	680	927	1450	224	493
	Receiver		110	166	221	351	203	306	405	652	107	207
	Reactor		516	516	516	516	960	960	960	960	516	1042

Tab. 3-5: Energy Delivery/Requirement of Heliostatfield, Receiver and Reactor in MWh

Module:

lengt	35 m
diameter	6 m
volume of containment	990 m³
mass of storing material	2,100 t
volume of storing material	700 m³
total heat exchange area	39,584 m²

Bricks:

material	Mg O
height of brick*	5 cm
width of brick*	5 cm
width of gas channel*	2,1 cm
density	3 kg/dm³
specific heat	1,26 kJ/kg K
heat conductivity	10,5 W/mK

*Detail of storage geometry:

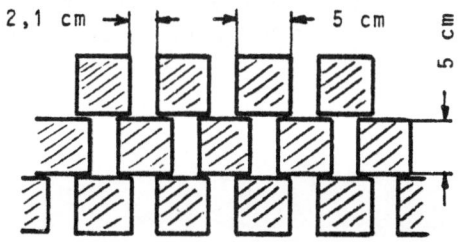

Tab. 3-6: Main data of storage moduls
         (according to / 6 /)

Number of Moduls	Storage outlet temp. cycle at the end of charging cycle (°C)	Stored energy (MWh)
1	777 (777)	115.0 (121)
2	687	207.6
3	627 (632)	245.3 (236)
4	577	261.4
5	567 (561)	272.1 (270)
6	537	277.4
7	527 (527)	282.6 (287)

Tab. 3-7:   Calibration of simplified M A N code
RECIR by comparison of gas outlet temperature
of storage and stored energy calculated
by RECIR (in brackets) with DFVLR results

reduced storage length	storage temperature (°C) at start of charging	at end
. 05	779	791 (790)
. 15	770	791 (789)
. 25	755	787 (787)
. 35	731	785 (783)
. 45	696	777 (776)
. 55	646	766 (764)
. 65	581	748 (746)
. 75	516	723 (721)
. 85	468	688 (686)
. 95	451	644 (644)

Tab. 3-8: Calibration of simplified M A N code
RECIR by comparison of the storage
temperatures at different distances
to the charging gas inlet calculated
by RECIR (in brackets) with DFVLR
results starting from the same
temperature profile

Concept	Storage Moduls	Heat Ex. Area (m²)	Storage Load (MWh)	$T_{Sto,out}$	$T_{Blow,in}$	ε
2-1	1	2500	69	566°C	482°C	1,0
2-2	1.5	2500	138	649°C	500°C	0,974
2-3	4	5000	312	636°C	500°C	0,952
4-1	1.5	2500	126	576°C	504°C	1.0
4-2	3	5000	261	599°C	522°C	1.0
4-3	2x3	5000	2x307	641°C	500°C	0.996

Tab. 3-9: Comparison of Storage Outlet and Blower Inlet Temperatures
for different plant concepts at 21th* of September

* on Sept. 21st appr. the same solar energy is collected as on a mean day
of the year

Plant concept	1	2.1	2.2	2.3	3	4.1	4.2	4.3
Number of storage moduls	-	1	1.5	4	-	1.5	3	2x3
Heat exchanger area of recuperator /m²/	-	2500	2500	5000	-	2500	5000	2x5000
Rated cooling power /MW/	-	-	4.5	14	-	-	-	2x3,5
Rated blower power /MW/	2.5	2.8	4.3	6.9	2.9	3.4	4.7	2x4,2

Tab. 3 - 10 : Main components of primary loop

| Plant | 2.1 | 2.2 | 2.3 | | | | 4.1 | 4.2 | 4.3¹⁾ | |
Date	21.09.	21.09.	21.03.	21.06.	21.09.	21.09.21.12.	21.09.	21.09.	21.03.	21.09.
Receiver	230.9	308.9	527.8	540.1	489.4	350.5	426.9	566.9	491.1	455.4
Blower	16.23	23.67	45.30	47.41	41.70	29.91	19.92	26.78	25.57	23.46
Reactor²⁾	168.9	178.5	195.6	218.2	189.3	156.8	316.1	333.5	181.8	176.0
Storage	68.79	137.7	333.7	330.9	311.9	214.3	126.3	260.6	335.6	307.3
Cooler	0	7.75	36.55	31.09	23.81	0.65	0	0	5.68	1.93
Loss factor ε	1.000	0.974	0.931	0.944	0.952	0.998	1.000	1.000	0.988	0.996

Tab. 3-11: Daily energy (MWh) of the investigated plants with storage during
receiver operation

1) half of the plant

2) from receiver direct into reactor

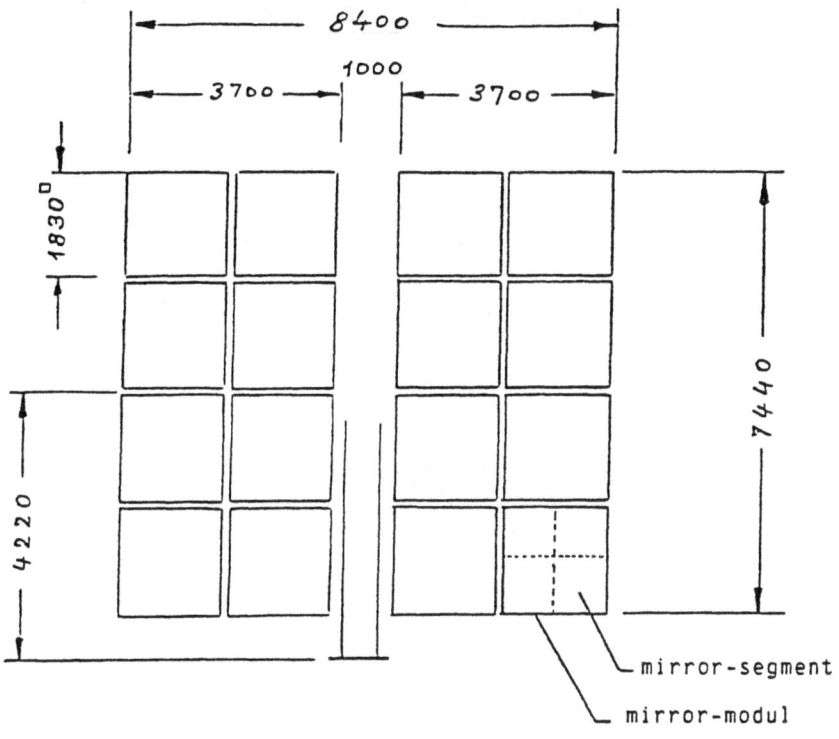

8400
1000
3700
3700
1830
7440
4220

mirror-segment
mirror-modul

reflectivity	0,85
beam quality	2,24 mrad
mirror area of the heliostat	51.84 m²
mirror-modules each consisting of 4 mirror segments	900 x 900 mm
gap between the mirror-segments	10 mm
dimensions of a mirror-modul	1830 x 1830 mm
gap between the mirror-modules	40 mm
ground-clearance	500 mm

<u>Fig. 3-1</u> : Main dimension of the heliostat assumed for the
field design (GAST configuration)

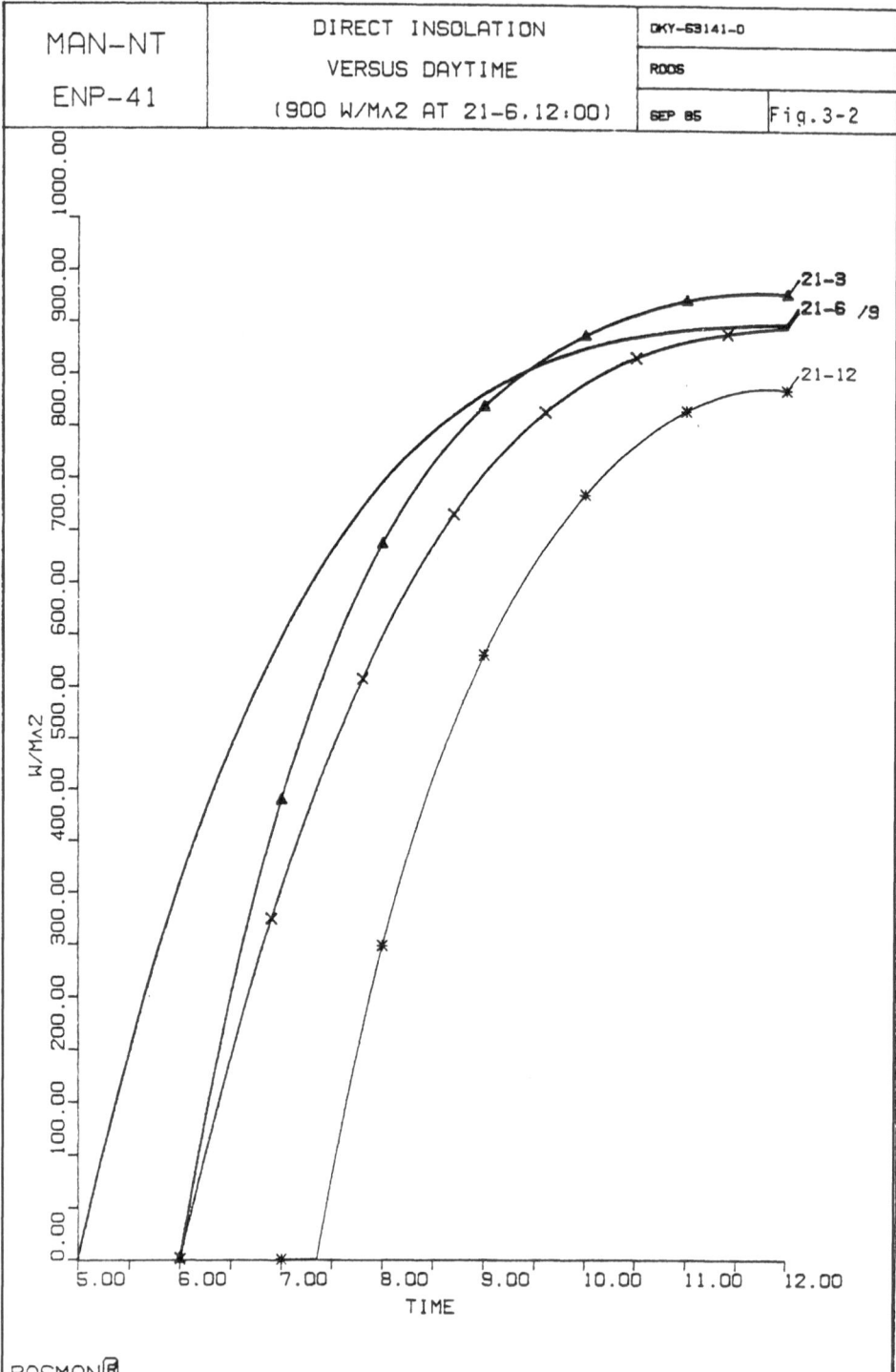

MAN-NT	DIRECT INSOLATION	DKY-63141-0	
ENP-41	VERSUS DAYTIME	ROOS	
	(900 W/M∧2 AT 21-6.12:00)	SEP 85	Fig.3-2

BASMAN⒝

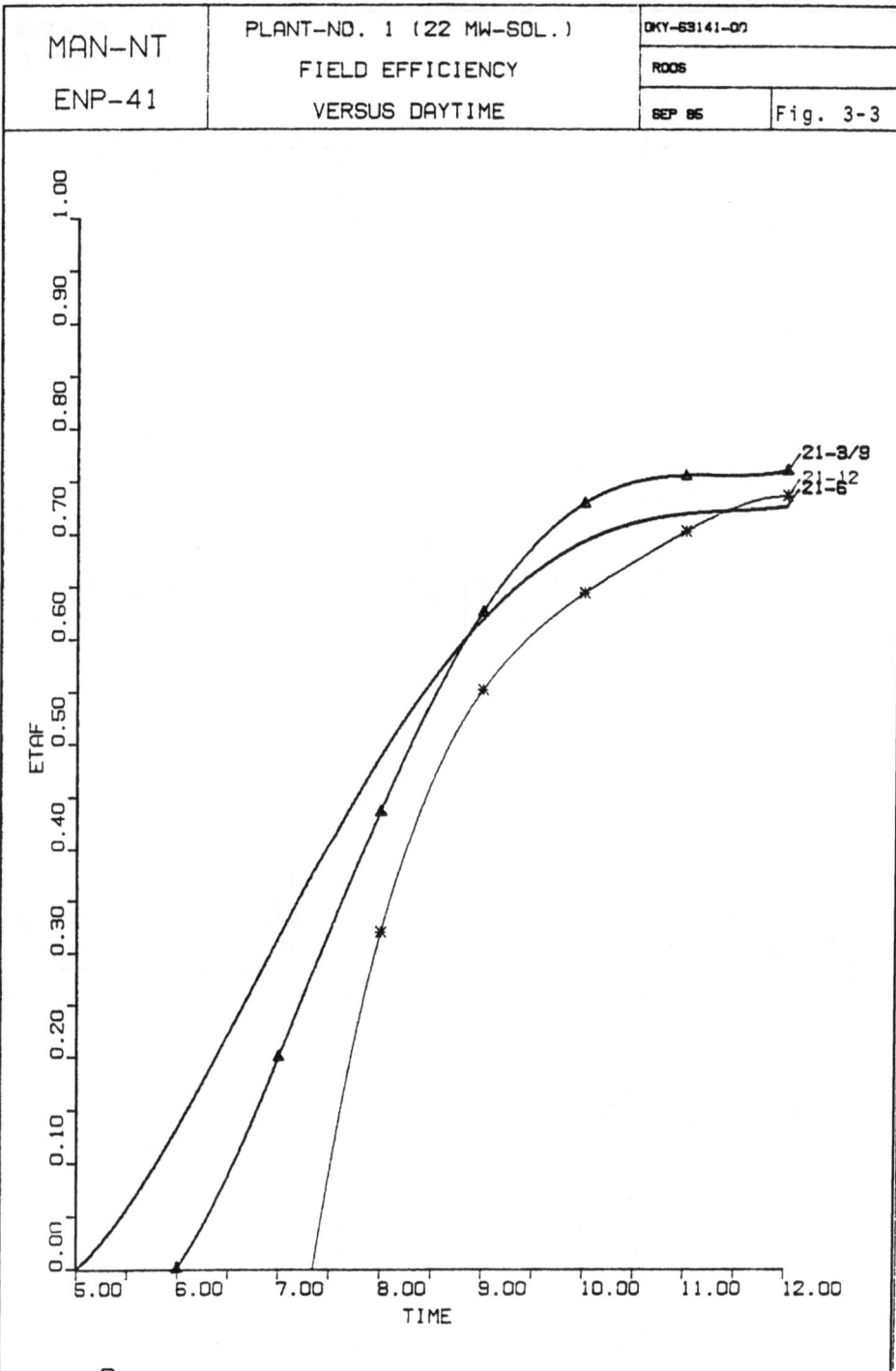

MAN–NT	PLANT–NO. 1 (22 MW–SOL.)	DKY-63141-00	
ENP–41	FIELD EFFICIENCY	ROOS	
	VERSUS DAYTIME	SEP 85	Fig. 3-3

21-3/9
21-12
21-6

ETAF

TIME

BASMAN Ⓡ

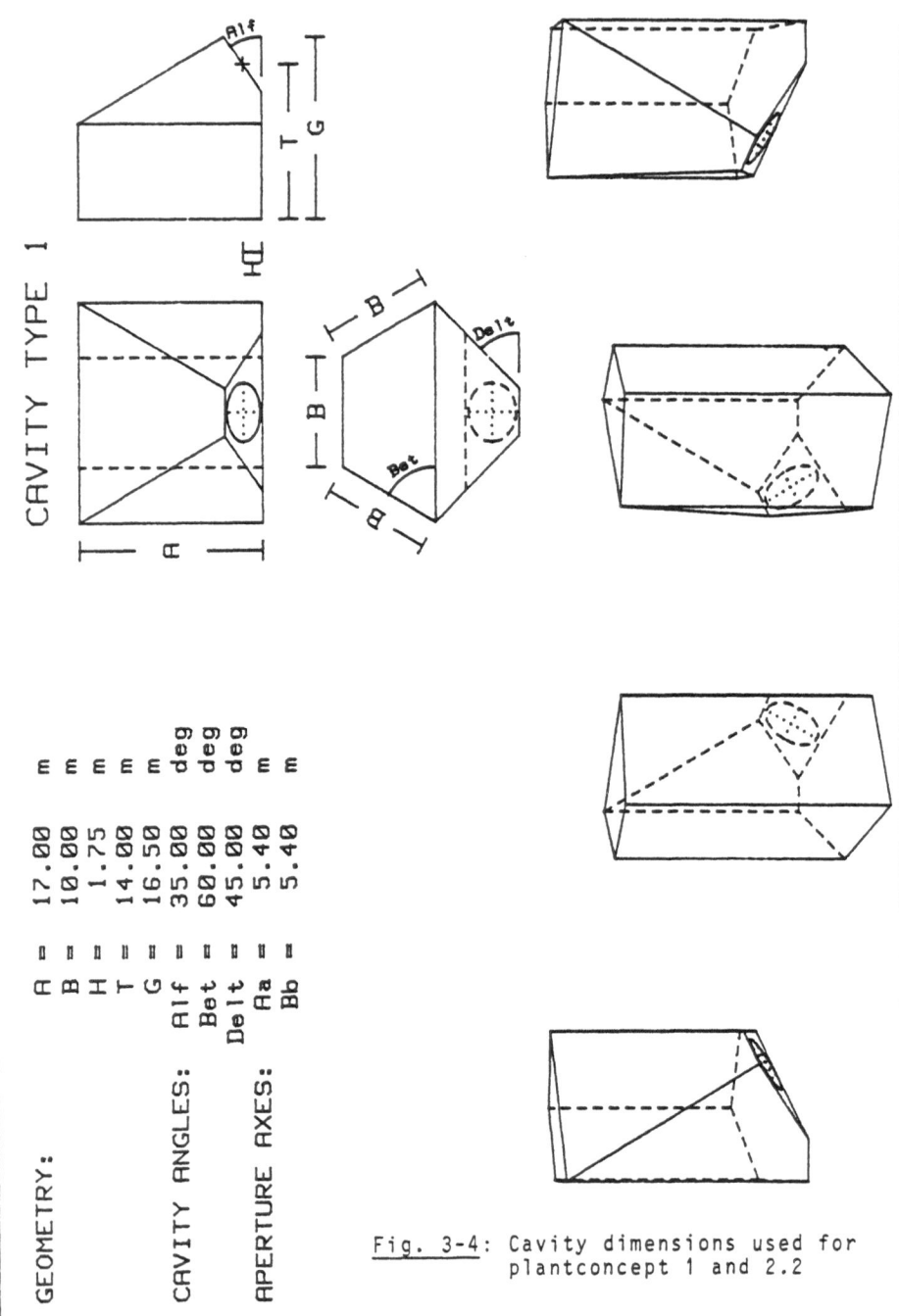

CAVITY TYPE 1

GEOMETRY:			
	A	=	17.00 m
	B	=	10.00 m
	H	=	1.75 m
	T	=	14.00 m
	G	=	16.50 m
CAVITY ANGLES:	Alf	=	35.00 deg
	Bet	=	60.00 deg
	Delt	=	45.00 deg
APERTURE AXES:	Aa	=	5.40 m
	Bb	=	5.40 m

Fig. 3-4: Cavity dimensions used for plantconcept 1 and 2.2

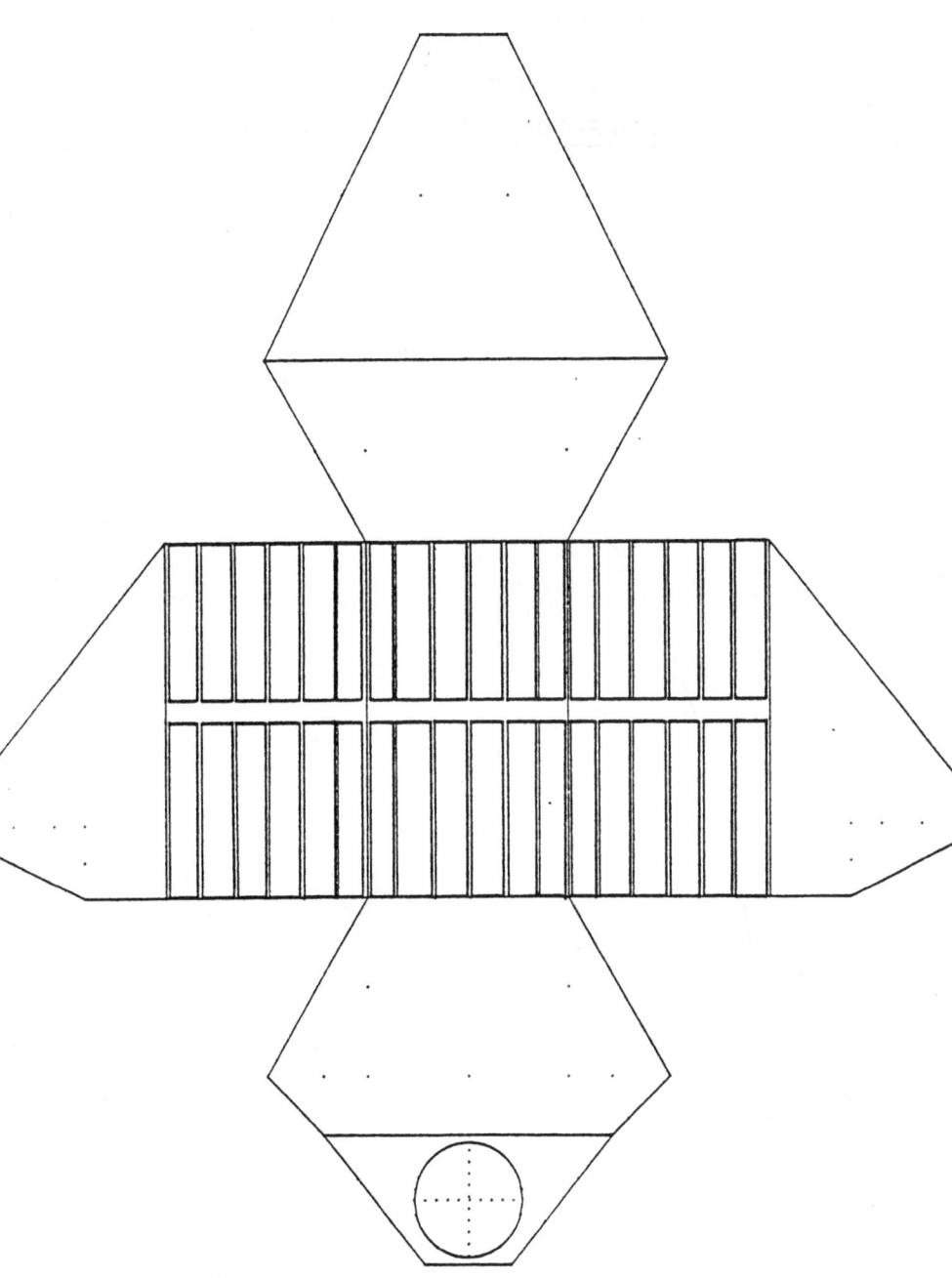

Fig. 3-5:   Arrangement of panels in the cavity of plant 1

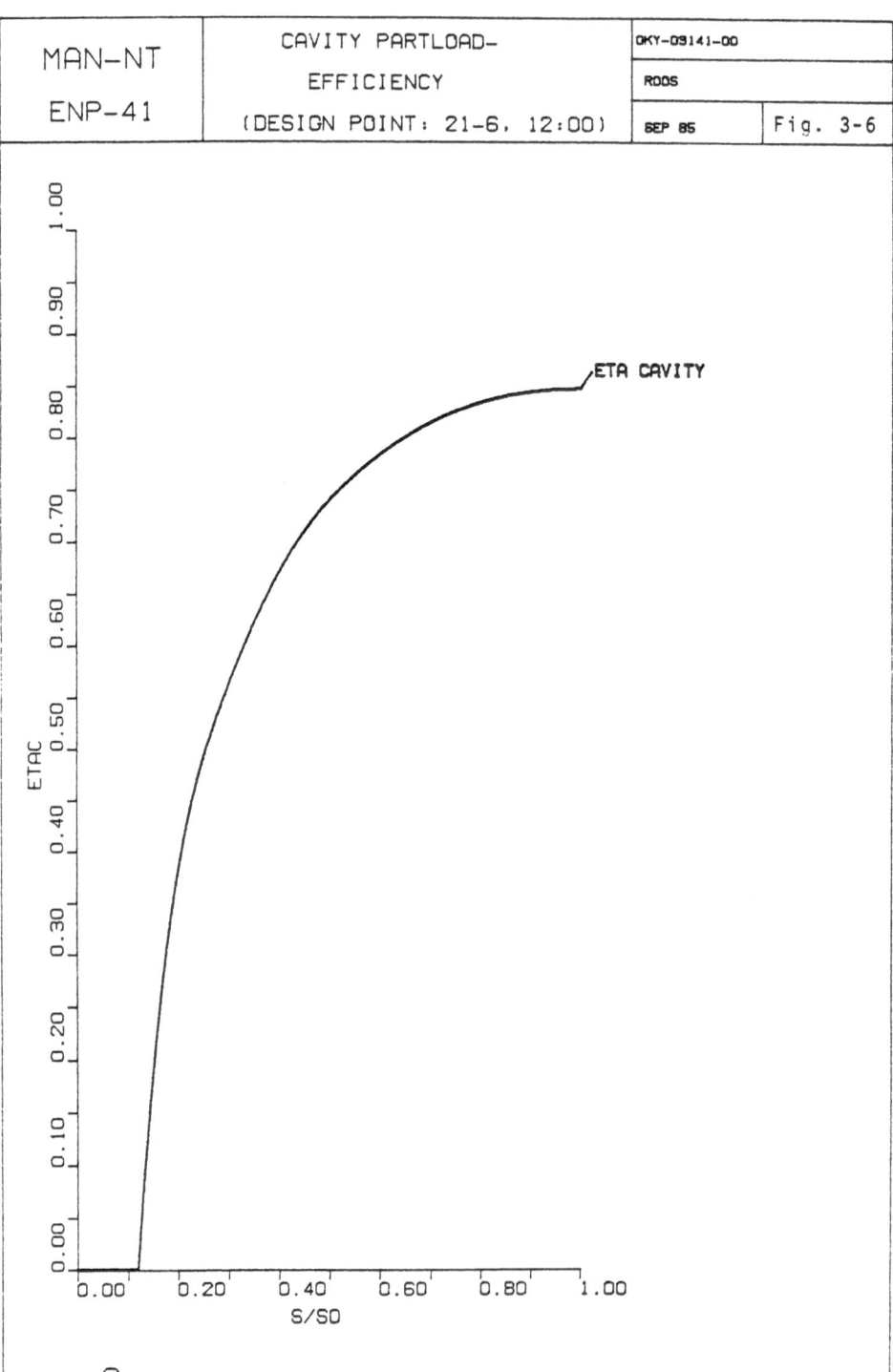

MAN—NT	CAVITY PARTLOAD—	OKY–09141–00	
ENP—41	EFFICIENCY	RODS	
	(DESIGN POINT: 21–6, 12:00)	SEP 85	Fig. 3-6

ETA CAVITY

BASMAN®

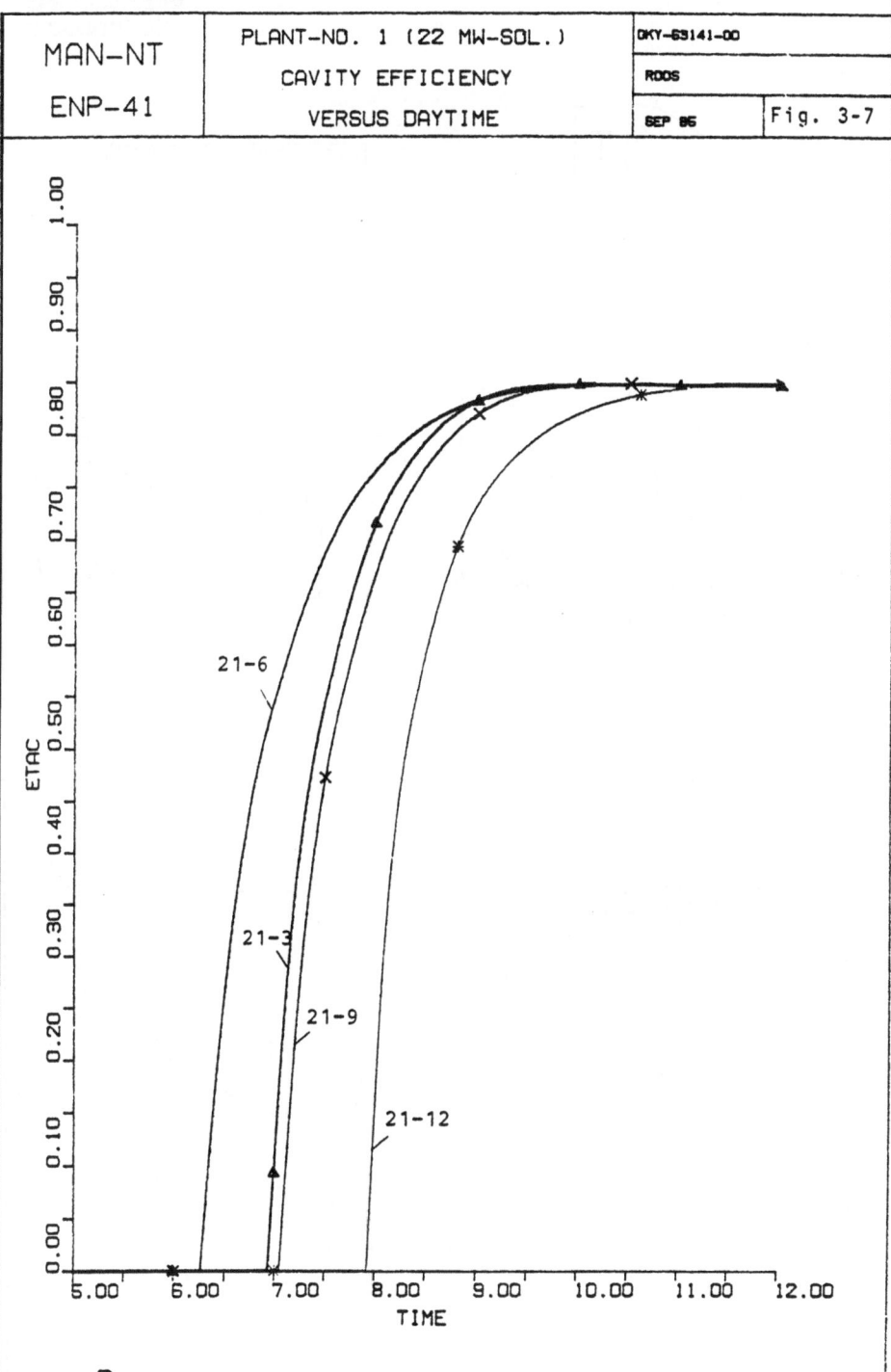

| MAN-NT ENP-41 | PLANT-NO. 1 (22 MW-SOL.) CAVITY EFFICIENCY VERSUS DAYTIME | DKY-63141-00 RODS SEP 86 | Fig. 3-7 |

BASMAN®

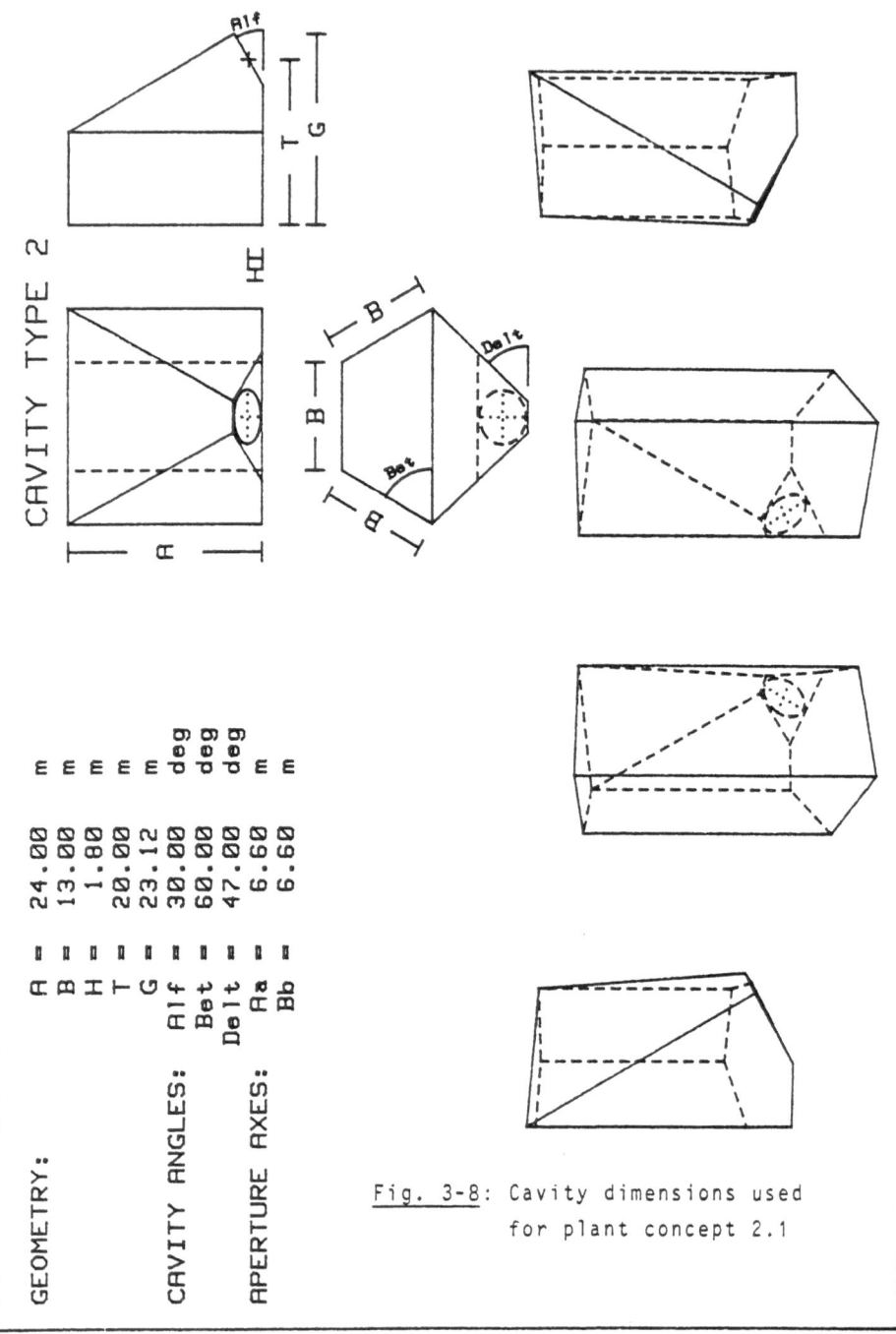

CAVITY TYPE 2

GEOMETRY:

A	=	24.00	m
B	=	13.00	m
H	=	1.80	m
T	=	20.00	m
G	=	23.12	m

CAVITY ANGLES:

Alf	=	30.00	deg
Bet	=	60.00	deg
Delt	=	47.00	deg

APERTURE AXES:

Aa	=	6.60	m
Bb	=	6.60	m

Fig. 3-8: Cavity dimensions used
for plant concept 2.1

- 364 -

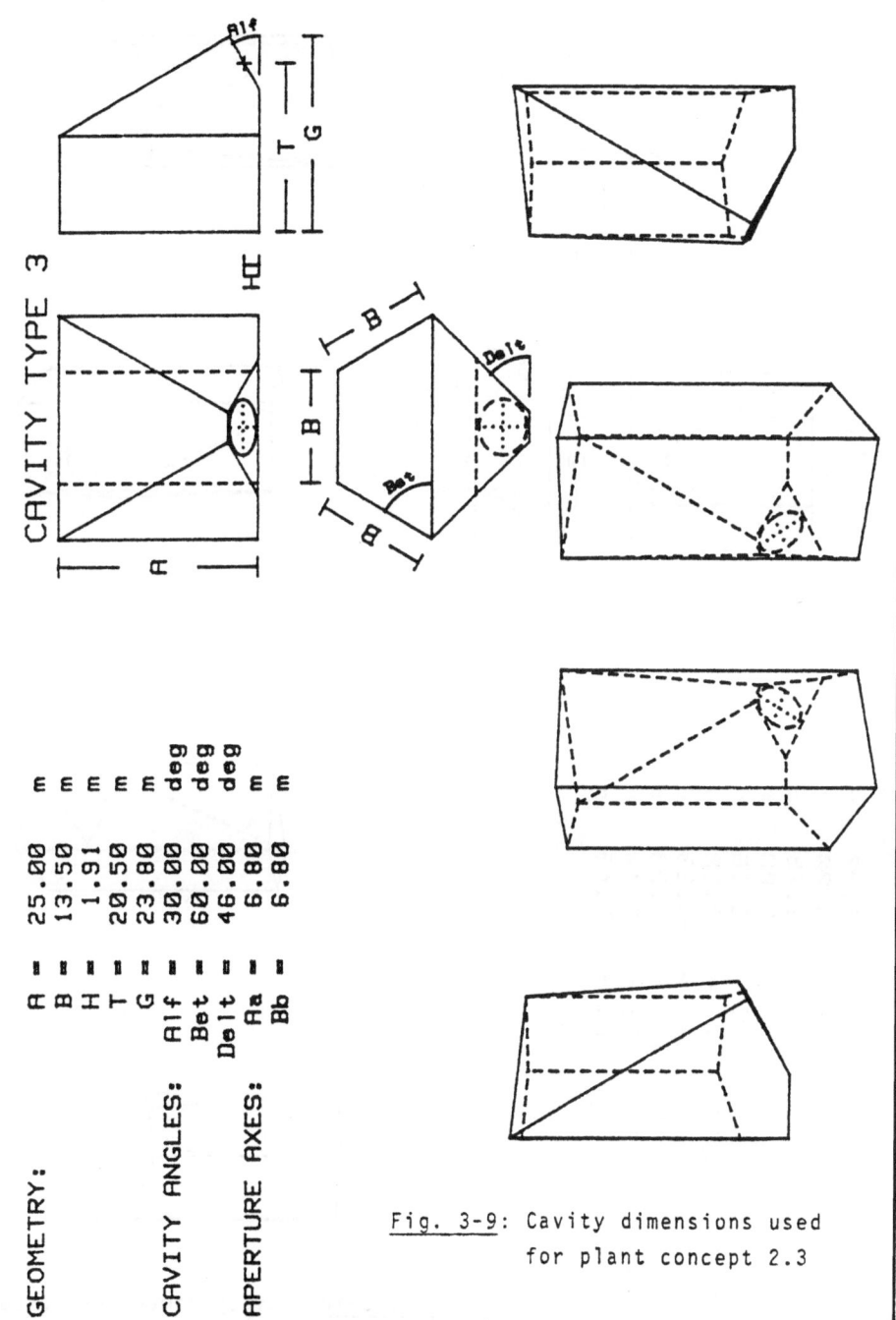

CAVITY TYPE 3

GEOMETRY:

A   = 25.00   m
B   = 13.50   m
H   = 1.91   m
T   = 20.50   m
G   = 23.80   m

CAVITY ANGLES:

Alf = 30.00   deg
Bet = 60.00   deg
Delt = 46.00 deg

APERTURE AXES:

Aa = 6.80   m
Bb = 6.80   m

Fig. 3-9: Cavity dimensions used
for plant concept 2.3

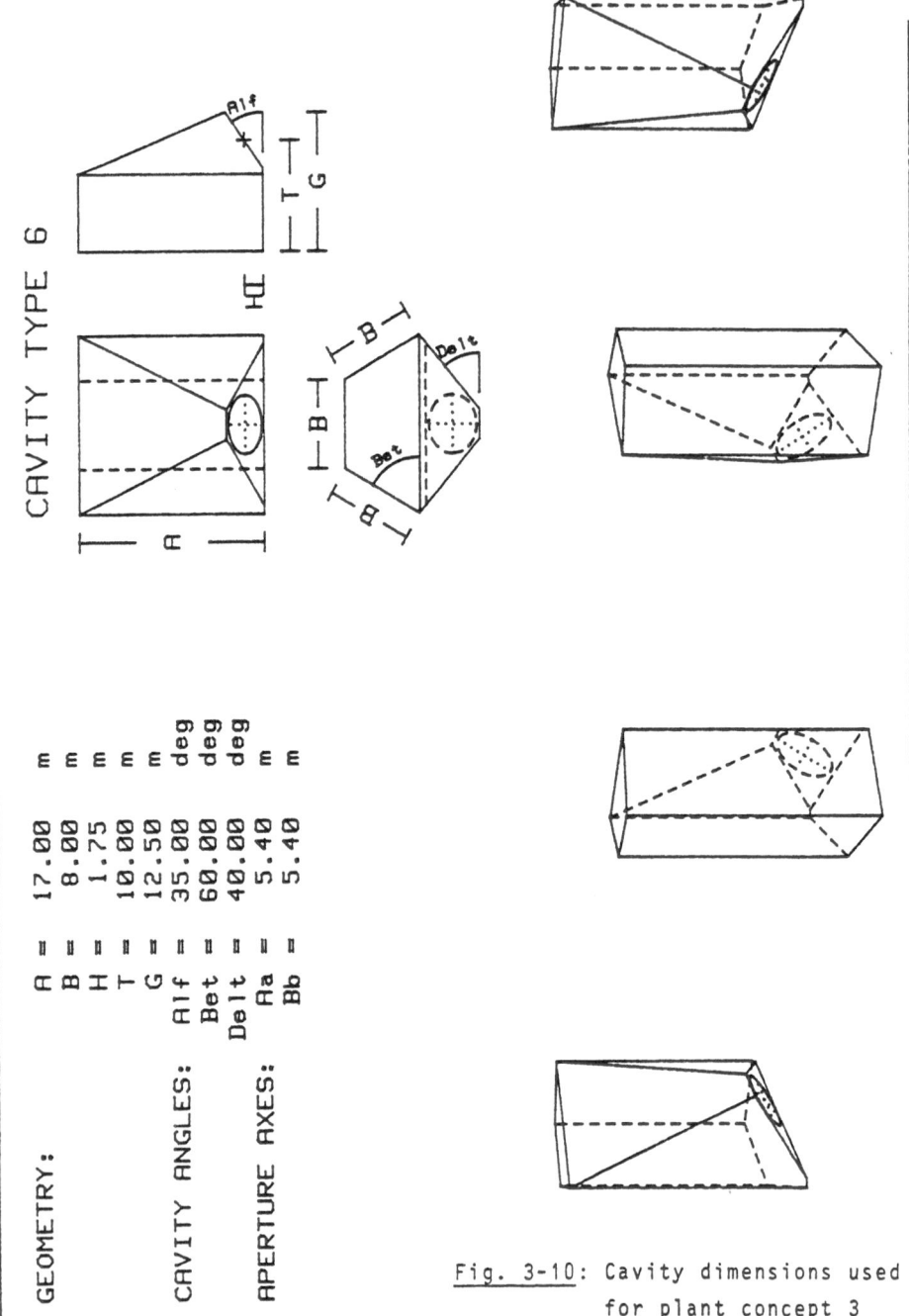

CAVITY TYPE 6

GEOMETRY:
A	=	17.00	m
B	=	8.00	m
H	=	1.75	m
T	=	10.00	m
G	=	12.50	m

CAVITY ANGLES:
Alf	=	35.00	deg
Bet	=	60.00	deg
Delt	=	40.00	deg

APERTURE AXES:
Aa	=	5.40	m
Bb	=	5.40	m

Fig. 3-10: Cavity dimensions used for plant concept 3

- 366 -

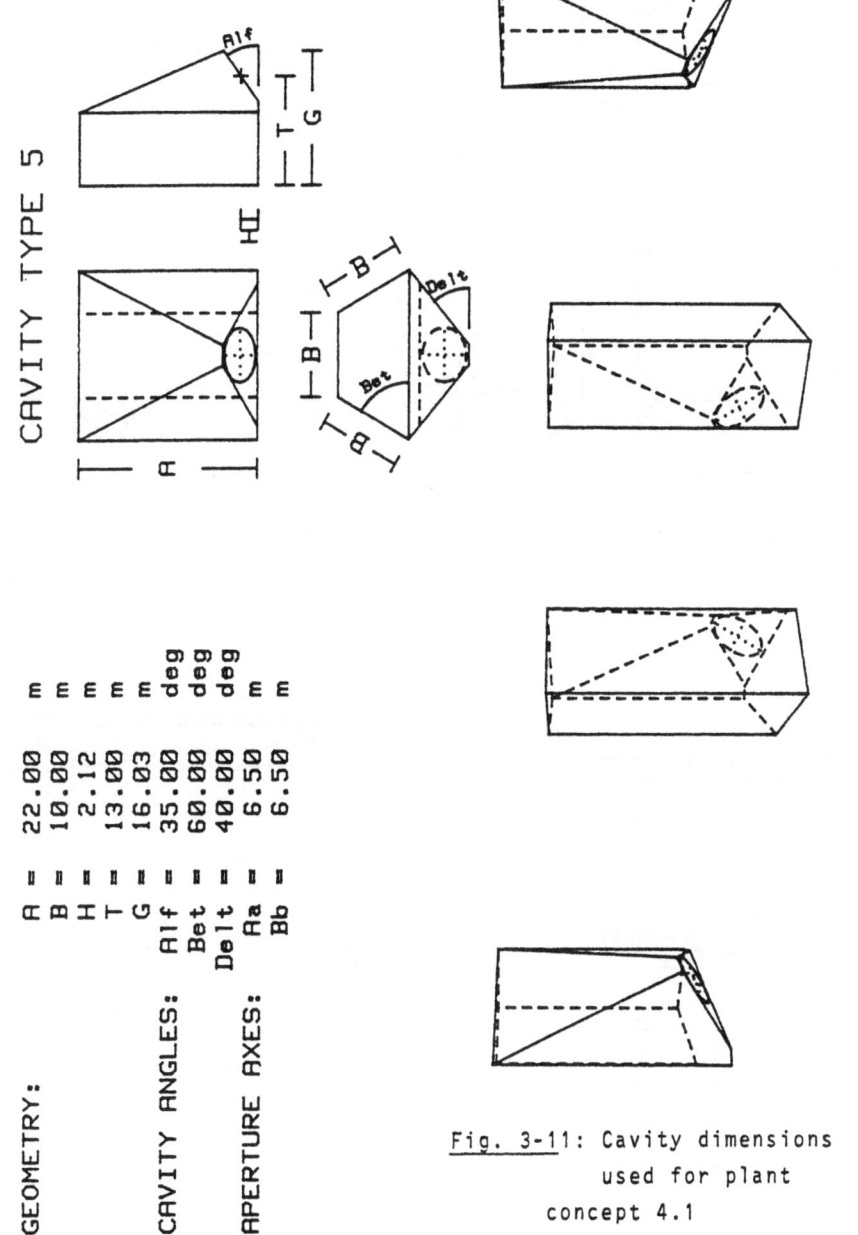

CAVITY TYPE 5

GEOMETRY:

A	=	22.00	m
B	=	10.00	m
H	=	2.12	m
T	=	13.00	m
G	=	16.03	m

CAVITY ANGLES:

Alf	=	35.00	deg
Bet	=	60.00	deg
Delt	=	40.00	deg

APERTURE AXES:

Aa	=	6.50	m
Bb	=	6.50	m

Fig. 3-11: Cavity dimensions
used for plant
concept 4.1

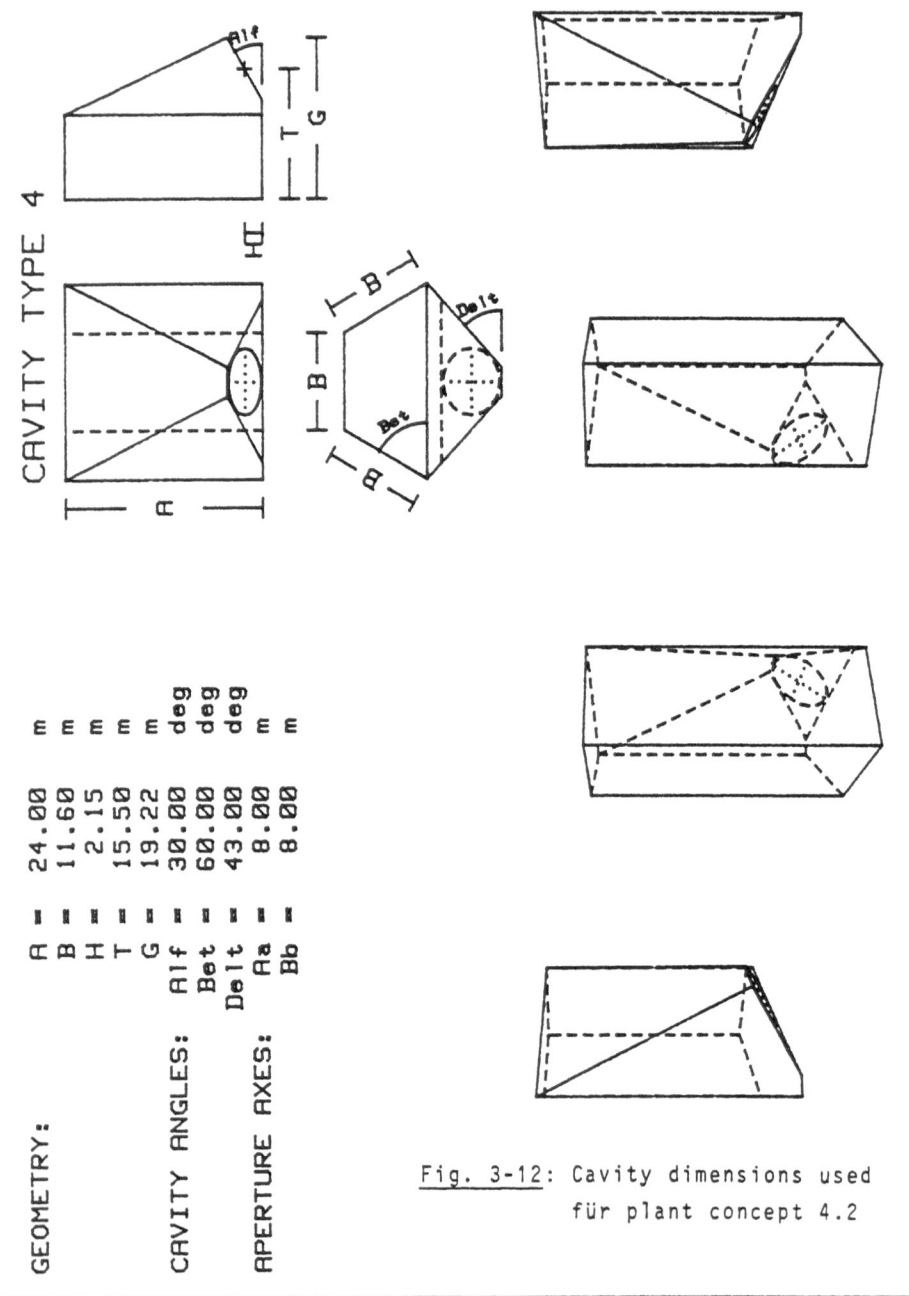

CAVITY TYPE 4

GEOMETRY:

A	=	24.00	m
B	=	11.60	m
H	=	2.15	m
T	=	15.50	m
G	=	19.22	m

CAVITY ANGLES:

Alf	=	30.00	deg
Bet	=	60.00	deg
Delt	=	43.00	deg

APERTURE AXES:

Aa	=	8.00	m
Bb	=	8.00	m

<u>Fig. 3-12</u>: Cavity dimensions used
für plant concept 4.2

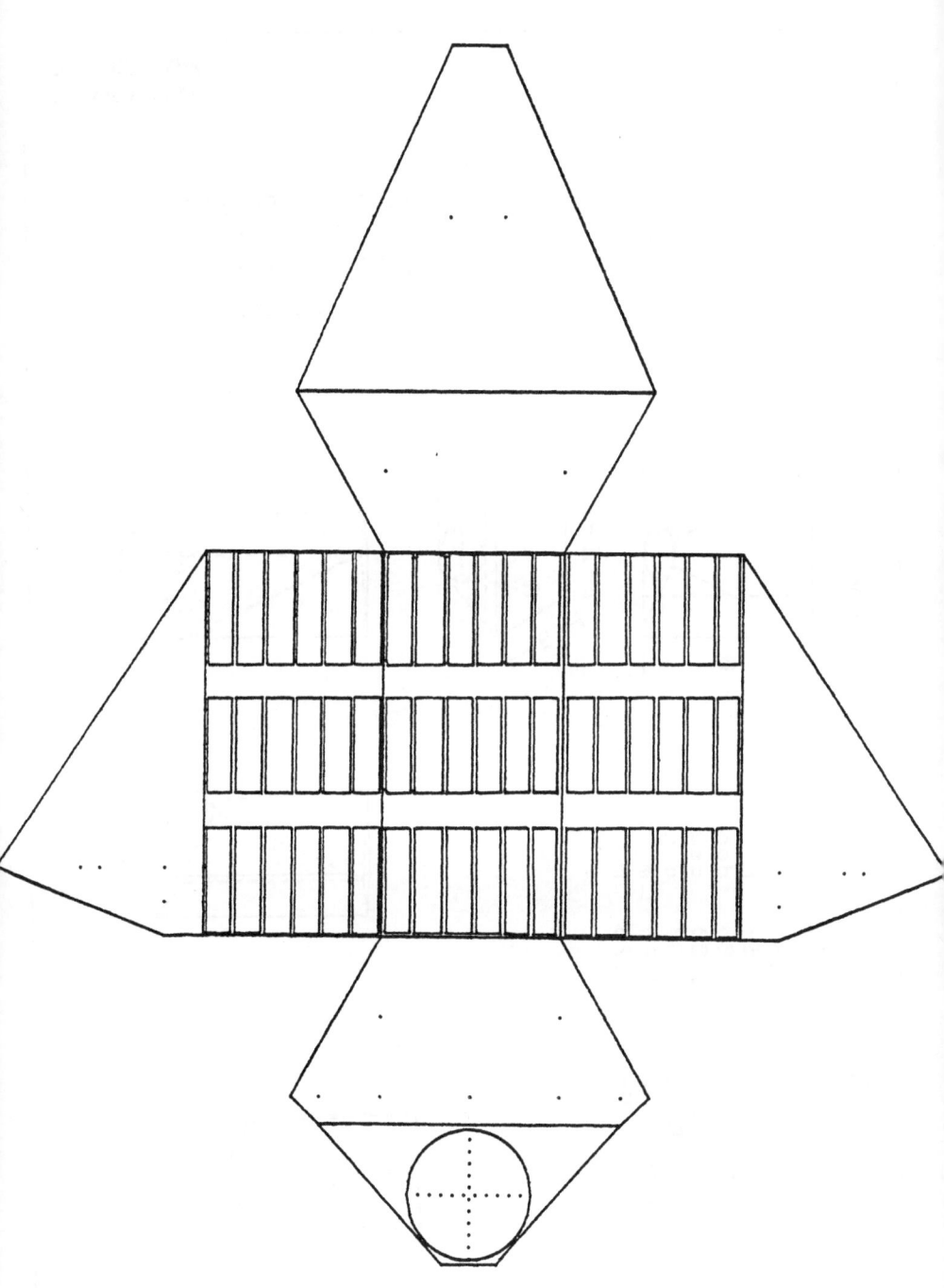

<u>Fig. 3-13:</u>   Arrangement of panels in the cavity of plant 4.2

- 369 -

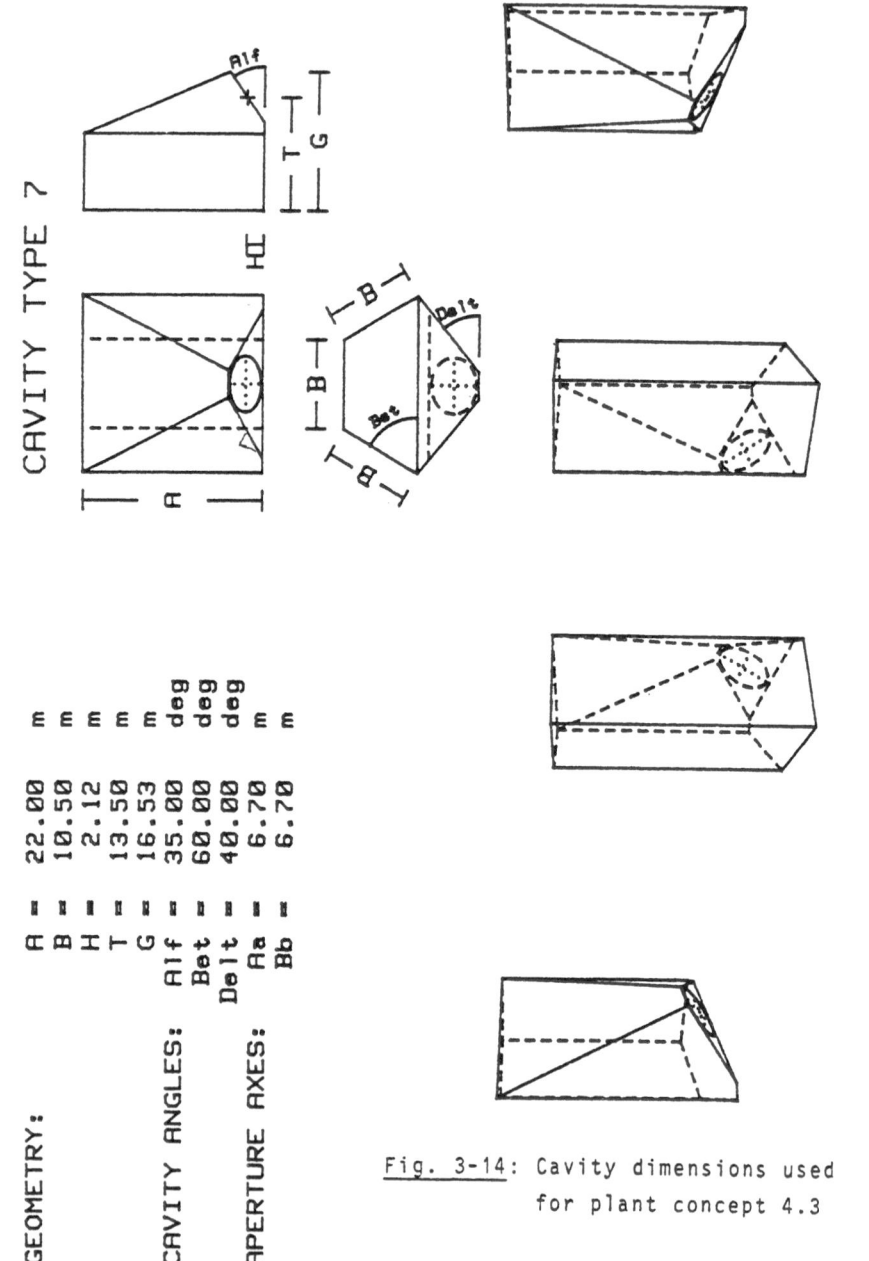

CAVITY TYPE 7

GEOMETRY:

A	=	22.00	m
B	=	10.50	m
H	=	2.12	m
T	=	13.50	m
G	=	16.53	m

CAVITY ANGLES:

Alf	=	35.00	deg
Bet	=	60.00	deg
Delt	=	40.00	deg

APERTURE AXES:

Aa	=	6.70	m
Bb	=	6.70	m

Fig. 3-14: Cavity dimensions used
for plant concept 4.3

- 370 -

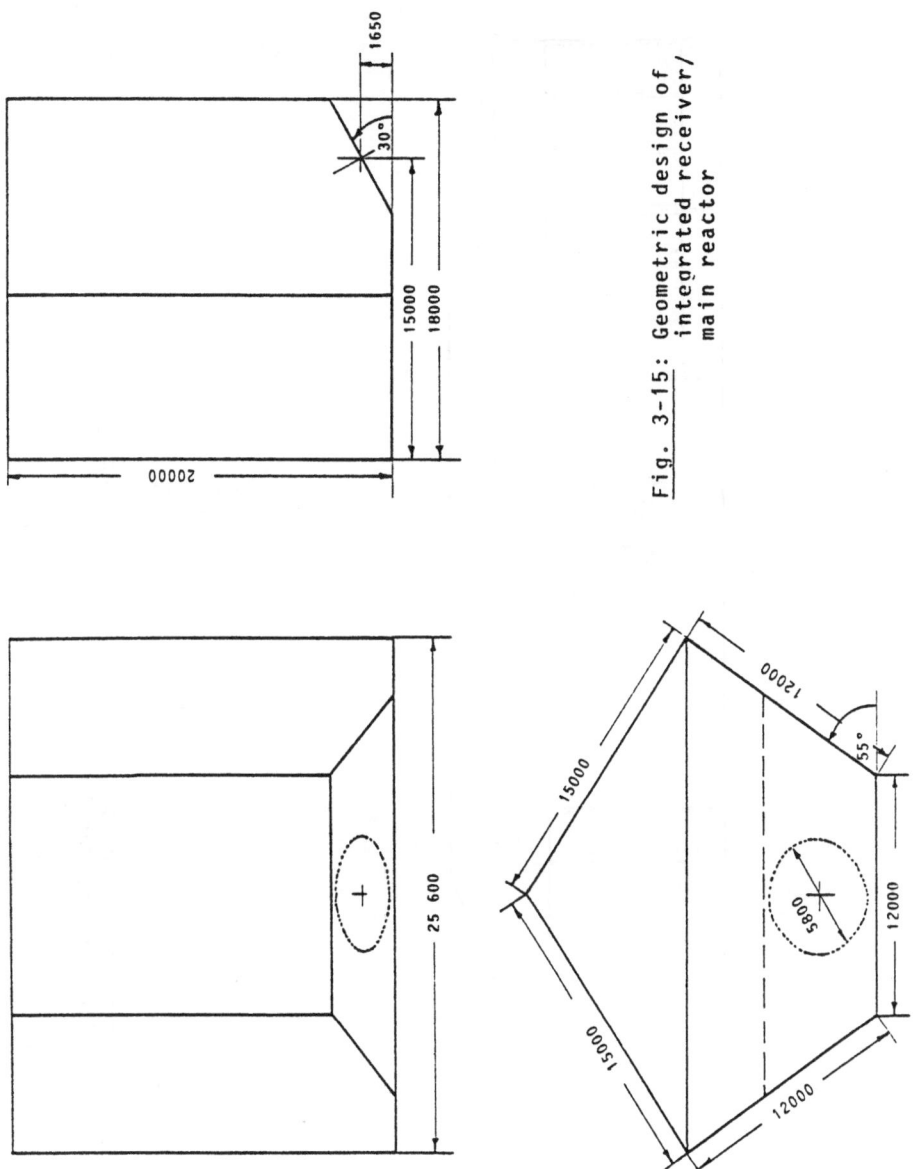

Fig. 3-15: Geometric design of integrated receiver/ main reactor

- 371 -

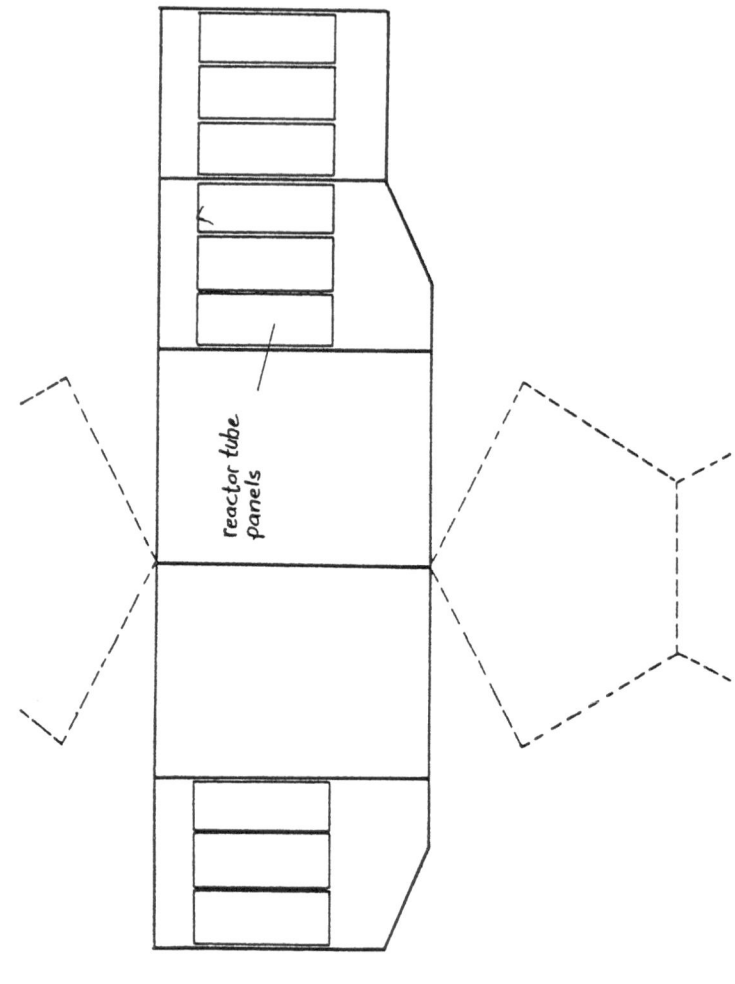

reactor tube panels

Fig. 3-16: Arrangement of Reactor Tube Panels of Concept 6 (Integrated Receiver/Main Reactor)

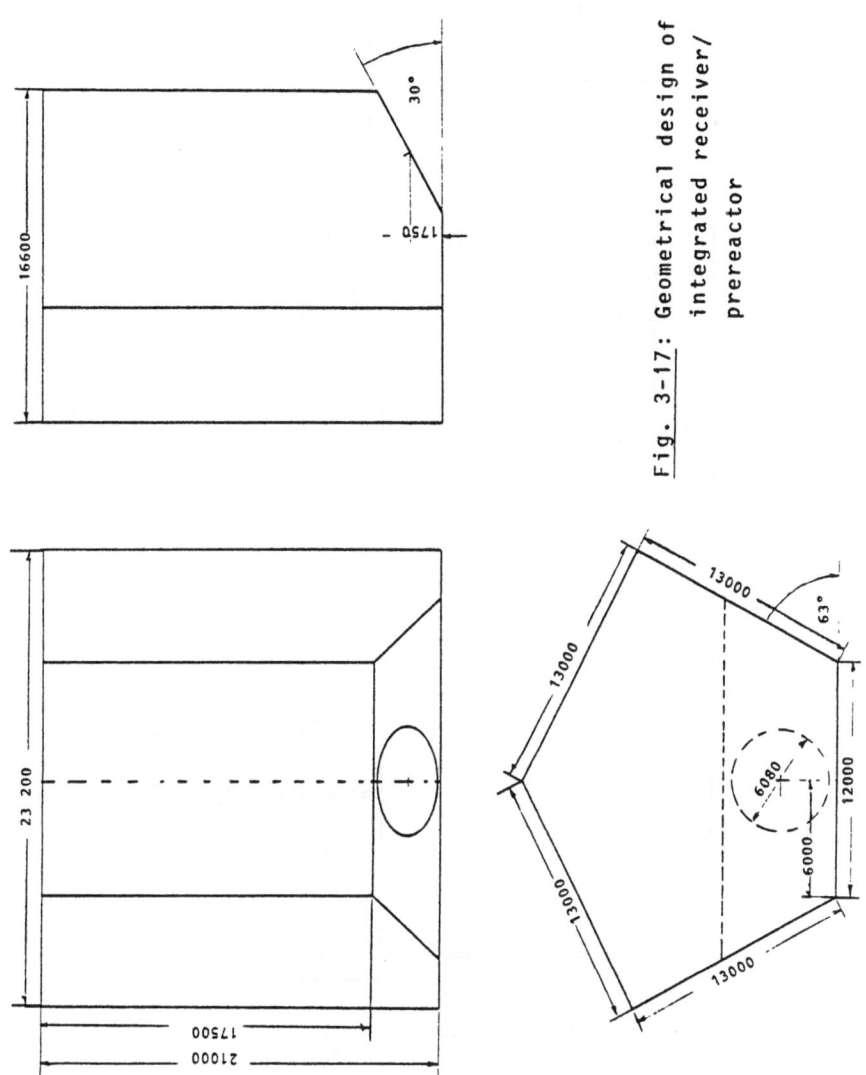

Fig. 3-17: Geometrical design of integrated receiver/ prereactor

- 373 -

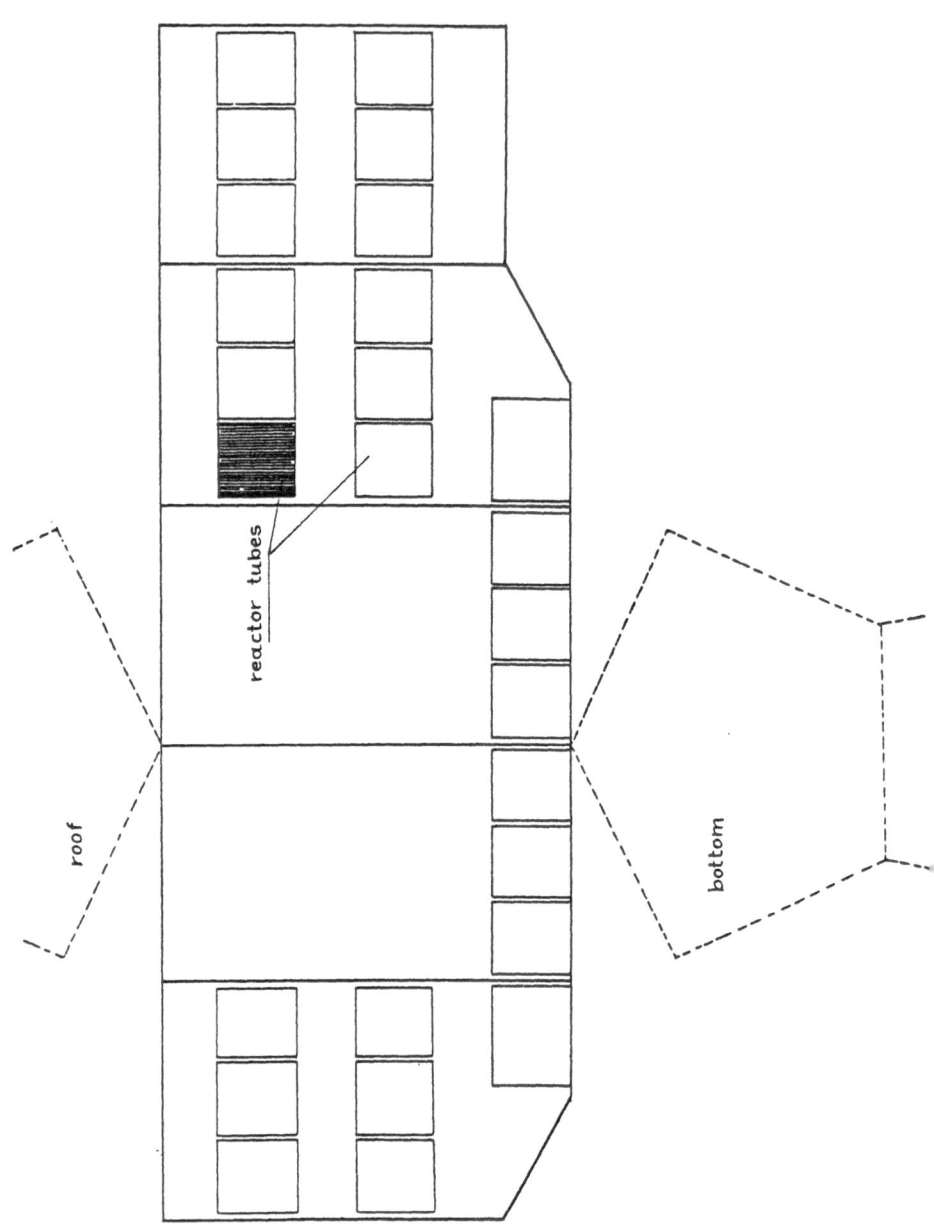

<u>Fig. 3-18</u>:   Arrangement of tubes in the IRC-type receiver (plant concept 5)

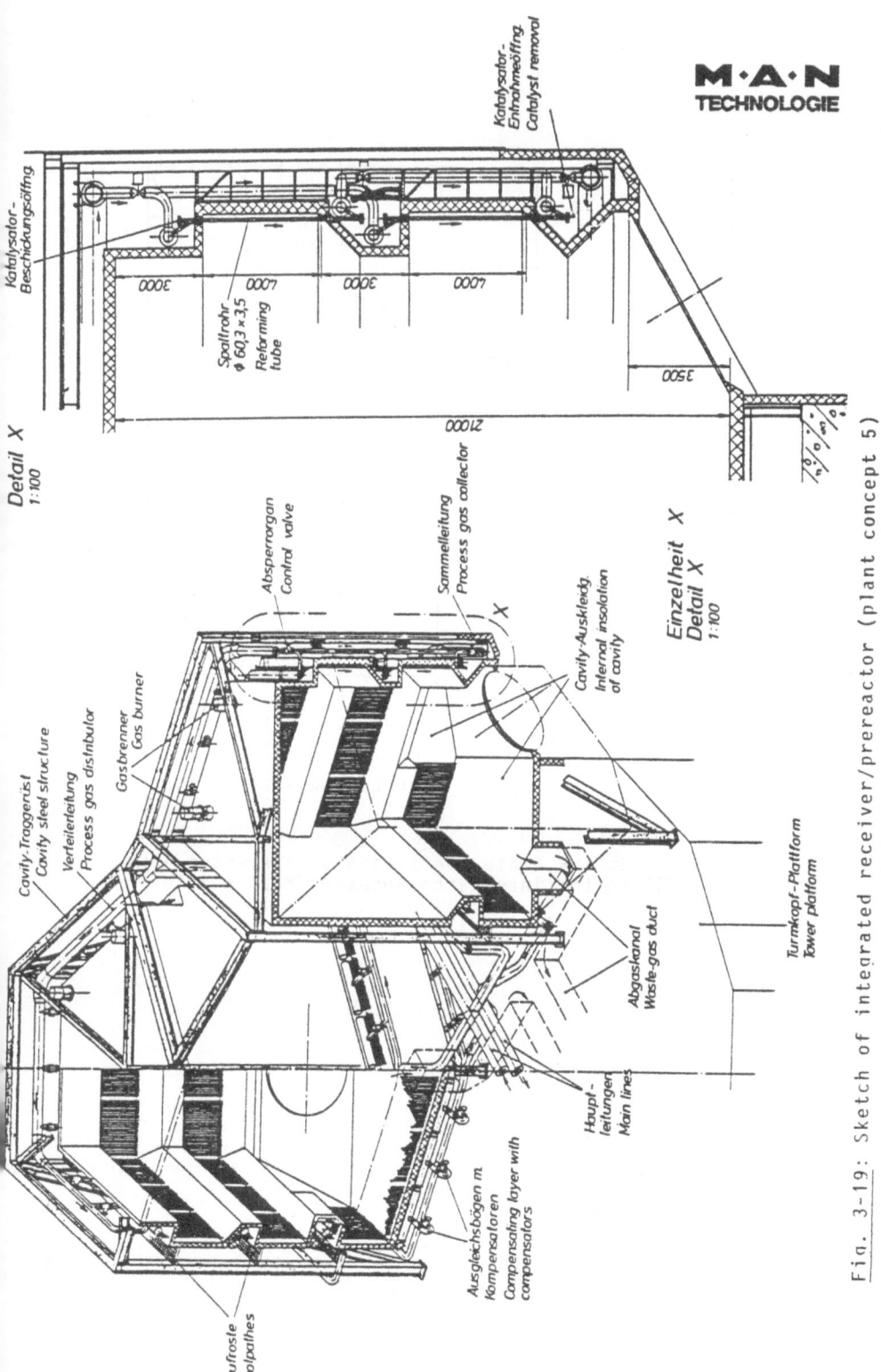

M·A·N
TECHNOLOGIE

Detail X
1:100

Katalysator-
Beschickungsöffng.

Spaltrohr
⌀ 60,3 × 3,5
Reforming
tube

3000 4000 3000 4000 3000

21000

3500

Katalysator-
Entnahmeöffng.
Catalyst removal

Absperrorgan
Control valve

Sammelleitung
Process gas collector

Cavity-Auskleidg.
Internal insolation
of cavity

Einzelheit X
Detail X
1:100

Cavity-Traggerüst
Cavity steel structure

Verteilerleitung
Process gas distributor

Gasbrenner
Gas burner

Abgaskanal
Waste-gas duct

Turmkopf-Plattform
Tower platform

Haupt-
leitungen
Main lines

Ausgleichsbögen m
Kompensatoren
Compensating layer with
compensators

Laufroste
Foolpathes

Fig. 3-19: Sketch of integrated receiver/prereactor (plant concept 5)

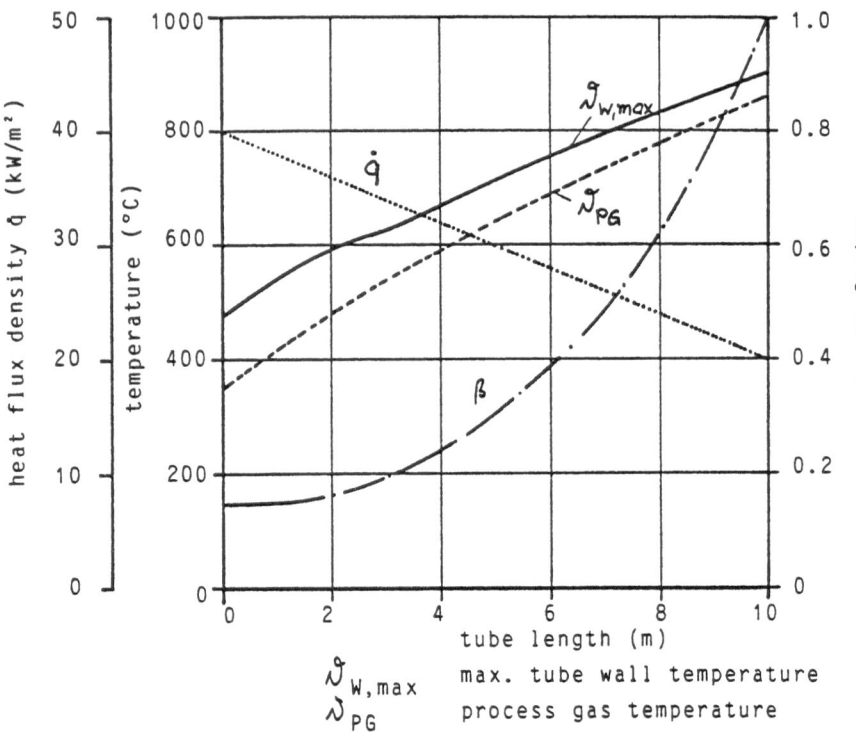

$\vartheta_{W,max}$ max. tube wall temperature
ϑ_{PG} process gas temperature

Fig. 3-20: Steady State Conditions of a Reactor Tube
           in the Integrated Receiver/Main Reactor

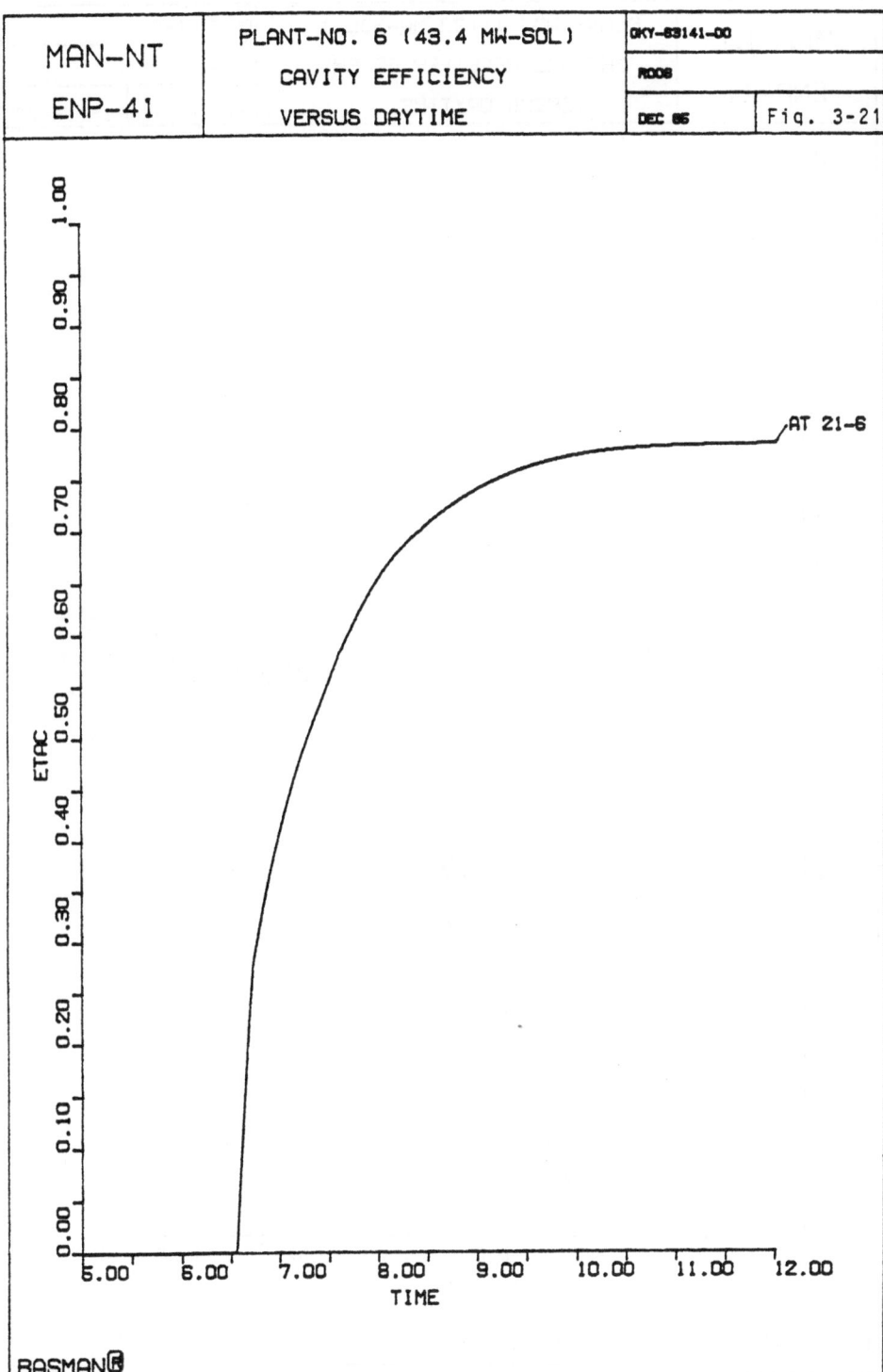

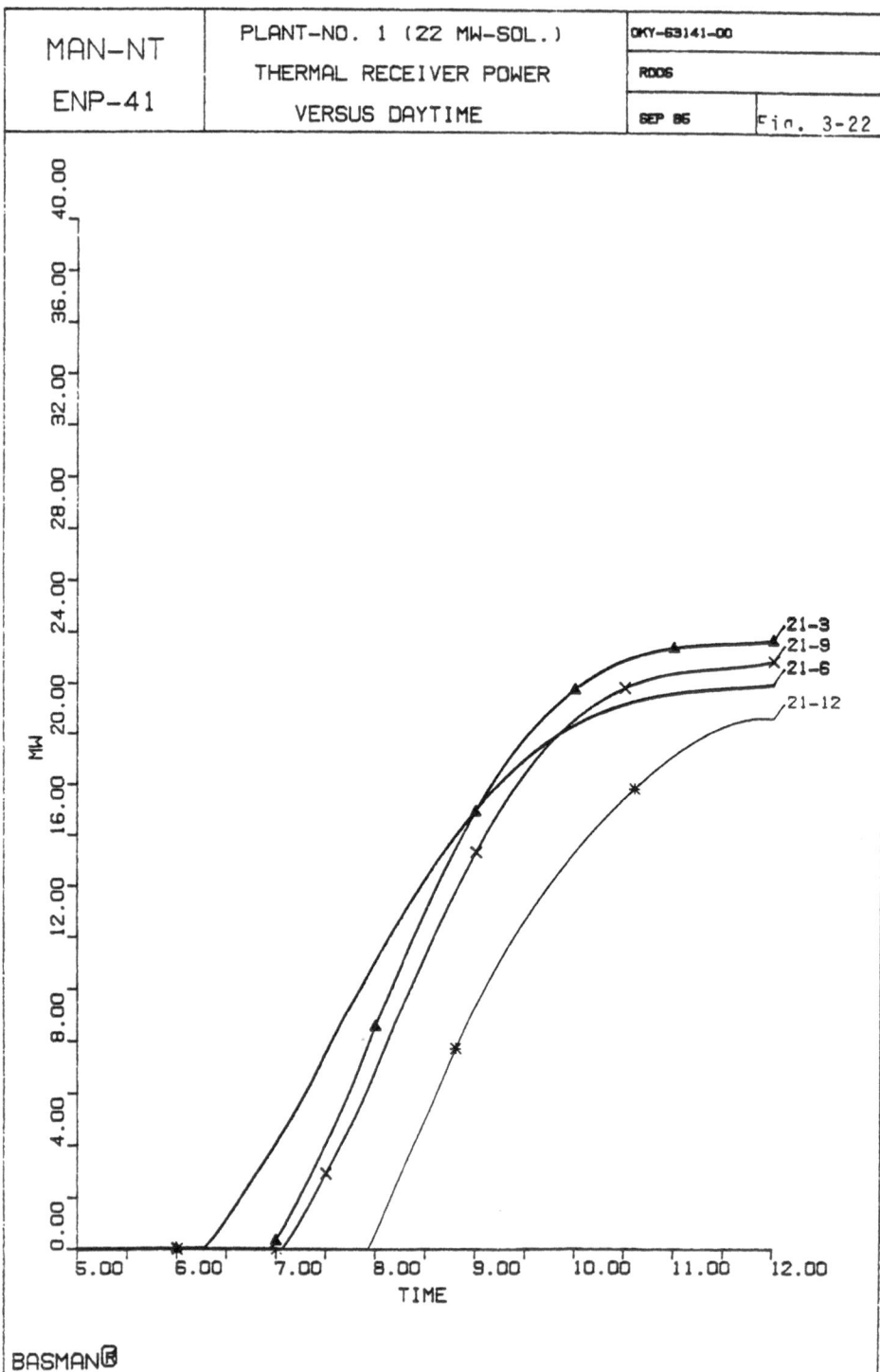

| MAN-NT ENP-41 | PLANT-NO. 1 (22 MW-SOL.) THERMAL RECEIVER POWER VERSUS DAYTIME | OKY-63141-00 RDOS SEP 86 Fig. 3-22 |

BASMAN®

- 378 -

| MAN–NT ENP-41 | PLANT-NO. 1 (22 MW-SOL.) ENERGY BALANCE FOR THE 21TH OF MARCH | OKY–63141–00 ROOS SEP 86 | Fig. 3-23 |

SOL.INPUT (338 MWH)

SOL.OUTPUT (166 MWH)

PROCESS (516 MWH)

MW

TIME

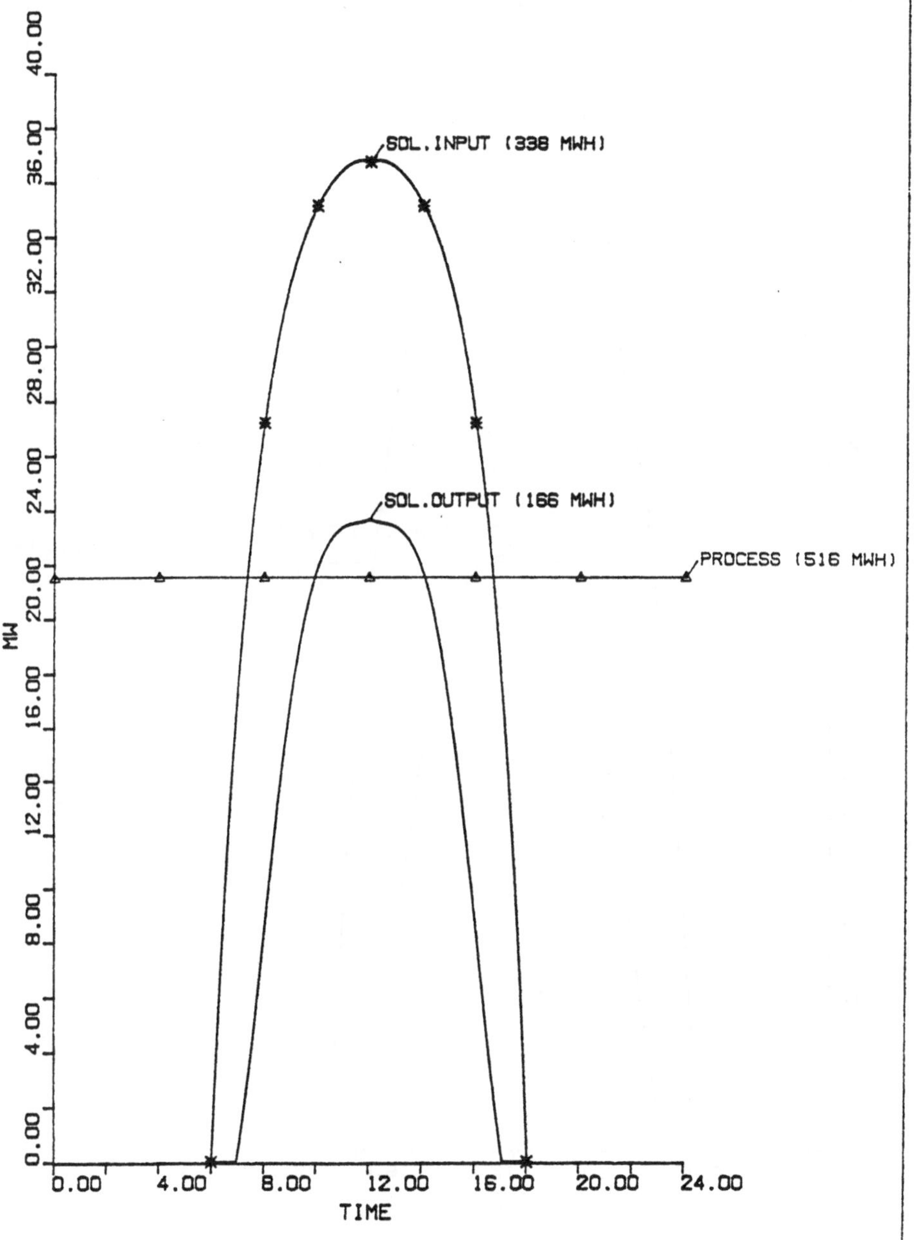

BASMAN®

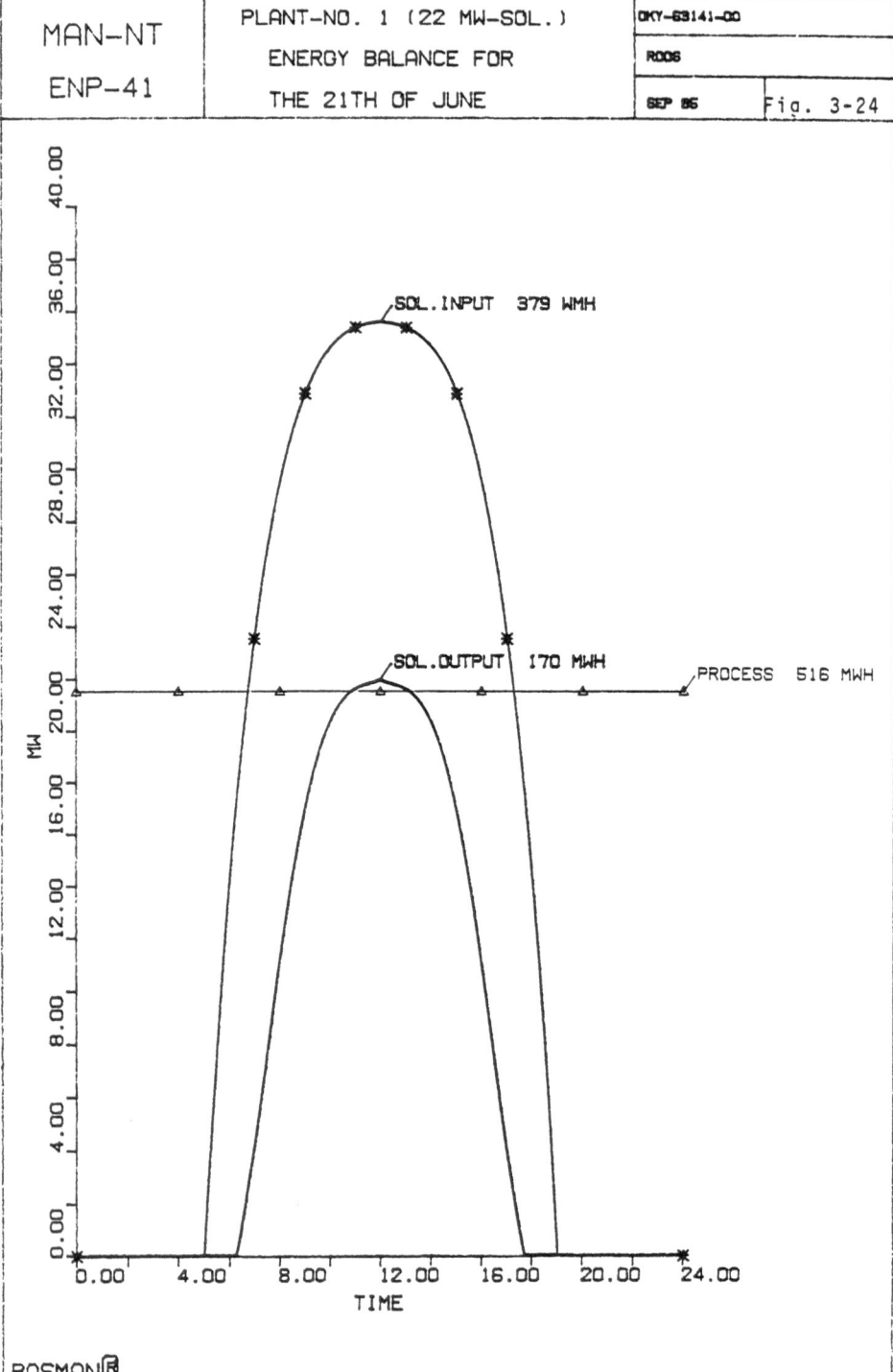

SOL.INPUT   379 WMH

SOL.OUTPUT   170 MWH

PROCESS   516 MWH

MW

TIME

BASMAN®

- 380 -

SOL.INPUT (310 MWH)

SOL.OUTPUT (154 MWH)

PROCESS (516 MWH)

BASMAN®

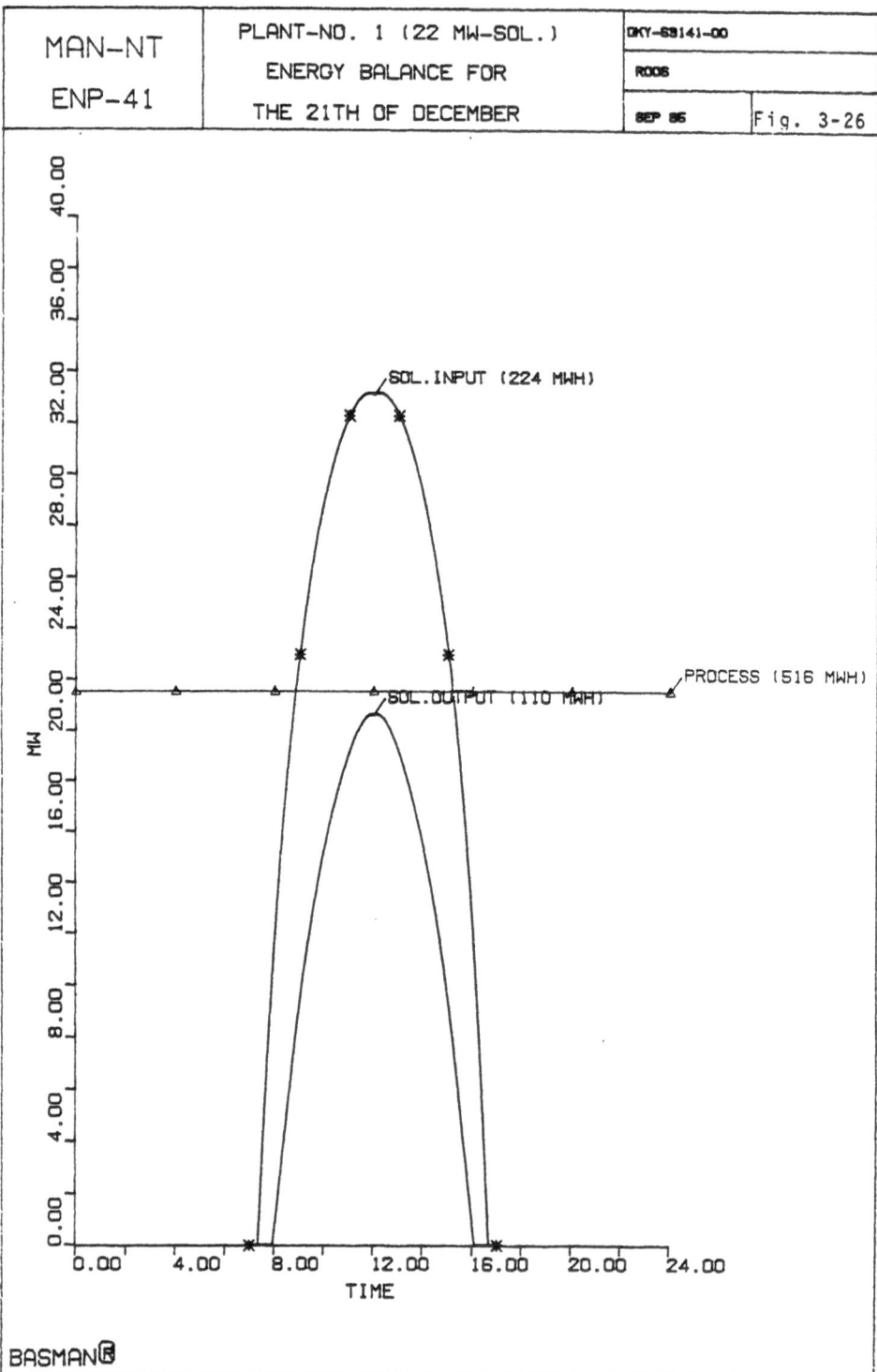

BASMAN®

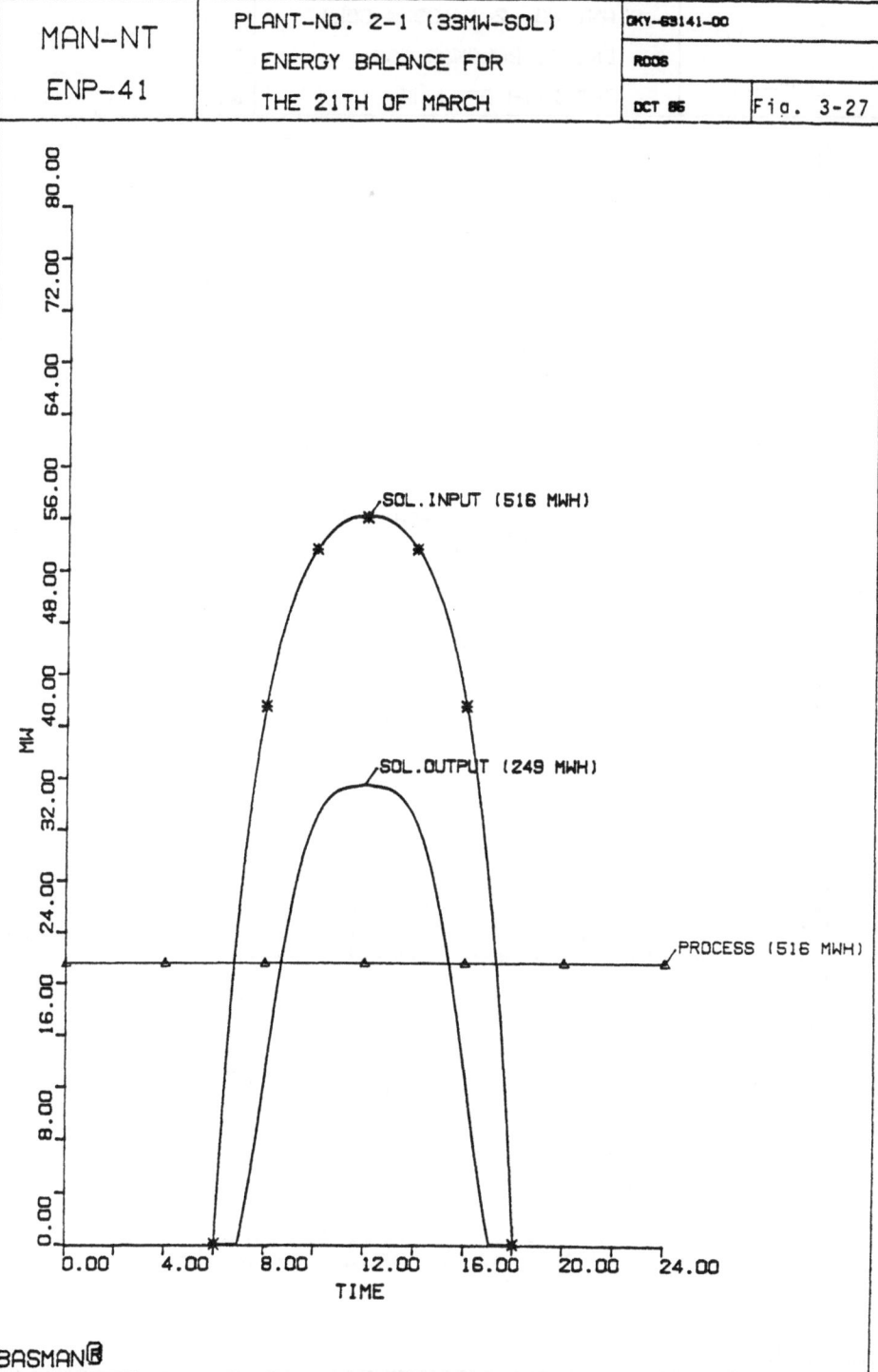

MAN-NT	PLANT-NO. 2-1 (33MW-SOL)	DKY-63141-00	
	ENERGY BALANCE FOR	ROOS	
ENP-41	THE 21TH OF MARCH	OCT 86	Fig. 3-27

SOL.INPUT (516 MWH)

SOL.OUTPUT (249 MWH)

PROCESS (516 MWH)

MW

TIME

BASMAN⑮

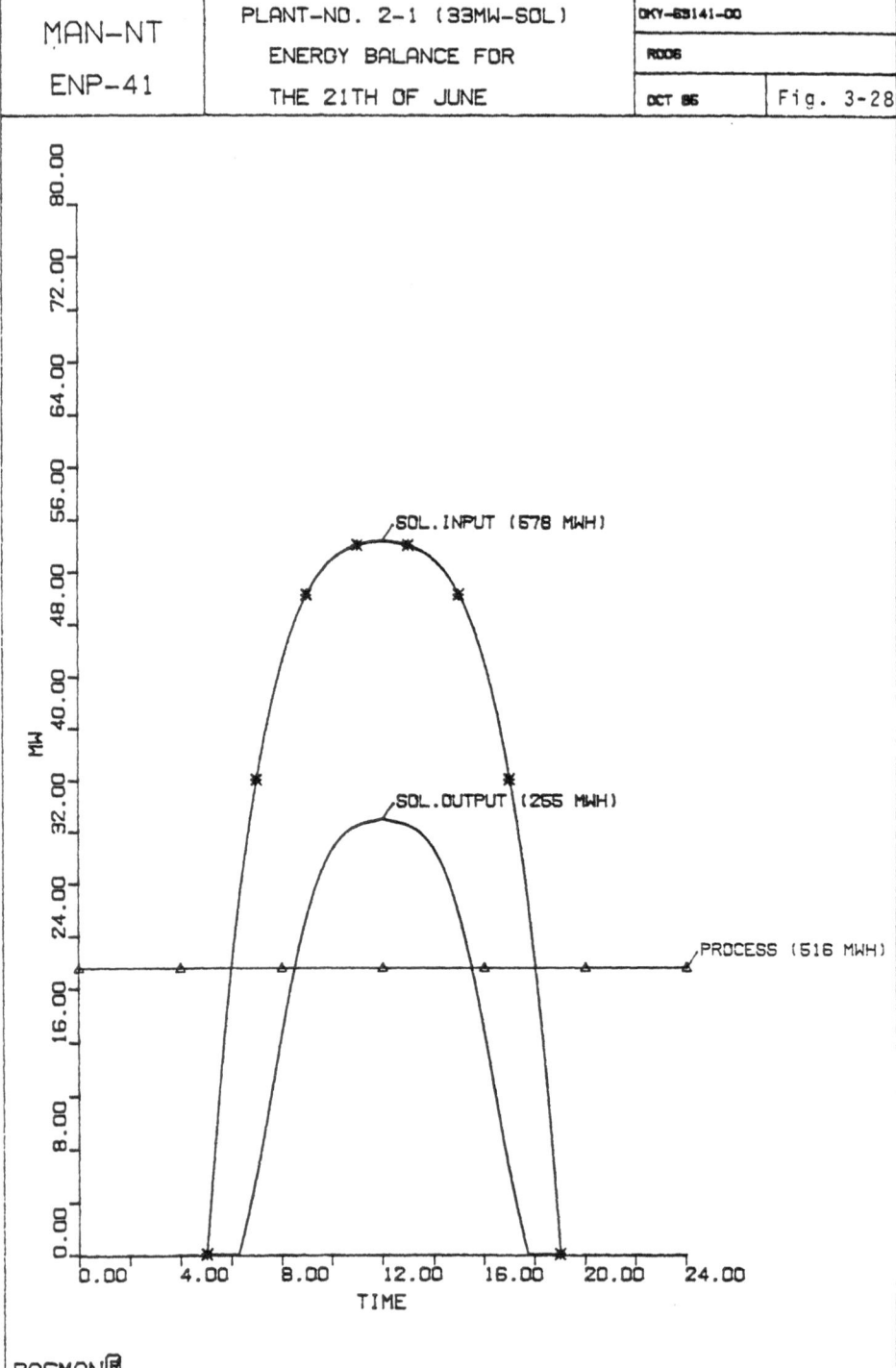

BASMAN®

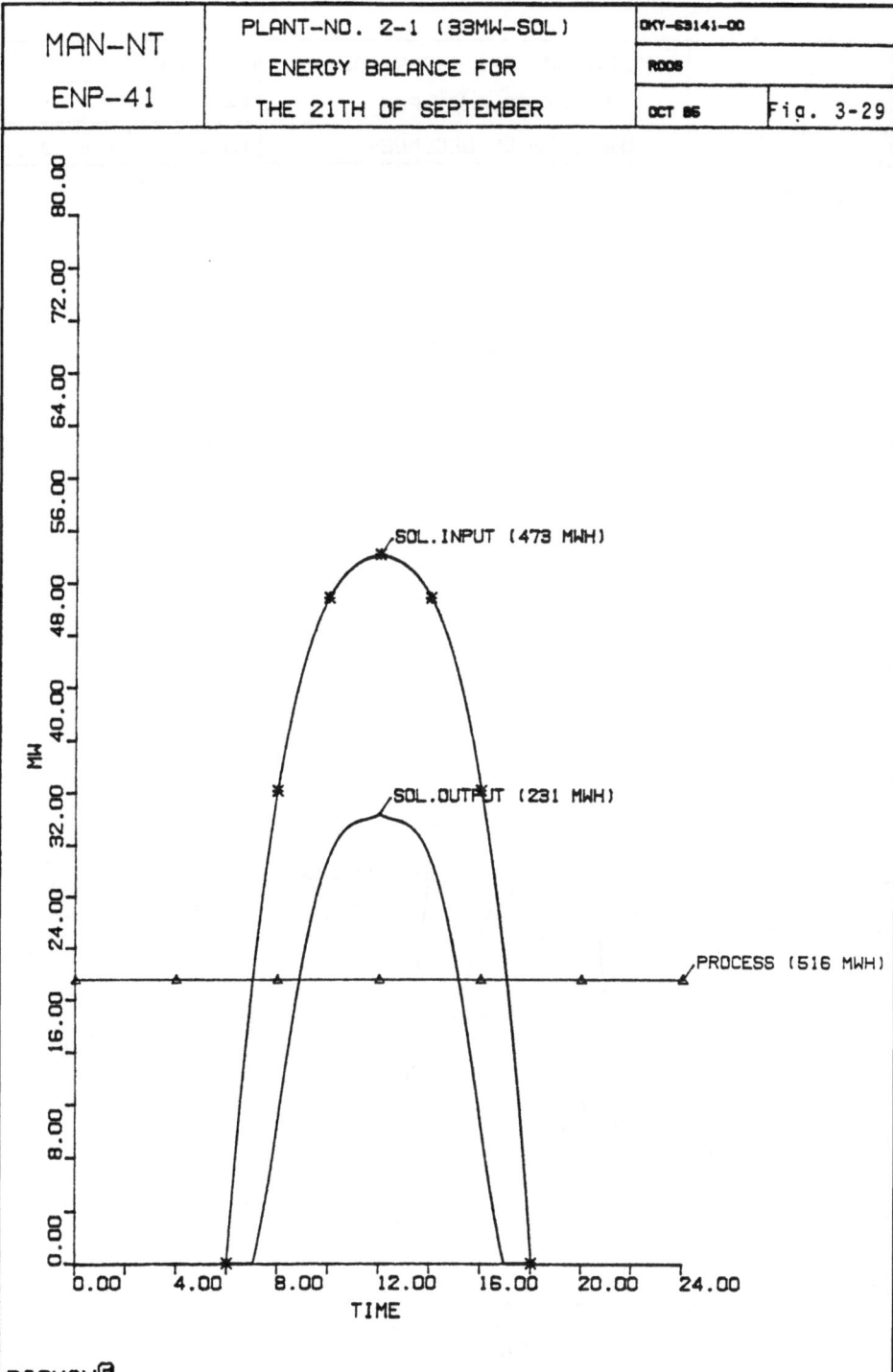

MAN–NT	PLANT–NO. 2–1 (33MW–SOL)	DKY–63141–00	
ENP–41	ENERGY BALANCE FOR	ROOS	
	THE 21TH OF SEPTEMBER	OCT 86	Fig. 3-29

SOL.INPUT (473 MWH)

SOL.OUTPUT (231 MWH)

PROCESS (516 MWH)

MW

TIME

BASMAN®

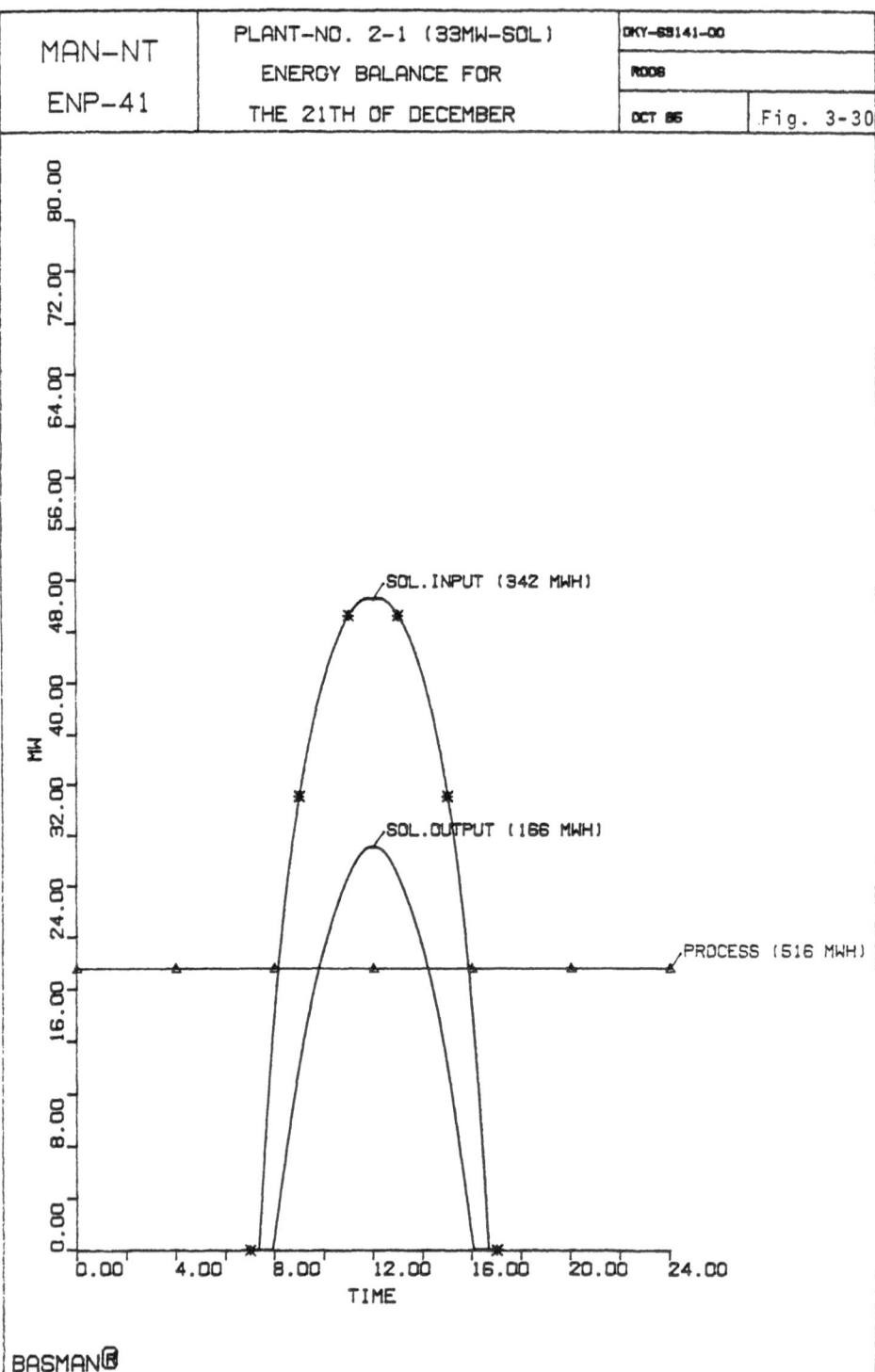

SOL.INPUT (342 MWH)

SOL.OUTPUT (166 MWH)

PROCESS (516 MWH)

MW

TIME

BASMAN®

MAN—NT ENP—41	PLANT-NO. 2-2 (44MW-SOL) ENERGY BALANCE FOR THE 21TH OF MARCH	DKY-63141-00	
		ROOS	
		OCT 85	Fig. 3-31

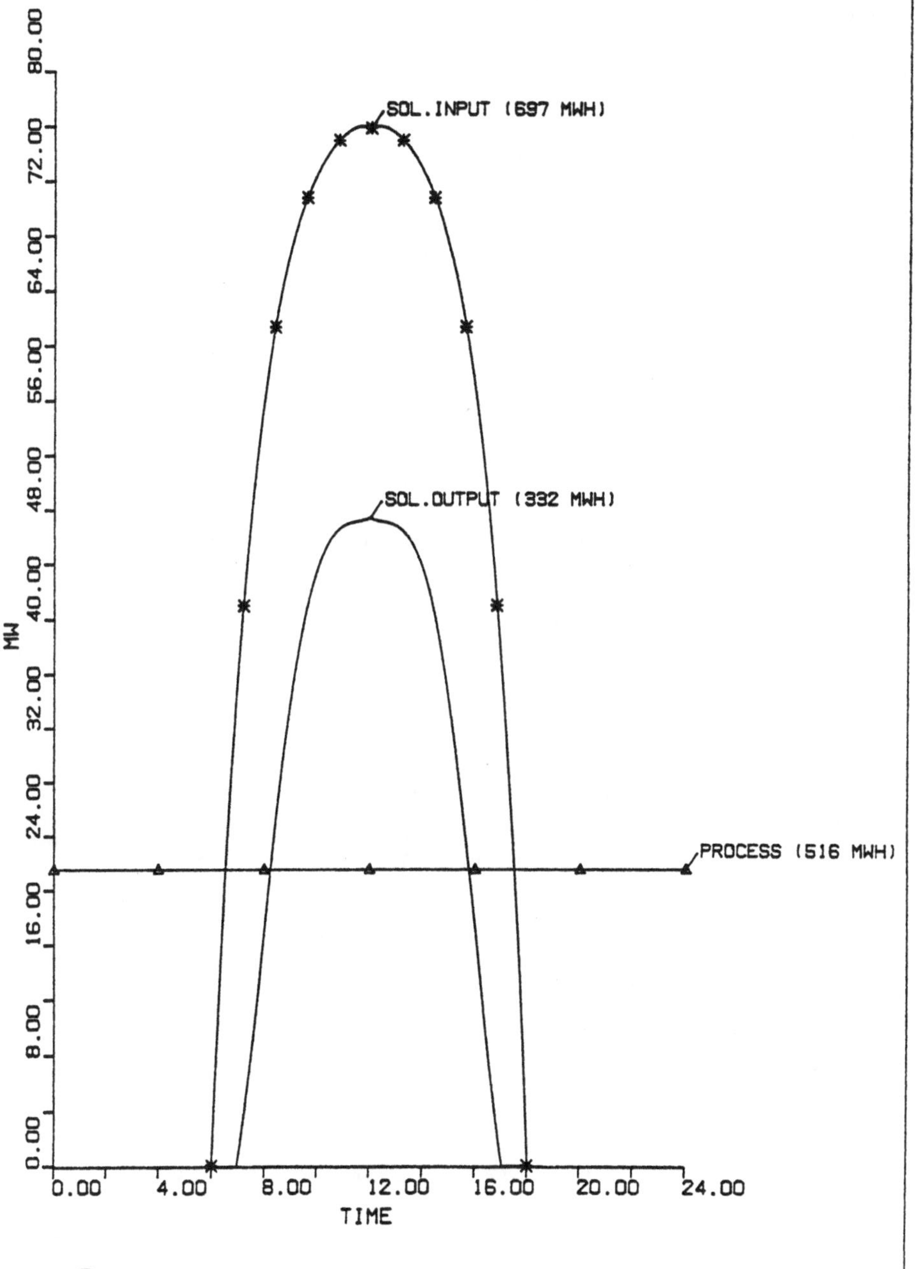

SOL.INPUT (697 MWH)

SOL.OUTPUT (332 MWH)

PROCESS (516 MWH)

MW

TIME

BASMAN℞

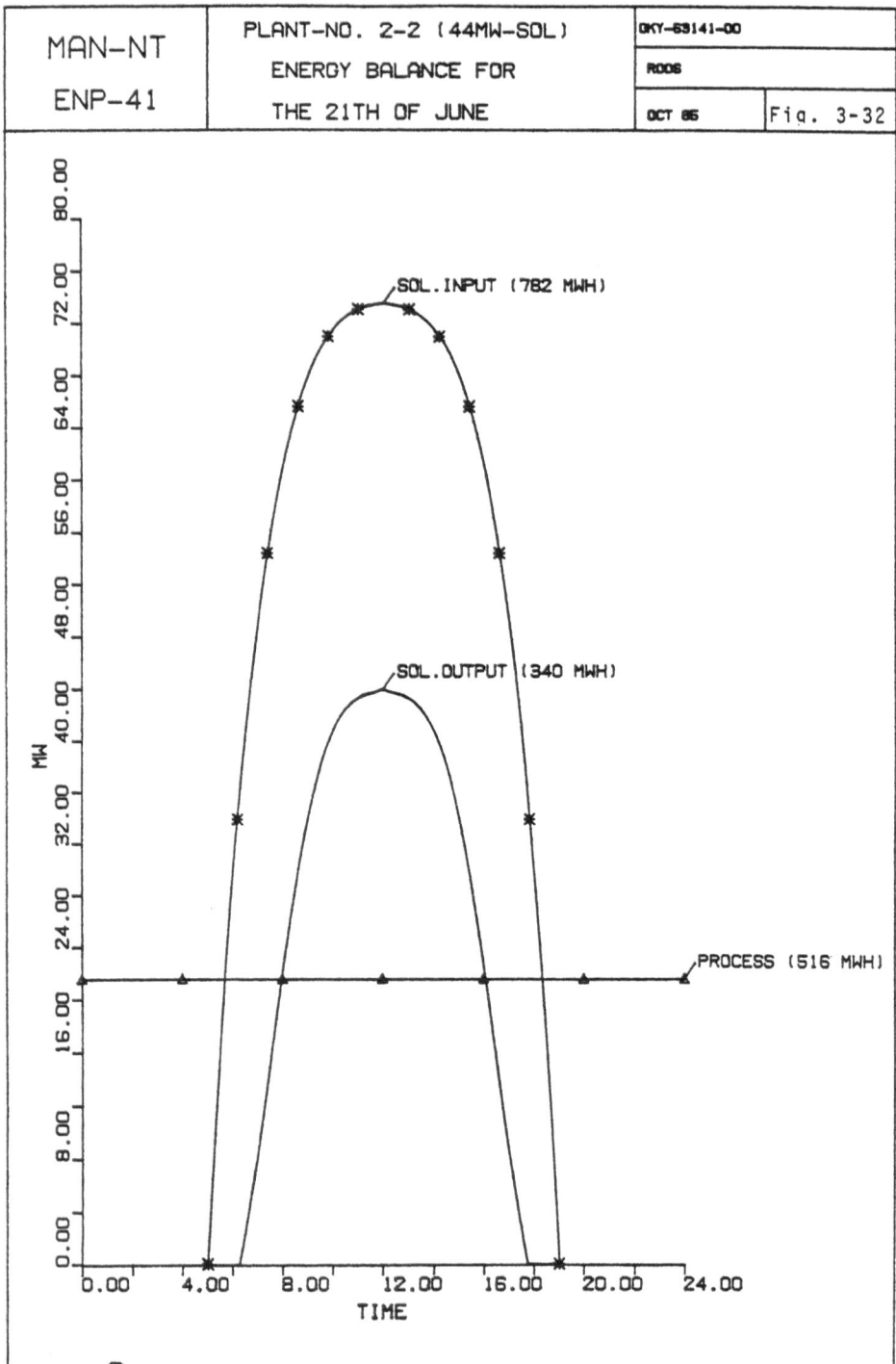

MAN-NT ENP-41	PLANT-NO. 2-2 (44MW-SOL) ENERGY BALANCE FOR THE 21TH OF JUNE	OKY-63141-00
		RODS
		OCT 85 — Fig. 3-32

SOL.INPUT (782 MWH)

SOL.OUTPUT (340 MWH)

PROCESS (516 MWH)

MW

TIME

BASMAN®

- 388 -

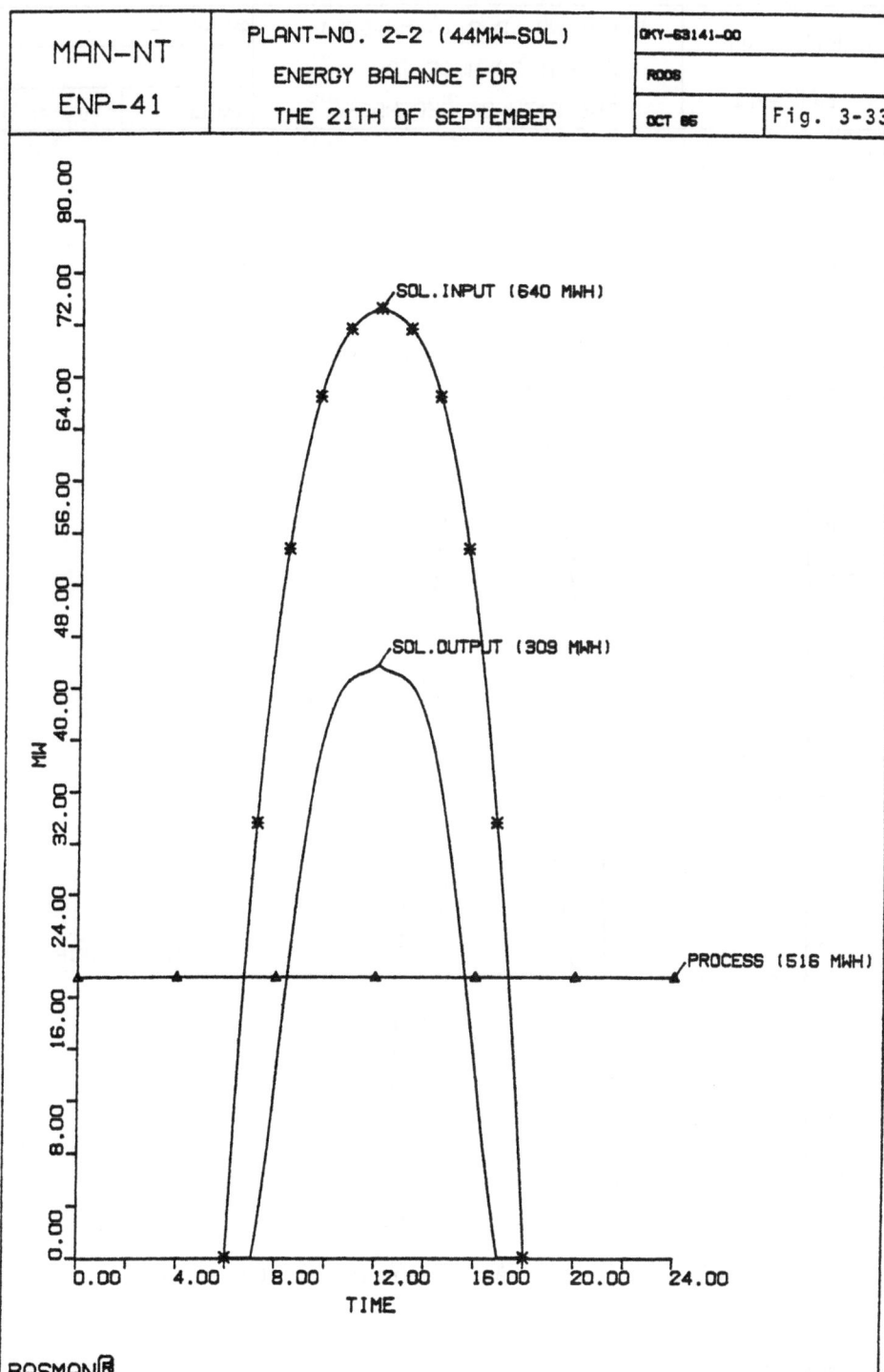

MAN–NT	PLANT–NO. 2–2 (44MW–SOL)	OKY–53141–00	
	ENERGY BALANCE FOR	ROOS	
ENP–41	THE 21TH OF SEPTEMBER	OCT 85	Fig. 3-33

SOL.INPUT (640 MWH)

SOL.OUTPUT (309 MWH)

PROCESS (516 MWH)

MW

TIME

BASMAN

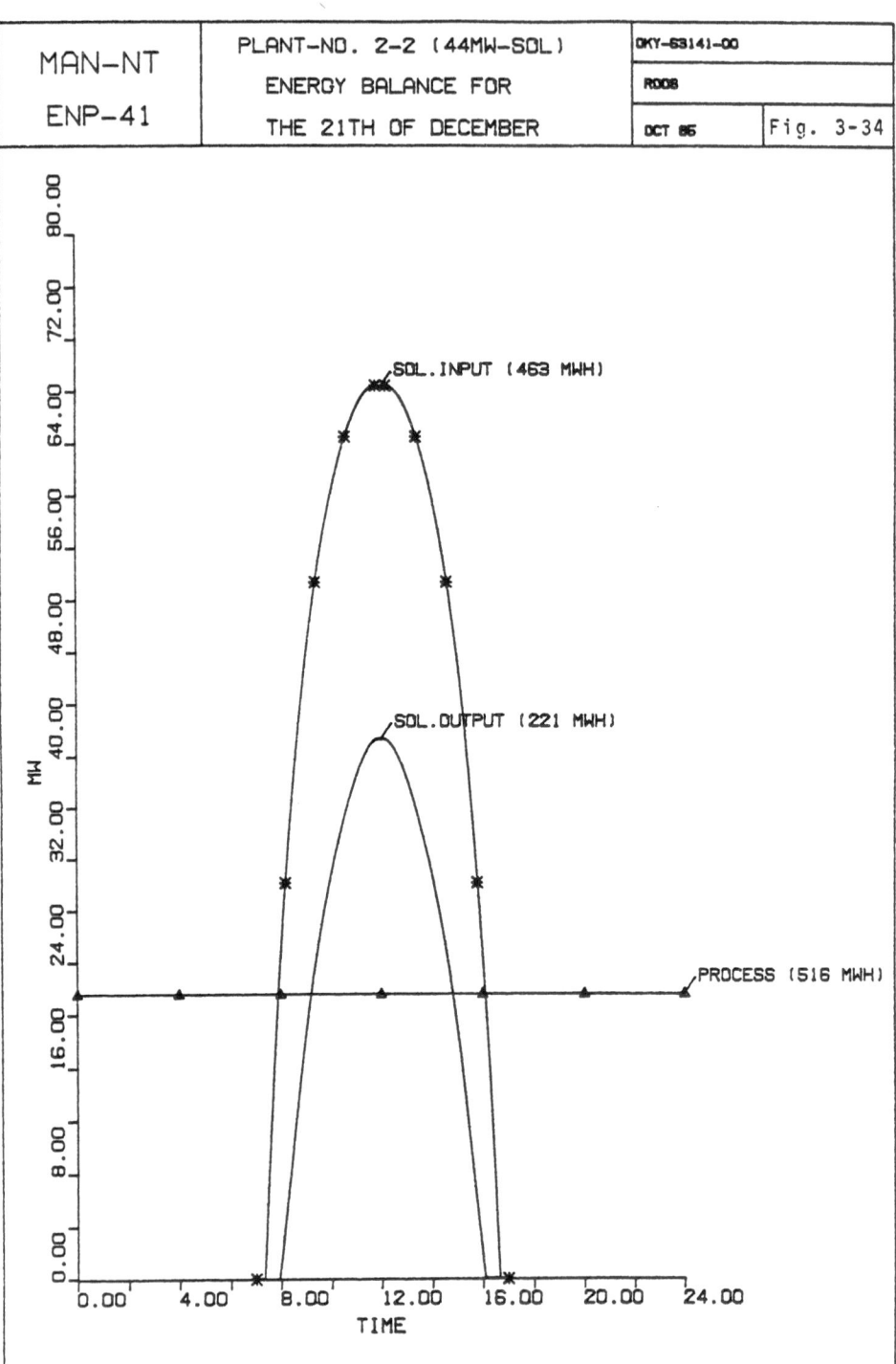

MAN-NT ENP-41	PLANT-NO. 2-2 (44MW-SOL) ENERGY BALANCE FOR THE 21TH OF DECEMBER	OKY-53141-00 ROOS	
		OCT 85	Fig. 3-34

SOL.INPUT (463 MWH)

SOL.OUTPUT (221 MWH)

PROCESS (516 MWH)

MW

TIME

BASMAN®

- 390 -

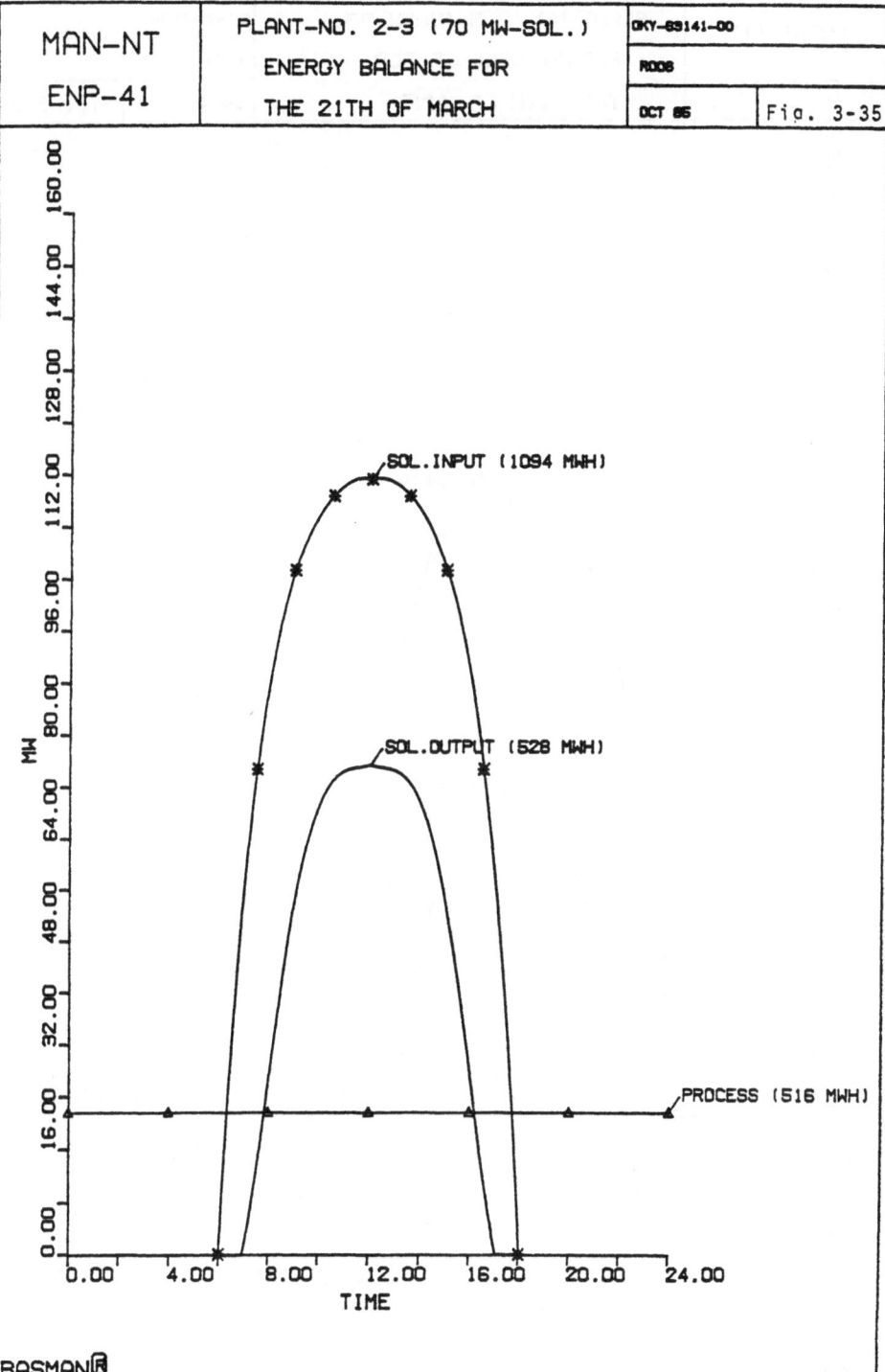

MAN–NT ENP-41	PLANT–NO. 2-3 (70 MW–SOL.) ENERGY BALANCE FOR THE 21TH OF MARCH	OKY-69141-00	
		ROOS	
		OCT 85	Fig. 3-35

SOL.INPUT (1094 MWH)

SOL.OUTPUT (528 MWH)

PROCESS (516 MWH)

MW

TIME

BASMAN®

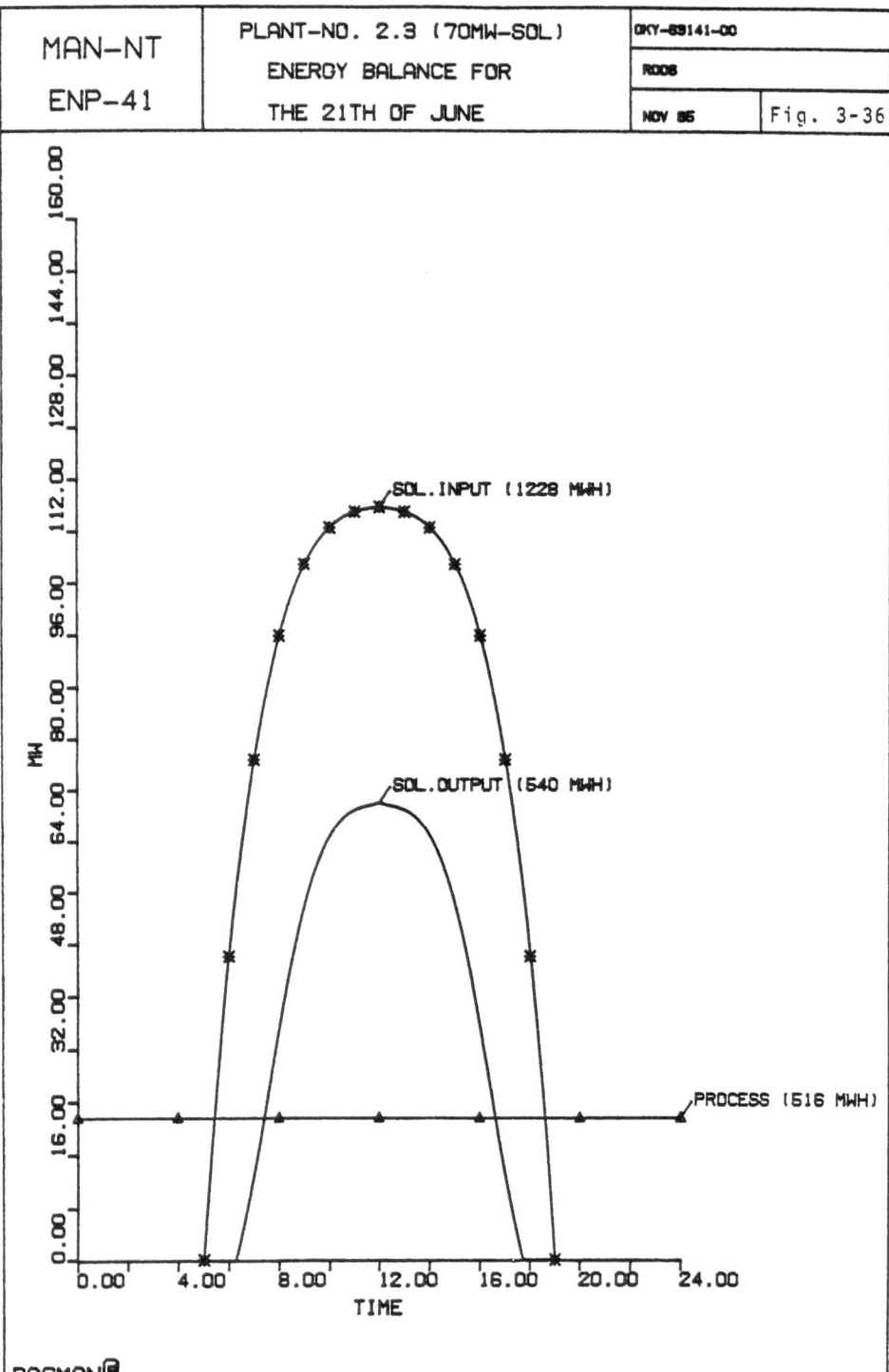

MAN–NT	PLANT–NO. 2.3 (70MW–SOL)	OKY–69141–00	
	ENERGY BALANCE FOR	RODS	
ENP–41	THE 21TH OF JUNE	NOV 85	Fig. 3-36

SOL.INPUT (1228 MWH)

SOL.OUTPUT (540 MWH)

PROCESS (516 MWH)

MW

TIME

BASMAN®

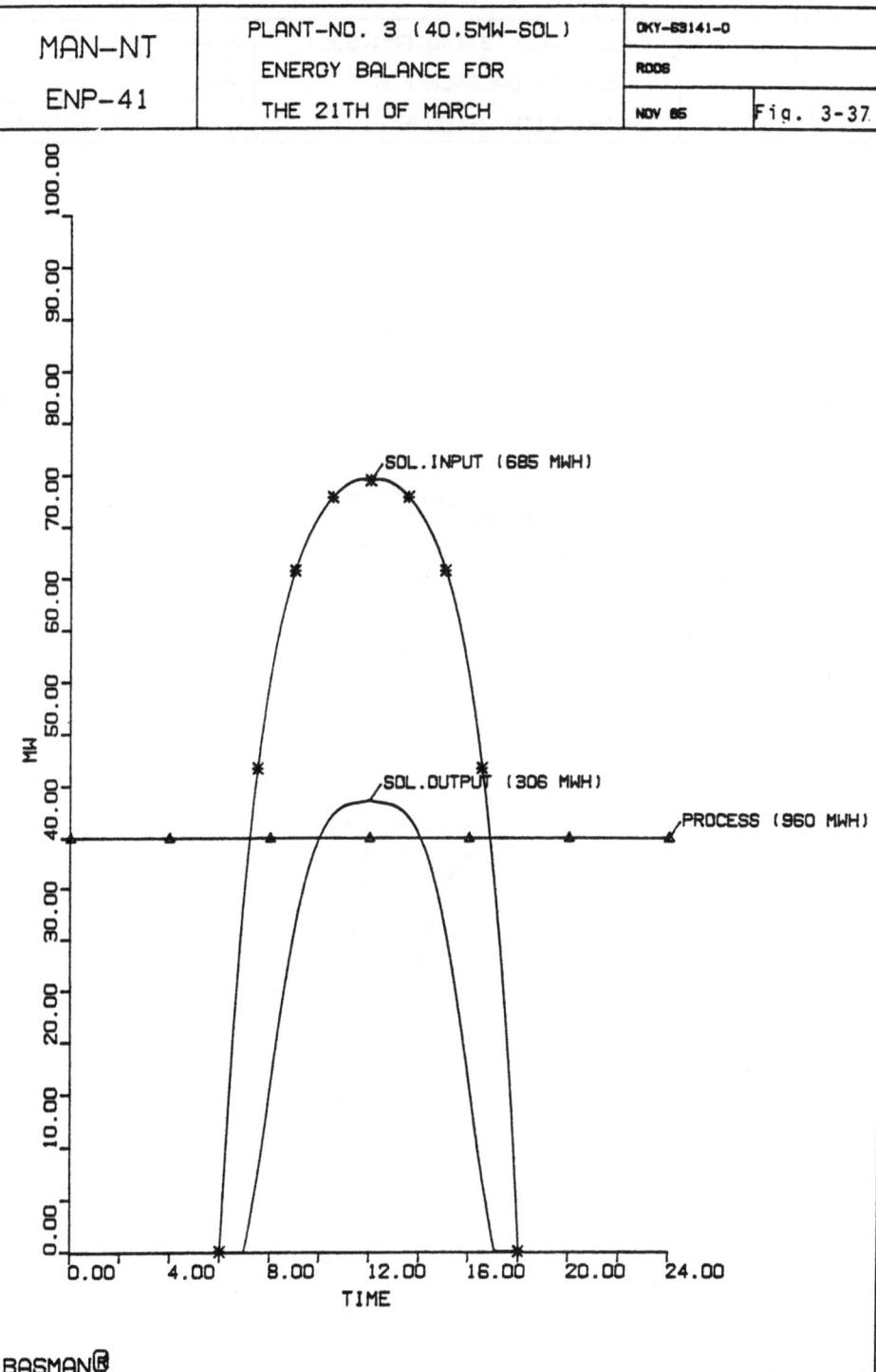

MAN–NT	PLANT–NO. 3 (40.5MW–SOL)	DKY–63141–0	
ENP–41	ENERGY BALANCE FOR	RODS	
	THE 21TH OF MARCH	NOV 85	Fig. 3-37

SOL.INPUT (685 MWH)

SOL.OUTPUT (306 MWH)

PROCESS (960 MWH)

MW

TIME

BASMAN®

MAN-NT ENP-41	PLANT-NO. 3 (40.5MW-SOL) ENERGY BALANCE FOR THE 21TH OF JUNE	OKY-63141-0 ROOS NOV 85 — Fig. 3-38

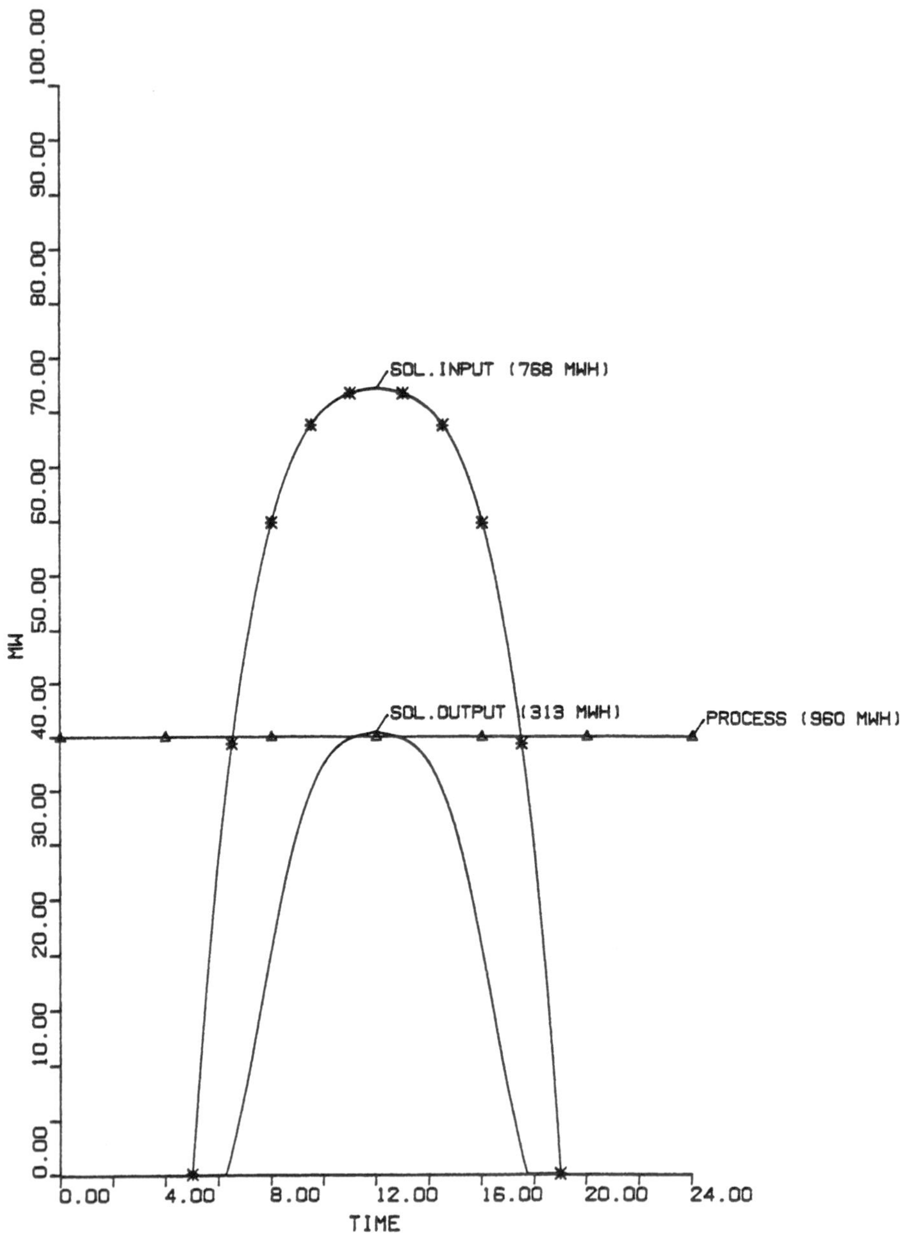

SOL.INPUT (768 MWH)
SOL.OUTPUT (313 MWH)
PROCESS (960 MWH)
MW
TIME
0.00
10.00
20.00
30.00
40.00
50.00
60.00
70.00
80.00
90.00
100.00
0.00
4.00
8.00
12.00
16.00
20.00
24.00

BASMAN®

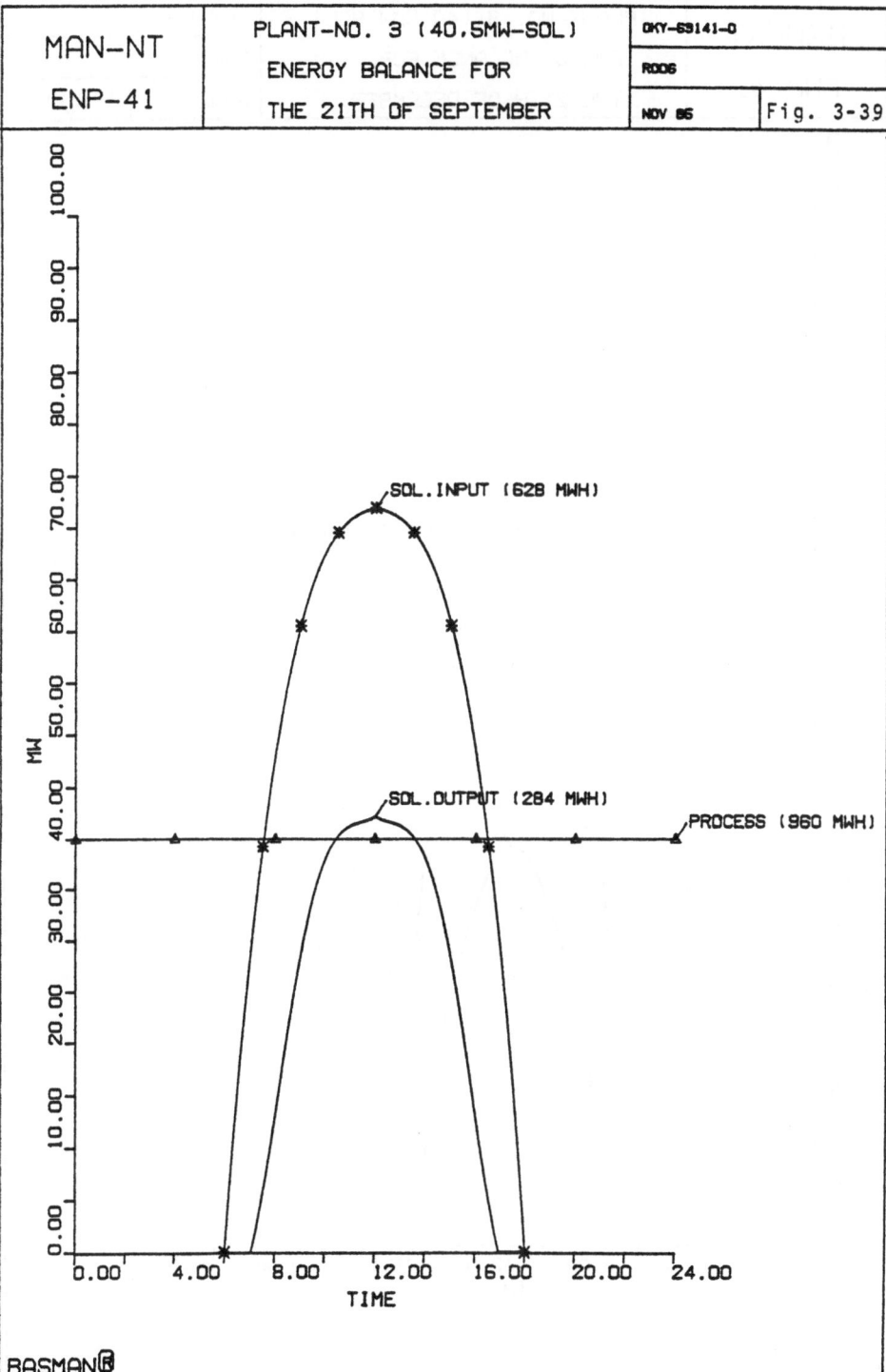

| MAN–NT ENP–41 | PLANT–NO. 3 (40,5MW–SOL) ENERGY BALANCE FOR THE 21TH OF SEPTEMBER | DKY–69141–0 ROOS NOV 86 | Fig. 3-39 |

SOL.INPUT (628 MWH)

SOL.OUTPUT (284 MWH)

PROCESS (960 MWH)

MW

TIME

BASMAN®

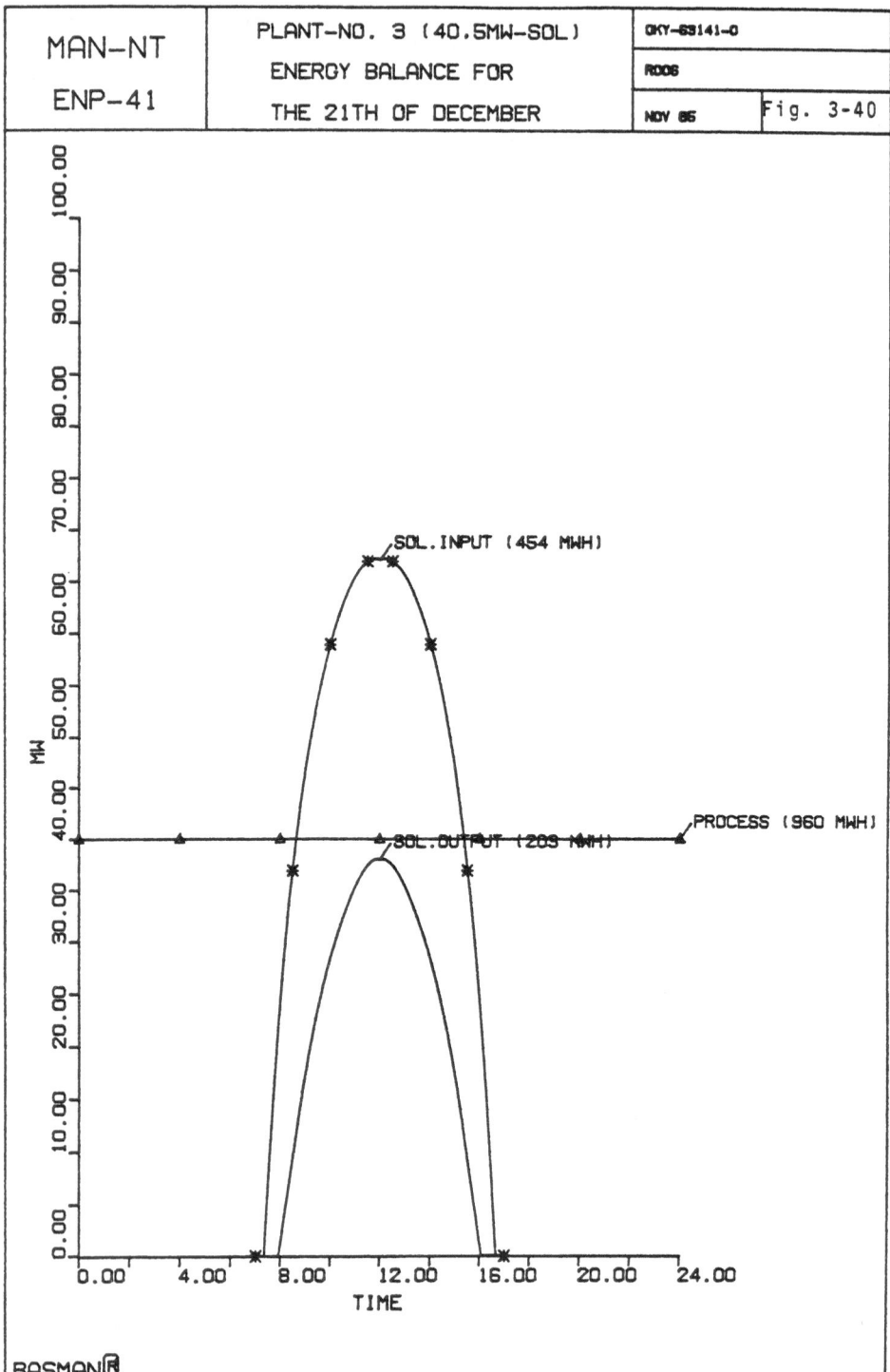

MAN-NT	PLANT-NO. 3 (40.5MW-SOL)	OKY-53141-0	
	ENERGY BALANCE FOR	ROO6	
ENP-41	THE 21TH OF DECEMBER	NOV 86	Fig. 3-40

SOL.INPUT (454 MWH)

PROCESS (960 MWH)

SOL.OUTPUT (203 MWH)

MW

TIME

BASMAN®

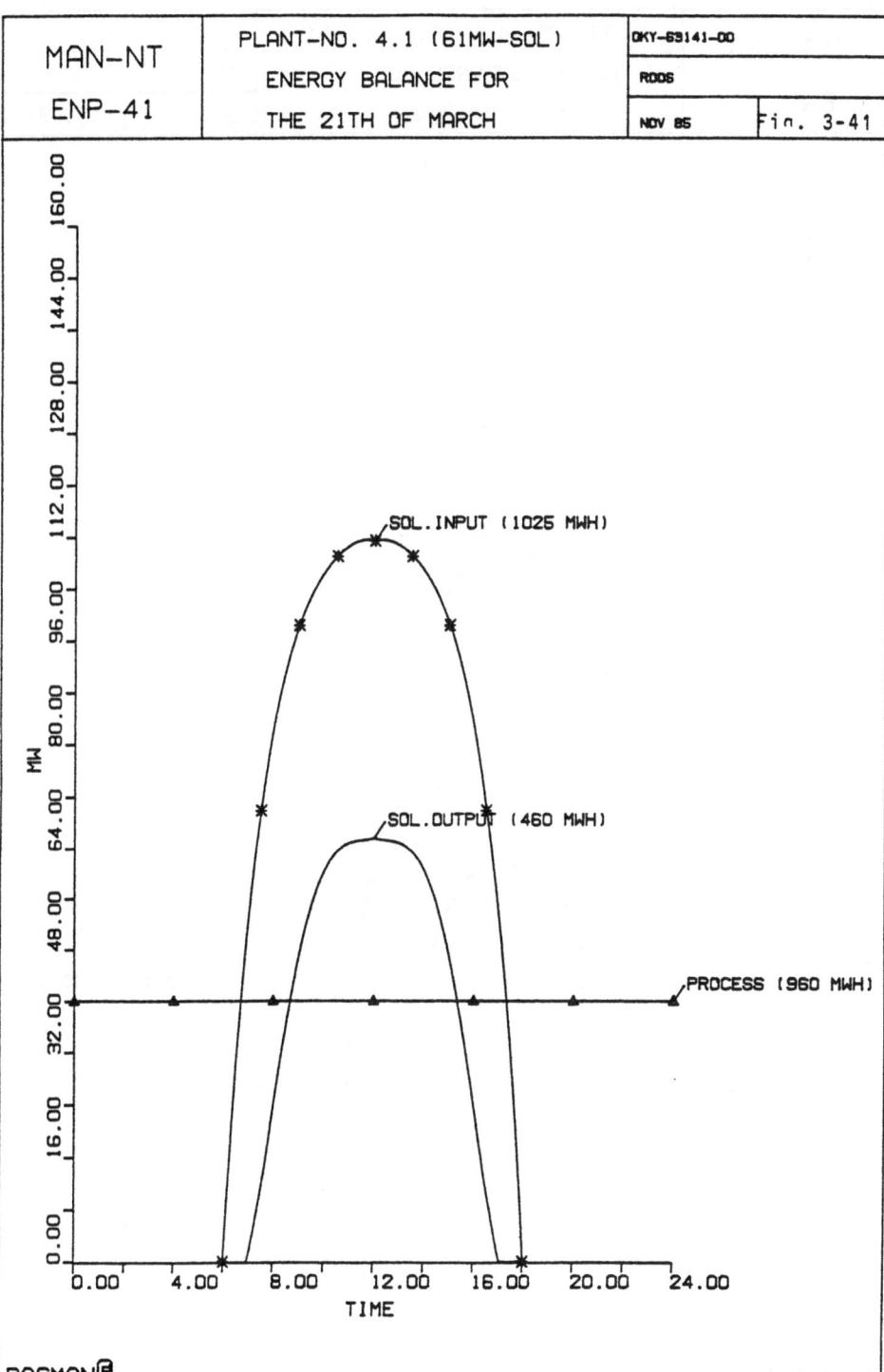

MAN-NT	PLANT-NO. 4.1 (61MW-SOL)	OKY-69141-00	
ENP-41	ENERGY BALANCE FOR	RODS	
	THE 21TH OF MARCH	NOV 85	Fig. 3-41

SOL.INPUT (1025 MWH)

SOL.OUTPUT (460 MWH)

PROCESS (960 MWH)

BASMAN®

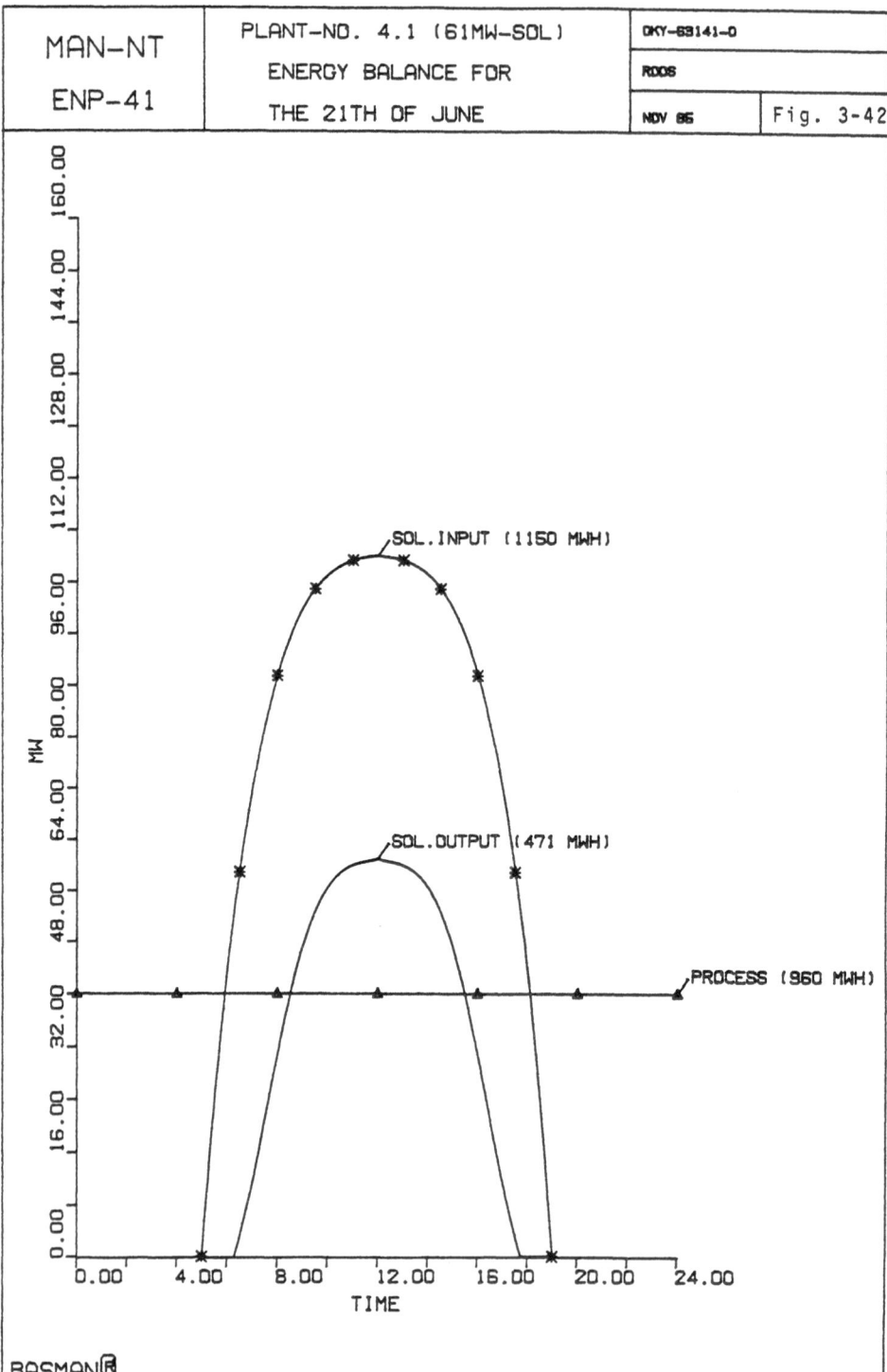

MAN-NT ENP-41	PLANT-NO. 4.1 (61MW-SOL) ENERGY BALANCE FOR THE 21TH OF JUNE	OKY-63141-0 RODS	
		NOV 86	Fig. 3-42

SOL.INPUT (1150 MWH)

SOL.OUTPUT (471 MWH)

PROCESS (960 MWH)

MW

TIME

BASMAN®

- 398 -

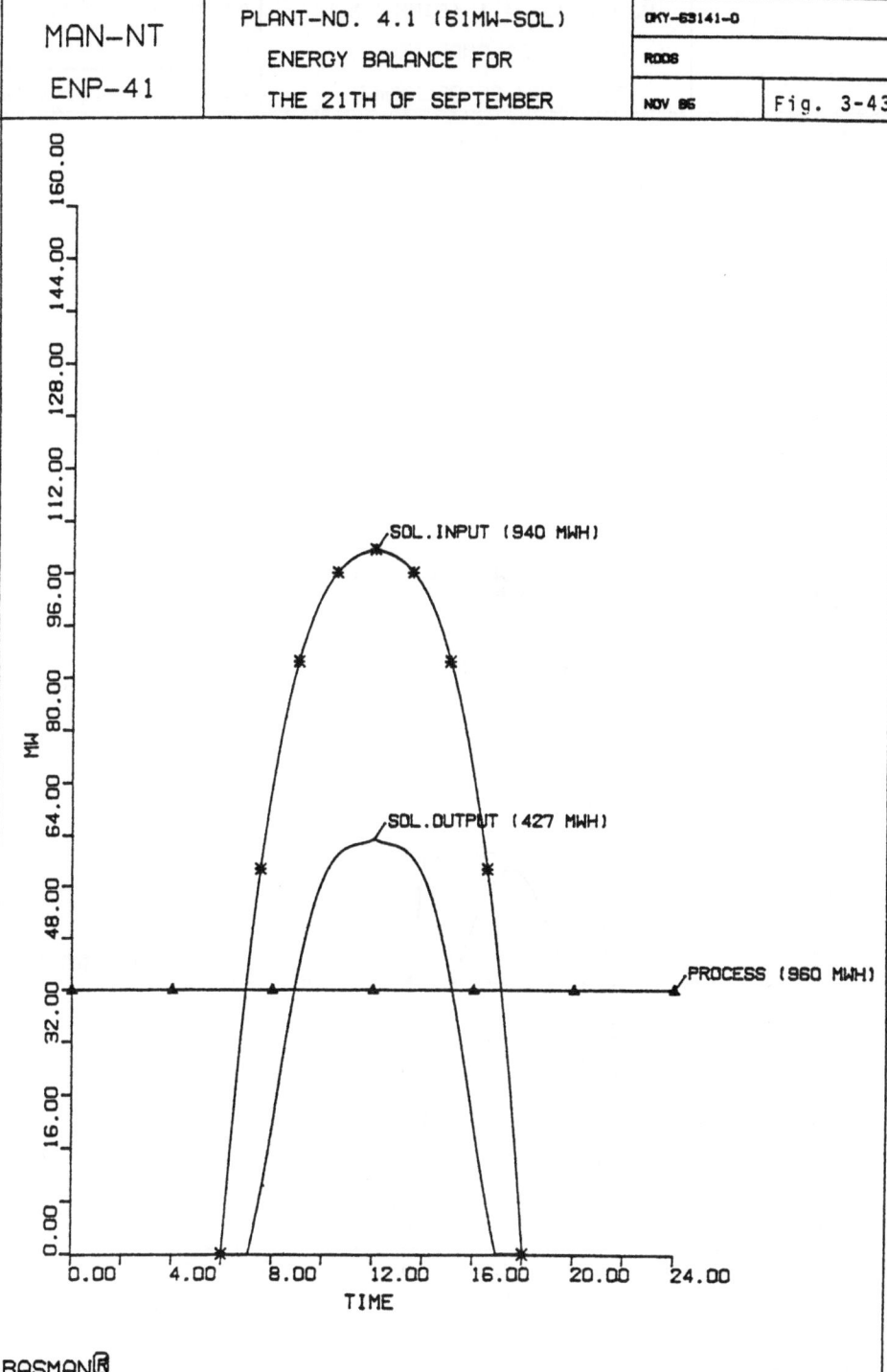

MAN—NT	PLANT—NO. 4.1 (61MW—SOL)	DKY-63141-0	
ENP—41	ENERGY BALANCE FOR	ROOS	
	THE 21TH OF SEPTEMBER	NOV 86	Fig. 3-43

SOL.INPUT (940 MWH)

SOL.OUTPUT (427 MWH)

PROCESS (960 MWH)

MW

TIME

BASMAN

SOL.INPUT (680 MWH)

SOL.OUTPUT (306 MWH)

PROCESS (960 MWH)

MW

TIME

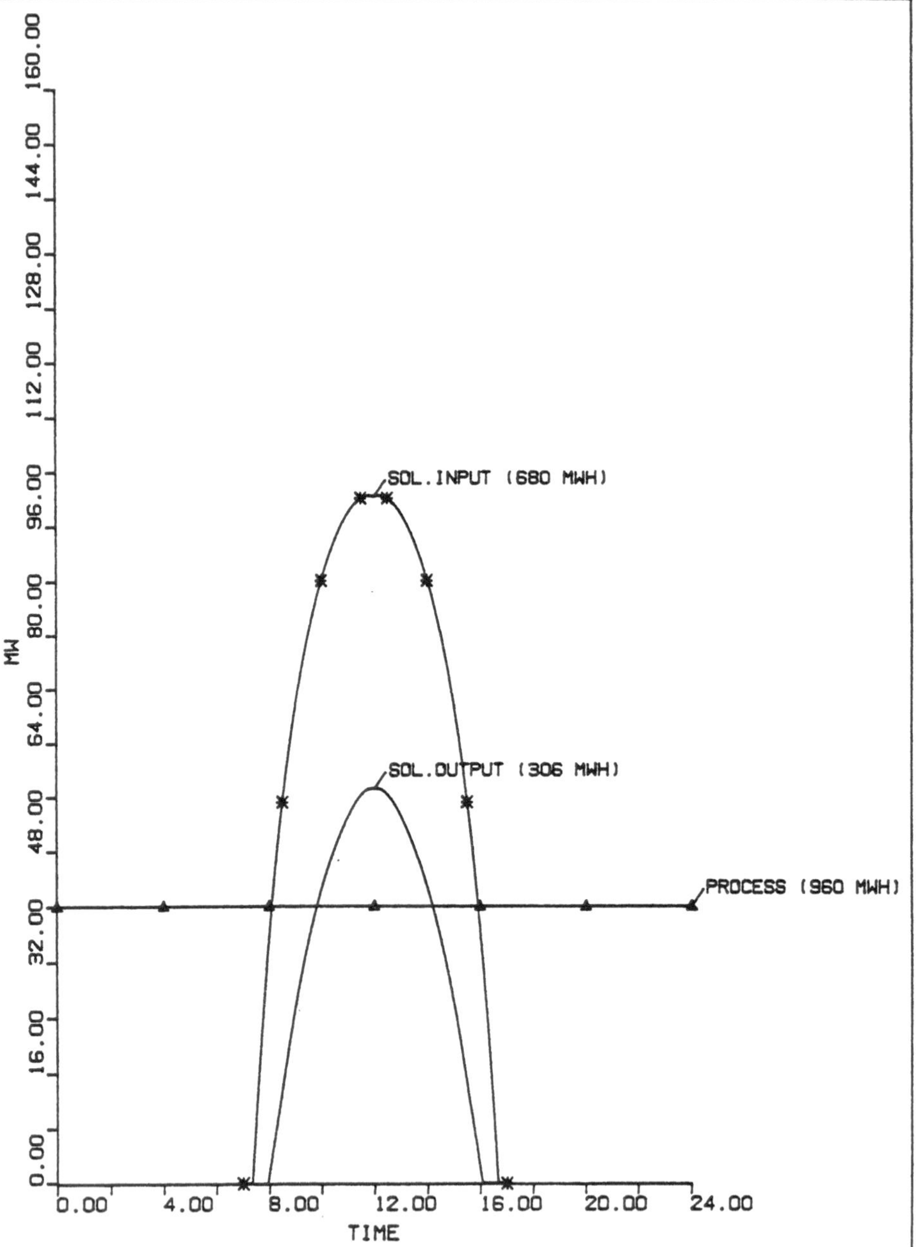

BASMAN®

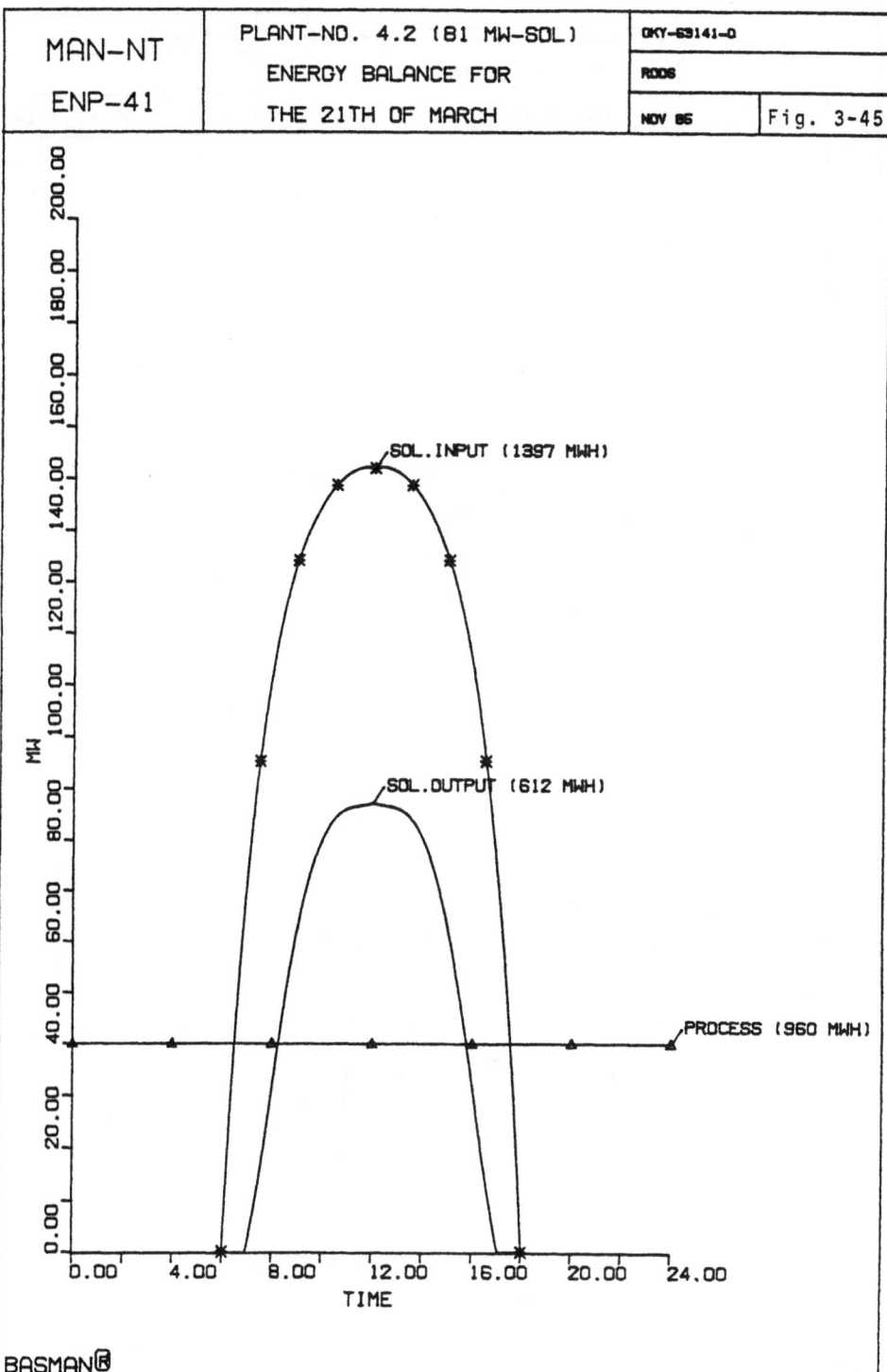

MAN-NT	PLANT-NO. 4.2 (81 MW-SOL)	OKY-69141-0	
ENP-41	ENERGY BALANCE FOR	ROOS	
	THE 21TH OF MARCH	NOV 86	Fig. 3-45

SOL.INPUT (1397 MWH)

SOL.OUTPUT (612 MWH)

PROCESS (960 MWH)

MW

TIME

BASMAN®

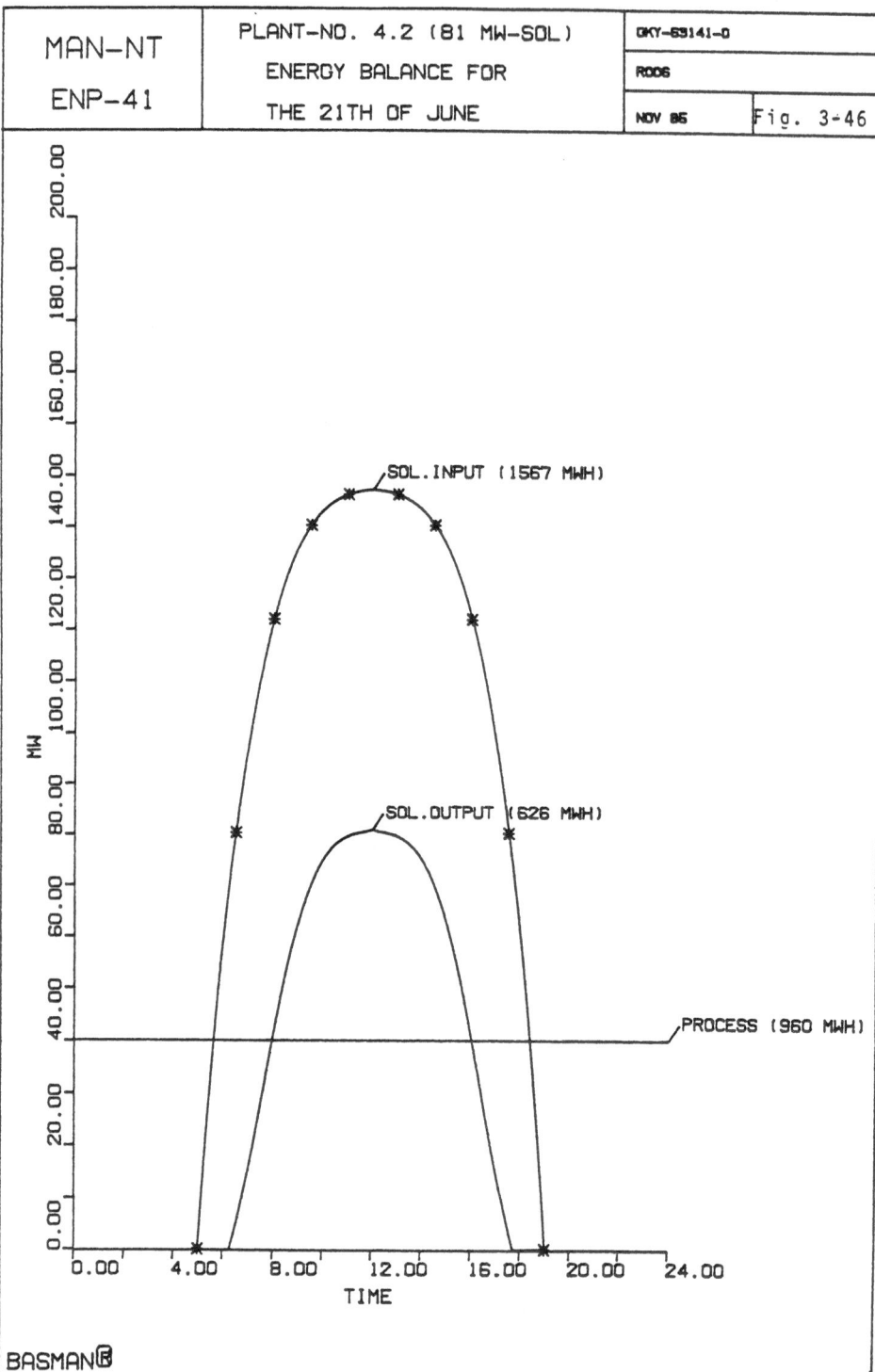

MAN-NT ENP-41	PLANT-NO. 4.2 (81 MW-SOL) ENERGY BALANCE FOR THE 21TH OF JUNE	DKY-63141-0
		ROOS
		NOV 86 — Fig. 3-46

SOL.INPUT (1567 MWH)

SOL.OUTPUT (626 MWH)

PROCESS (960 MWH)

MW

TIME

BASMAN Ⓡ

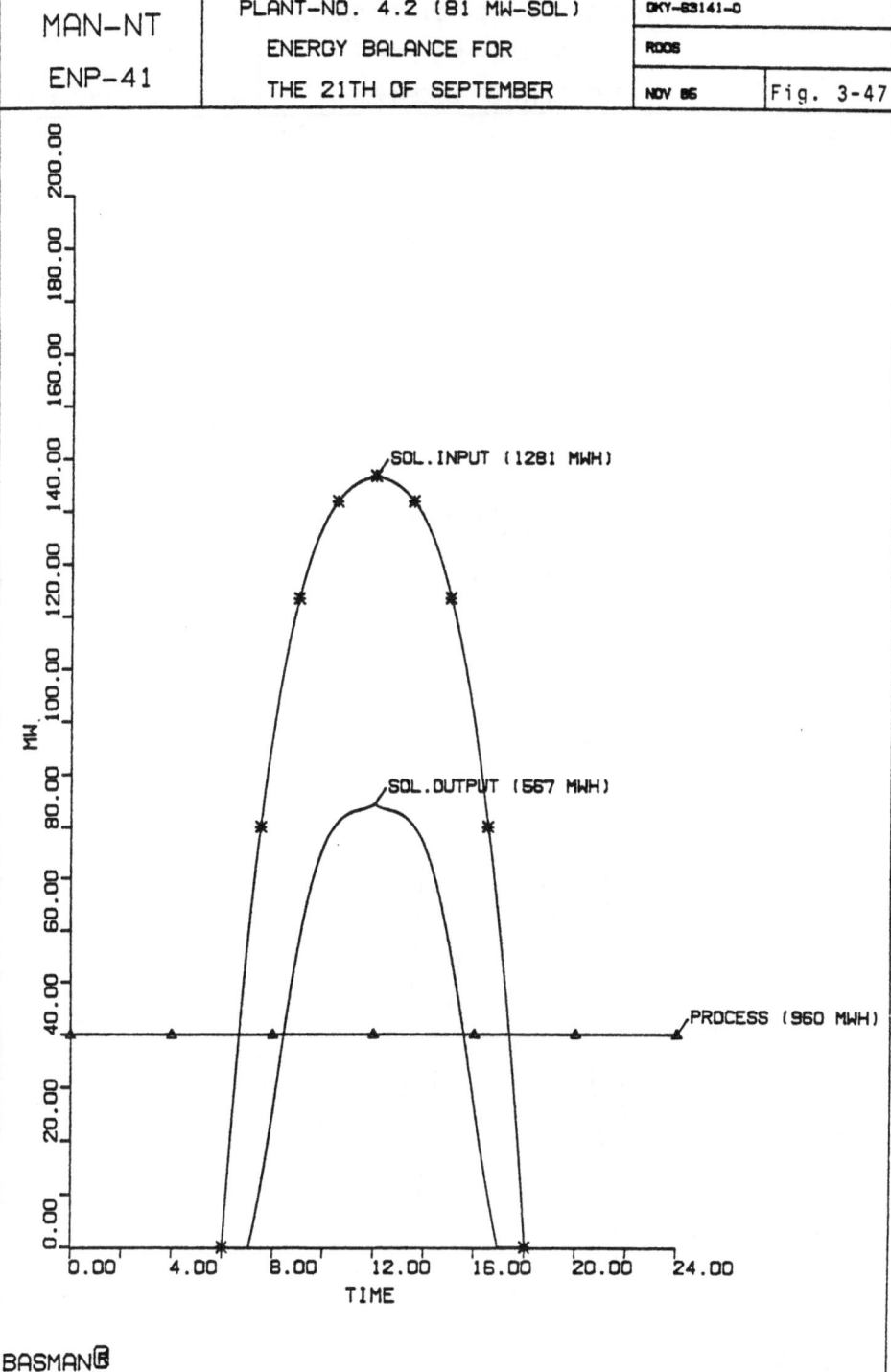

MAN-NT	PLANT-NO. 4.2 (81 MW-SOL)	DKY-63141-0	
ENP-41	ENERGY BALANCE FOR	ROOS	
	THE 21TH OF SEPTEMBER	NOV 86	Fig. 3-47

SOL.INPUT (1281 MWH)

SOL.OUTPUT (567 MWH)

PROCESS (960 MWH)

MW

TIME

BASMAN®

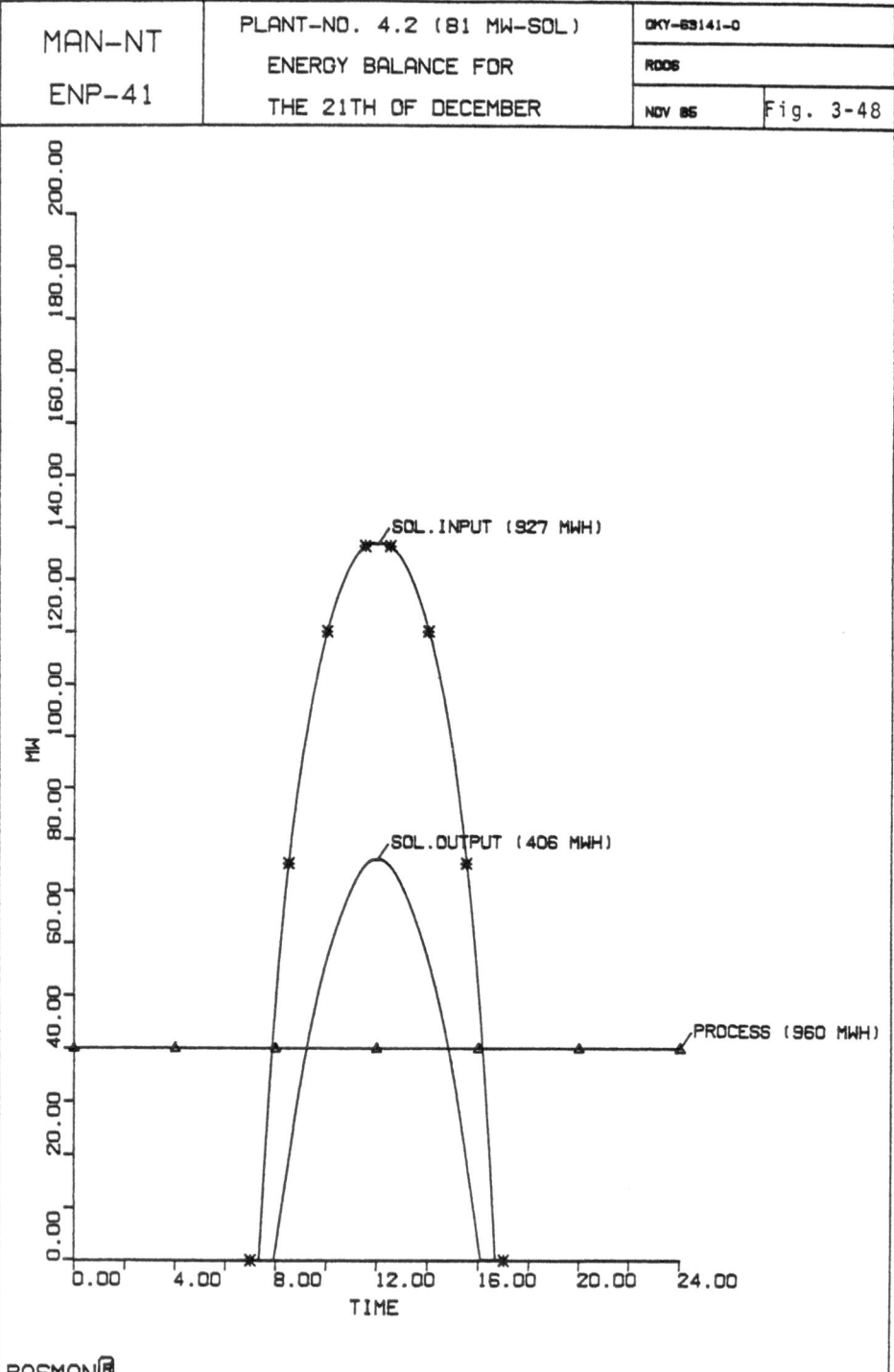

MAN-NT ENP-41	PLANT-NO. 4.2 (81 MW-SOL) ENERGY BALANCE FOR THE 21TH OF DECEMBER	DKY-63141-0
		ROOS
		NOV 86 Fig. 3-48

SOL.INPUT (927 MWH)

SOL.OUTPUT (406 MWH)

PROCESS (960 MWH)

MW

TIME

0.00   4.00   8.00   12.00   16.00   20.00   24.00

BASMAN

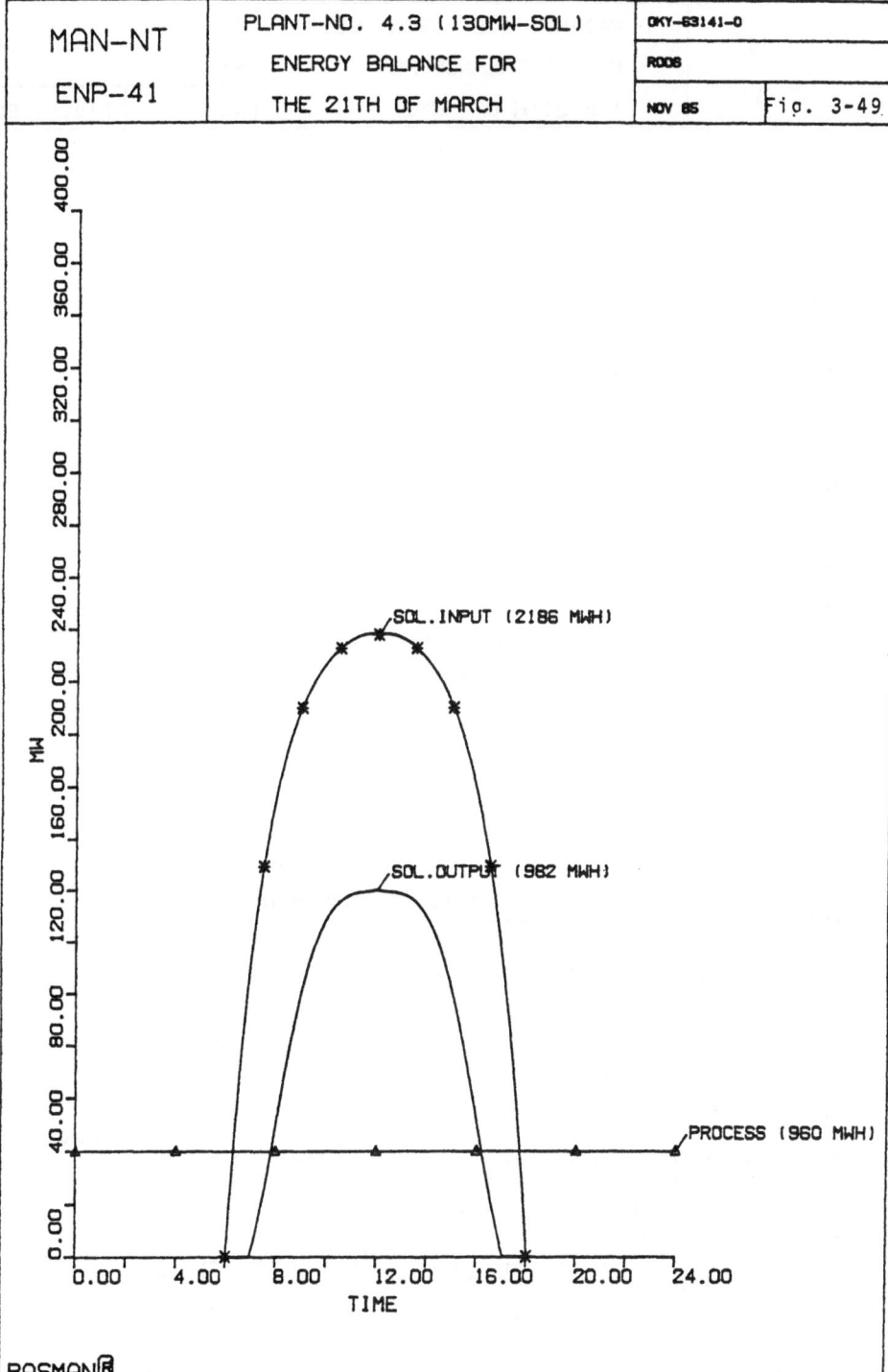

SOL.INPUT (2186 MWH)

SOL.OUTPUT (982 MWH)

PROCESS (960 MWH)

MW

TIME

BASMAN®

SOL.INPUT (2452 MWH)

SOL.OUTPUT (1005 MWH)

PROCESS (960 MWH)

MW

TIME

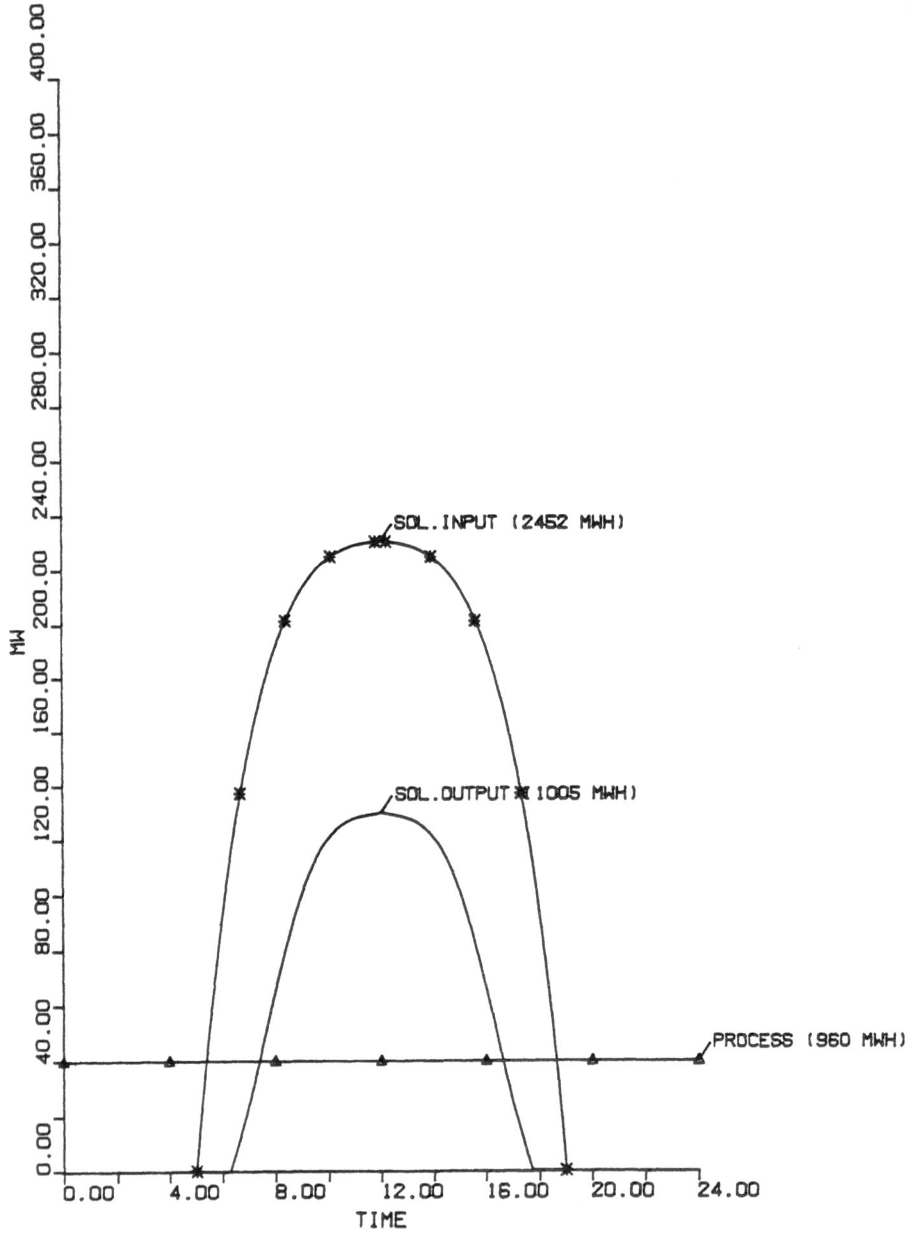

BASMAN

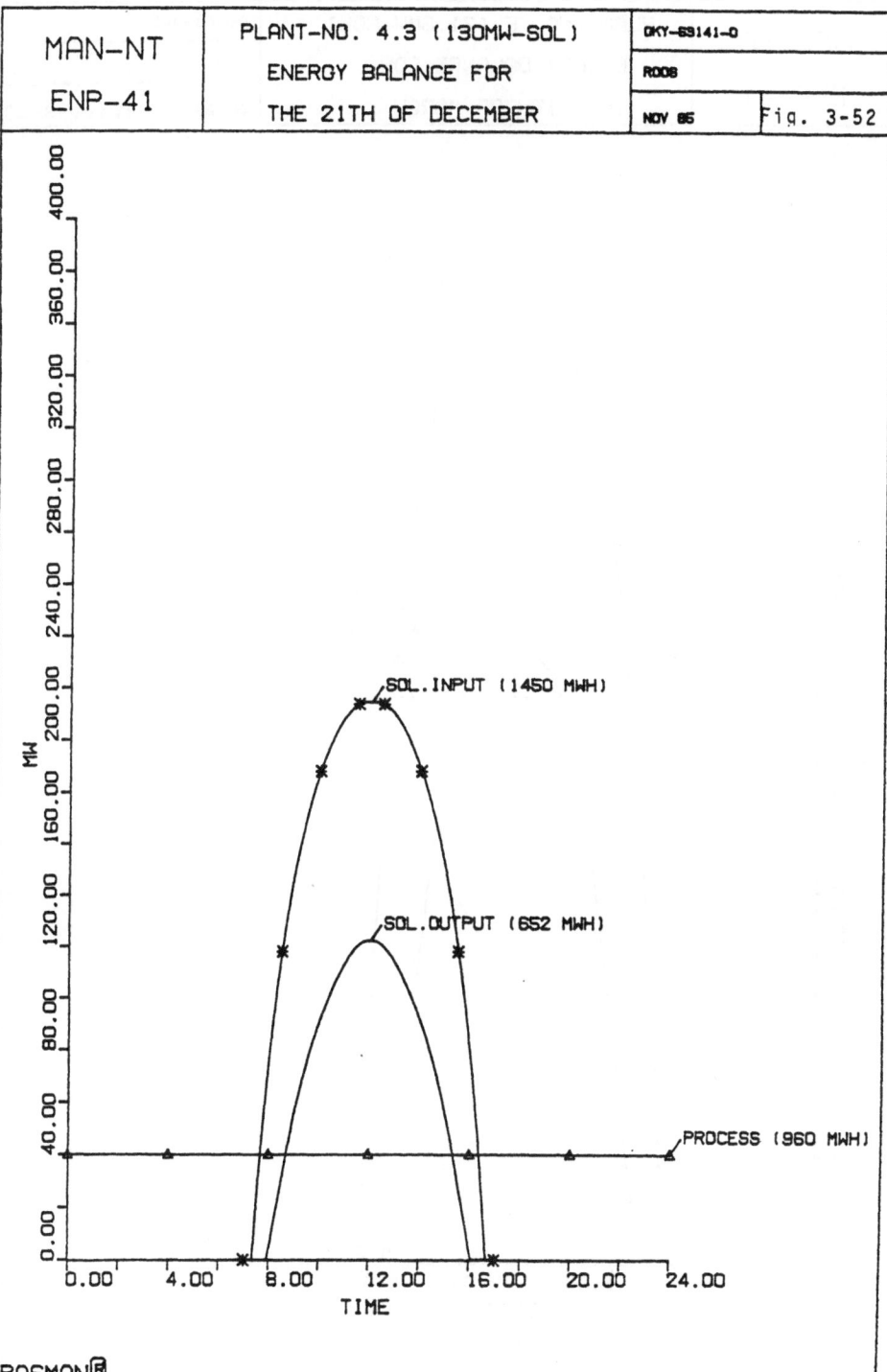

MAN–NT	PLANT–NO. 4.3 (130MW–SOL)	DKY–SS141–O	
ENP–41	ENERGY BALANCE FOR	ROOB	
	THE 21TH OF DECEMBER	NOV 85	Fig. 3-52

SOL.INPUT (1450 MWH)

SOL.OUTPUT (652 MWH)

PROCESS (960 MWH)

MW

TIME

BASMAN®

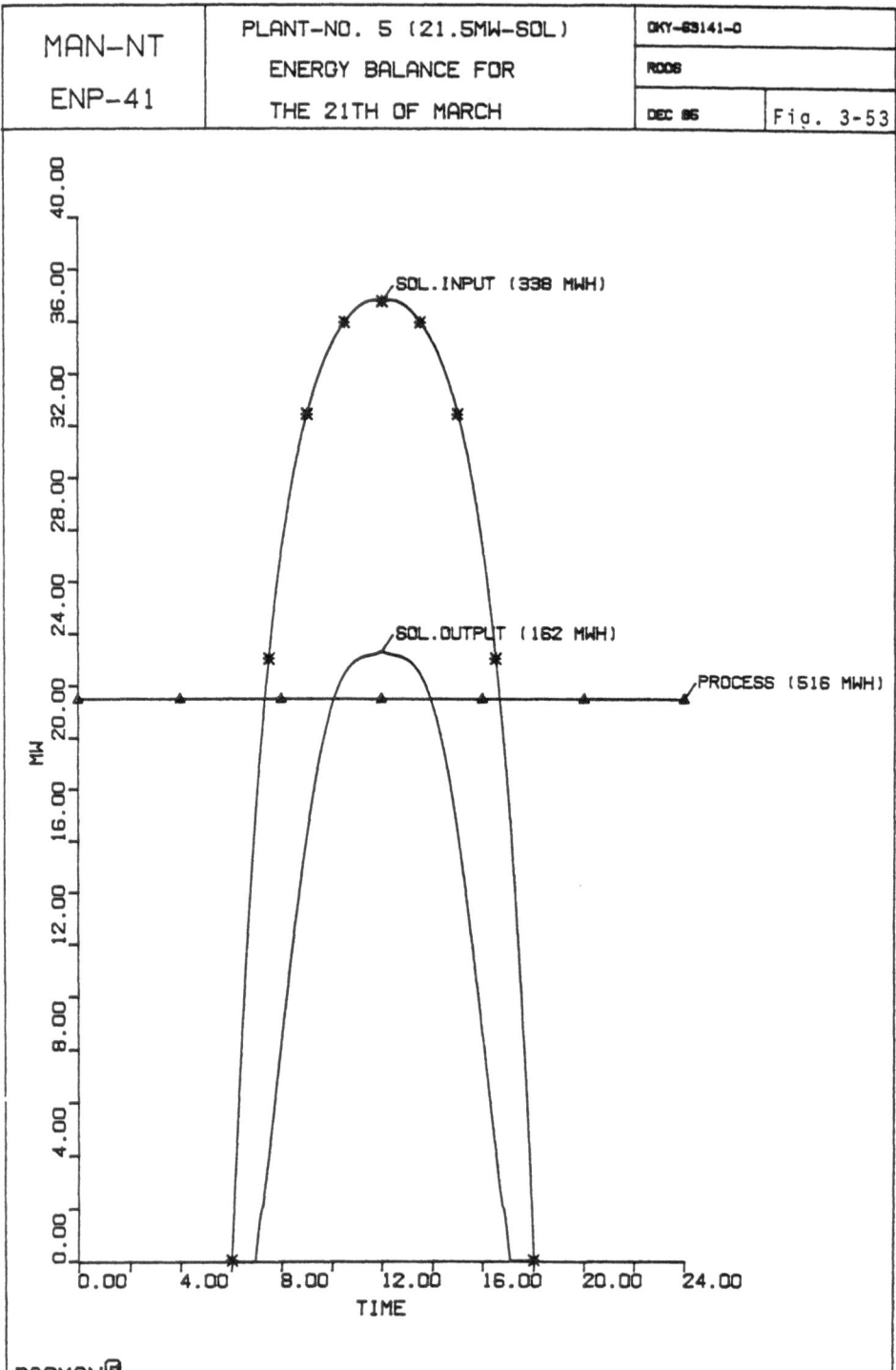

BASMAN®

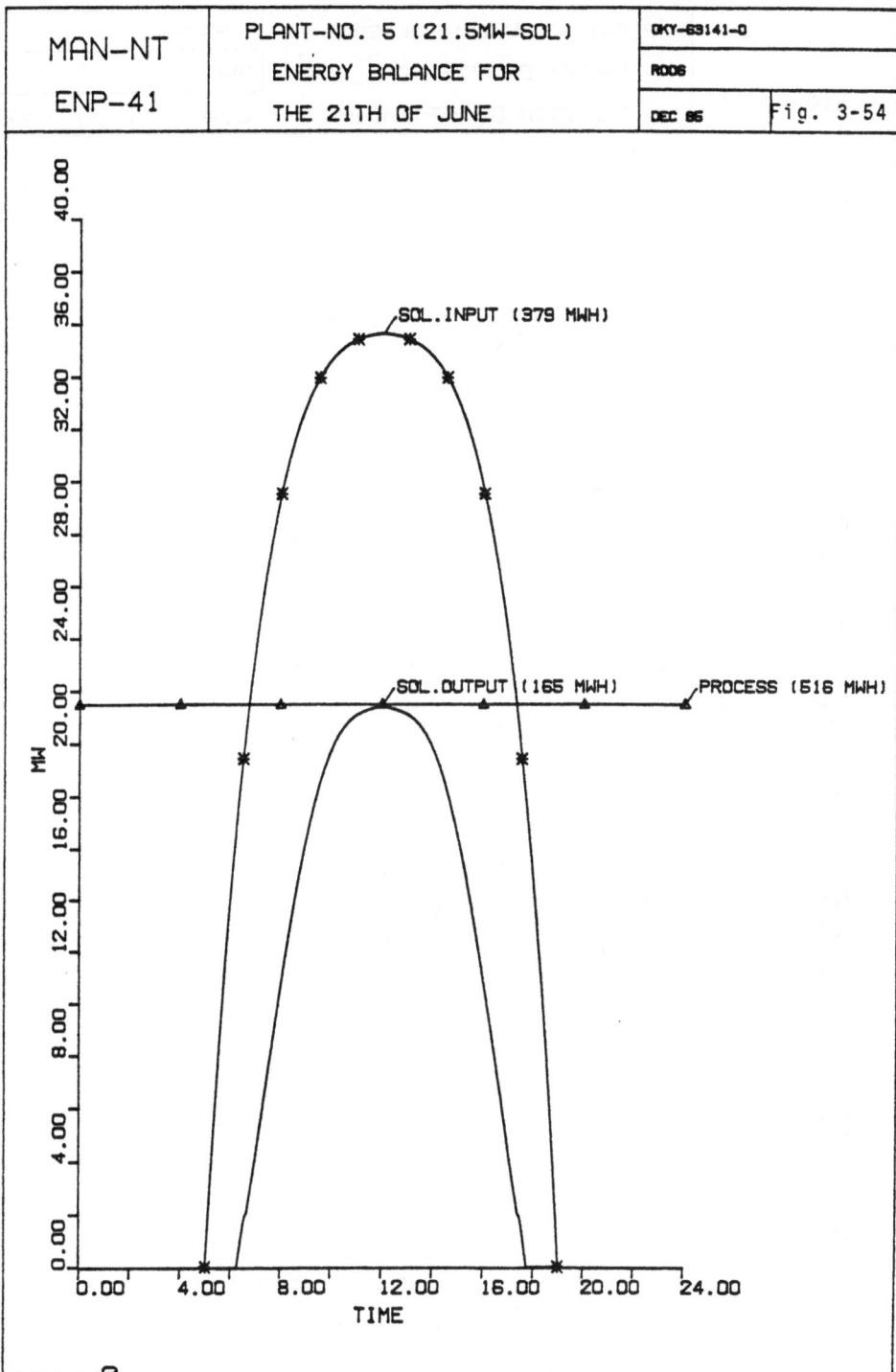

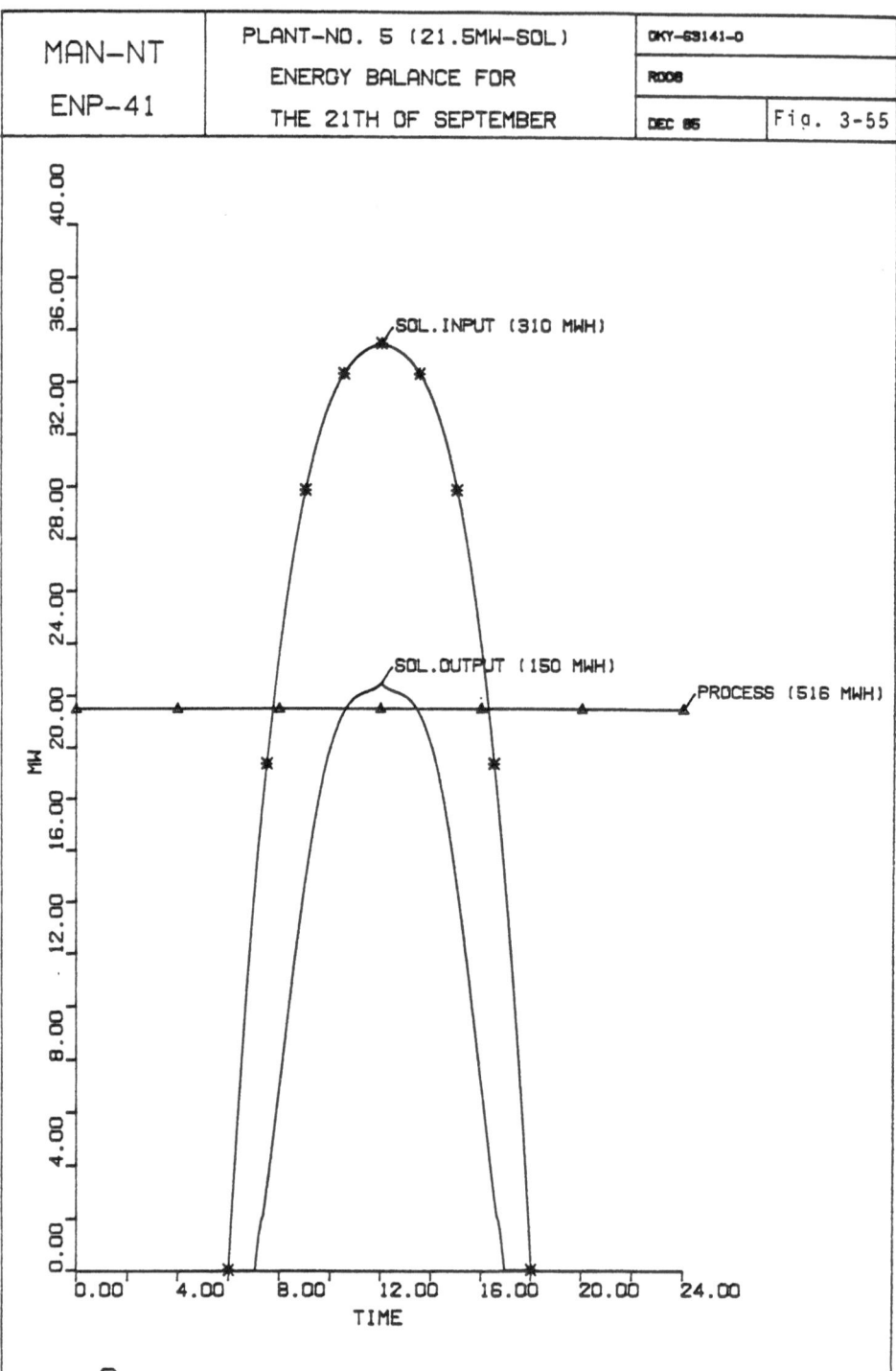

| MAN-NT ENP-41 | PLANT-NO. 5 (21.5MW-SOL) ENERGY BALANCE FOR THE 21TH OF SEPTEMBER | DKY-69141-0 ROO8 DEC 85 | Fig. 3-55 |

SOL.INPUT (310 MWH)

SOL.OUTPUT (150 MWH)

PROCESS (516 MWH)

MW

TIME

BASMAN®

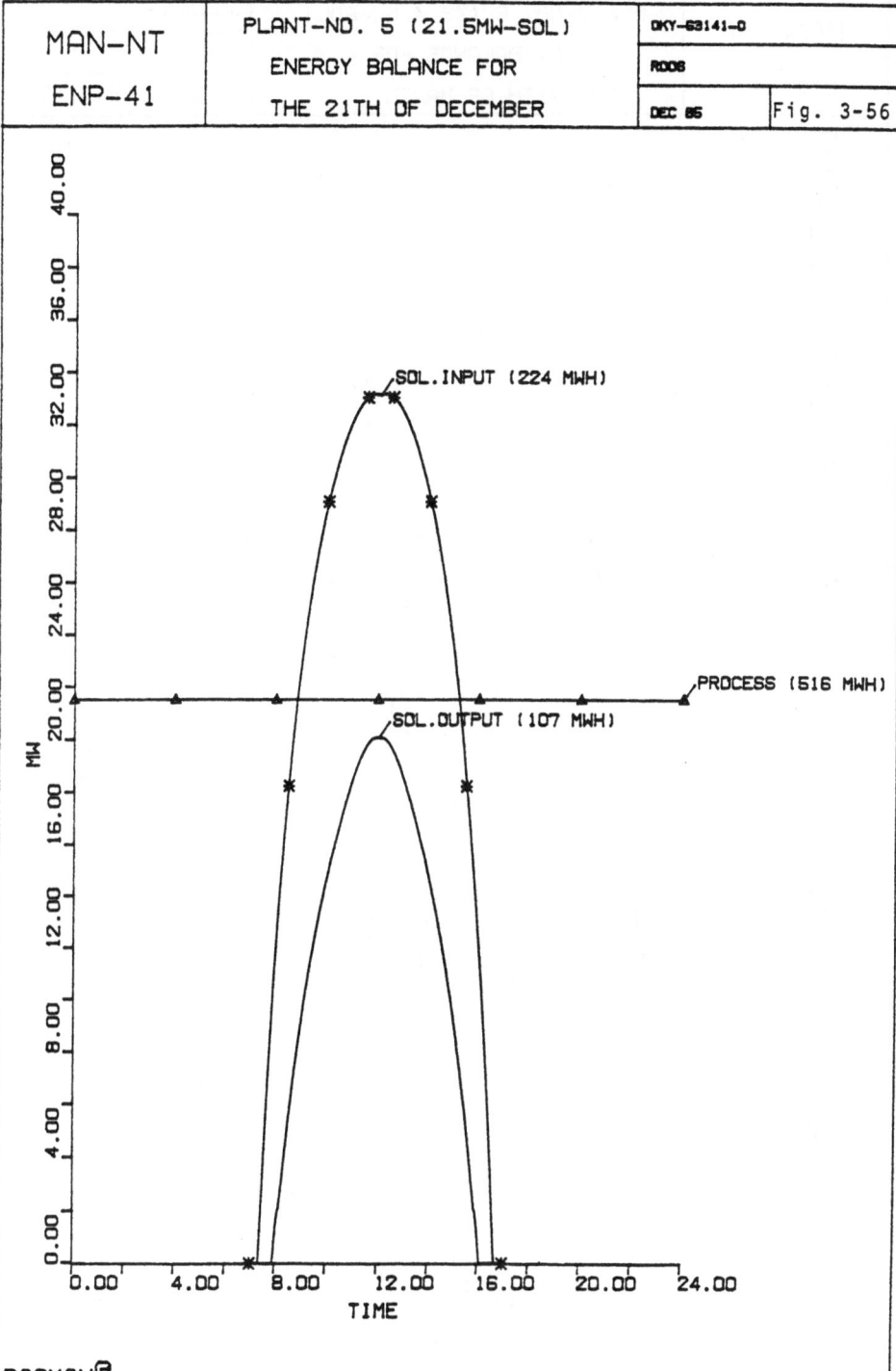

MAN-NT	PLANT-NO. 5 (21.5MW-SOL)	DKY-63141-0	
	ENERGY BALANCE FOR	ROOB	
ENP-41	THE 21TH OF DECEMBER	DEC 85	Fig. 3-56

SOL.INPUT (224 MWH)

PROCESS (516 MWH)

SOL.OUTPUT (107 MWH)

MW

TIME

BASMAN®

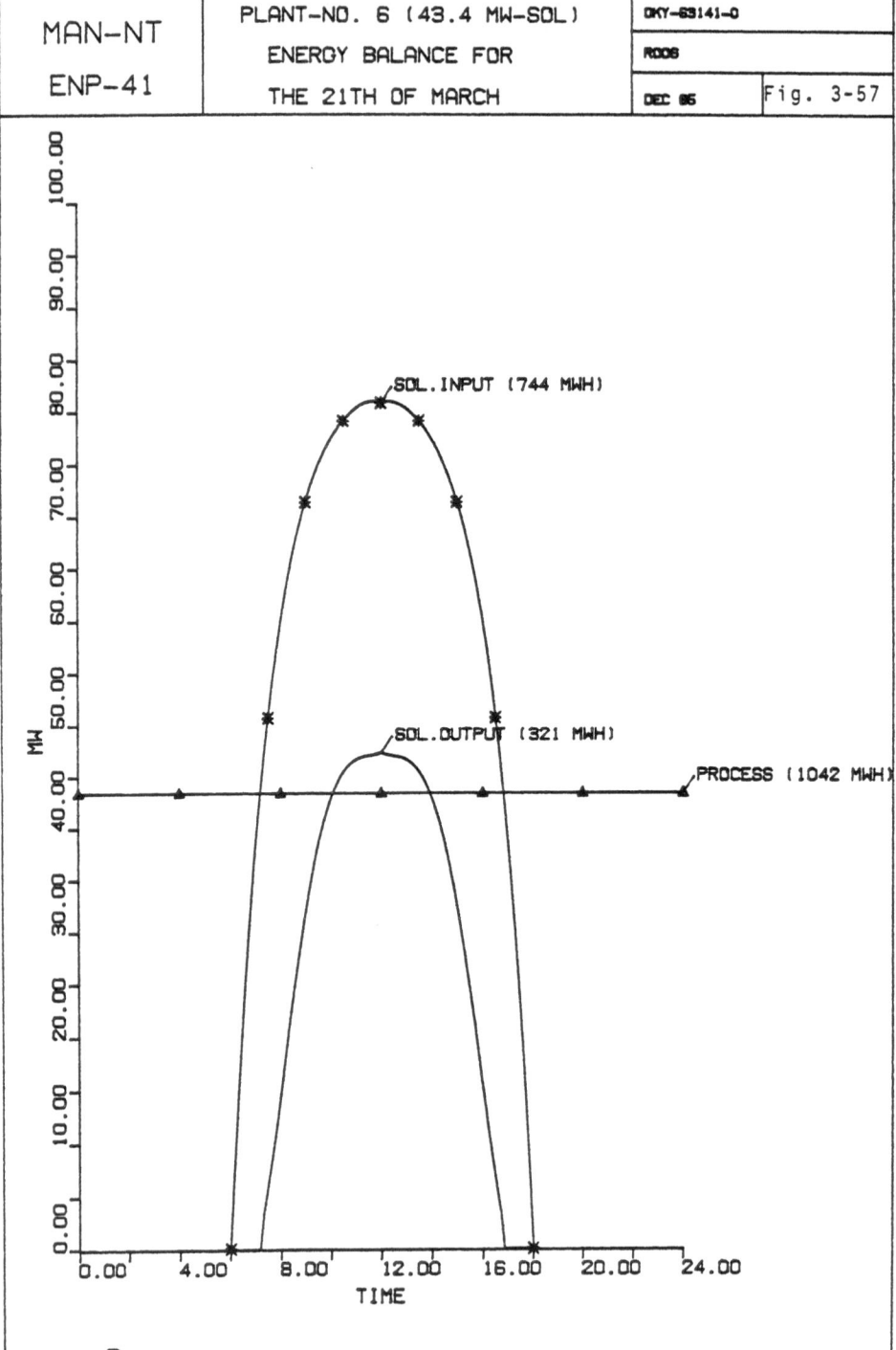

MAN-NT	PLANT-NO. 6 (43.4 MW-SOL)	DKY-63141-0	
	ENERGY BALANCE FOR	ROOS	
ENP-41	THE 21TH OF MARCH	DEC 85	Fig. 3-57

SOL.INPUT (744 MWH)

SOL.OUTPUT (321 MWH)

PROCESS (1042 MWH)

MW

TIME

BASMAN®

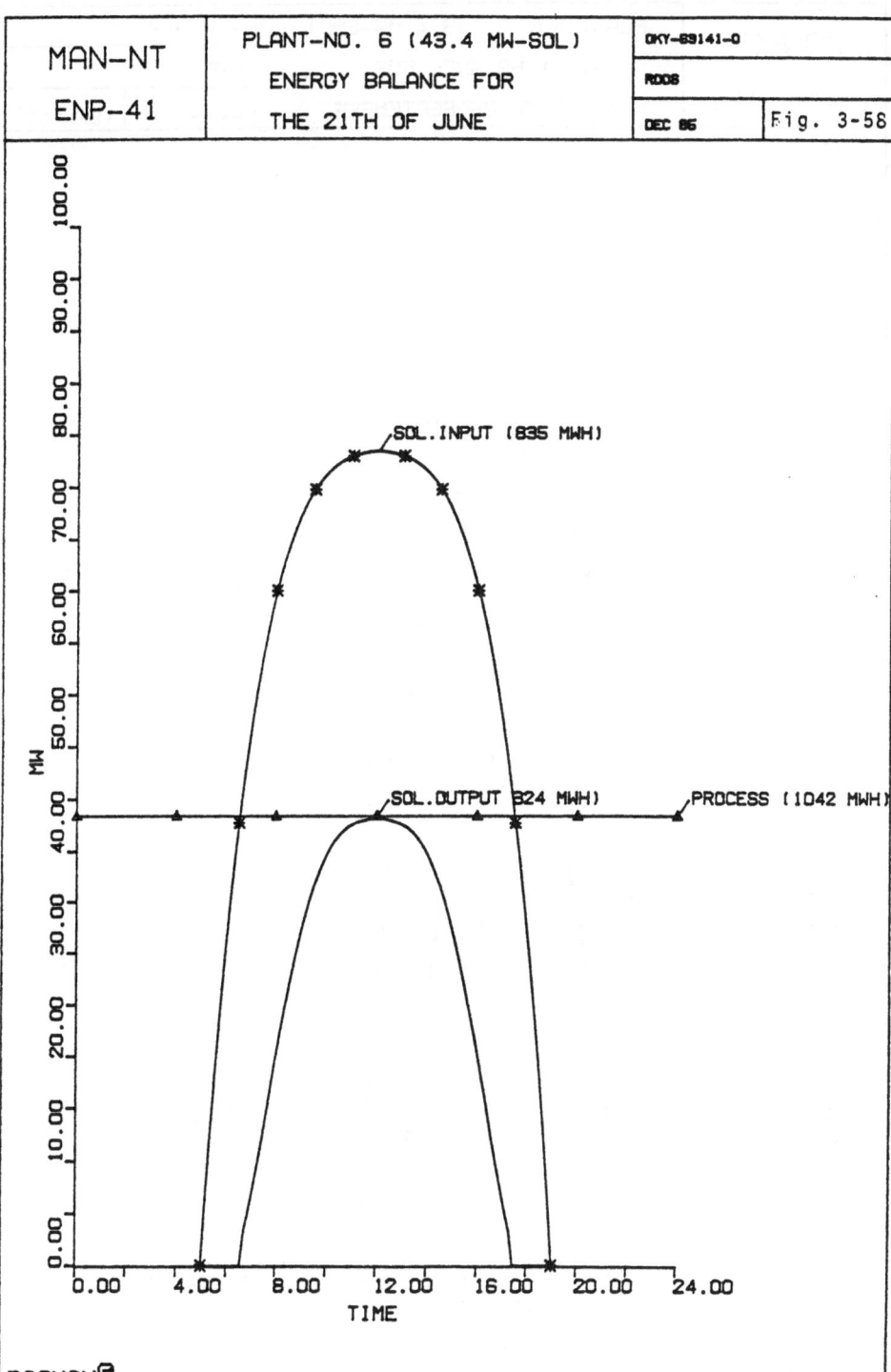

| MAN-NT ENP-41 | PLANT-NO. 6 (43.4 MW-SOL) ENERGY BALANCE FOR THE 21TH OF JUNE | OKY-69141-0 ROOS DEC 86 | Fig. 3-58 |

SOL.INPUT (835 MWH)

SOL.OUTPUT (824 MWH) PROCESS (1042 MWH)

MW

TIME

BASMAN®

- 413 -

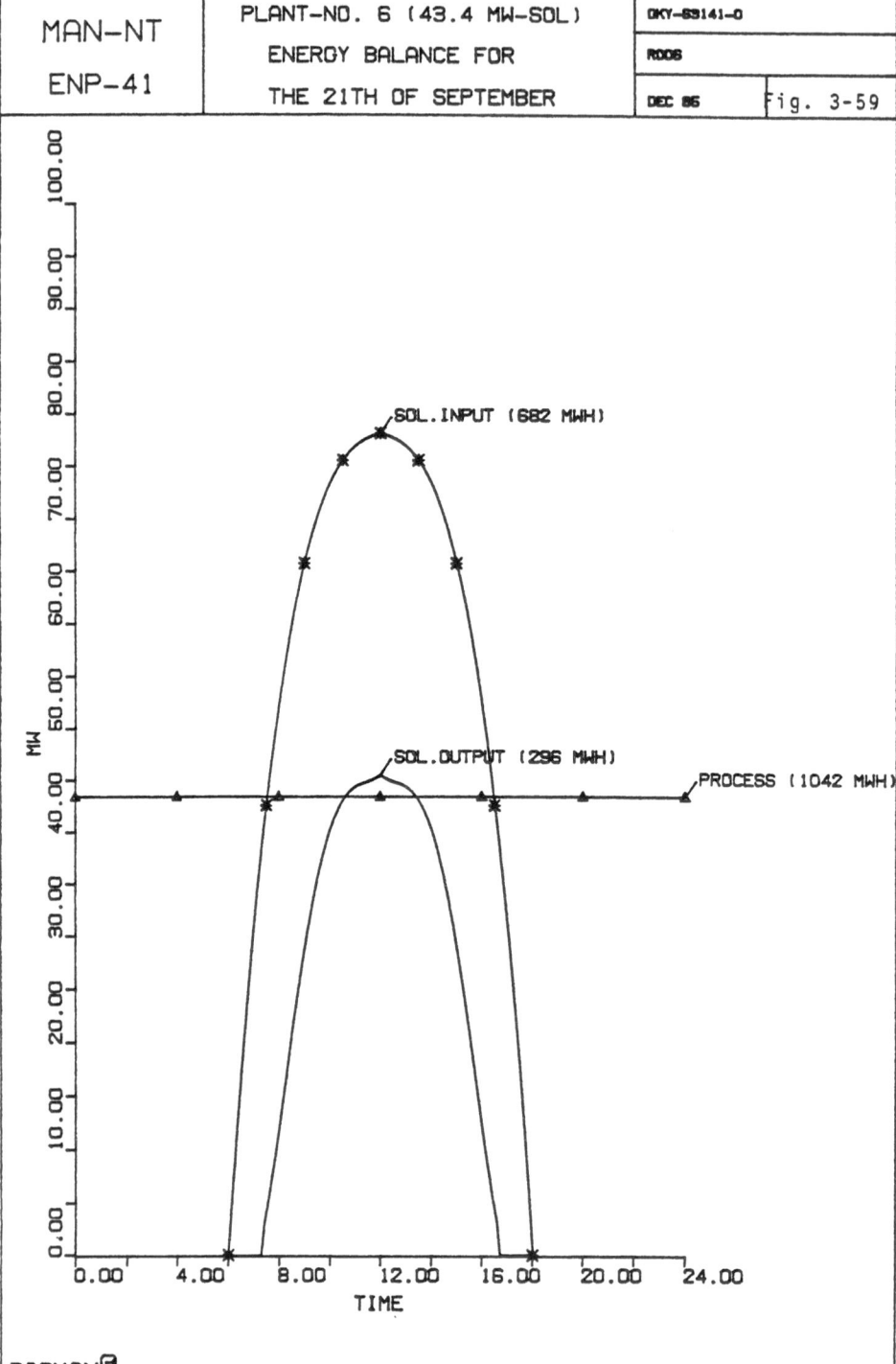

BASMAN®

- 414 -

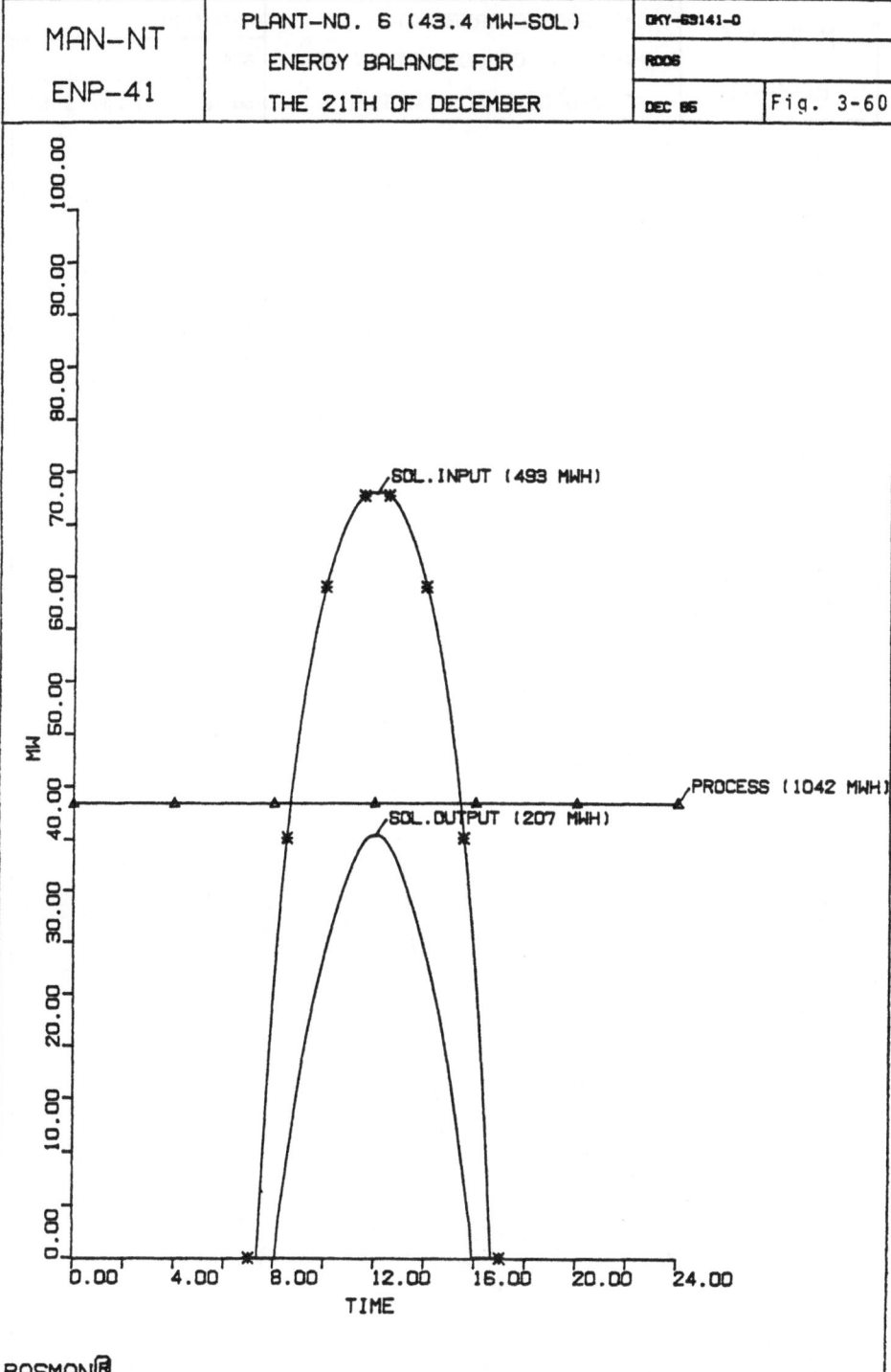

MAN-NT	PLANT-NO. 6 (43.4 MW-SOL)	OKY-69141-0	
ENP-41	ENERGY BALANCE FOR	RODS	
	THE 21TH OF DECEMBER	DEC 86	Fig. 3-60

SOL.INPUT (493 MWH)

PROCESS (1042 MWH)

SOL.OUTPUT (207 MWH)

TIME

MW

BASMAN®

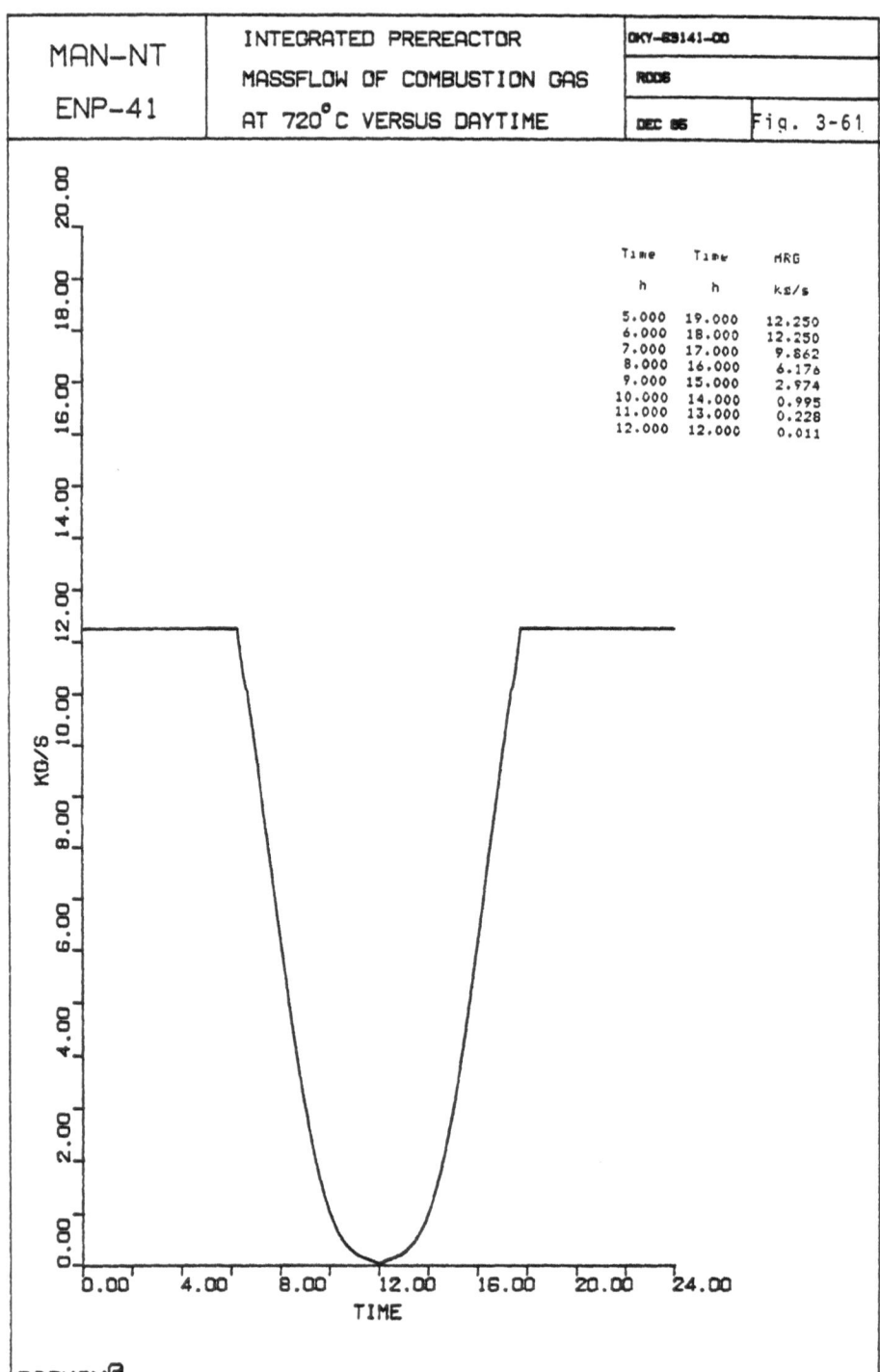

MAN—NT ENP-41	INTEGRATED PREREACTOR MASSFLOW OF COMBUSTION GAS AT 720°C VERSUS DAYTIME	DKY—83141—00	
		RDOS	
		DEC 86	Fig. 3-61

Time h	Time h	MRG kg/s
5.000	19.000	12.250
6.000	18.000	12.250
7.000	17.000	9.862
8.000	16.000	6.176
9.000	15.000	2.974
10.000	14.000	0.995
11.000	13.000	0.228
12.000	12.000	0.011

BASMAN

- 416 -

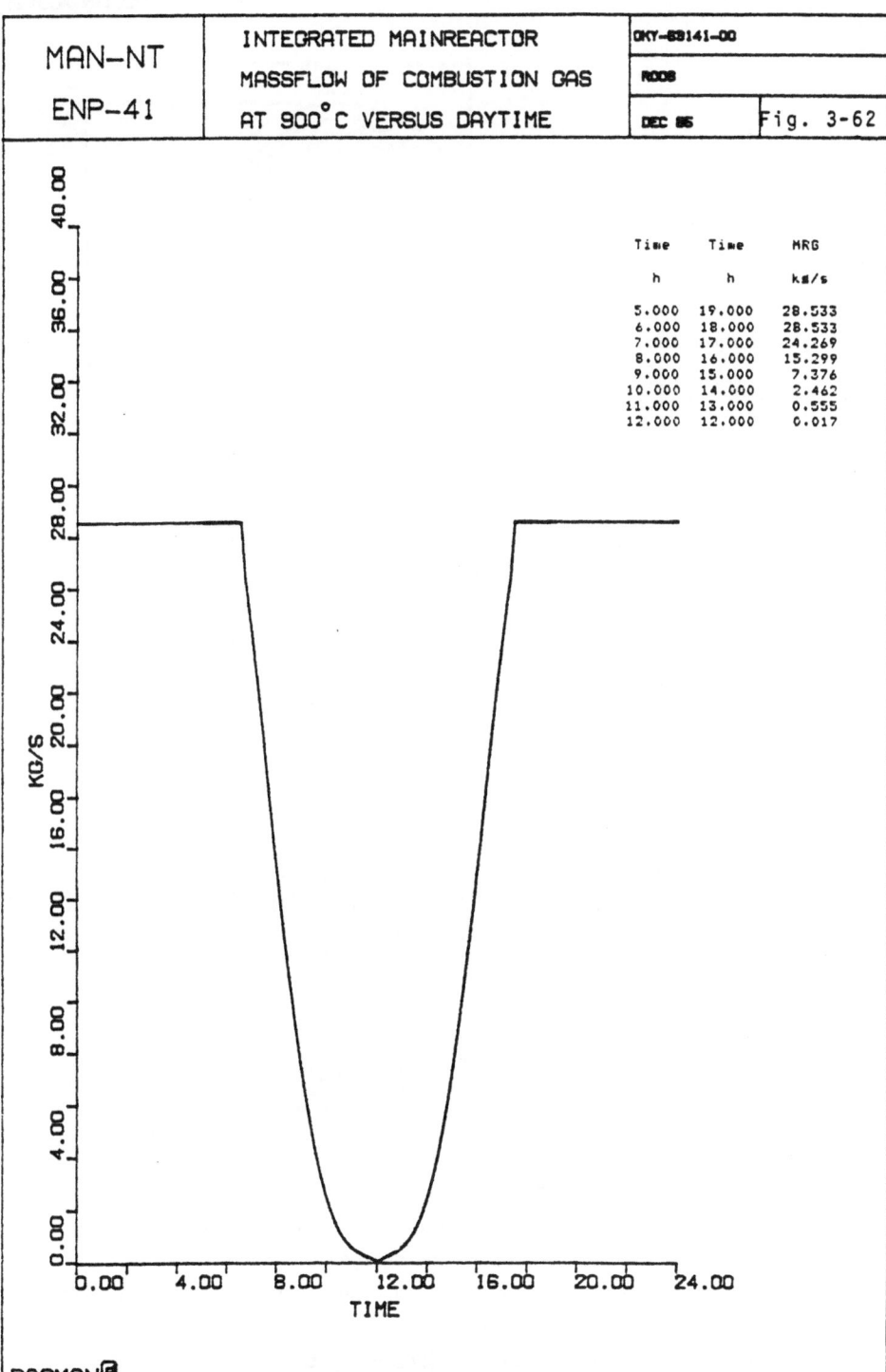

MAN–NT ENP–41	INTEGRATED MAINREACTOR MASSFLOW OF COMBUSTION GAS AT 900°C VERSUS DAYTIME	OKY–89141–00	
		ROOB	
		DEC 86	Fig. 3-62

Time	Time	MRG
h	h	kg/s
5.000	19.000	28.533
6.000	18.000	28.533
7.000	17.000	24.269
8.000	16.000	15.299
9.000	15.000	7.376
10.000	14.000	2.462
11.000	13.000	0.555
12.000	12.000	0.017

BASMAN®

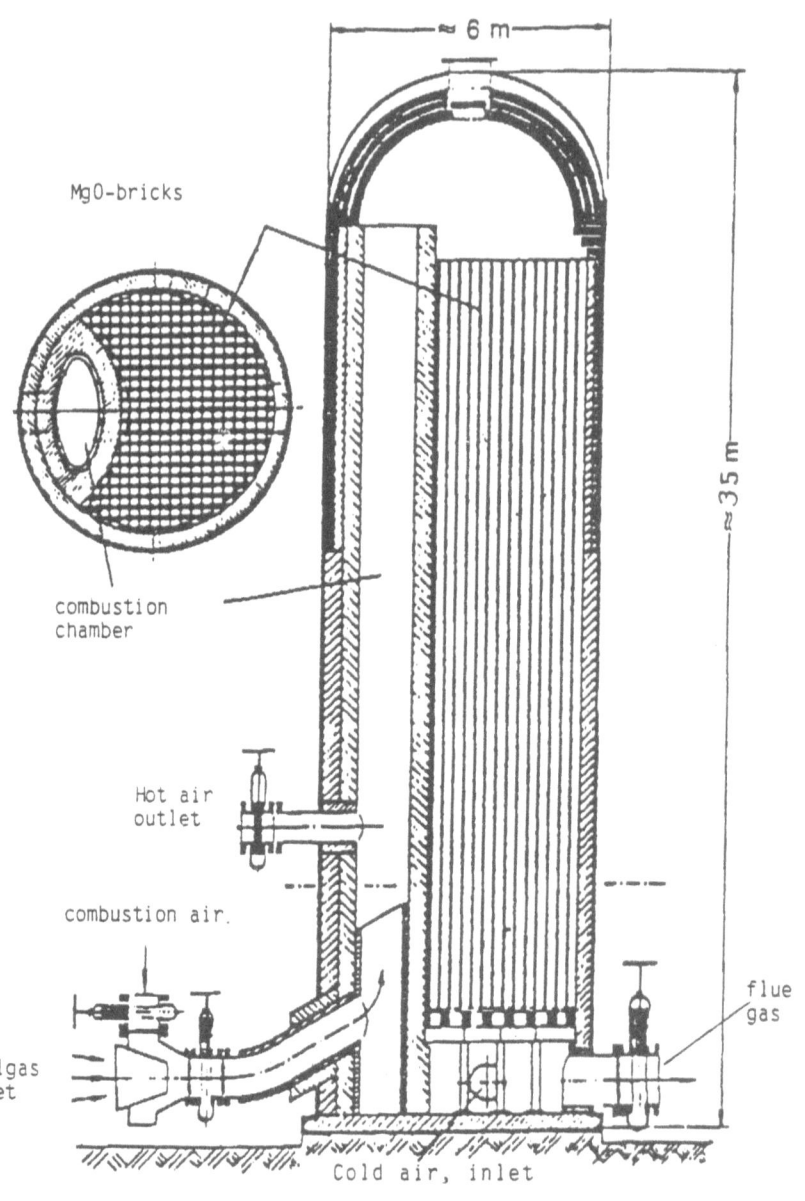

Fig.3-63:Sketch of storage module

(Any combination of the fossil fired energy source with the storage
module isn't foreseen yet in the design)

Source: Bohn, Bitterlich: Grundlagen der Energie- und Kraftwerkstechnik/
Handbuchreihe Energie, Band 1
Technischer Verlag Resch 1982 (according to /6/)

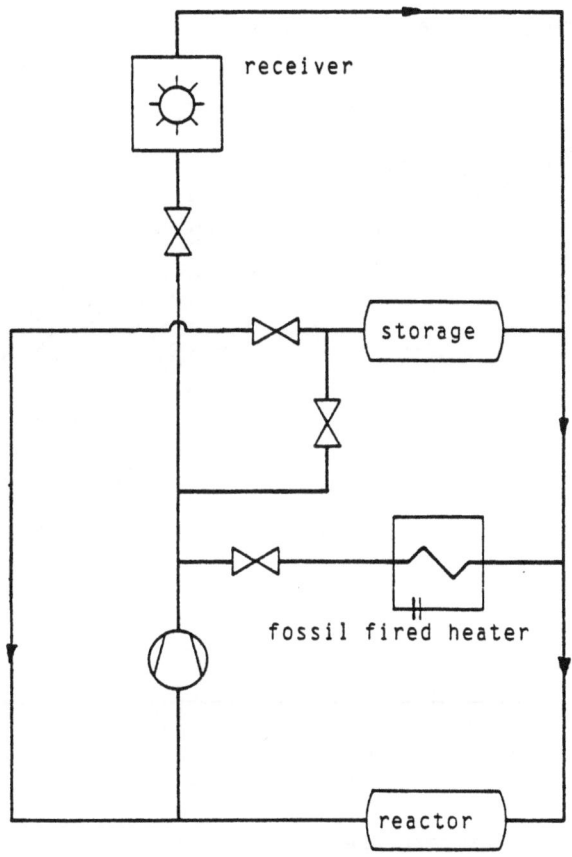

receiver

storage

fossil fired heater

reactor

Fig. 3-64: Flow scheme of primary loop without
recuperation

Temperature profiles during charging and discharging cycles

Charging energy: 327,2 MWh

Number of storage moduls: 3

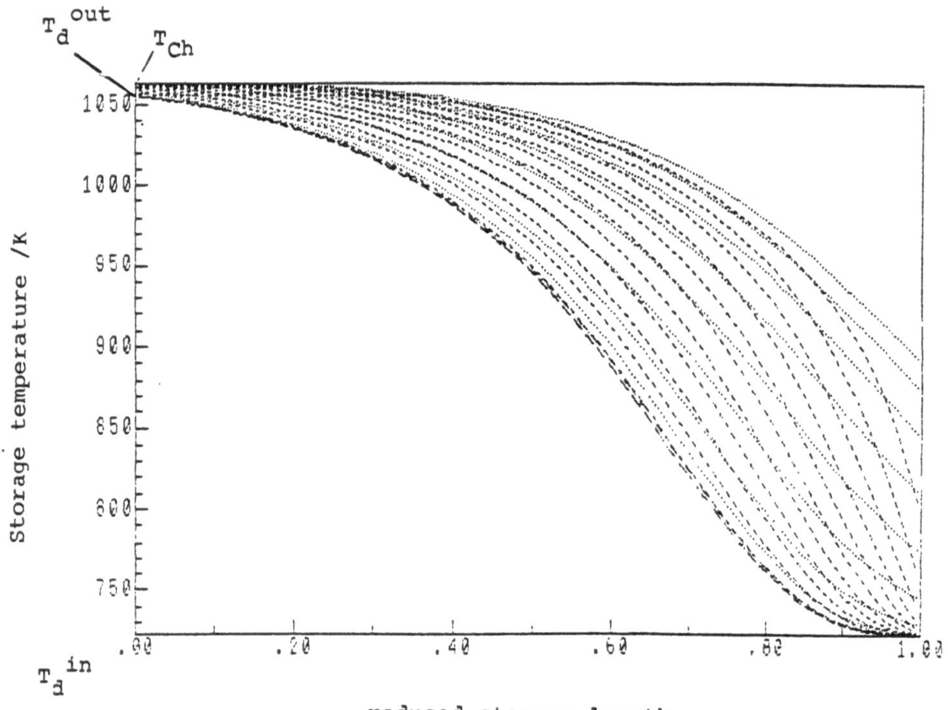

reduced storage length

T_{Ch}	–	temperature of charging gas at storage inlet	=	1063 K	=	790°
T_d^{out}	–	required gas temperature during discharging cycle	=	1053 K	=	780°
T_d^{in}	–	temperature of discharging gas at storage inlet	=	723 K	=	450°

Fig. 3-65: Storage temperatures during charging and discharging cycles (plant concept 2.3)
(according to /6/)

Charging energy: 327,2 MWh

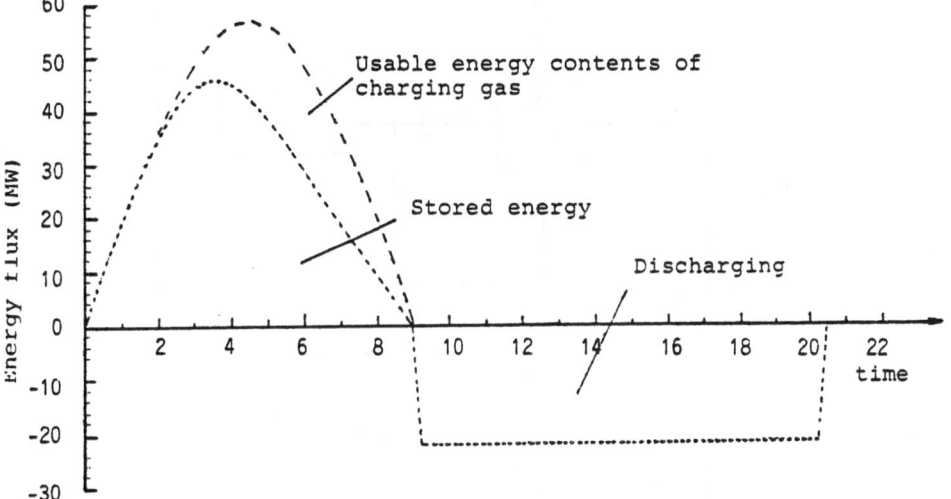

Fig. 3-66:   Energy flux during charging and discharging cycle
             (according to /6/)

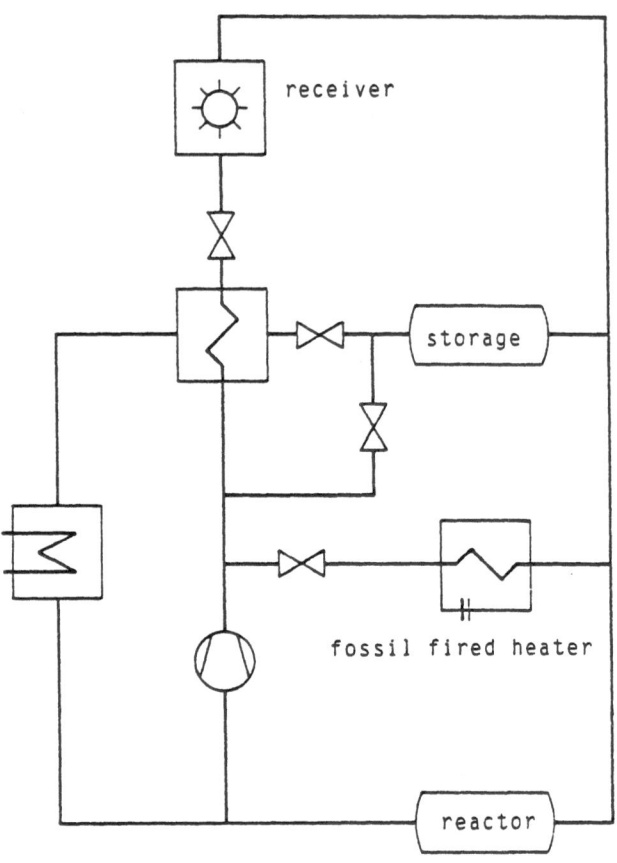

receiver

storage

fossil fired heater

reactor

Fig. 3-67: Flow scheme of primary loop with recuperation

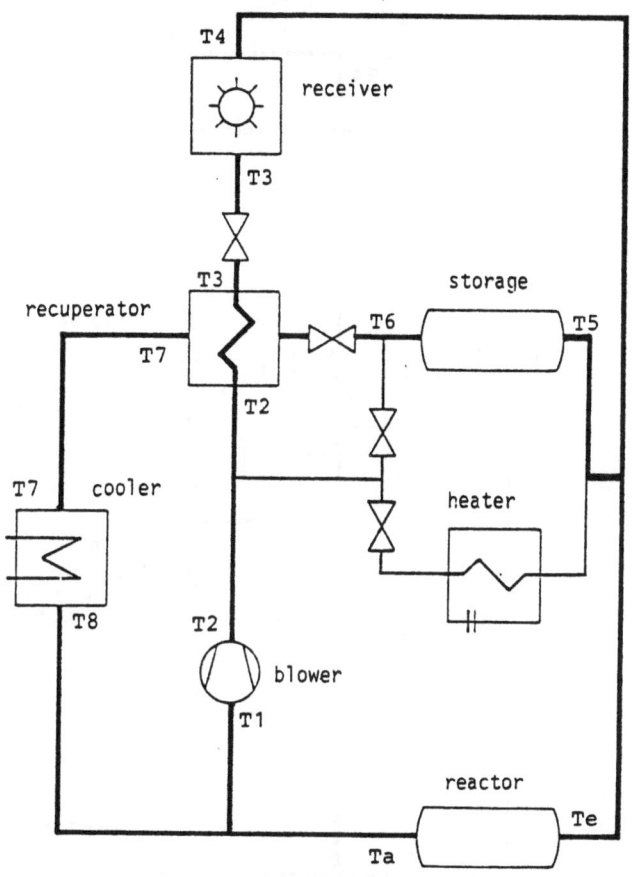

Fig. 3-68: Direct receiver and charging mode .

- 423 -

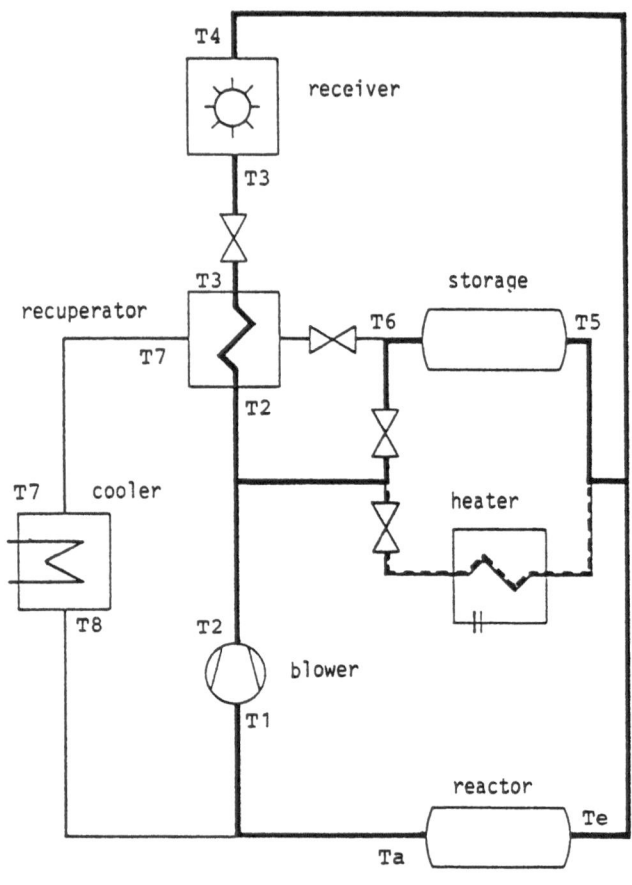

Fig. 3-69: Direct receiver and discharging mode

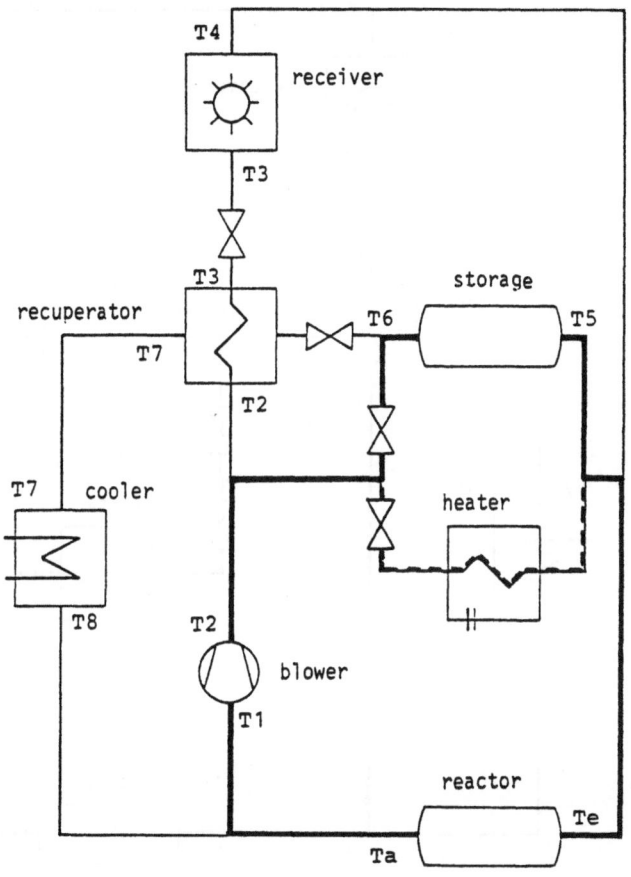

Fig. 3-70: Discharging mode

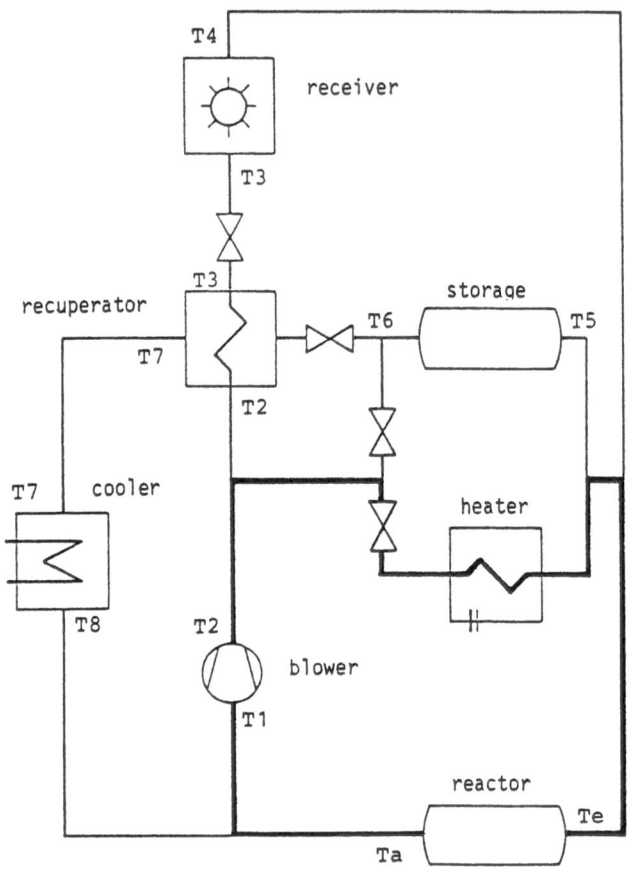

Fig. 3-71: Pure heater mode

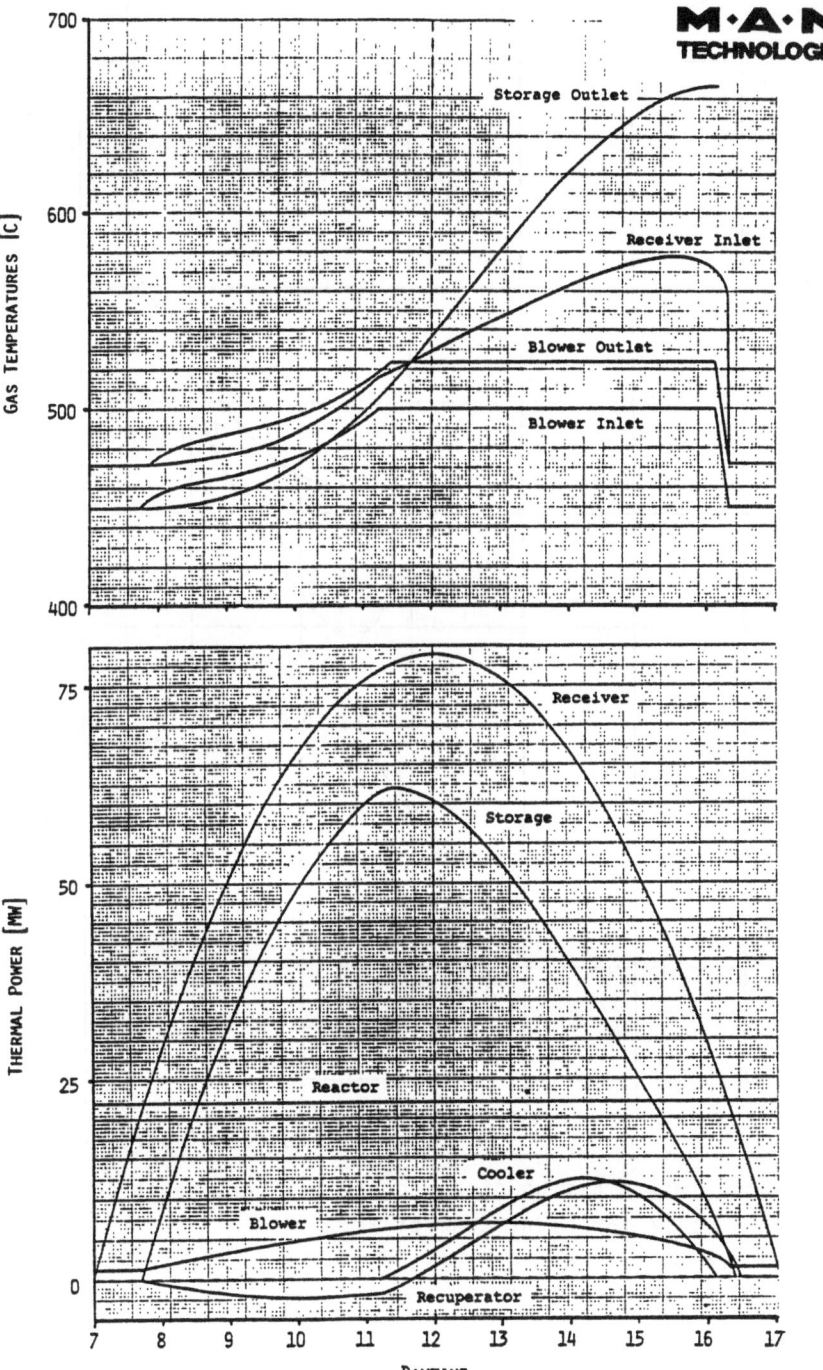

Fig. 3-72: Gas Temperatures and Thermal Power of Different Components
            of Concept 2.3 at 21th March (Loading Mode)

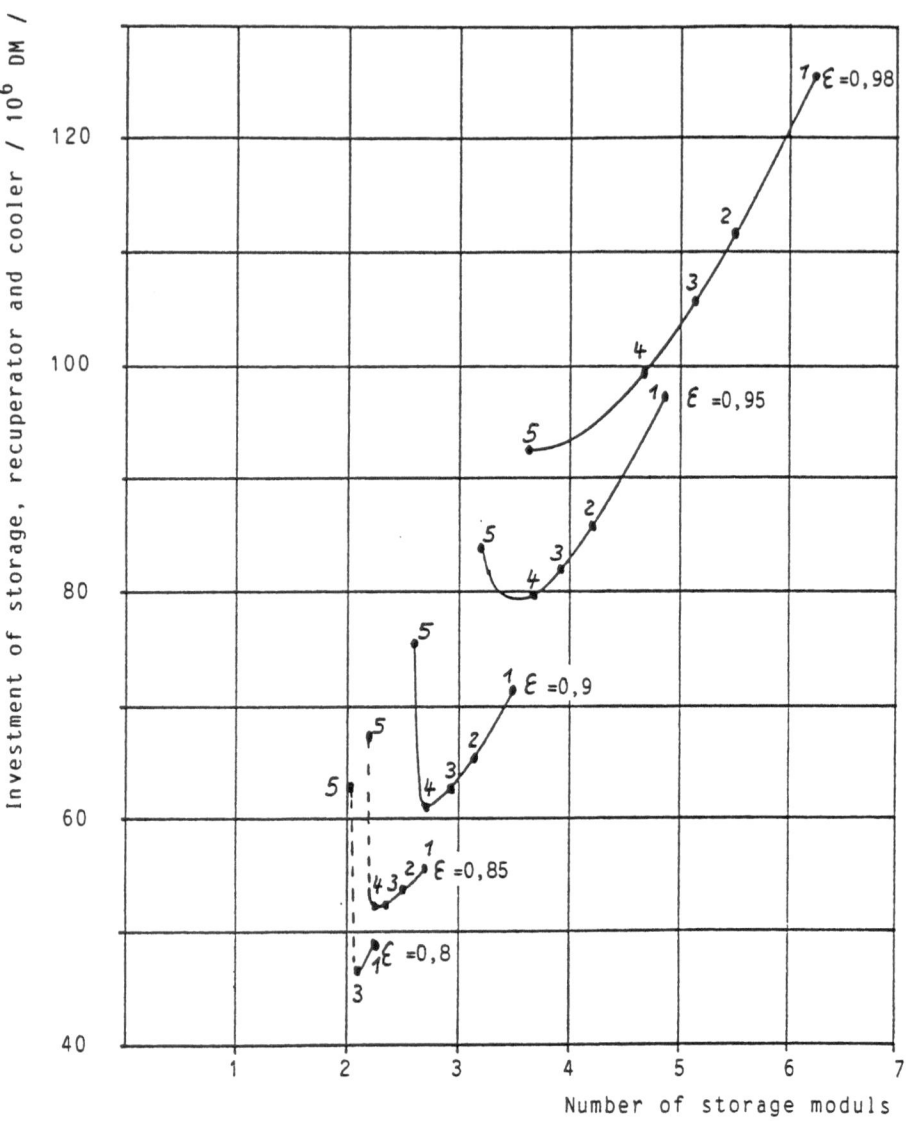

Fig. 3-73:   Determination of an economical reasonable storage
             size for plant concept 2.3
(Heat exchanger area of recuperator: 1 - 0 m², 2 - 2500 m²,
 3 - 5000 m², 4 - 10000 m², 5 - 40000 m²)

4.   Design and Construction of Separated Solar Reactors

4.1   Design Considerations

4.1.1   General Purpose

The design conditions for solar systems depend from many
factors. Each of the tabulated systems A to D or ...-1
to ...-6 principally requires a special design for the solar reactor. Due to the principle standardization, which
has been chosen, reactors will not differ very much but
only in minor details from each other,independent whether
they are constructed for a system A, B, C or D.

For the purpose of complete cost calculation it was necessary within this study, to fix the constructural details
of the solar supported reactors to a certain degree. The
results will be found in the attached drawings. By agreement it was fixed that reactors for the productin of methanol synthesis gas should be taken for detailed design.
Integrated reactors are handled separately within the solar chapter 3.

4.1.2   Choice of Appropriate Temperature Difference

In separated systems the design conditions are set by the
inlet and outlet temperatures of the media (heat carrier
and reactants), by the pressure of both sides and of
course by the energy rate to be exchanged. Of importance
also will be the quality and shape of the catalyst as it
will strongly influence the heat exchange within the tubes.
In the prereactor systems the upper temperature of the
heat carrier, which was fixed to 780 $^\circ$C at the inlet of
the reactor, will govern all secondary criteria, for the
main reactor systems the maximum process temperature,
which will have to be reached, will set such parameters.
The general problem of separated reactors can be seen by
Fig. 4 - 1 to Fig. 4 - 5: The heat requirement from the
process side is not a linear function with temperature but
is staped in such a manner that the temperature difference
between heat carrier and reactants will be smallest in the

- 429 -

middle of the reactor and can be less than 50 % of the
temperature differences at the ends, depending from the
choice of approach. It is therefore essential for the construction and cost of such plants that an optimal approach
should be chosen.

The following explanations will give some reasons for the
chosen parameters.

4.1.2.1 Principle_Dependencies

By the following mathematical combinations it will be
shown that specific mean logarithmic temperature differences must not be underrun, to still achieve reactors, which
technically can be built. This especially is essential
also, when considering systems A'-7 and A'-8, where near
approches in temperature are most desirable for good
efficiency.

Under the assumption of a high heat transfer coefficient
on the outside of a tube (at least $\alpha_{outside} = 10 \times \alpha_{inside}$)
and very thin walled tubes the following formulas will be
correct:

$$1) \quad d' \;\; = \;\; A \cdot m^{0,75}$$

$$2) \quad Fa \;\; = \;\; \pi \cdot Da \cdot L \cdot z$$

$$3) \quad \Delta p \;\; = \;\; B' \cdot L \cdot m^{1,9}$$

$$4) \quad Q \;\; = \;\; Fa \cdot \alpha' \cdot \Delta \vartheta$$

$$5) \quad m \;\; = \;\; G/F_h$$

$$6) \quad F_n \;\; = \;\; z \cdot \pi/4 \cdot Di^2$$

because of

$$k = \cfrac{1}{\cfrac{1}{\alpha_a} + \cfrac{Da}{\alpha_i \cdot D_i} + \cfrac{S}{\lambda} \cdot \cfrac{Di+Da}{2 \cdot D_i} + F} \approx \alpha_i \cdot \frac{Di}{Da} = \alpha'$$

when all terms except the second tend versus zero.

Wherein

α' = heat transfer coefficient of tube inside, related to outer surface of the tube (W/m^2K)

G = mass flow in z tubes (kg/sec)

Fa = outer surface of z tubes (m^2)

Da = outer diameter of tube (m)

Di = inner diameter of tube (m)

L = Length of heat transferring tube(s) (m)

z = number of tubes (-)

Δp = differential pressure of catalyst filled tube(s) (bar)

Q = transferred energy of z tubes (W)

F_n = $\pi \cdot Di^2/4$ = flow area (m^2)

$\Delta \vartheta$ = average logarithmic temperature difference (K)

k = overall heat transfer coefficient (W/m^2K)

α_a = heat transfer coefficient at outside of tube (W/m^2K)

α_i = actual heat transfer coefficient at inside of tube (W/m^2K)

s = wall thickness of tube (m)

λ = heat conductivity of material (W/mK)

F = fouling factors (m^2K/W)

A, B', B, S   specific constants with different dimensions, only valid for a defined heat exchanger

V_{cat} = Catalyst volume in z tubes (m^3)

By substitution it can be derived:

7) $Q/\Delta\vartheta = S \cdot z^{2,15} \cdot Da \cdot Di^{2,3} \cdot \Delta p \cdot G^{-1,15}$

8) $L = B \cdot G^{-1,9} \cdot z^{1,9} \cdot Di^{3,8} \cdot \Delta p$

9) $V_{cat} = z \cdot \pi \cdot Di^2 \cdot L/4$

With these correlations it is possible to describe the
effect of certain parameter variations, as may show the
following table.

This table has been established with the side conditions
as set out before (high outer heat transfer coefficient,
thin tubes, no fouling) and for an energy transfer rate
of

$$Q = 35,0 \text{ MW}$$

and a process gas mass flow of

$$G = 12,0 \text{ kg/sec.}$$

$\Delta\vartheta$ (K)	z (-)	Δp (bar)	D_a (m)	D_i (m)	L (m)	V_{cat} (m^3)
150	800	1,0	0,12	0,10	10	62,8
100	966	1,0	0,12	0,10	14,3	108,6
50	1334	1,0	0,12	0,10	26,4	276,5
150	580	2,0	0,12	0,10	10,8	49,3
80	776	2,0	0,12	0,10	18,9	115,2
50	966	2,0	0,12	0,10	28,6	217,1
50	908	1,0	0,15	0,13	34,5	415.3
50	658	2,0	0,15	0,13	37,4	326,1
80	2250	2,0	0,060	0,050	10,2	45,2
50	2800	2,0	0,060	0,050	15,5	85,3

With effective heat transfer coefficients at the outside
of the tubes being only about two times as high as in the
tube and with non negligible wall thickness of the tubes,
negative aspects are more evident even, which means:longer
tubes, more catalyst, higher pressure drop for the same $\Delta\vartheta$
and tube number.
Reactors of the proposed construction are limited in
height, weight and diameter for reasons of fabrication or
transport. If one sets a reasonable limit by experience
of about 15 m as the maximum allowable tube length, many
of the configurations of the table could be realized only
by admitting series of reactions, one behind the other,
what would be extremely undesirable. It comes out that

mean logarithmic temperature differences of less than 80K
should be avoided from such principle considerations.
What can be derived from the formulas and table:

a) the mean log. temperature difference should be as high
    as possible

b) smaller tube diameters are desirable, as they reduce
    the length of the reactor tubes and the volume of ne-
    cessary catalyst (But with smaller tubes systems with
    helical coil heat recovery heat recovery becomes im-
    possible!)

c) higher admissible differential pressures will reduce
    exchange area and needed catalyst volume but will in-
    crease the tube length

When also taking into account
    a) a finite value of the heat transfer coefficient at
       the outside

    b) a limited allowed height of catalyst because of
       limited strength

    c) a correlation between such admissible catalyst
       height and allowed pressure drop in the catalyst

    d) aging of catalyst

it becomes clear that below certain values of the log.
temperature difference no solution can be found, as such
real parameters will demand for not feasible tube sizes,
diameters or the like.

When considering such real effects it comes out that on
average for prereactors a $\Delta \vartheta$ of 50 K, for main reactors
a $\Delta \vartheta$ of 70 K should not be underrun. These values are
near to what has been the basis for the shown constructions.

4.1.2.2 Application for Prereactors

Fig. 4 - 1 shows the energy requirements of the prereactor for a specific natural gas quantity in dependency of temperature, to bring the process gas into equilibrium. The heat carrier is shown with constant specific heat, which is valid with very little deviation. Due to the curvature of the process gas function the average logarithmic temperature increment changes with temperature for the same amount of transferred energy. (The log. temp. increment has been indicated for each 200 kW energy increment). Fig. 4 - 2 shows,how the construction will be affected by such combinations, taking into account also different shapes of catalysts, which will influence the heat transfer properties at the inside of the reactor tubes. Much more length is necessary for the transfer of a fixed amount of energy in the middle of the reaction zone than at the ends.

The effect will be more distinct even with the mainreactor concept.

4.1.2.3 Application for Main Reactors

Fig. 4 - 3, similarly to Fig. 4 - 1, shows the energy requirements of the main reactor for a specific natural gas quantity in dependency of temperature under the assumption of thermodynamic equilibrium at each point of the reaction tube. The heat carrier is shown with constant specific heat, which also for the extended range of temperature still is valid with good approximation. Due to the curvature of the process gas function,the average logarithmic temperature increment changes with temperature for the same amount of transferred energy. (The log. temperature increment has been indicated for each 250 kW energy increment).
Fig. 4 - 4 shows the effects as Fig. 4 - 2. The small log. temperature differences in the middle of the reactor are a burden, which enlarges the reactor by increasing the necessary tube length.

A nearer approach than chosen hardly will be possible.
For comparison also the relevant line for the carrier gas
in case of the prereactor is indicated in Fig. 4 - 3. The
lines for the heat carrier are not identic, because the
specific recycle flow rate per energy quantity is smaller
with the main reactor than with the prereactor, which with
regard to secondary energy needs is in favour of the main
reactor system.

4.1.2.4 Application for Main Reactors at Low Pressure

As has basically already been pointed out in chapter 2.1.4
it is possible only to achieve maximum temperatures of
about 800 °C at the inlet of a separated reactor due to
available techniques with metallic receivers. Higher temperatures would require not yet available ceramic tubing
in the receiver and relevant fabrication procedures. Purposely it therefore was the aim of this study also to investigate, whether it would be possible with such limitation
in temperature to design a workable system.

That such a system would be possible, already has been
shown by Fig. 2 - 5 in chapter 2.1.4, where also the gas-
composition for this system A' - 7 has been given. Further
details have been given with Fig. 2 - 14 in chapter
2.2.2.2.

To sum up the results of these aforegoing and later following chapters:

Due to the high energy consumption of the reformed gas compression to synthesis pressure the specific natural gas
consumption is nearly as high as for a purely fossil
operated plant. Because of the additional investment of
the solar plant, the product prices will be compatible only around the year 2060. Other, with nowadays techniques
presentable solar systems will result in cheaper product
costs than with the compared system A' - 7.

Therefore the design considerations have not been extended
for system A' - 7 beyond some principle investigations.
Fig. 4 - 5 shows the correlations of energy input and temperature in analogy to Fig. 4 - 1 and 4.3. It becomes
evident that a smaller temperature approach between heat
carrier and reactant technically cannot be chosen. It
shows that an exit temperature of the heat carrier from
the reactor of 480 ^{o}C for this reason cannot be underrun.

By this lower temperature and the fixed upper one the recycle rate for the heat carrier is fixed. As a result it
comes out that also the specific energy requirements for
driving the separated recycle are not advantageous. The
relative figures for the energy transport compare as
follows:

Separated Prereactor, at high pressure   = 1,00

Separated Main Reactor, at high pressure = o,75

Separated Main Reactor, at low pressure   = 1,35

Because of the tight temperature approach, as it is shown
by Fig. 4 - 5, it can be expected that also the reactor
will be very expensive in comparison to higher pressure
versions, especially as with the low pressure of the reactants special precautions have to be respected to stay
within reasonable differential pressures, which condition
will require additional heat exchange area and catalyst.
On the background of such results no special efforts have
been made to work out detailed constructional drawings of
the reactor for system A' - 7.

4.2   Construction Drawings

4.2.1   Construction of Prereactor
         Two alternatives are presented by drawings
                 D 6314100-OL-00001, Fig. 4 - 6
                 D 6314100-OL-00002, Fig. 4 - 7

Fig. 4 - 6 shows an internally clad, vertical vessel. By
the internal cladding the temperature of the pressure
bearing casing is only slightly above ambient. The heat
carrier enters radial at the lower end and is forced
around the catalyst filled tubes on a cross-countercurrent
flow. The tubes are arranged in a "no-tubes-in-the-win-
dow"-principle, allowing high heat transfer coefficients
at the outside of the tubes with little pressure drop.
The holes in the baffles are slightly increased above minimum for better heat transfer (axial component being increased) and easier mounting of the tubes. The cooled air
(450°C) leaves the reactor below the tube plate.

The reactants enter from top and are distributed to the
tubes by differential pressure only. The tubes are filled
with highly active catalyst at the top for about 3,5 m.
The catalyst is hold by special supports in each tube, to
be brought into the tube after filling the lower part catalyst. It is supposed to leave about 10 times the diameter
of the tubes free of catalyst below these supports for reasons
of better heat transfer below. (Generation of turbulences
and high heat transfer in the first part of a catalytic
zone). In the lower part for a height of about - 5,5 m
normal steam reformer catalyst in form of rings will be
found. The support of the catalyst is shown as by inert
ceramic balls in the outlet cone. The reacted gas leaves
the reactor at the bottom flange.

Fig. 4 - 7 shows a reactor with cooled wall. By this principle of construction the pressure bearing casing is kept
at the temperature of the leaving heat carrier. The heat
carrier (air) enters from top and is passed through a tube
to the lower tube plate, where it enters the "shell side"
of the heat-exchanger-reactor. The arrangement of tubes
and baffles is similar to Fig. 4 - 6. At the upper tube
plate the heat carrier is directed towards the wall of the
reactor and is flowing downwards between casing and an insulated guide plate. The heat carrier leaves the reactor
lateral at the bottom.

The reactants enter the reactor from top,too, but lateral.
The distribution to the single tube is effected by differential pressure only as in the case before. The catalyst
arrangement is identic as in Fig. 4 - 6, but the lower
catalyst is being supposed to be supported by gratings for
every tube, as they are usual for synthesis reactors of
similar nature. The reacted gas leaves by the central
lower flange.

4.2.2   Comparison of Constructions

The aforementioned constructions were developed also on
the basis of earlier concepts of solar studies. At the
time of origination it could not be decided, which type
of reactor would be preferable. This still is valid at
present, as the overall dimensions and the weight of both
reactors are rather equal and also the quoted prices are
differing only by about 4 %, which is within the margin
of error. As both types of reactors are common in chemical industry the final choice of one type would depend on
the overall layout of the plant and the specific arrangement in relation to other equipment.

4.2.3   Construction of Main Reactor

The proposed construction is shown by drawing

    D 6314100-0L-00003, Fig. 4 - 8

The construction is identic with that one of Fig. 4 - 7,
but differes in dimensions due to the altered load and
different temperature conditions.

The choice in favour of this second alternative was made,
because the construction is better apt for long walls, as
they are needed here. Then the internal cladding at high
temperatures becomes difficult, to prevent cracks by which
hot gases could reach the pressure bearing wall and could
generate troubels by hot spots.

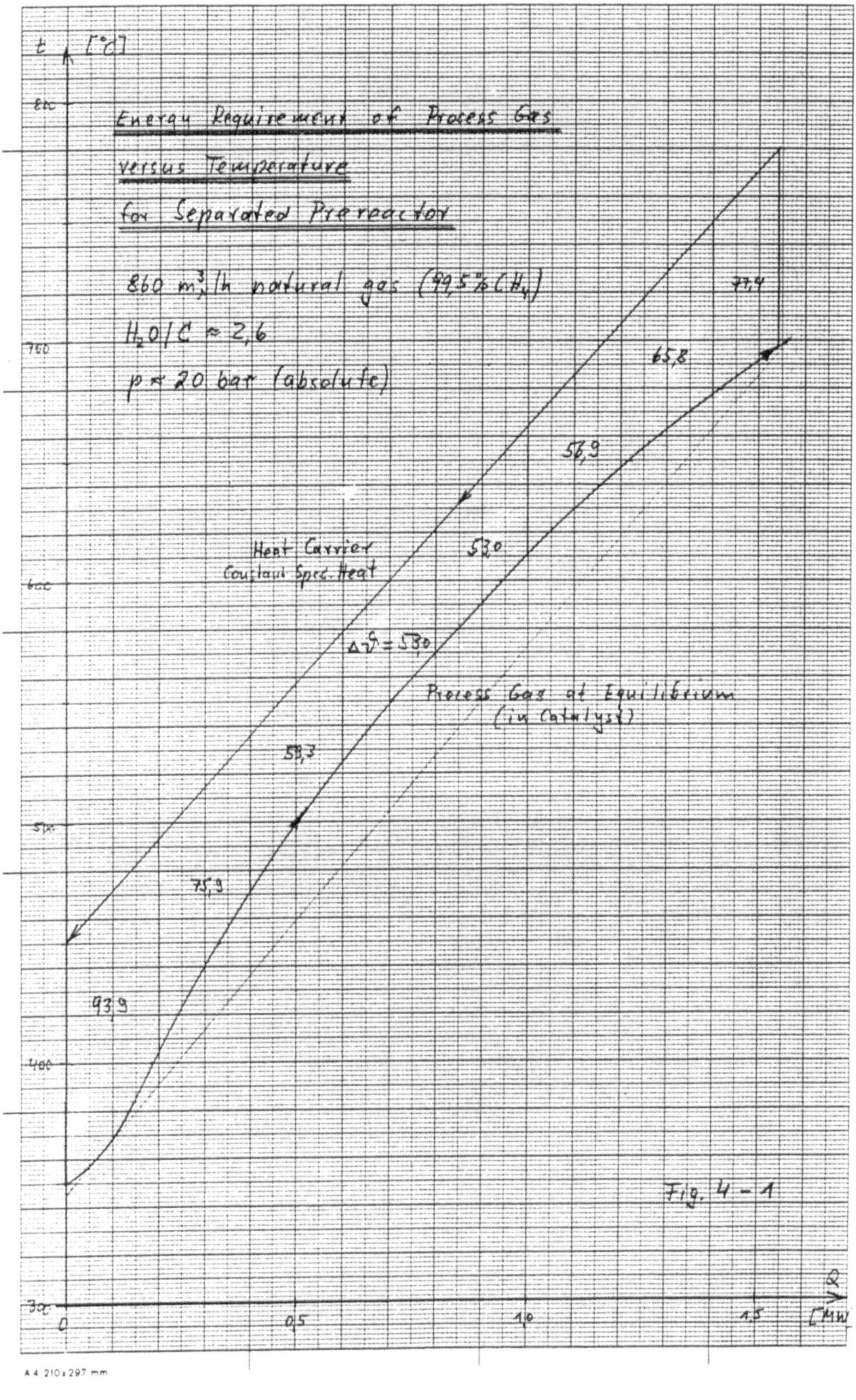

t [°C]

800

Energy Requirement of Process Gas
versus Temperature
for Separated Pre-reactor

860 m³/h natural gas (99,5% CH_4)

$H_2O/C \approx 2,6$

$p \approx 20$ bar (absolute)

700

77,4

65,8

58,9

Heat Carrier
Constant Spec. Heat

53,0

$\Delta \hat{r}^2 = 530$

Process Gas at Equilibrium
(in Catalyst)

58,3

500

75,3

Fig. 4 - 1

93,9

400

300

0   0,5   1,0   1,5   [MW]

A 4 210 x 297 mm

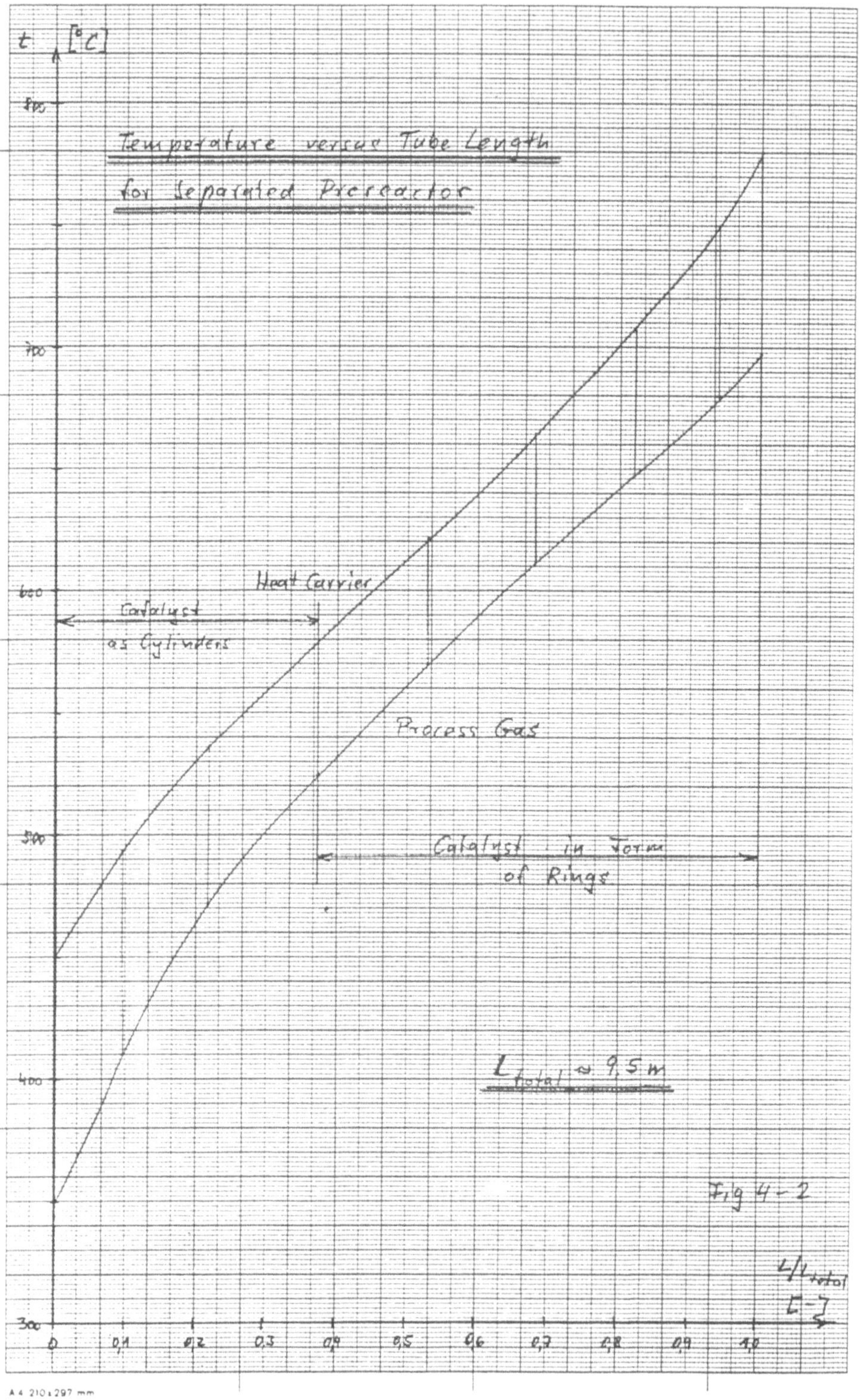

Temperature versus Tube Length for Separated Preheater

t [°C]

800

700

600 — Heat Carrier

Catalyst as Cylinders

Process Gas

500 — Catalyst in Form of Rings

400

$L_{total} \approx 9.5\,m$

Fig 4 - 2

L/L_{total} [-]

300

0   0.1   0.2   0.3   0.4   0.5   0.6   0.7   0.8   0.9   1.0

A 4 210 x 297 mm

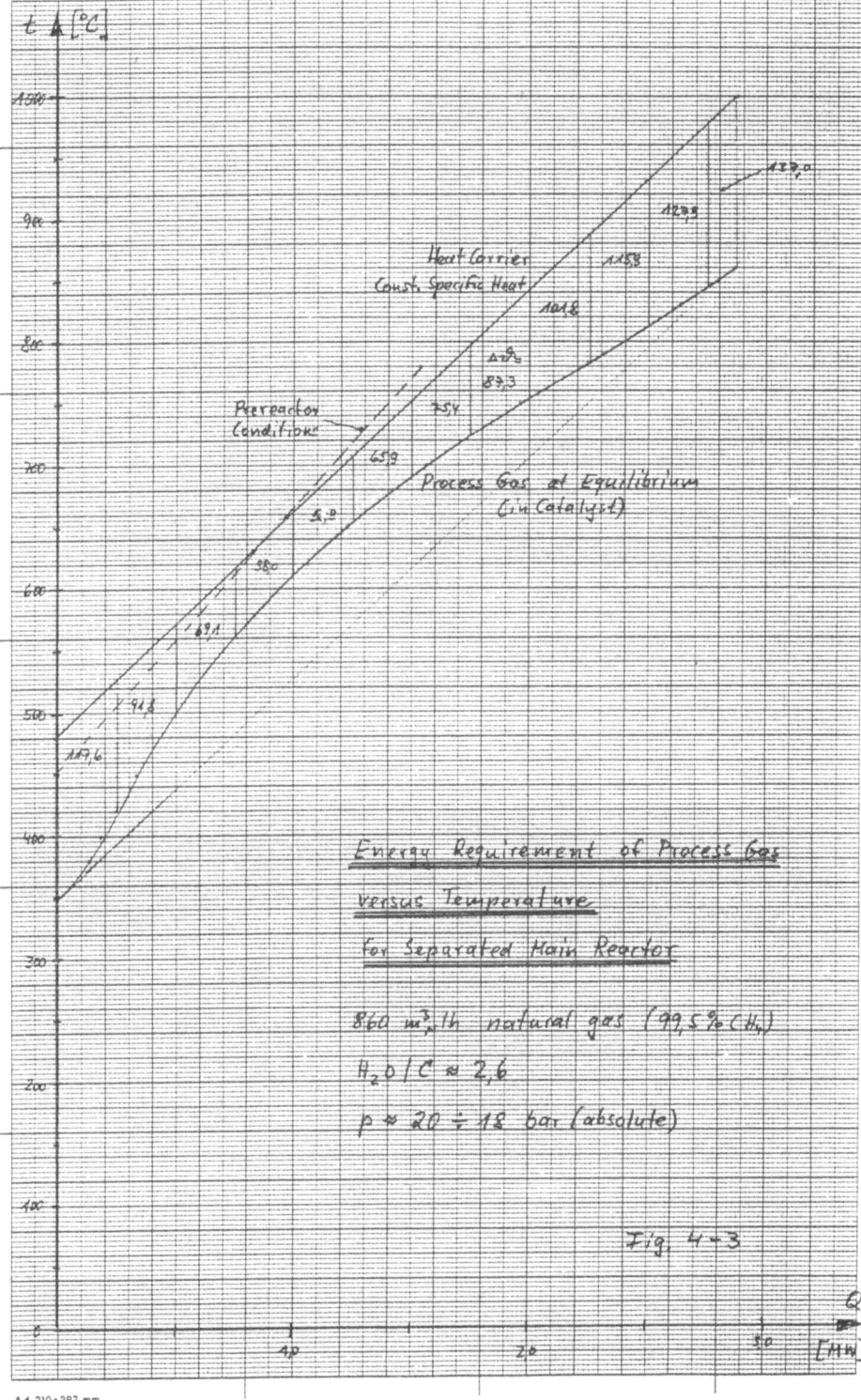

Fig. 4-3

Energy Requirement of Process Gas versus Temperature for Separated Main Reactor

$860 \, m^3/h$ natural gas (99,5 % CH_4)

$H_2O/C \approx 2,6$

$p \approx 20 \div 18$ bar (absolute)

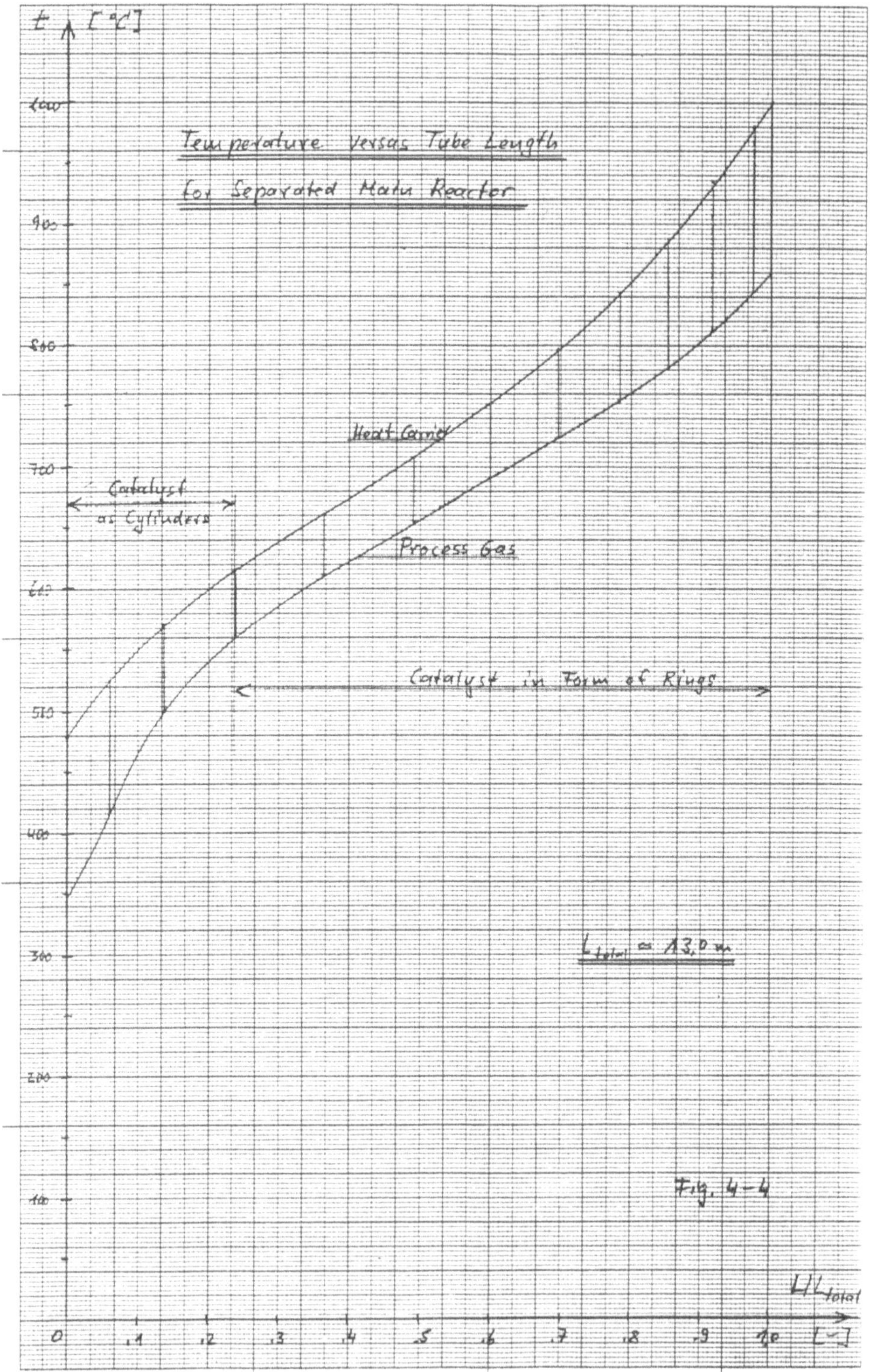

Temperature versus Tube Length for Separated Main Reactor

Fig. 4-4

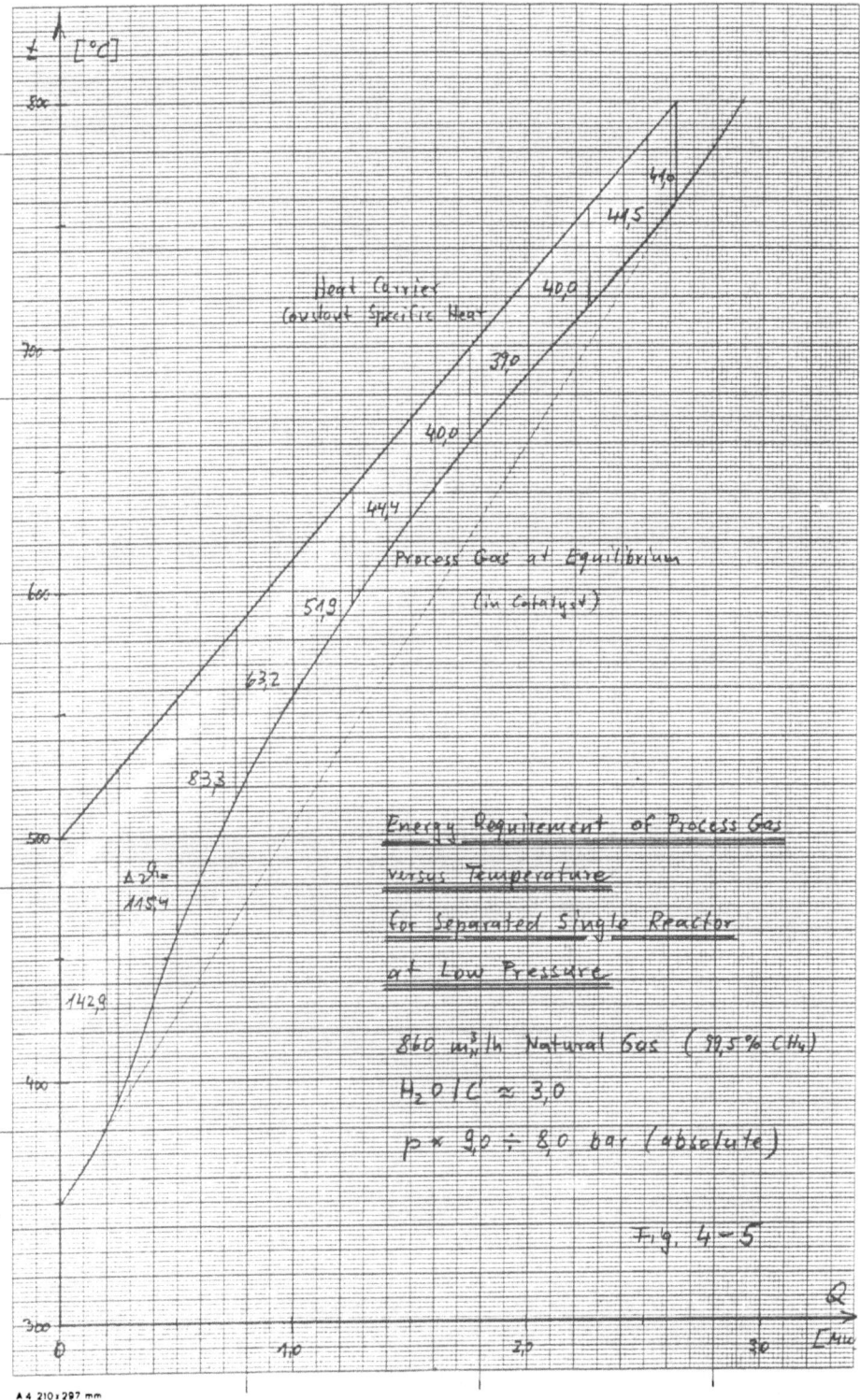

t [°C]

800

700

600

500

400

300

Heat Carrier
Constant Specific Heat

Process Gas at Equilibrium
(in Catalyst)

41,0
41,5
40,0
39,0
40,0
44,4
51,9
63,2
83,3

$\Delta z \frac{q_o}{a} =$
115,4

142,9

Energy Requirement of Process Gas
versus Temperature
for Separated Single Reactor
at Low Pressure

860 m_N^3/h Natural Gas (99,5% CH_4)
$H_2O/C \approx 3,0$
$p \approx 9,0 \div 8,0$ bar (absolute)

Fig. 4-5

0   1,0   2,0   3,0   Q [MW]

A 4 210 x 297 mm

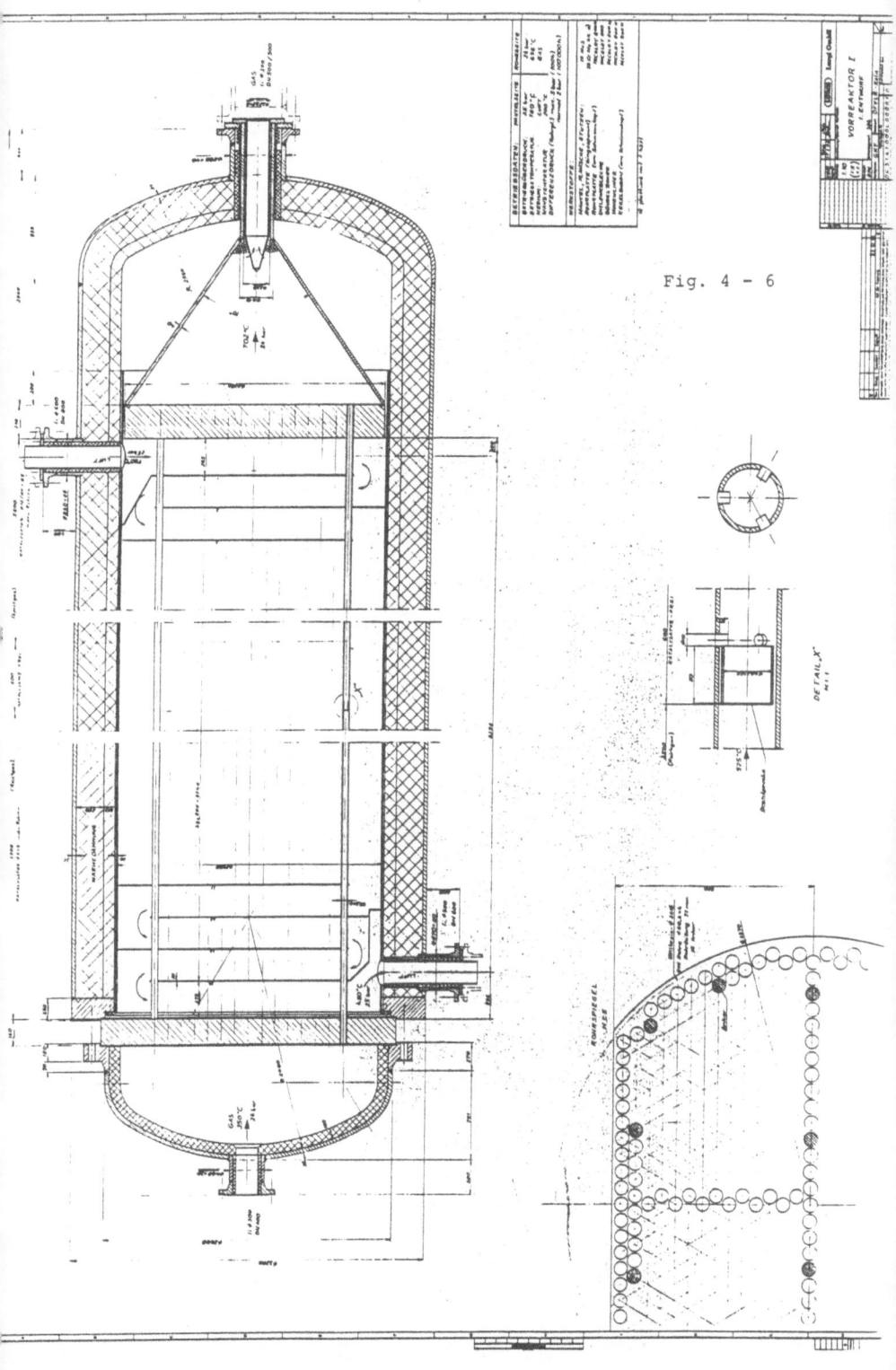

Fig. 4 - 6

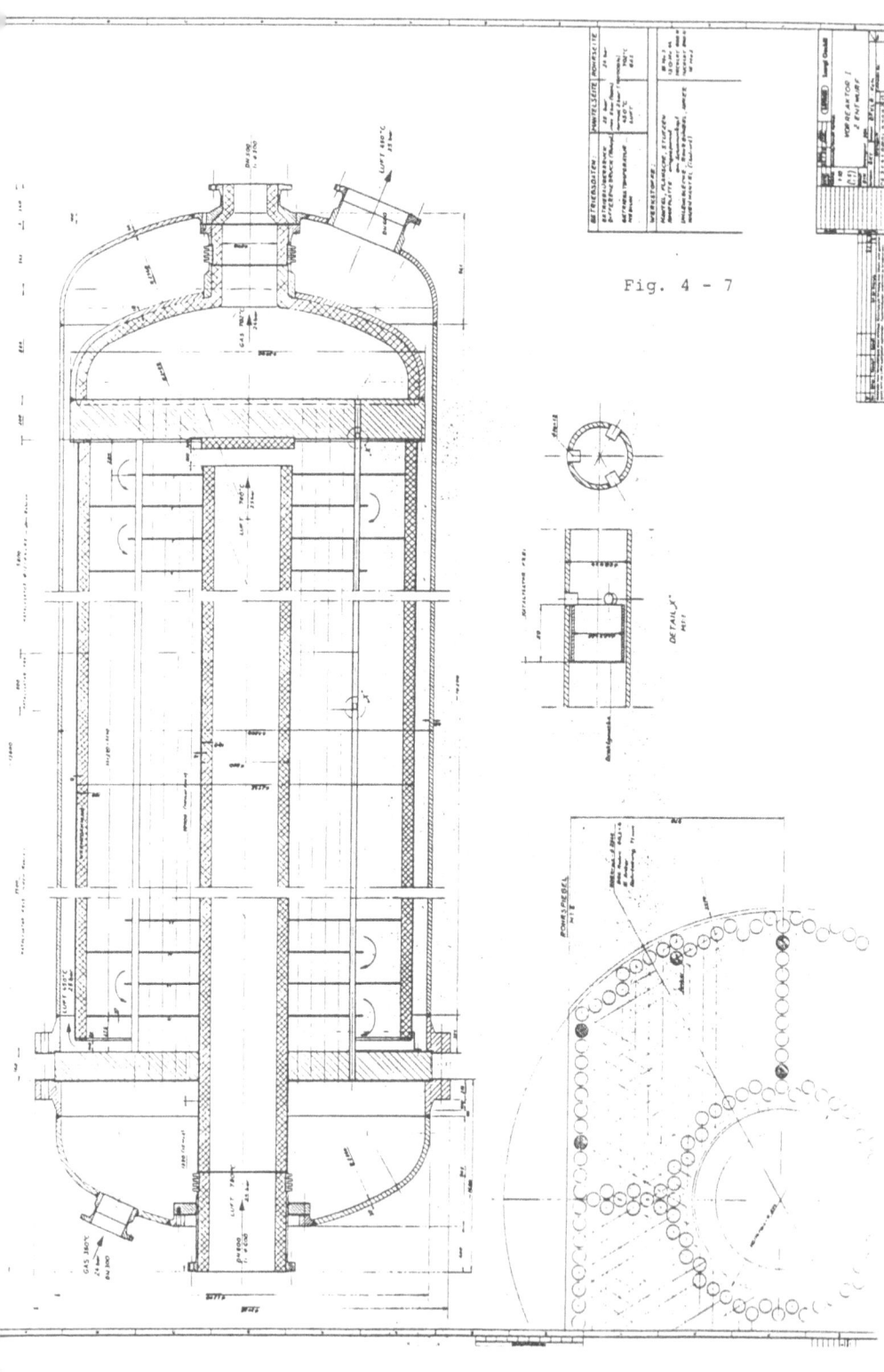

Fig. 4 - 7

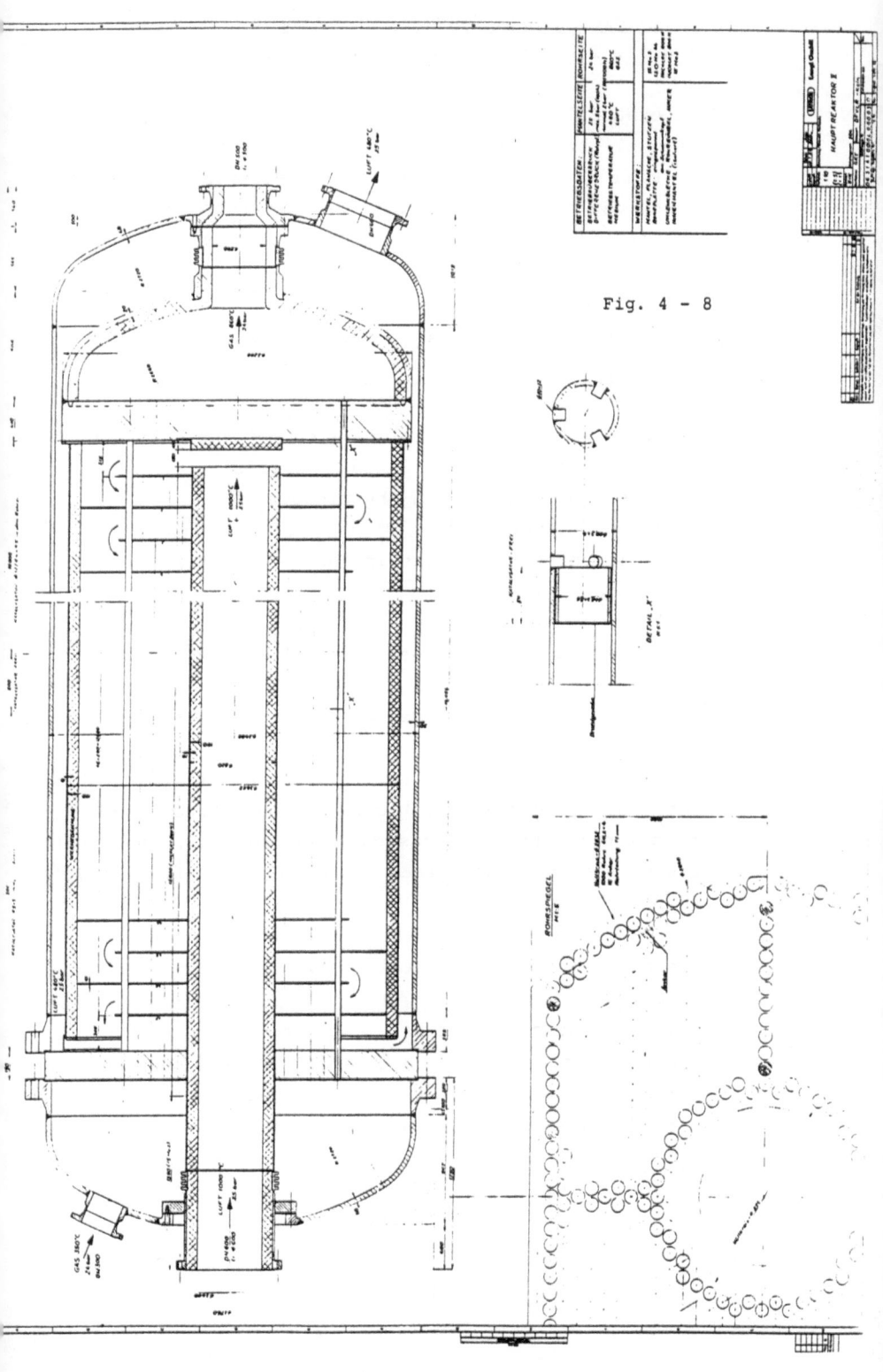

Fig. 4 - 8

5.   Annual Solar Energy Output, Solar Energy Rates in
        Products, Investments and Product Costs

        This chapter mainly is made up from tables and graphs.
        There will be given general comments for the purpose of
        each kind of table/graph and some specifics with the de-
        tailed results. It is inevitable that there have to be
        made many crosswise references to avoid constant repeti-
        tions.

5.1   Summary of Costs and Performance Data of the Solar
        Section of the Plant

        The comparison of the solar operated methane reforming
        plants is based on the following criteria
        - share of solar energy in the product,
        - specific product costs,
        - specific investment costs in the year 2030 and
        - development risk.

        With the design results described in Section 3 for the
        solar part of the plant, the costs of all subsystems of
        this part of the plant, the solar energy transferred to
        the reactor and the own energy requirements of the plant
        can be determined. Together with the assumptions relating
        to economic marginal conditions and the data of the
        chemical plant to be discussed in Section 5.2, the first
        three of the above specified criteria can be fulfilled
        with these data.
        Tab. 5.1-1 provides an overview of the costs and
        performance data of the total of ten solar plants under
        examination.
        The investment costs are based on estimations from other
        projects, particularly from the GAST project and the
        relevant literature which are transferred in a suitable
        way to the plants under examination in this study.

        Heliostat field costs: For the heliostat field, costs of
        750 DM/m² are estimated for the heliostat field ready for

operation. This assumption has been adopted from /1/, where estimations are based on a heliostat with a surface area of 51.84 m².

Receiver costs: The costs both for the 800 °C receiver and for the 1000 °C receiver are derived from the GAST 20 MW_{el} plants. Results from /2/ were adapted for the costs of the integrated solutions; these costs contain the additional cavity firing. The receiver costs are based on the following structure:

- Cavity costs - were recalculated corresponding to the surface conditions

- Costs of ring header and distributor lines - were recalculated, while taking into consideration length, mass flow and pressure

- Heating surface costs - were recalculated in relationship to the heating surfaces

- Measuring, control and regulation systems - were assumed as being equal for all receivers

- Spare parts, transport, installation, equipment and engineering were estimated, based on the GAST receivers.

Tower costs: The estimation of the tower costs took into consideration the tower height, receiver mass and receiver size. Data from /1/, /3/ and /4/ were used for cost estimation.

Primary circuit - the blower and line costs were defined on the basis of estimated prices of a blower manufacturer and a manufacturer of internally insulated hot gas lines. The recuperator and cooler costs were adoptped from other projects and adapted via the heat exchanger surface to the equipment forming the basis of this study.

Accumulator - the estimated costs specified in /5/ (20
million DM/module) of DFVLR Stuttgart were adopted.
Others - includes the control and electrical equipment
necessary for the operation of the solar section of the
plant, the location costs (site procurement and
development as well as infrastructure) and constructional
services without the tower. For these costs, data were
used from /1/ which were recalculated in relationship to
the services.

With the investment costs specified in Tab. 5.1-1,
specific investment costs averaging 4900 DM/kW are
obtained for the examined systems, referred to the
nominal receiver output with a fluctuation range of
approx. +/-10 %.

The annual operating costs include the costs for
operating personnel, maintenance and repair costs for the
solar section of the plant. These costs were estimated as
being 2 %/a of the investment costs. This cost range is
also used in /1/ (1.8 %/a), and as part of own
examinations a cost rate of 2.7 %/a was determined for a
small plant.

The performance data were defined in Section 3 and are
listed in Tab. 5.1-1.
The annual energy requirements of the reactors at
constant plant capacity amount to approx. 190 GWh for the
plants with a prereactor (plants 1, 2 and 5) and to
approx. 350 GWh for the plants with a main reactor
(plants 3, 4 and 6). Of these energy requirements,
between approx. 30 % (plant 1) and approx. 97 % (plant
4.3) can be covered by process heat generated by solar
systems. These values are based on the assumption of only
clear days over the entire year and start-up and shut-
down losses have not been taken into consideration.
The solar section of the plant has defined own power
requirements for making available the process heat at the
reactor which, in the case of the plants with separate

heat transfer circuit (plants 1 to 4), are clearly
dominated by the consumption of the blower required for
the circulation of the heat transfer medium. The
consumption specified in Tab. 5.1-1 is already given in
the form of a thermal energy equivalent which should be
equated with the consumption of a corresponding amount of
natural gas, whereby a degree of conversion efficiency of
the heating energy contained in the natural gas into
mechanical energy was assumed as being 35 %. It can be
seen that between 65 % (plant 1) and 25 % (plant 4.3) of
the solar energy must be provided when air is used as the
heat transfer medium for transporting the process heat
from the receiver to the reactor. The estimation of the
energy required for the circulation circuit was therefore
also carried out for helium, which provided considerably
more favourable results (approx. 25 % for plant 1 and 12
% for plant 4.3).

Further costs were calculated under the assumption that
helium is used in the primary circuit as the heat
transfer medium.
Based on the investment costs, the annual operating costs
(personnel, maintenance, repair and own consumption) and
the amount of process heat transferred annually to the
reactor, the specific energy costs can be estimated for
process heat produced by the solar plant at a temperature
of 800 °C to 1000 °C and a pressure of 20 bar. Further
marginal conditions required for this estimation are
listed in Tab. 5.1-2.

The dynamic present value method is used for calculating
the specific energy production costs. The specific energy
costs determined for 1995 when using helium as the heat
transfer medium are specified in Tab. 5.1-1. The
development of these costs between 1995 and 2020 referred
to the base case in Tab. 5.1-2 are specified in Fig. 5.1-
1 to 5.1-10. In accordance to the above mentioned criteria
on which the comparison of the plant concepts is based
the plant concepts 4.3 and 6 were selected as most
promising (see below). Therefore for the plants 4.3 and
6, the cost progression was also determined for the best

case and the worst case in accordance with Tab. 5.1-2 and
illustrated in Figs. 5.1-11 and 5.1-12.
It should be mentioned at this point that the costs and
performance data specified in this study were compiled
particularly from the point of view of mutual comparability of the 10 examined plants. Deviations from the
values specified in this study can occur during detailed
design and precise cost calculations and particularly in
the case of more precise assumptions relating to
insolation conditions.

Summarizing, however, it should be stated once again that
with the storage systems in accordance with concept 4 up
to approx. 90 % (under idealized insolation conditions)
of the natural gas required to heat a reformer can be
substituted by process heat produced by a solar system.
The costs of the process heat (800°C - 1000°C, 22 bar)
should then be between 10 DPf/kWh and 15 DPf/kWh.
In /6/ specific energy costs of 7.6 US Cent/kWh were
estimated for air heated systems (800°C - 1100°C, 10
bar).

Bibliography to Section 5.1

/1/ Potential Gasgekühlter Sonnenturmkraftwerke
Hrgb.: Becker, M., H. Dunker, H.N. Sharan,
DFVLR Köln, 1985

/2/ Steam Reforming of Methane Utilizing Solar Heat
Abschlußbericht der Firmen LURGI und MAN-Neue
Technologie, Februar 1984

/3/ Battelson, K.W. "Solar Power Tower Design Guide:
Solar Thermal Central Receiver Power Systems,
A Source of Electricity and/or Process Heat"
SAND 81.8005, 1981

/4/ Schlußbericht der ARGE GAST für die Vorentwurfs-
phase
GAST 1 ARG BSA 11 000 003, 1980

/5/ Klaiß, H., G. Merkel, M. Geyer
"Speicheranalyse im Rahmen der Studie
'Vergleichende Untersuchung und Bewertung ver-
schiedener solarer Methanreformierungs-Systeme
mit Folgeanlage'"
DFVLR EN-TT, IB 444 005/86, 1986

/6/ De Laquil, C.L. Yang, J.E. Noring
"Solar Central Receiver High Temperature Process
Air Systems"
SAND 82-8254, 1983

Cost and Performance Compilation of Solar Related Part of the Plant

		Plant Concept									
		1	2.1	2.2	2.3	3	4.1	4.2	4.3	5	6
Investment / 10^6 DM /	Heliostat field	30	47	57	99	56	84	115	180	30	61
	Receiver	22	32	41	64	52	77	95	163	29	55
	Tower	4	6	11	27	11	25	48	53	5	15
	Primary loop	17	27	32	49	26	41	54	85	9	13
	Storage system	-	20	30	80	-	30	60	120	-	-
	Others	34	36	38	43	37	41	45	63	34	38
	Total Investment	107	168	209	362	182	298	417	675	107	182
Performance	Annual operational cost (staff, main-tenance, repair) / 10^6 DM/a /	2	3	4	7	4	6	8	14	2	4
	Annual solar thermal energy to reactor /GWh/	56.0	78.4	109.4	172.1	102.9	150.6	205.8	339.3	53.3	104.7
	Annual internal consumption in a thermal energy equiva-lent / GWh/a / — Coolant Air	36.1	39.7	44.2	66.1	44.6	52.7	61.2	84.1		
	Coolant He	15.2	17.3	19.5	29.3	20.6	25.1	29.6	41.0	2.2	4.2
	Specific cost of solar thermal energy / DM/kWh / Coolant He	0,145	0,159	0,141	0,153	0,133	0,144	0,147	0,142	0,14	0,121
	Start of operation: 1995										

Tab. 5.1-1: Cost and performance compilation

	Base case	Best case	Worst case
Cost base	1986	1986	1986
Time of erection	4a	4a	4a
Service life	20a	20a	15a
Plant start-up	1995-2020	1995-2020	1995-2020
Cost of natural gas	45 DM/MWh	50 DM/MWh	40 DM/MWh
Real escalation rate of natural gas	3 %/a	4 %/a	2 %/a
Inflation rate	5 %/a	6 %/a	3 %/a
Rate of interest	8 %/a	6 %/a	8 %/a
Operational cost incl. insurance rate (excl.) internal consumption)	2,3 %/a	2 %/a	4 %/a
Decreasing rate of solar specific investment (over 20 years)	1,25 %/a	1,75 %/a	0,5 %/a
Uncertainty in investment	J[*]	0,8·J	1,2·J
Annual operational time	8160 h/a	8160 h/a	7800 h/a

[*]J = investment as indicated in Tab. 5.1-1

Tab. 5.1-2: Non-technical assumptions for energy cost
            estimations

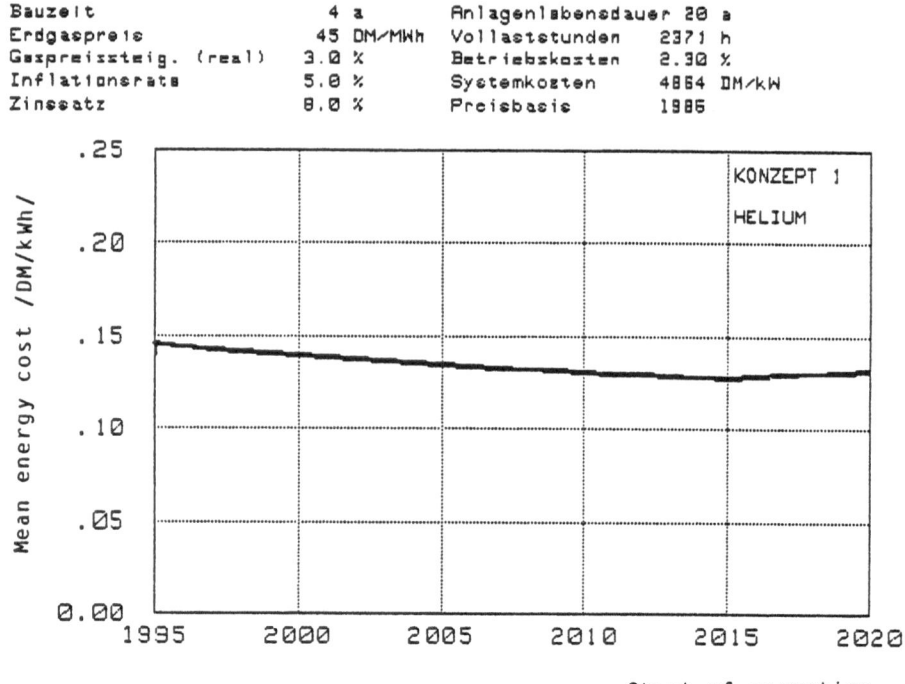

Bauzeit	4 a	Anlagenlebensdauer	20 a
Erdgaspreis	45 DM/MWh	Vollaststunden	2371 h
Gaspreissteig. (real)	3.0 %	Betriebskosten	2.30 %
Inflationsrate	5.0 %	Systemkosten	4864 DM/kW
Zinssatz	8.0 %	Preisbasis	1986

KONZEPT 1

HELIUM

Start of operation

Fig. 5.1-1: Plant concept 1 - specific cost of solar
produced thermal energy delivered to the reactor

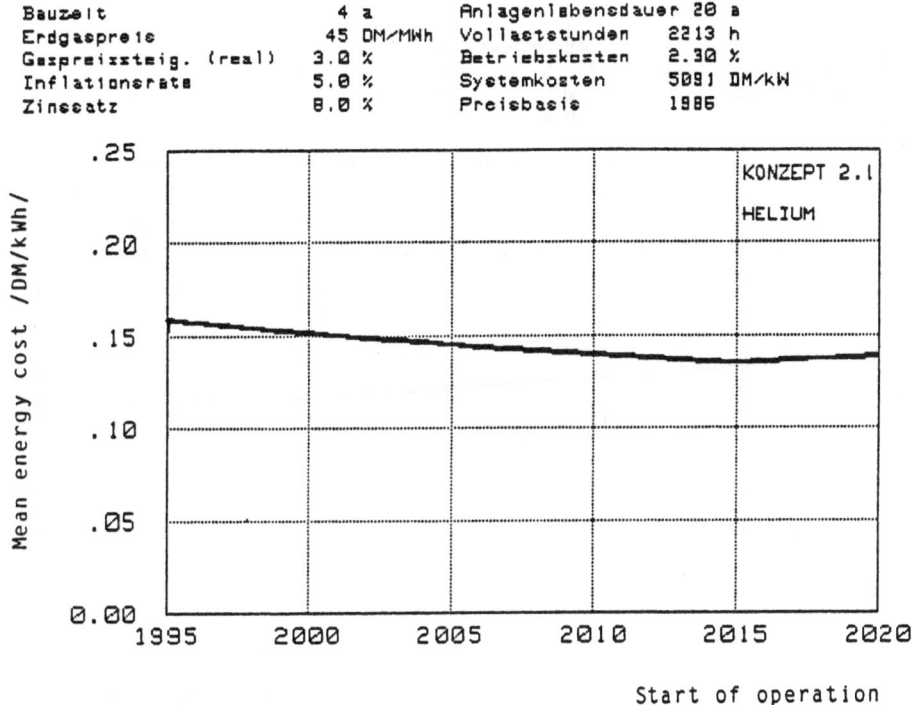

Bauzeit	4 a	Anlagenlebensdauer	20 a
Erdgaspreis	45 DM/MWh	Vollaststunden	2213 h
Gaspreissteig. (real)	3.0 %	Betriebskosten	2.30 %
Inflationsrate	5.0 %	Systemkosten	5091 DM/kW
Zinssatz	8.0 %	Preisbasis	1986

Fig. 5.1-2: Plant concept 2.1 - specific cost of solar
produced thermal energy delivered to the reactor

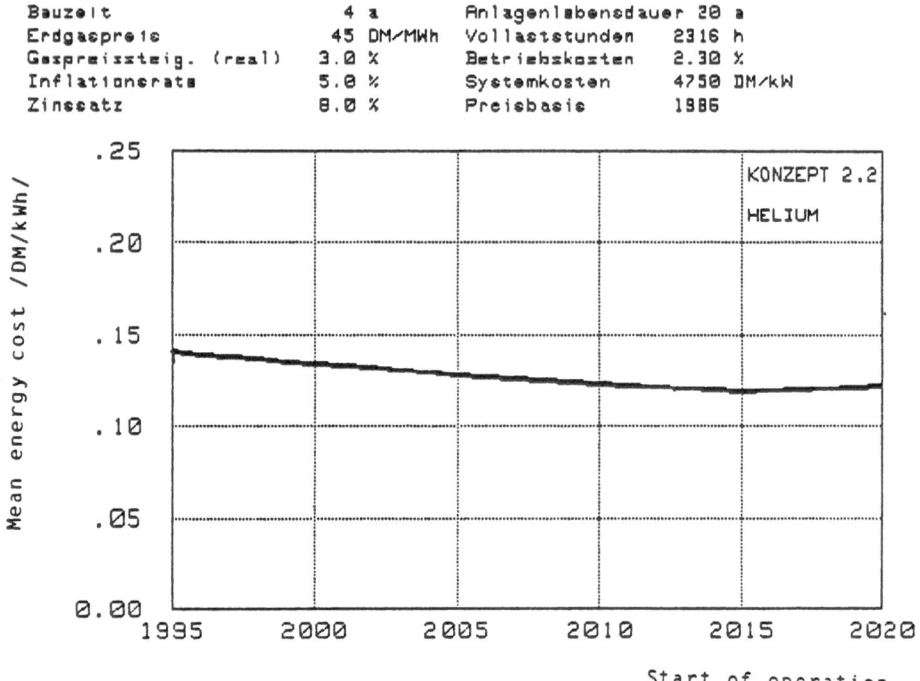

Bauzeit	4 a	Anlagenlebensdauer	20 a
Erdgaspreis	45 DM/MWh	Vollaststunden	2316 h
Gaspreissteig. (real)	3.0 %	Betriebskosten	2.30 %
Inflationsrate	5.0 %	Systemkosten	4750 DM/kW
Zinssatz	8.0 %	Preisbasis	1986

Fig. 5.1-3:   Plant concept 2.2 - specific cost of solar
produced thermal energy delivered to the reactor

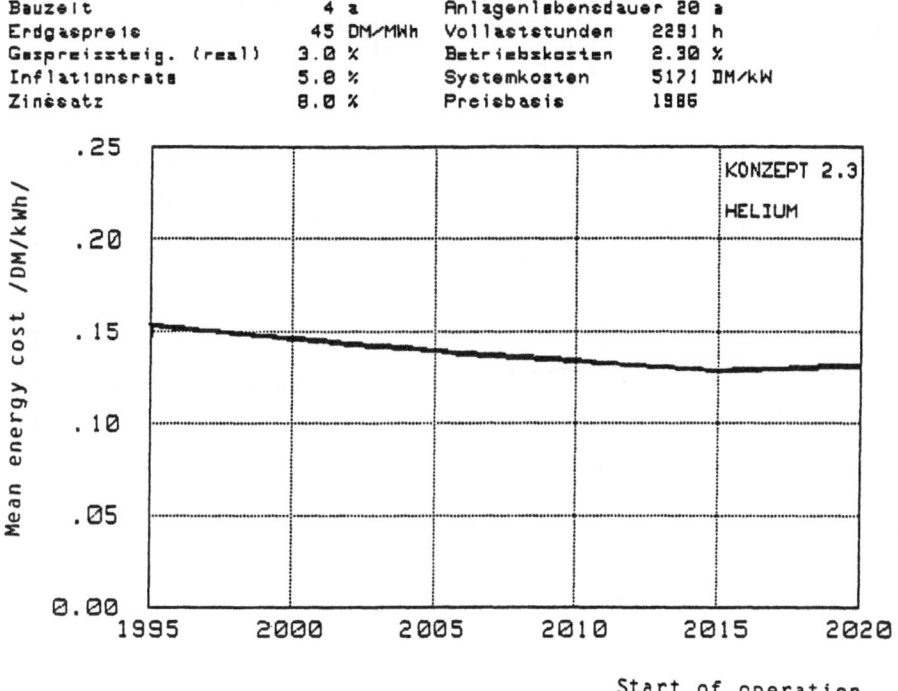

Bauzeit	4 a	Anlagenlebensdauer	20 a
Erdgaspreis	45 DM/MWh	Volllaststunden	2291 h
Gaspreissteig. (real)	3.0 %	Betriebskosten	2.30 %
Inflationsrate	5.0 %	Systemkosten	5171 DM/kW
Zinssatz	8.0 %	Preisbasis	1986

Fig. 5.1-4: Plant concept 2.3 - specific cost of solar
produced thermal energy delivered to the reactor

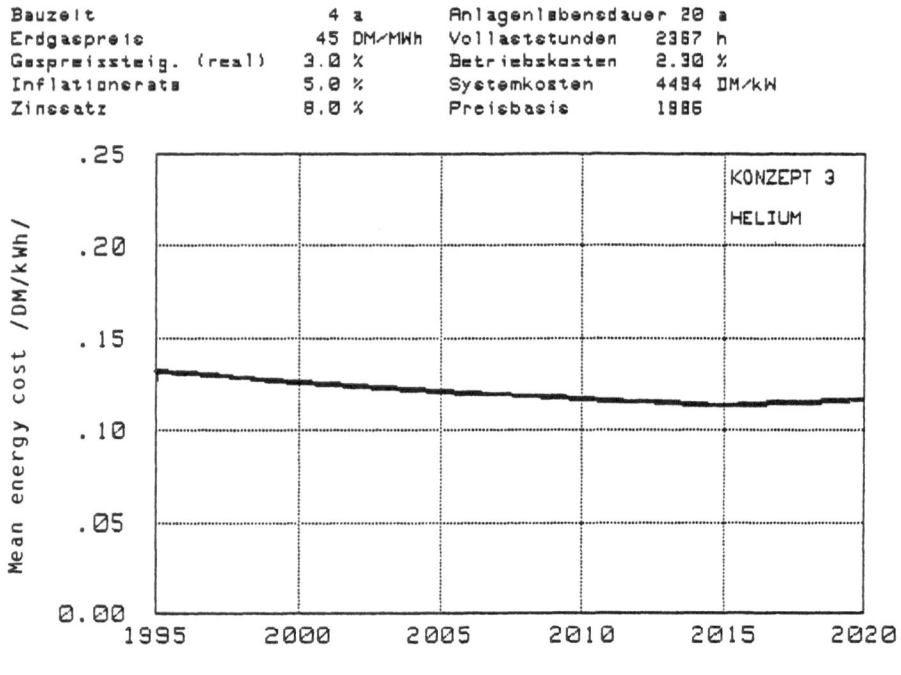

Bauzeit	4 a	Anlagenlebensdauer 20 a
Erdgaspreis	45 DM/MWh	Vollaststunden 2367 h
Gaspreissteig. (real)	3.0 %	Betriebskosten 2.30 %
Inflationsrate	5.0 %	Systemkosten 4494 DM/kW
Zinssatz	8.0 %	Preisbasis 1986

Start of operation

Fig. 5.1-5: Plant concept 3 - specific cost of solar produced thermal energy delivered to the reactor

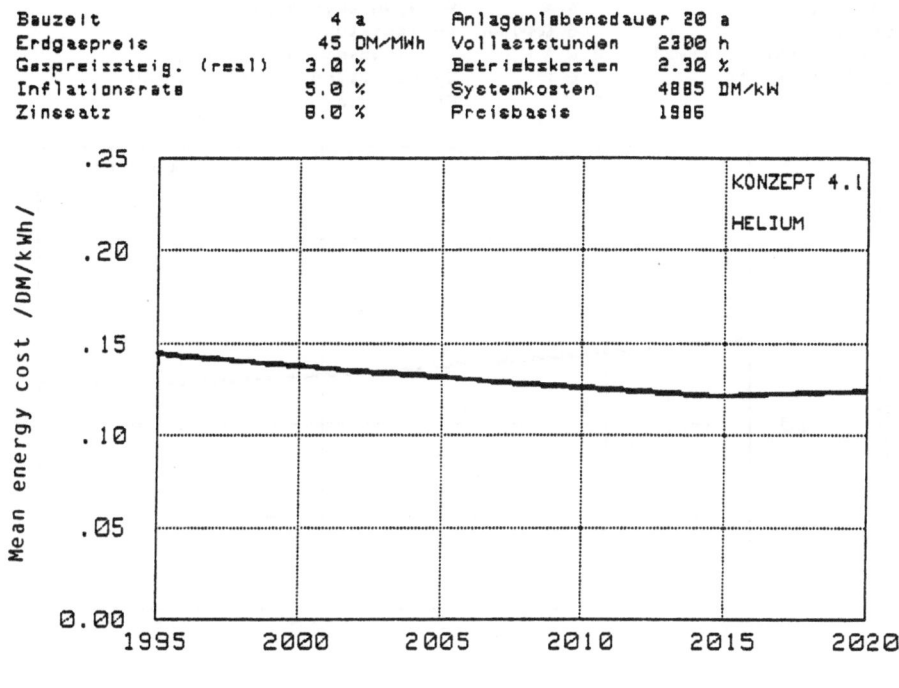

Fig. 5.1-6: Plant concept 4.1 - specific cost of solar produced thermal energy delivered to the reactor

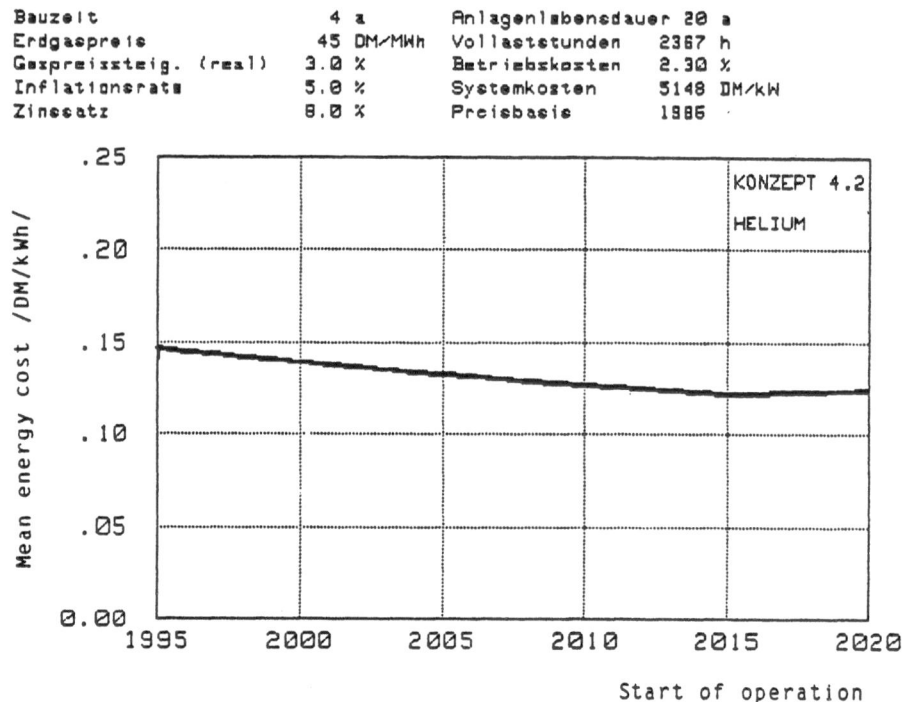

Bauzeit	4 a	Anlagenlebensdauer	20 a
Erdgaspreis	45 DM/MWh	Vollaststunden	2367 h
Gaspreissteig. (real)	3.0 %	Betriebskosten	2.30 %
Inflationsrate	5.0 %	Systemkosten	5148 DM/kW
Zinssatz	8.0 %	Preisbasis	1986

Fig. 5.1-7: Plant concept 4.2 - specific cost of solar produced thermal energy delivered to the reactor

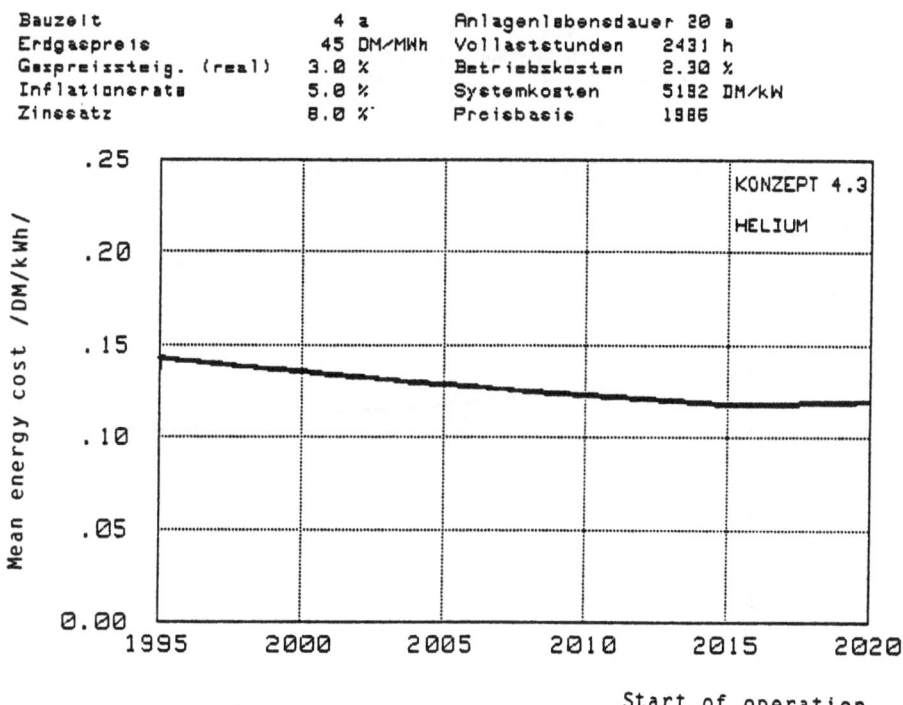

Fig. 5.1-8: Plant concept 4.3 - specific cost of solar
produced thermal energy delivered to the reactor

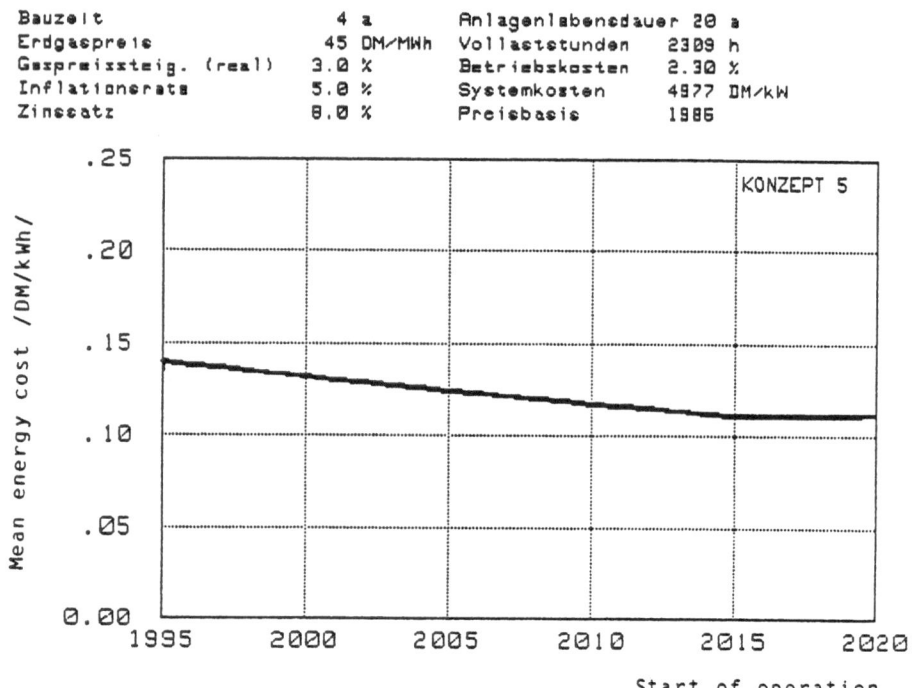

Bauzeit	4 a	Anlagenlebensdauer	20 a
Erdgaspreis	45 DM/MWh	Vollaststunden	2309 h
Gaspreissteig. (real)	3.0 %	Betriebskosten	2.30 %
Inflationsrate	5.0 %	Systemkosten	4977 DM/kW
Zinssatz	8.0 %	Preisbasis	1986

Fig. 5.1-9: Plant concept 5 - specific cost of solar
produced thermal energy delivered to the reactor

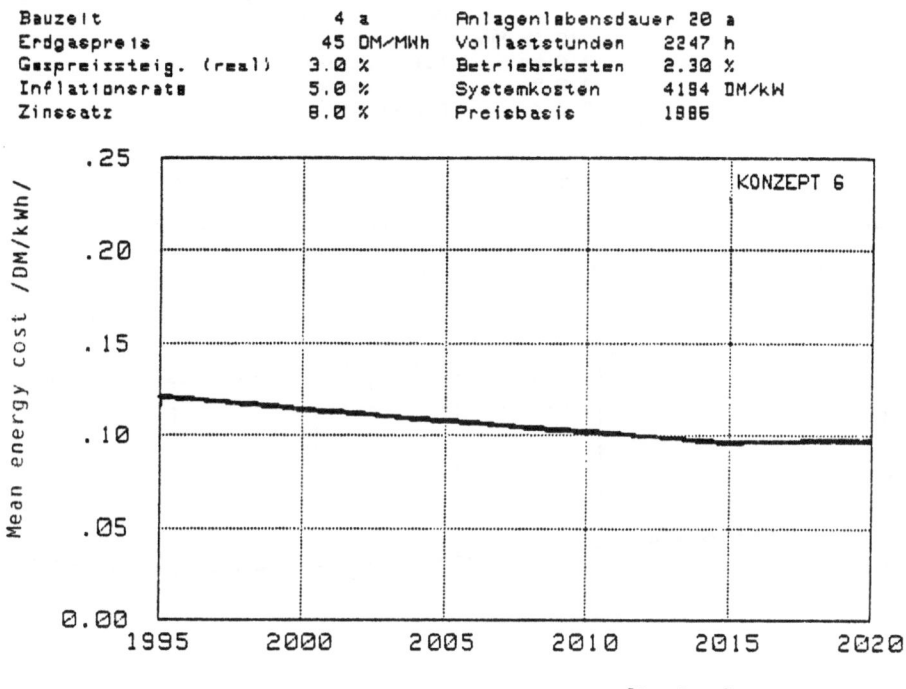

Bauzeit	4 a	Anlagenlebensdauer 20 a
Erdgaspreis	45 DM/MWh	Vollaststunden 2247 h
Gaspreissteig. (real)	3.0 %	Betriebskosten 2.30 %
Inflationsrate	5.0 %	Systemkosten 4194 DM/kW
Zinssatz	8.0 %	Preisbasis 1986

KONZEPT 6

Mean energy cost /DM/kWh/

Start of operation

Fig. 5.1-10: Plant concept 6 - specific cost of solar
produced thermal energy delivered to the reactor

- 465 -

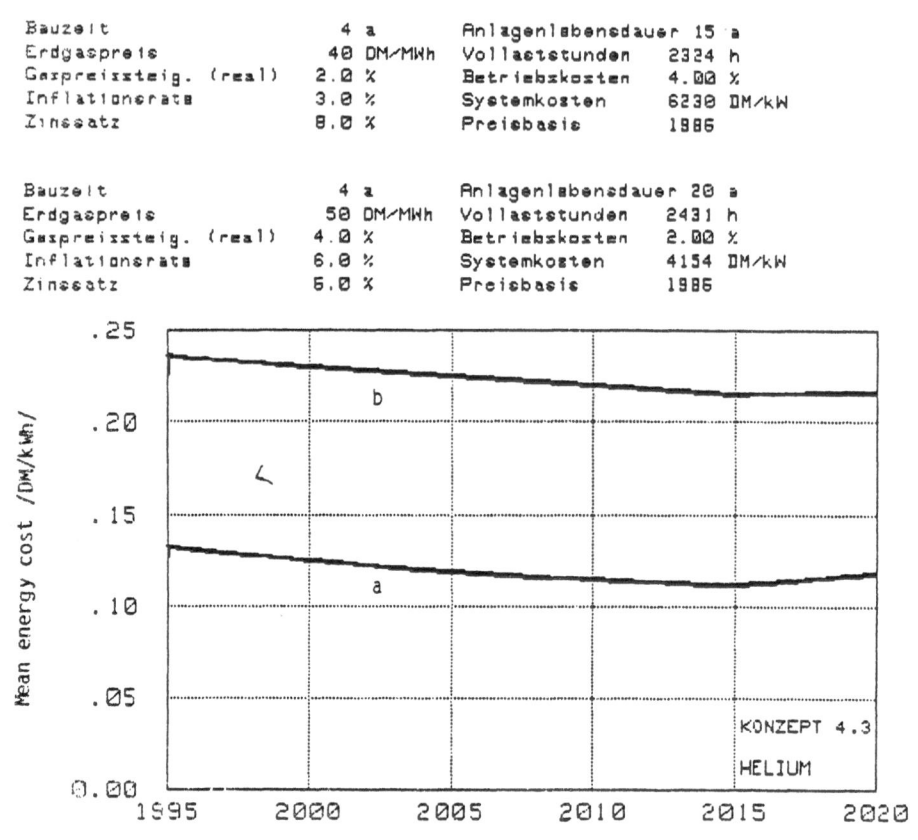

```
Bauzeit                      4  a    Anlagenlebensdauer 15 a
Erdgaspreis                 40 DM/MWh Vollaststunden    2324 h
Gaspreissteig. (real)      2.0 %    Betriebskosten     4.00 %
Inflationsrate             3.0 %    Systemkosten       6230 DM/kW
Zinssatz                   8.0 %    Preisbasis         1986

Bauzeit                      4  a    Anlagenlebensdauer 20 a
Erdgaspreis                 50 DM/MWh Vollaststunden    2431 h
Gaspreissteig. (real)      4.0 %    Betriebskosten     2.00 %
Inflationsrate             6.0 %    Systemkosten       4154 DM/kW
Zinssatz                   5.0 %    Preisbasis         1986
```

Fig. 5.1-11:Plant concept 4.3 - specific cost of solar produced thermal energy delivered
tc the reactor (a - best case, b - worst case)

- 466 -

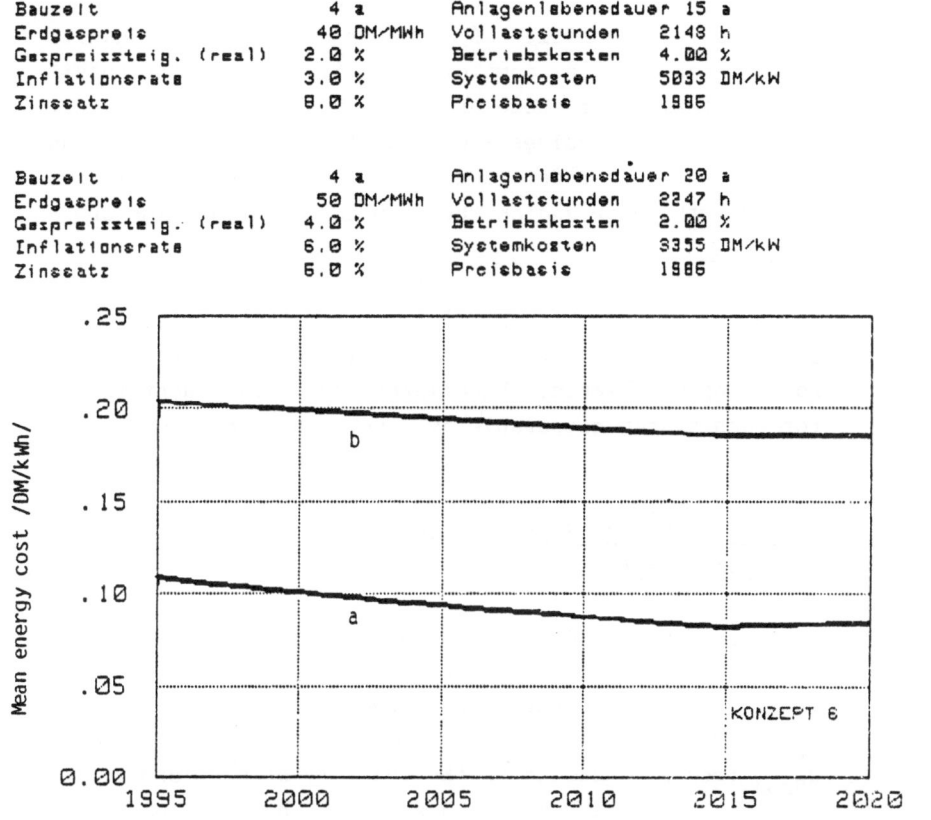

```
Bauzeit              4 a        Anlagenlebensdauer 15 a
Erdgaspreis          40 DM/MWh  Vollaststunden     2149 h
Gaspreissteig. (real) 2.0 %     Betriebskosten     4.00 %
Inflationsrate       3.0 %      Systemkosten       5033 DM/kW
Zinssatz             8.0 %      Preisbasis         1986


Bauzeit              4 a        Anlagenlebensdauer 20 a
Erdgaspreis          50 DM/MWh  Vollaststunden     2247 h
Gaspreissteig. (real) 4.0 %     Betriebskosten     2.00 %
Inflationsrate       6.0 %      Systemkosten       3355 DM/kW
Zinssatz             5.0 %      Preisbasis         1986
```

5.1-12: Plant concept 6 - specific cost of solar produced thermal energy delivered to the reactor (a - best case, b - worst case)

- 467 -

5.2   Energetic Data for Investigated Chemical Systems

5.2.1   Basic Energy Data for Systems A, A', B, C and D

The tabulations for the four investigated chemicals (tab.
5 - 2, 5 - 4, 5 - 6, 5 - 8) are made up identically. They
indicate the quantities of media or energies to be handled
within each chemical complex, when fitted out with fossil
steam reformers or those of specific solar design. Most
of the indicated values are related to an hourly basis as
it will come out under idealized, favourable conditions
all over the year and on the assumed operating time of
8160 h/a.

The received actual quantities of the tables 5 - 2,
5 - 4, 5 - 6 and 5 - 8 are the basis for the always
following next table, which makes the actual quantities
for each system comparable by generating specific data.

5.2.2   Abbreviations and Significance of Tables 5 - 2, 5 4,
        5 - 6 and 5 - 8

$V_{E\ Proc}$ (row 1) indicates the quantitiy of natural gas
                being used for process reaction. It cannot
                be substituted by solar means.

G_{Prod} (row 2) indicates the quantity of pure product be-
                ing released from the chemical complex,
                being methanol with A, ammonia with B and
                C, Octamix-Fuel-Methanol with D.

V_{H2} (row 3) indicates the quantity of pure hydrogen
                being produced simultaneously with the main
                product. If there is no such pure hydrogen
                separated within the process sequences (with
                systems B and C), row (3) was left open to
                easen comparison of the tabulations with
                each other.

$V_{E\ fuel}$ (row 4) indicates the quantity of natural gas being
used as fuel within the reforming plant.
It is the same figure as it was developed
in tables of chapter 2.

$V_{Esubst.}$ (row 5) represents the quantity of natural gas,
which probably could be saved in natural
gas fuel consumption, if further solar means
were introduced into the system. Such solar
means would be for instance an additional
solar plant with storage systems that would
provide energy on a temperature level of
about $600^{\circ}C$, to produce high pressure superheated steam for the purpose of driving compressor turbines.
Such additional means were not studied
within this work. The relevant figures were
taken from the energy balance. If there was
a demand of natural gas with solar operation
and night operation also at the same location within the plant, it was suggested,
that this energy equivalent could be substituted by solar means. The figures were developed only for comparison reasons to show
the potential of a process for further
acceptance of solar energy. Especially
costs of such additional equipment have not
been taken into consideration by any means.

$V_{E\ sol\ He}$ (row 6) indicates the equivalent amount of natural
gas that will be used to serve the energy
requirements of the solar plant. It mainly
consists of the power for driving the recycle compressor of separated systems, when
Helium is the heat carrier medium. An
efficiency of 35 % (thermal) has been assumed.

$V_{E\ Sol\ L}$ (row 7) indicates the relevant quantity of natural
gas, when **air** is being used as heat carrier.

$Q_{solar\ max}$ (row 8) is the design energy of the solar plant
to be supplied/made available for chemical
reaction.

Q_{solar} (row 9) is the average energy supply of the solar
plant on the basis of the annual solar ener-
gy output, as being found in tab. 5 - 1.
Details for computation have already been
given in chapter 2.2.3.1.

ζ_{sol} (row 10) is the quotient of $Q_{solar}/Q_{solar\ max}$. See
also 2.2.3.1.

$Q_{E\ fuel}$ (row 11) is the thermal energy equivalent of $V_{E\ fuel}$
row (4).

$Q_{tot\ fuel}$ (row 12) adds the thermal energy equivalent of
the residual gases, which are used as fuel,
to the value of $Q_{E\ fuel}$. $Q_{tot\ fuel}$ represents the total energy equivalent used as
fuel in the reforming unit

$Q_{E\ subst.}$ (row 13) is the thermal energy equivalent of
$V_{E\ subst.}$, row (5)

$V_{E\ util.}$ (row 14) represents the quantity of natural gas,
which will be needed to produce the neces-
sary utilities within the studied complex.
The figure was produced by evaluating the
main electric consumers and by further adding a certain percentage for smaller consumptions.
The figure additionally was made up by using
a monetary equivalent for other utilities
(as cooling water and raw water) and considering these values as natural gas consump-

tion. The value of $V_{E\ util}/G_{Prod}$ was
thought to be fix. While the original figure for $V_{E\ util.}$ was calculated for the
fossil case, the figures for the solar cases
with different product output were prorated
by such specific values.

$Q_{E\ util.}$ (row 15) is the thermal energy equivalent of
$V_{E\ util.}$, row (14).

5.2.3 Abbreviations and Significance of Tables 5 - 3, 5 - 5, 5 - 7 and 5 - 9

The values in brackets indicate by what division
"row No. .../row No. ... of the always preceeding tabulation" the new value is gained.

$V_{E\ Proc}$ (row 1) represents the specific natural gas volume
per weight unit of product (1/2)

V_{h2} (row 2) is the specific pure hydrogen sideproduct
volume per weight unit of product (3/2)

$V_{E\ fuel}$ (row 3) represents the specific natural gas fuel
volume per weight unit of product (4/2)

$V_{E\ Sol\ He}$ (row 4) is the specific natural gas volume per
weight unit of product, which is needed to
drive the solar auxiliary circuits, if the
heat carrier is Helium (6/2)

$V_{E\ Sol\ L}$ (row 5) represents a value as before, if instead
of helium air will be taken (7/2)

$V_{E\ util.}$ (row 6) is the specific natural gas consumption
for the generation/valuation of utilities
per weight unit of product (14/2)

$v_{E\ tot\ He}$(row 7) is the addition of rows (1) + (3) + (4)
+ (6) for the concerned line and represents
the total specific natural gas consumption
per weight unit of product for the indicated
fossil or solar production system, when
using helium as heat carrier (if any) in
the solar circuit.

$v_{E\ tot\ L}$(row 8) is the addition of rows (1) + (3) + (5)
+ (6) for the concerned line and represents
the total specific natural gas consumption
per weight unit of product as before, when
air is being used as heat carrier (if any)
in the solar circuit.

$q_{sol\ max}$(row 9) is the solar specific energy input at design condition per weight unit of product
(8/2)

q_{sol} (row 10) is the average specific energy input on an
annual basis per weight unit of product
(9/2)

$q_{E\ tot\ He}$(row 11) represents the specific overall energy
requirement, which has to be provided by
natural gas, per weight unit of product for
the indicated fossil or solar production
system. The figure corresponds on a thermal
basis with row (7), which means: the solar
auxiliary circuit shall be operated with
helium. (A relevant figure for air operation
has not been provided as this in any case
will be less economic).

$v_{E\ subst.}$(row 12) is the specific quantity of natural gas
per weight unit of product, which may addi-
tionally be substituted by solar energy
(see $V_{E\ subst}$, chapter 5.2.2). (5/2)

$q_{E\ subst}$(row 13) is the thermal equivalent of row (12)

$$(13/2)$$

$v_{e\ optim}$(row 14) represents probably best achievable specific consumption figure for natural gas
per weight unit of product. It reduces the
lowest calculated specific consumption figure (row 7) by the value of possible additional substitution (row 12)

$$(14) = (7) - (12)$$

$q_{E\ optim}$(row 15) is the thermal equivalent of row (14).
A lower heating value of 9,915 kWh/m^{3}N natural gas generally has been adopted for
such correlations.

5.2.4 Discussion of Results of Tables 5 - 2 to 5 - 9

Attention should be paid to rows (11) and (15) of the specific energy tables. There it can be seen that the highest
fossil energy requirement is linked with the fossil plant
and that savings gradually can be achieved when proceeding
from prereactor-systems to those with main reactors and
that the introduction of storage systems always improves
the situation. Thus a good storage system for the prereactor system yields better specific numbers than a single
main reactor system.

System A' - 8 has higher specific energy requirements than
the normal fossil one, which exclude A' - 8 from interest
of development. A' - 7 is slightly better than the fossil
operated plant, but it has to be checked, at what costs
such improvements can be gained. This generally is valid
for all solar comparable systems, which was the reason for
further investigation on costs and economics.

Some of the figures, displayed by the aforementioned
tables, only will be used for valuations in later chapters
of this study.

In this table the specific total natural gas consumption
figures $v_{e\ tot\ He}$ and $v_{E\ opt}$ have been listed again for
the different chemical systems. By multiplying these specific figures with the annual product quantity of each
plant, one arrives at the actual total natural gas consumption per year.

Table 5 - 10 enlists the natural gas savings of each plant
type (with the earlier fixed energy input and other related
side conditions) versus the purely fossil operated plant
at normal reforming conditions.

The table shows that - in accordance with expectations -
the savings of natural gas are fixed with nearly no excep-
tions by the type of solar front plant only, when considering the "studied plant concepts". From the savings in natural gas the cost of investments for the solar plant (including related over expense in the reforming unit) have
to be paid from. Later chapter will enlarge on this point.
The constant savings in natural gas for any solar system
explain, why a system   - 4.3 gets economic from a certain
point of time onward, independent whether it is in connection with a methanol production or with an ammonia production.

While it is of little interest, what production route is
following, when looking at savings with natural gas reforming only, the scene gets quite different, when regarding
further energy substitutions and savings coupled therewith.

In earlier chapters it has already been explained that
methanol systems with solar aid are balanced or even over-
supplied energywise. This is different with the other
systems (B, C and D), where further important natural gas
savings are deemed possible, bringing the specific natural
gas consumption down to values near to the mere process

consumption. While this already is achieved for A-systems, this might well be worthwile to look at in more detail withthe other chemical systems.

As the substitutable energy is proposed to be supplied at
a temperature level of about 600°C, the applicable techniques will be different, for instance heat carrier as
molten salt and storage of such liquid might be possible.
As the energy costs of such other systems will not be much
lower than the calculated ones in this study, the time of
economic parity will not be much different with such enlarged systems, than what will be given later. But it
might be expected that the savings and profitableness will
be much better for such systems, once the parity has been
surpassed.

5.2.6   Comparison of Specific Energy Data, Tab. 5 - 11

From Tab. 5 - 10 it followed that the annual natural gas
savings are more or less independent from the chemical
system, which will follow a solar aided reforming plant.
Yet it is not clarified, where a high solar portion of
energy will be found in a product or what criterium else
would identify a chemical plant, to preferably be equipped
with solar reforming plants of a certain performance.

The attempt for such classification was made with Tab.
5 - 11. It contains specific energy data without dimensions in generating relations between specific energy data
of tables 5 - 3, 5 - 5, 5 - 7 and 5 - 9.

One of the specific reference numbers definitely should
be the natural gas consumption for process use or its equivalent in thermal terms. - $q_{E\ Proc}$.

$q_{E\ Proc}(row\ 1) = v_{E\ Proc} \cdot H_u$
        (Hu = lower heating value of nat.gas =
        9,915 kWh/m^3N)

q_{Sol}(repeatedly) - as with chap. 5.2.3

- 475 -

It is felt that the natural gas for process use is a good connecting link between the processes, better apt for comparison than a product quantity, which is influenced by other factors that are not connected with the sun at all (for instance addition of nitrogen!).

The figures of table 5 - 11 are explained by combination of specific figures. Yet one arrives at the same dimensionless figures, if one starts from the absolute values as the weight unit reference disappears by division.

$q_{Sol}/q_{E\ Proc.}$ (row 1) indicates the part of solar energy in relation to the process bound natural gas energy. It is highest with system D, where the smallest quantity of natural gas for process operation is used with a given solar energy quantity.

Another reference should be the thermal equivalent of the overall natural gas consumption $-q_{E\ tot\ He}$

$q_{E\ tot\ He}$ - as with chapter 5.2.3

$q_{Sol}/q_{E\ tot\ He}$ indicates the rate of solar energy in relation to the residual total energy requirement for natural gas.

As natural gas for process operation cannot be substituted it may be interesting to learn also, what part of the remaining natural gas for the other purposes has been replaced, for which row (3) has been introduced. There the reference number is $q_{aux.foss.}$, meaning the energy equivalent of the auxiliary uses of the fossil operated plant. The numbers are the specific figures for the sum of $V_{E\ fuel}$ and $V_{E\ util}$ in chapter 5.2.2 for the purely fossil operated plants.

$$q_{aux.foss.} = (V_{E\ fuel} + V_{E\ util})_{foss} \cdot 9,915/G_{Prod.} (kWh/t)$$

The reasons and possibilities of further energy substitution have been explained in detail before. It seems adequate also to investigate the influence of such additional solar power addition. The sum of solar energy for reforming and of solar additional substitution should be calculated in relation to the auxiliary natural gas consumption of the purely fossil operated plant.
This has been done by row (5).

$q_{E\ subst}$ (row 5) as with chapter 5.2.3

Rows (7) and (8) are related to the total natural gas consumption of the fossil operated plant. These lines indicate the relative savings or rates of solar energy
(row 8) or solar energy + solar substitutable energy
(row 7) in relation to the present status of technology.
In row (8) the solar net energy credit has been considered.

$q_{E\ optim}$ (row 7) - as with chapter 5.2.3

$q_{E\ tot\ fossil}$ (row 7, row 8) - corresponds to $q_{E\ tot\ He}$ of chapter 5.2.3 for the purely fossil operated plant.

$q_{Sol\ He(E)}$ (row 8) corresponds to $v_{E\ Sol\ He}$ of chapter 5.2.3 on a thermal basis.

$$q_{Sol\ He\ (E)} = v_{E\ Sol\ He} \cdot H^u \text{(natural gas)}$$

5.2.7 Valuation of Data from Tab. 5 - 11

In rows (4) and (6) the summation of various specific data has been performed. Though these numbers have no physical or actual meaning they are an indication, which of the systems specifically accepts more or less solar energy, as these summations can be regarded as average numbers.

When considering the solar supported plants only as they are investigated within this study, the A-systems accept the highest portion of solar energy (highest figures in row 4 and also in row 8).
This was the reason that more detailed investigations (construction, investment, economics) have basically been made for all A-systems and that many data of B-, C- and D-systems have been adopted in comparison with the gained A-figures.

If in future investigations also the effect of additional solar energy substitution shall be found out in more detail, one would have to consider the B-systems as they reach highest value of summations in rows (5) and (7).

It should be stated another time that the purpose of this present study best is met with the A-systems as they leave hardly any capacity for further solar energy acceptance.

5.3   Investment of Chemical Plants

In tables 5 - 12 to 5 - 15 the investments of the chemical plants are enlisted.

They are based on 1986 prices and include delivery of the complete plant sections, engineering, erection and civil work.

The figures are thought to be accurate within +/- 20 %. They are budgetary figures and not binding the originators of the study by any means but have been developed on best available comparative calculation data for single apparatus and/or plants of similar size and type. As the limits between plant sections are fluctuating in any case the total, summarized value of the plant systems should be used as references only.

The prices of tables 5 - 12 to 5 -15 are additive to the costs of MAN-tabulations (tab. 5 - 1) for the solar related plant parts.

5.4   Present and Future Product Costs

With the consumption figures and investment costs it is possible to calculate the product costs, if other cost-related factors are fixed.

Such cost-related factors will be found in Tab. 5 - 16, for instance: time of erection, service life, cost of natural gas rate of interest, cost of operation and maintenance in relation to the total investment, costs of insurance.

Other conditions are related to the probable development of costs in the future. They are: the inflation rate of investments and the escalation rate of the natural gas beyond the inflation rate (real escalation).

The conditions of tab. 5 - 16 are referred to as standard conditions.

They are the basis for the cost calculations of tables 5- 18 to 5 - 21 and the relevant figures.
Investigations have also been made for conditions, deviating from these standard conditions. They will be found behind the standard calculation/presentations.

5.4.1   Product Cost Calculations

For the calculation of the product costs of the various production routes a dynamic method for the net present values was applied.

Cost of Product = (Net value of investment
 + net value of the operating costs
 + net value of natural gas costs)
 divided by
 Product Quantity in total lifetime.

The product costs have been calculated for the time of start-up in prices of the time, when the plant was started to be built (reference time).
The scheme of calculation, Tab. 5 - 17, has been applied for all such calculations. Each row in tables 5 - 18 to 5 - 21 has its own calculation sheet of the form, presented with fig. 5 - 17, which will be explained in some more detail.

5.4.1.1 Cost Calculation Scheme, Tab. 5 - 17

The data as being found in such tables were fed to the stores of a pocket computer and the relevant costs and prices were calculated by an appropriate program, the results being available from other stores.

The following lines/rows still may need explanations:

A) Input-Information

Investment Io in 1986

 These are the investment costs of each plant type, being composed from solar related prices and chemicals plant prices. The prices are on 1986 cost basis.

Solar energy available

 comes from tables 5 - 1

Natural Gas to Process

 as in 5.2.2.

Natural Gas Consumption

means that equivalent portion of natural gas, which is needed to drive the auxiliaries of the solar related part. It corresponds with $V_{E\ Sol\ He}$ of chapter 5.2.2

N.G. to combustion

corresponds with $V_{E\ fuel}$ of 5.2.2

Other N.G. Consumers

corresponds with $V_{E\ util.}$ of 5.2.2

Year of start-up

is the first year, which has been investigated, usually 1990. The other years of start-up are chosen by "intervals of next T".

Part of Operating Costs of Io

is the percentage of costs, related to the total investment. This figure comprises annual operating costs and insurance.

Increase of Value H_2/N,G , Factor F

For systems A and D, pure hydrogen is produced as by-product. The credit for such hydrogen has been taken into account by the following relation:

$$\text{Price Hydrogen} = F \cdot \frac{HuH2}{HuN.G} \cdot \text{Price N.G.}$$

with 45 DM/MWh and F = 1,4 a price of about 0,20 DM/m^{3}N hydrogen results for nowadays plants, which seems reasonable.

A change of figure F will not change the point of time, when solar plants are equal in price with fossil ones, but will change the margin of profit after such cross-over points. Therefore F = 1,4 has been considered

as adequate and constant and has not been changed
within this study.

Annual Price Reduction etc.

The four last lines of the input data deal with
possible savings in the solar related part of the plant
with future installations. It is expected that the
solar related parts will specifically decrease in
costs, when more installation of standardized sizes
will be built. Especially for mirrors this has proven
to be correct in the past.

By 1,25 % cost decrease per year, the non-inflated investment will amount to aobut 78 % of the present
value only after 20 years. It was assumed that
further reductions would be speculative only, which
was the reason for mathematical limitations in this
field.

B)   From the General Results only "Total N.G. consumptions under consideration of H_2-credit" needs explanation in so
far, as this figure has no technical background at all,
but is reducing the natural gas consumption by a quantity
equivalent in price with the produced hydrogen. The indicated figure of this row is the basis for the net value
of natural gas overall consumption.

C)   Specific Results

The hydrogen costs are determined by the above cited
formula, depending from F = 1,4 and the basic natural gas
price

The product costs are found out by the procedure as detailed in 5.4.1.0.

Reduced Investment 86 refers to the reduction of prices
as contemplated above (last paragraph of A). For purely
fossil operated plants no such reductions have been adopted, but with solar operated plants the lowest investement,
reduced to 1986, will be found only from the year 2015 onwards.

Io is the escalated or real investment at the time of reference. As inflation goes with

$$I_{oT} = I_{1986} \cdot (1 + p)^{(T - 1986)}$$

there is a fixed relation between prices of startup or reference and basic year (1986). This relation is given
also in the pen ulitmate row (cash value T/cash value 86).
This factor is used for all values referring to 1986, as
otherwise by inflation the feeling for the actual value
gets lost. Thus Natural Gas Price g_1 is escalated from
1986 onwards only by the real excalation factor.

Cash Value of Investment refers to the actual value at
time of starting construction, the startup time to be 4
years later. The given value, divided by penultimate row,
gives the net value of investment for 1986, which is lower
than the value of row (4), due to staggered payments
during construction period.
Cash Value of Operating Costs refers to the summation of
operating costs during the total lifetime of the plant.
The indicated costs are for reference time.
Cash Value of Natural Gas refers to the summation of the
natural gas costs during the total lifetime of the plant,
changing year by year. The indicated costs refer to the
reference time.
Total Costs Cash Value is the summation of the three net
values for investment, operation and natural gas for the
reference time. By division through the penultimate figure
of the line one arrives at costs, related to 1986 prices

By division of the total costs by the total product quantity, produced during the total life of the plant, one

arrives at the specific product costs of row (2). The ulitmate row indicates the annual costs of natural gas in the year of startup, related to the time of reference. This annual value will increase by the real escalation factor and inflation rate in course of the plant's life.

5.4.1.2 Tables of Product Costs

From the detailed calculation schemes for every type of plant only the product costs as function of the startup time have been transferred to tables 5 - 18 to 5 - 21. For these tables the "standard conditions" of tab. 5 - 16 are valid and also F= 1,4 for hydrogen price bonus (see chapter 5.4.1.1).

A final line in tables 5 - 18 t 5 - 21 shows the year, when equal product prices are gained from the purely fossil operated plant and the plant type of respective reference.

5.4.1.3 Curves of Costs

Fig. 5 - 1 to Fig. 5 -   show the results of tables 5.18 to 5 - 21 in a different form.

Fig. 5 - 1 shows the cost of natural gas under the assumption of 3 % real escalation rate. The inflation rate is eliminated to stay with the cost basis of 1986.

Fig. 5 - 2 The credit for byproduct hydrogen on the cost basis of 1986 is shown in this figure. As by the equation of chapter 5.4.1.1 the price of hydrogen directly was related to that one of natural gas. To arrive at real hydrogen credits at time of startup it would be necessary to additionally consider the inflation rate (as with natural gas price aslo).

Fig. 5 - 3 and Fig. 5 - 4 show the development of prices for
methanol from different solar systems in comparison with fossil plants under the scenario
of the "standard-conditions" of table 5 - 16.

The points of parity with the purely fossil
operated plant are marked. Therefrom it
comes that the earliest time of reaching economic production with a solar supported plant
would be in the year 2006 by the system
A - 6 and that the latest time of parity
would be reached around the year 2031 for
system A - 1. The highest profil in the year
2040 (or the lowest methanol costs) would be
reached by the solar system A - 4.3.

Fig. 5 - 5 to 5 - 10 are the equivalents of figures 5 - 3 and
figures 5 - 4 for other chemical products, as
explained just before. For reasons, as already emphasized in connection with table
5 - 10 (chapter 5.2.5) the time of parity with
purely fossil operated plants does not changes
with the product and also the savings in energy (or costs) are related only to the type of
solar reforming plant.

5.4.1.4 Graphs of Costs for Changed Operation Conditions

Fig. 5 - 11 shows the results for the systems A' - 7 and
A' - 8 in comparison with the normal, entirely fossil operated methanol plant. While
systems A' - 7 and A' - 8 are operated at low
pressure only, for reasons as explained in
chapter 2.1.4,the comparison of A-fossil-plant
remains with its outlet conditions as in the
comparison for the other solar plants as no
reason can be seen to worsen an available
technique.

- 485 -

The curves show that parity in costs with the
fossil plant is reached for system A'- 7 only
by the year 2062 under the "standard conditions", but no corss-over can be found at all
for system A' - 8. In the latter the savings
in natural gas are that small that they cannot compensate the additional costs of equipment and operation even with a very high natural gas price.

The curves also show that the costs of product always are higher for systems A' - 7 and
A' - 8 than with any other investigated solar
system.

Therefore the development of any system at
low pressure is not deemed interesting nor
advisable.

Fig. 5 - 12   shows the effects for operating the plant
A - 3 with fluctuations in capacity, following solar radiation. For the partial load
curve it has been assumed that the capacity
of the plant will be adjusted with the availability of solar energy and that, if 15 % of
nominal solar energy supply are underrun, the
plant capacity is stabilized at this 15 % by
fossil heating.

With such an operation the consumption of natural gas goes down, as does overall production rate, but the investment is used by a
smaller portion only. Therefore the overall
costs are increased, especially as long as
natural gas will be available at relatively
low cost.

The graph shows that parity of costs with the
entirely fossil operated plant is reached for
the variable load   - 3 plants only by about

10 years later, than if the plant A - 3 is
operated on 100 % load throughout its life,
allowing up to 100 % fossil firing during
night. Therefore such fluctuations are not
found to be a good solution for solar operated plants. The application of adequately
sized storage systems will anyhow eliminate
the need for such mixed operation.

5.4.2   Product Costs with Changes in Scenario

Figures 5 - 13 to 5 - 17 investigate the influence of
changed basic parameters, deviating from the standard conditions of table 5 - 16. The comparison always has been
made between the systems A-fossil and A - 4.3, the latter
having a point of parity for standard conditions just in
the middle region of time, not at the edges and also accep-
ting a high portion of solar energy.

The extreme values of the differentiation and the values
of the standard conditions compare as follows:

Changing topic	Fig-No	Standard	Considered extremes	Delta time
Price of na- tural gas	5 - 13	45 DM/MWh	35 to 55	15 a
Real escala- tion of na- tural gas	5 - 14	3,0 %/a	2,0 to 4,0	32 a
Operating life of the plant	5 - 15	20 a	10 to 20	17 a
Inflation and Interest	5 - 16	5%/8%	6%/6% to 3%/8%	14 a
Investment	5 - 17	100 %	80 % to 120 %	14 a

The "standard conditions" meet average for all combinations
except for the life time of the plant. There it was felt
that a high service life time for the solar plant should
be acceptable and possible. Chemical plants normally assume payout times of less than ten years. But even with

life time of ten years only, fossil operated plants and
A - 4.3-system still tender equal prices within the investigated time frame.

What comes from the above list:

The point of time for equal prices on average changes by
+/- 9,2 years within this investigation. When adjusting
the life time of the plants to 15 years instead of 20
years, what might be a bit more on the real side of economic considerations, the time of parity will be extended
by about 8 years. Then the variation band should be about
+/- 10 years on average.

As it must not be expected that all solar-positive and all
solar-negative factors will accumulate at the same time,
it should prove true that system A - 4.3 should become economic around the year 2020 +/- 10 years, mainly depending
from the development of the natural gas price and its
availability, as the other factors as accuracy of the calculated investment and inflation and interest rate should
be of minor influence. It is expected that for the other
solar systems the time variation for economic parity will
vary within the same band.

BASIC ENERGY DATA   A - SYSTEMS   TAB. 5 - 2

Row System/Value Dimension	1 $V_{E\,PROC}$ m³N/H	2 G_{PROD} T/H	3 $V\,H2$ m³N/H	4 $V_{E\,FUEL}$ m³N/H	5 $V_{E\,SUBST}$ m³N/H	6 $V_{ESOL\,HE}$ m³N/H	7 $V_{ESOL\,L}$ m³N/H	8 $Q_{SOL\,MAX}$ MW	9 Q_{SOLAR} MW	10 η_{SOL} -	11 $Q_{E\,FUEL}$ MW	12 $Q_{TOT\,FUEL}$ MW	13 $Q_{E\,SUBST}$ MW	14 $V_{E\,UTIL}$ m³N/H	15 $Q_{E\,UTIL}$ MW
A-FOSSIL	1200	13.75	9420	3578	0	0	0	0	0	0	35.5	57.5	0	550	5.45
A - 1	1200	13.75	9420	2892	22	175	416	21.5	6,392	0.2973	28.7	50.7	0.25	550	5.45
A - 2.1	1200	13.75	9420	2616	22	199	457	21.5	8,951	0.4163	25.9	47.9	0.25	550	5.45
A - 2.2	1200	13.75	9420	2234	22	225	509	21.5	12,489	0.5809	22.2	44.2	0.25	550	5.45
A - 2.3	1200	13.75	9420	1462	22	337	761	21.5	19,647	0.9138	14.5	36.5	0.25	550	5.45
A - 3	1200	13.75	9420	2312	203	237	513	40.5	11,748	0.2937	22.9	44.9	2.3	550	5.45
A - 4.1	1200	13.75	9420	1725	27	289	607	40.0	17,192	0.4298	17.1	39.1	0.3	550	5.45
A - 4.2	1200	13.75	9420	1045	0	341	705	40.0	23,492	0.5873	10.4	32.4	0	550	5.45
A - 4.3	1200	13.75	9420	0	0	472	968	40.0	38,732	0.9683	0[1]	22.0[1]	0	550	5.45
A - 5	13020	14.90	10220	3228	0	25	25	21.5	6,085	0.2830	32.1	56.0	0	597	5.92
A - 6	13020	14.90	10220	2605	0	48	48	43.4	11,952	0.2754	25.8	49.7	0	597	5.92
A'- 7	15125	16.35	11195	1205	878	904	1900	40.0	39,576	0.9894	11.9	47.1	10.1	790	7.83
A'- 8	15125	16.35	11195	3800	0	52	52	46.1	12,696	0.2754	37.7	72.9	0	790	7.83

- 489 -

SPECIFIC ENERGY DATA   A - SYSTEMS   TAB. 5 - 3

Row / System / Value Dimension	1 $V_{E\ PROC}$ M³N/T	2 $V\ H_2$ M³N/T	3 $V_{E\ FUEL}$ M³N/T	4 $V_{ESOL\ HE}$ M³N/T	5 $V_{E\ SOL\ L}$ M³N/T	6 $V_{E\ UTIL}$ M³N/T	7 V_{ETOTHE} M³N/T	8 $V_{ETOT\ L}$ M³N/T	9 Q_{SOLMAX} MWH/T	10 Q_{SOL} MWH/T	11 Q_{ETOTHE} MWH/T	12 $V_{E\ SUBST}$ M³N/T	13 $Q_{E\ SUBST}$ MWH/T	14 V_{EOPTIM} M³N/T	15 Q_{EOPTIM} MWH/T
A-FOSSIL	872.7	685.1	260.2	0	0	40.0	1172.9	1172.9	0	0	11.63	0	0	1172.9	11.63
A - 1	872.7	685.1	210.3	12.7	30.3	40.0	1135.7	1153.3	1.564	0.465	11.26	1.600	0.016	1135.7	11.26
A - 2.1	872.7	685.1	190.3	14.5	33.2	40.0	1117.5	1136.2	1.564	0.651	11.08	1.600	0.016	1117.5	11.08
A - 2.2	872.7	685.1	162.5	16.4	37.0	40.0	1091.6	1112.2	1.564	0.908	10.82	1.600	0.016	1091.6	10.82
A - 2.3	872.7	685.1	106.3	24.5	55.3	40.0	1043.5	1074.3	1.564	1.429	10.35	1.600	0.016	1043.5	10.35
A - 3	872.7	685.1	168.1	17.2	37.3	40.0	1098.6	1118.1	2.909	0.854	10.89	14.8	0.146	1083.2	10.74
A - 4.1	872.7	685.1	125.5	21.0	44.1	40.0	1059.2	1082.3	2.909	1.250	10.50	2.0	0.019	1059.2	10.50
A - 4.2	872.7	685.1	76.0	24.8	51.3	40.0	1013.5	1040.0	2.909	1.708	10.05	0	0	1013.5	10.05
A - 4.3	872.7	685.1	0	(34,3)	(70,4)	30.7	903.4	939.5	2.909	2.817	8.96	0	0	903.4	8.96
A - 5	873.8	685.9	217.4	1.7	1.7	40.0	1132.9	1132.9	1.443	0.408	11.23	0	0	1132.9	11.23
A - 6	873.8	685.9	174.8	3.2	3.2	40.0	1091.8	1091.8	2.913	0.802	10.83	0	0	1091.8	10.83
A' - 7	925.1	684.7	73.7	55.3	116.2	48.3	1102.4	1163.3	2.446	2.421	10.93	53.7	0.532	1048.7	10.40
A' - 8	925.1	684.7	232.4	3.2	3.2	48.3	1209.0	1209.0	2.821	0.777	12.0	0	0	1209.0	12.0

BASIC ENERGY DATA   B-SYSTEMS   TAB. 5 - 4

Row System/Value Dimension	1 V_E PROC m³N/H	2 G_{PROD} T/H	3 V H2 m³N/H	4 V_E FUEL m³N/H	5 V_E SUBST m³N/H	6 V_{ESOL} HE m³N/H	7 V_{ESOL} L m³N/H	8 Q_{SOL} MAX MW	9 Q_{SOLAR} MW	10 η_{SOL} -	11 Q_E FUEL MW	12 $Q_{TOT\ FUEL}$ MW	13 Q_E SUBST MW	14 V_E UTIL m³N/H	15 Q_E UTIL MW
B-FOSSIL	14325	22.76		8385	3745	0	0	0	0	0	83.1	92.4	37.1	915	9.1
B - 1	14325	22.76		7695	3440	175	416	21.5	6.392	0.2973	76.3	85.6	34.1	915	9.1
B - 2.1	14325	22.76		7420	3440	199	457	21.5	8.951	0.4163	73.6	82.9	34.1	915	9.1
B - 2.2	14325	22.76		7038	3440	255	509	21.5	12.489	0.5809	69.8	79.1	34.1	915	9.1
B - 2.3	14325	22.76		6265	3440	337	761	21.5	19.647	0.9138	62.1	71.4	34.1	915	9.1
B - 3	14325	22.76		7117	3745	237	513	40.0	11.748	0.2937	70.6	79.9	37.1	915	9.1
B - 4.1	14325	22.76		6530	3745	289	607	40.0	17.192	0.4298	64.7	74.0	37.1	915	9.1
B - 4.2	14325	22.76		5850	3745	341	705	40.0	23.492	0.5873	58.0	67.3	37.1	915	9.1
B - 4.3	14325	22.76		4205	3745	472	968	40.0	38.732	0.9683	41.7	51.0	37.1	915	9.1
B - 5	15543	24.695		8386	3608	25	25	21.5	6.085	0.2830	83.1	93.2	35.8	993	9.8
B - 6	15543	24.695		7808	2063	48	48	43.4	11.952	0.2754	77.4	87.5	20.5	993	9.8

SPECIFIC ENERGY DATA   B - SYSTEMS   TAB. 5 - 5

Row System/Value Dimension	1 $V_{E\,PROC}$ m^3N/T	2 $V\,H2$ m^3N/T	3 $V_{E\,FUEL}$ m^3N/T	4 $V_{ESOL\,HE}$ m^3N/T	5 $V_{E\,SOL\,L}$ m^3N/T	6 $V_{E\,UTIL}$ m^3N/T	7 V_{ETOTHE} m^3N/T	8 $V_{ETOT\,L}$ m^3N/T	9 Q_{SOLMAX} MWH/T	10 Q_{SOL} MWH/T	11 Q_{ETOTHE} MWH/T	12 $V_{E\,SUBST}$ m^3N/T	13 $Q_{E\,SUBST}$ MWH/T	14 V_{EOPTIM} m^3N/T	15 Q_{EOPTIM} MWH/T
B-FOSSIL	629.4		368.4	0	0	40.2	1038.0	1038.0	0	0	10.29	164.5	1.63	873.5	8.66
B - 1	629.4		338.1	7.7	18.3	40.2	1015.4	1026.0	0.9446	0.2808	10.07	151.1	1.50	864.3	8.57
B - 2.1	629.4		326.0	8.7	20.1	40.2	1004.3	1015.7	0.9446	0.3932	0.96	151.1	1.50	853.2	8.46
B - 2.2	629.4		309.2	9.9	22.4	40.2	988.7	1001.2	0.9446	0.5487	9.80	151.1	1.50	837.6	8.30
B - 2.3	629.4		275.3	14.8	33.4	40.2	959.7	978.3	0.9446	0.8632	9.52	151.1	1.50	808.6	8.02
B - 3	629.4		312.7	10.4	22.5	40.2	992.7	1004.8	1.7575	0.5162	9.84	164.5	1.63	828.2	8.21
B - 4.1	629.4		286.9	12.7	26.7	40.2	969.2	983.2	1.7575	0.7554	9.61	164.5	1.63	804.7	7.98
B - 4.2	629.4		257.0	15.0	31.0	40.2	941.6	957.6	1.7575	1.0322	9.34	164.5	1.63	777.1	7.70
B - 4.3	629.4		184.8	20.7	42.5	40.2	875.1	896.9	1.7575	1.7018	8.68	164.5	1.63	710.6	7.05
B - 5	629.4		339.6	1.0	1.0	40.2	1010.2	1010.2	0.8706	0.2464	10.02	146.1	1.45	864.1	8.57
B - 6	629.4		316.2	1.9	1.9	40.2	987.7	987.7	1.7574	0.4840	9.79	83.5	0.83	904.2	8.97

BASIC ENERGY DATA   C - SYSTEMS   TAB. 5 - 6

Row System/Value Dimension	1 $V_{E\,PROC}$ m³N/H	2 G_{PROD} T/H	3 V_{H2} m³N/H	4 $V_{E\,FUEL}$ m³N/H	5 $V_{E\,SUBST}$ m³N/H	6 $V_{E\,SOL\,HE}$ m³N/H	7 $V_{E\,SOL\,L}$ m³N/H	8 $Q_{SOL\,MAX}$ MW	9 Q_{SOLAR} MW	10 η_{SOL} -	11 $Q_{E\,FUEL}$ MW	12 $Q_{TOT\,FUEL}$ MW	13 $Q_{E\,SUBST}$ MW	14 $V_{E\,UTIL}$ m³N/H	15 $Q_{E\,UTIL}$ MW
C-FOSSIL	11395	17.625		4105	1060	0	0	0	0	0	40.7	71.6	10.5	640	6.3
C - 1	11395	17.625		3416	755	175	416	21.5	6.392	0.2973	33.9	64.8	7.5	640	6.3
C - 2.1	11395	17.625		3140	755	199	457	21.5	8.951	0.4163	31.1	62.0	7.5	640	6.3
C - 2.2	11395	17.625		2758	755	255	509	21.5	12.489	0.5809	27.3	58.2	7.5	640	6.3
C - 2.3	11395	17.625		1985	755	337	761	21.5	19.647	0.9138	19.7	50.6	7.5	640	6.3
C - 3	11395	17.625		2777	0	237	513	40.0	11.748	0.2937	27.5	58.4	0	640	6.3
C - 4.1	11395	17.625		2162	0	289	607	40.0	17.192	0.4298	21.4	52.3	0	640	6.3
C 4.2	11395	17.625		1450	0	341	705	40.0	23.492	0.5873	14.4	45.3	0	640	6.3
C - 4.3	11395	17.625		0	0	472	968	40.0	38.732	0.9683	0	30.9	0	367	3.6
C - 5	12365	19.125		3742	925	25	25	21.5	6.085	0.2830	37.1	70.6	9.2	695	6.9
C - 6	12365	19.125		3165	0	48	48	43.4	11.952	0.2754	31.4	64.9	0	695	6.9

SPECIFIC ENERGY DATA — C-SYSTEMS — TAB. 5 - 7

Row System/Value Dimension	1 V_E PROC M³N/T	2 V H2 M³N/T	3 V_E FUEL M³N/T	4 V_ESOL HE M³N/T	5 V_E SOL L M³N/T	6 V_E UTIL M³N/T	7 V_ETOTHE M³N/T	8 V_ETOT L M³N/T	9 q_{SOLMAX} MWH/T	10 q_{SOL} MWH/T	11 q_{ETOTHE} MWH/T	12 V_E SUBST M³N/T	13 q_E SUBST MWH/T	14 V_{EOPTIM} M³N/T	15 q_{EOPTIM} MWH/T
C-FOSSIL	646.5		232.9	0	0	36.3	915.7	915.7	0	0	9.08	60.1	0.60	855.6	8.48
C - 1	646.5		193.8	9.9	23.6	36.3	886.5	900.2	1.220	0.363	8.79	42.8	0.42	843.7	8.37
C - 2.1	646.5		178.2	11.3	25.9	36.3	872.3	886.9	1.220	0.508	8.65	42.8	0.42	829.5	8.22
C - 2.2	646.5		156.5	12.8	28.9	36.3	852.1	868.2	1.220	0.709	8.45	42.8	0.42	809.3	8.02
C - 2.3	646.5		112.6	19.1	43.2	36.3	814.5	838.6	1.220	1.115	8.08	42.8	0.42	771.7	7.65
C - 3	646.5		157.6	13.4	29.1	36.3	853.8	869.5	1.220	0.667	8.47	0	0	853.8	8.47
C - 4.1	646.5		122.7	16.4	34.4	36.3	821.9	839.9	2.270	0.976	8.15	0	0	821.9	8.15
C - 4.2	646.5		82.3	19.3	40.0	36.3	784.4	805.1	2.270	1.333	7.78	0	0	784.4	7.78
C - 4.3	646.5		0	26.8	54.9	20.8	694.1	722.2	2.270	2.198	6.88	0	0	694.1	6.88
C - 5	646.5		195.7	1.	1.3	36.3	879.8	879.8	1.124	0.318	8.72	52.5	0.52	827.3	8.20
C - 6	646.5		165.5	2.5	2.5	36.3	850.8	850.8	2.269	0.625	8.44	0	0	850.8	8.44

BASIC ENERGY DATA   D-SYSTEMS   TAB. 5 - 8

Row System/Value Dimension	1 V_E PROC m³N/H	2 G_{PROD} T/H	3 V H2 m³N/H	4 V_E FUEL m³N/H	5 V_E SUBST m³N/H	6 V_{ESOL} HE m³N/H	7 V_{ESOL} L m³N/H	8 Q_{SOL} MAX MW	9 Q_{SOLAR} MW	10 η_{SOL} -	11 Q_E FUEL MW	12 Q_{TOT} FUEL MW	13 Q_E SUBST MW	14 V_E UTIL m³N/H	15 Q_E UTIL MW
D-FOSSIL	10290	8.625	11.280	7040	2480	0	0	0	0	0	69.8	79.9	24.6	785	7.8
D - 1	10290	8.625	11.280	6350	2179	175	416	21.5	6.392	0.2973	63.0	73.1	21.6	785	7.8
D - 2.1	10290	8.625	11.280	6074	2179	199	457	21.5	8.951	0.4163	60.2	70.3	21.6	785	7.8
D - 2.2	10290	8.625	11.280	5692	2179	225	509	21.5	12.489	0.5809	54.4	64.5	21.6	785	7.8
D - 2.3	10290	8.625	11.280	4920	2179	337	761	21.5	19.647	0.9138	48.8	58.9	21.6	785	7.8
D - 3	10290	8.625	11.280	5772	2537	237	513	40.0	11.748	0.2937	57.2	67.3	25.2	785	7.8
D 4.1	10290	8.625	11.280	5185	2537	289	607	40.0	17.192	0.4298	51.4	61.5	25.2	785	7.8
D 4.2	10290	8.625	11.280	4505	2537	341	705	40.0	23.492	0.5873	44.7	54.8	25.2	785	7.8
D - 4.3	10290	8.625	11.280	2859	2537	472	968	40.0	38.732	0.9683	28.3	38.4	25.2	785	7.8
D - 5	11170	9.36	12.240	6927	2350	25	25	21.5	6.085	0.2830	68.7	79.7	23.3	852	8.4
D 6	11170	9.36	12.240	6350	850	48	48	43.4	11.952	0.2754	63.0	74.0	8.4	852	8.4

SPECIFIC ENERGY DATA   D - SYSTEMS   TAB. 5 - 9

Row System/Value Dimension	1 $V_{E\ PROC}$ m³N/t	2 V_{H2} m³N/t	3 $V_{E\ FUEL}$ m³N/t	4 $V_{ESOL\ HE}$ m³N/t	5 $V_{E\ SOL\ L}$ m³N/t	6 $V_{E\ UTIL}$ m³N/t	7 V_{ETOTHE} m³N/t	8 $V_{ETOT\ L}$ m³N/t	9 Q_{SOLMAX} MWH/t	10 Q_{SOL} MWH/t	11 Q_{ETOTHE} MWH/t	12 $V_{E\ SUBST}$ m³N/t	13 $Q_{E\ SUBST}$ MWH/t	14 V_{EOPTIM} m³N/t	15 Q_{EOPTIM} MWH/t
D-FOSSIL	1193.0	1307.8	816.2	0	0	91.0	2100.2	2100.2	0	0	20.82	287.5	2.85	1812.7	17.97
D - 1	1193.0	1307.8	736.2	20.3	48.2	91.0	2040.5	2068.4	2.493	0.741	20.23	252.6	2.50	1787.9	17.73
D - 2.1	1193.0	1307.8	704.2	23.1	53.0	91.0	2011.3	2041.2	2.493	1.038	19.94	252.6	2.50	1758.7	17.44
D - 2.2	1193.0	1307.8	659.9	26.1	59.0	91.0	1970.0	2002.9	2.493	1.448	19.53	252.6	2.50	1717.4	17.03
D - 2.3	1193.0	1307.8	570.4	39.1	88.2	91.0	1893.5	1942.6	2.493	2.278	18.77	252.6	2.50	1640.9	16.27
D - 3	1193.0	1307.8	669.2	27.5	59.5	91.0	1980.7	2012.7	4.638	1.362	19.64	294.1	2.92	1686.6	16.72
D - 4.1	1193.0	1307.8	601.2	33.5	70.4	91.0	1918.7	1955.6	4.638	1.993	19.02	294.1	2.92	1624.6	16.11
D - 4.2	1193.0	1307.8	522.3	39.5	81.4	91.0	1845.8	1887.7	4.638	2.724	18.30	294.1	2.92	1551.7	15.23
D - 4.3	1193.0	1307.8	331.5	54.7	112.2	91.0	1670.2	1727.7	4.638	4.491	16.56	294.1	2.92	1376.1	13.64
D - 5	1193.0	1307.8	740.1	2.7	2.7	91.0	2026.8	2026.8	2.297	0.650	20.10	251.1	2.49	1775.7	17.61
D - 6	1193.0	1307.8	678.4	5.1	5.1	91.0	1967.5	1967.5	4.637	1.277	19.51	90.8	0.90	1876.7	18.61

System/Value Dimension	V_E TOT HE M^3N/T	V_E OPTIM M^3N/T	$V_{EFOSSIL} - V_{ESOLHE}$ $10^6 \, M^3N/A^{1)}$	$V_{EFOSSIL} - V_E$ OTIM $10^6 \, M^3N/A^{1)}$
A-FOSSIL	1172.9	1172.9	0	0
A - 1	1135.7	1135.7	4,178	4,178
A - 2.1	1117.5	1117.5	6,216	6,216
A - 2.2	1091.6	1091.6	9,122	9,122
A - 2.3	1043.5	1043.5	14,519	14,519
A - 3	1098.0	1083.2	8,404	10,064
A - 4.1	1059.2	1059.2	12,757	12,757
A - 4.2	1013.5	1013.5	17,885	17,885
A - 4.3	903.4	903.4	30,238	30,238
A - 5	1132.9	1132.9	4,863	4,863
A - 6	1091.8	1091.8	9,860	9,860
A' - 7	1102.4	1048.7	9,406	16,570
A' - 8	1209.0	1209.0	4,816	4,816
B-FOSSIL	1038.0	873.5	0	30,551
B - 1	1015.4	864.3	4,197	32,260
B - 2.1	1004.3	853.2	6,259	34,321
B - 2.2	988.7	837.6	9,156	37,219
B - 2.3	959.7	808.6	14,542	42,605
B - 3	992.7	828.2	8,413	38,964
B - 4.1	969.2	804.7	12,778	43,329
B - 4.2	941.6	777.1	17,904	48,455
B - 4.3	875.1	710.6	30,254	60,805
B - 5	1010.2	864.1	5,602	35,042
B - 6	987.7	904.2	10,136	26,962
C-FOSSIL	915.7	855.6	0	8,644
C - 1	886.5	843.7	4,200	10,369
C - 2.1	872.3	829.5	6,242	12,397
C - 2.2	852.1	809.3	9,147	15,302
C - 2.3	814.5	771.7	14,555	20,710
C - 3	853.8	853.8	8,902	8,902
C - 4.1	821.9	821.9	13,490	13,490
C - 4.2	784.4	784.4	18,884	18,884
C - 4.3	694.1	694.1	31,871	31,871
C - 5	879.8	827.3	5,602	13,794
C - 6	850.8	850.8	10,127	10,127
D-FOSSIL	2100.2	1812.7	0	20,234
D - 1	2040.5	1787.9	4,202	21,980
D - 2.3	1893.5	1640.9	14,548	32,326
D - 3	1980.7	1686.6	8,410	29,109
D - 4.3	1670.2	1376.1	30,263	50,962
D - 5	2026.8	1775.7	5,605	24,780
D - 6	1967.5	1876.7	10,133	17,067

1) 1 A = 8160 H

ENERGETIC COMPARISONS

TAB. 5 - 11

Row System/Value	1 $q_{SOL}/q_{E\,PROC}$	2 $q_{SOL}/q_{E\,TOTHE}$	3 $q_{SOL}/q_{AUX.FOSS}$	4 $\sum 1 \div 3$	5 $\dfrac{q_{SOL}+q_{E\,SUBST}}{q_{AUX.FOSS}}$	6 $\sum 4 \div 5$	7 $1-\dfrac{q_{EOPTIM}}{q_{E\,TOT\,FOSS}}$	8 $\dfrac{q_{SOL}-q_{SOL\,HE}(E)}{q_{E\,TOT\,FOSSIL}}$
A-FOSSIL	0	0	0		0		0	0
A - 1	0.0537	0.0413	0.1562		0.1616		0.0317	0.0292
A - 2.3	0.1651	0.1381	0.4801		0.4855		0.1103	0.1020
A - 3	0.0987	0.0784	0.2869		0.3360		0.0765	0.0588
A - 4.3	0.3256	0.3145	0.9464		0.9464		0.2298	0.2130
A - 5	0.0471	0.0363	0.1371		0.1371		0.0330	0.0336
A - 6	0.0926	0.0758	0.2694		0.2694		0.0905	0.0662
Σ A	0.7828	0.6844	2.2761	3.7433	2.3360	6.0793	0.5718	0.5028
B-FOSSIL	0	0	0		0		0	0
B - 1	0.6450	0.0279	0.0693		0.4023		0.1585	0.0199
B - 2.3	0.1383	0.0907	0.2131		0.4396		0.1673	0.0696
B - 3	0.0827	0.0524	0.1274		0.5833		0.2210	0.0401
B - 4.3	0.2727	0.1961	0.4201		0.5298		0.3154	0.1454
B - 5	0.0395	0.0246	0.0608		0.4187		0.1675	0.0230
B - 6	0.0776	0.0494	0.1195		0.3243		0.1289	0.0452
Σ B	0.6558	0.4411	1.0102	2.1071	3.5204	5.6275	1.3607	0.3432
C-FOSSIL	0	0	0		0		0	0
C - 1	0.0566	0.0413	0.1360		0.2248		0.0656	0.0292
C - 2.3	0.1739	0.1381	0.4177		0.2934		0.0786	0.1020
C - 3	0.1041	0.0788	0.2499		0.5751		0.1573	0.0588
C - 4.3	0.3429	0.3194	0.8235		0.8235		0.2420	0.2128
C - 5	0.0496	0.0365	0.1191		0.3140		0.0965	0.0336
C - 6	0.0957	0.0741	0.2342		0.2342		0.0709	0.0661
Σ C	0.8246	0.6882	1.9804	3.4932	2.7149	6.2081	0.7785	0.5025
D-FOSSIL	0	0	0		0		0	0
D - 1	0.0626	0.0566	0.0824		0.3168		0.1369	0.0259
D - 2.3	0.1926	0.1213	0.2533		0.3603		0.1487	0.0908
D - 3	0.1151	0.0694	0.1514		0.5312		0.2187	0.0523
D - 4.3	0.3797	0.2712	0.4993		0.8239		0.3448	0.1896
D - 5	0.0550	0.0323	0.0723		0.3491		0.1545	0.0299
D - 6	0.1080	0.0655	0.1420		0.2420		0.1064	0.0589
Σ D	0.9150	0.5963	1.2007	2.7100	3.0993	5.8093	1.3069	0.4474

INVESTMENTS

CODE: A   TAB. 5 - 12

Row SYSTEM	1 A-FOSSIL	2 A - 1	3 A - 2	4 A - 3	5 A - 4	6 A - 5	7 A - 6
REFORMING PLANT INCL. FLUEGAS SYSTEM INCL. FOSSIL HEATER, WASTE HEAT R.	25	46	46	56	56	17	17
METHANOL SYNTHESIS PLANT AND -DISTILLATION INCL. SYNGAS COMPRESSEION	45	45	45	45	45	47	47
PRESSURE SWING ADSORPTION THERMAL TREATMENT OF CONDENSATES	8	8	8	8	8	8	8
BOILER FEED WATER PREPARATION	2	2	2	2	2	2	2
OTHERS GASHOLDERS CONTINGENCIES	4	4	6	4	6	4	6
SUMMATION	84	105	107	115	117	78	80

ALL PRICES IN MILLION DM
THE PRICES ARE ADDITIONAL TO THOSE OF SOLAR PLANT, TAB. 5 - 1

INVESTMENTS

CODE: __B__   TAB. 5 - 13

Row / System	1 B-FOSSIL	2 B - 1	3 B - 2	4 B - 3	5 B - 4	6 B - 5	7 B - 6
Reforming Plant, incl. Secondary Reactor incl. Fluegas System, incl. Fossil Heater Waste Heat Recovery	30	51	51	61	61	20	16
Synthesisgas Preparation, Thermal Condensate Treatment, Air Compression	25	25	25	25	25	27	27
Ammonia Synthesis, incl. Refrigeration Unit and Syngas Compression	90	90	90	90	90	92	92
Boiler Feed Water Preparation	3	3	3	3	3	3	3
Others, Contingencies	10	10	10	10	10	10	10
Summation	158	179	179	189	189	155	151

All prices in million DM
The prices are additional to those of Solar Plant, Tab. 5 - 1

INVESTMENTS   <u>CODE: C</u>   TAB. 5 - 14

Row System	1 C-FOSSIL	2 C - 1	3 C - 2	4 C - 3	5 C - 4	6 C - 5	7 C - 6
Reforming Plant, incl. Fluegas Syst., Fossil Heater, Waste Heat Rec. Shift Conversion	25	46	46	56	56	21	17
Pressure Swing Adsorpt., Thermal Treatment of Condensate	12	12	12	12	12	13	13
Air Separation Plant Nitrogen Compression	25	25	25	25	25	26	26
Ammonia Synthesis incl. Refrigeration Unit and Syngas Compression	86	86	86	86	86	90	90
Boiler Feed Water Preparation	3	3	3	3	3	3	3
Others, Contingencies Gasholders	10	10	10	12	12	10	12
Summation	161	182	182	194	194	163	169

All prices in million DM
The prices are additional to those of the Solar Plant, Tab. 5 - 1

INVESTMENTS

CODE: D   TAB. 5 - 15

Row	1	2	3	4	5	6	7
System	D-Fossil	D - 1	D - 2	D - 3	D - 4	D - 5	D - 6
Reforming Plant, incl. Fluegas Syst., Fossil Heater, Waste Heat Recovery	25	46	46	56	56	21	17
First CO_2-Removal, CO_2-Compression, Thermal Condensate Treatment	6	6	6	6	6	7	7
Pressure Swing Adsorpt., Compression of CO-Gas	10	10	10	10	10	12	12
Ocatmix-Synthesis Plant Stabilization Second CO_2-Removal	36	36	36	36	36	38	38
Boiler Feedwater Treatment	2	2	2	2	2	2	2
Others, Contigencies	4	4	4	4	4	4	4
Summation	83	104	104	114	114	84	80

All prices in million DM.
The prices are additional to those of the Solar Plant, Tab. 5 - 1

Tab. 5 - 16

Assumption for the Estimation of the Specific
Solar Energy Cost

Cost Basis	1986
Time of Erection	4 a
Service Life	20 a
Cost of Natural Gas	45 DM/MWh
Real Escalation Rate of Natural Gas	3 % /a
Inflation Rate	5 % /a
Rate of Interest	8 % /a
Cost of Operation and Maintenance	2 %/a
Insurance	0,3 %/a
Decreasing of Solar Specific Investment	1,25 % /a (over 20 years)
$\dfrac{\text{Value thermal Unit Hydrogen}}{\text{Value thermal Unit Natural Gas}}$	1,40

COSTS CALCULATED BY DYNAMIC CASH FLUCTUATION METHODS

DESCRIPTION OF SYSTEM:

$A - Fossil \cdot CH_3OH - Fossil$

INPUT-INFORMATION

INVESTMENT Io IN 1986		10^6DM	84			
SOLAR-ENERGY AVAILABLE	GWh/a	0		REAL INCREASE OF N.G.VALUE F 1/a		0.030
NATURAL GAS(N.G.)TO PROC.	mN³/h	12 000		BASIC PRICE OF N.G. Hu 10^6DM/GWh		0.045
NATURAL GAS CONSUMPTION	GWh/a	0		LOWER HEATING VALVE OF N.G. KWh/mN³		99 15
N.G.TO COMBUSTION	mN³/h	3582		PART OF OPERATING COSTS OF Io		0.023
OTHER N.G.CONSUMERS	mN³/h	550		INTERVALLS OF NEXT T		5
ANNUAL OPERATING HOURS	h/a	8160		PRODUCT-QUANTITY t/h		13,75
REFERENCE YEAR (=1986)		1986		PURE H2-QUANTITY mN³/h		9420
YEAR OF START UP T		1990		INCREASE OF VALVE H2/N.G.F		1.4
PERIOD OF CONSTRUCTION	a	4		ANN.PRICE REDUCT.OF SOL.INVEST.Io		0.0125
LIFE IN YEARS	a	20		START OF REDUCTION		1991
RATE OF INTEREST	1/a	0.08		NUMBER OF YEARS FOR REDUCTION		20
RATE OF INFLATION R	1/a	0.05		PART OF INVESTMENT FOR REDUCT.		0

GENERAL RESULTS

$v_E^i = 883,6 \ mN^3 N.G/t \ CH_3OH$

TOTAL N.G.CONSUMPTION	mN³/h	16132	
TOTAL PRODUCTION TIME	h	163 200	
TOTAL PRODUCT QUANTITY	t	2 244 000	

TOTAL N.G.CONSUMPT. UNDER CONSIDERATION OF H2-CREDIT mN³/h   12 149,7

YEAR OF START UP	PRODUCT COSTS 86	HYDROGEN COSTS 86	REDUCED INVESTM. FOR T IN 1986	Io (t)	NAT.GAS PRICE g_1 (T) for 1986	CASH VALUE OF INVESTM. AT T(BT)	CASH VALUE FOR OPERAING COSTS AT T(CT)	CASH VALUE COSTS FOR NAT.GAS AT T(DT)	TOTAL COSTS CASH VALUE AT T(ET)	CASH VALUE T/ CASH VALUE 86	COSTS O N.G.Go AT T
	DM/t	DM/mN³	10^6DM	10^6DM	10^6DM/GWh	10^6DM	10^6DM	10^6DM	10^6DM		10^6DM/
1990	449.5	0.1886	84.0	84.0	0.045	80.564	26.766	901.4	1008.8	1.0000	44.23
1995	513.5	0.2187	84.0	107.2	0.0522	102.8	34.2	1333.7	1470.7	1.2763	65.04
2000	597.7	0.2535	84.0	136.8	0.0605	131.2	43.6	1973.4	2148.2	1.6289	96.83
2005	673.7	0.2939	84.0	174.6	0.0701	167.5	55.6	2919.7	3142.8	2.0789	143.27
2010	773.4	0.3407	84.0	222.9	0.0813	213.8	71.0	4319.9	4604.7	2.6533	211.97
2015	888.9	0.3949	84.0	284.5	0.0942	272.8	90.6	6391.5	6755.0	3.3864	313.63
2020	1022.9	0.4578	84.0	363.0	0.1092	348.2	115.7	9456.6	9920.5	4.3219	464.04
2025	1178.2	0.5308	84.0	463.3	0.1266	444.4	147.6	13992	14584	5.5160	686.575
2030	1358.2	0.6153	84.0	591.4	0.1468	567.2	188.4	20702	21457	7.0400	1015.8
2035	1567.0	0.7133	84.0	754.7	0.1702	723.9	240.5	30629	31593	8.9850	1503.0
2040	1808.9	0.8269	84.0	963.3	0.1973	923.9	306.9	45318	46548	11.4674	2223.7
2045	2089.4	0.9586	84.0	1229.4	0.2287	1179.1	391.7	67050	68621	14.6356	3290.2
2050	2414.6	1.1113	84.0	1568.7	0.2651	1504.9	500.0	99205	101210	18.6792	4868.0
2060	3228.5	1.4935	84.0	2555.8	0.3563	2451	814.	217169	220435	30.4264	10658.5

Tab. 5-17

COSTS CALCULATED BY DYNAMIC CASH FLUCTUATION METHODS

DESCRIPTION OF SYSTEM:

$A - 4.3$ = Separated Main Reactor with Large Storage for Methanol Production

INPUT-INFORMATION

INVESTMENT Io IN 1986	10^6DM	790				
SOLAR-ENERGY AVAILABLE	GWh/a	339,3	REAL INCREASE OF N.G.VALUE	F 1/a		0,030
NATURAL GAS(N.G.)TO PROC.	mN^3/h	12 000	BASIC PRICE OF N.G.	Hu 10^6DM/GWh		0,045
NATURAL GAS CONSUMPTION	GWh/a	0	LOWER HEATING VALVE OF N.G.	(KWh)mN^3		9,915
N.G.TO COMBUSTION	mN^3/h	0	PART OF OPERATING COSTS OF Io			0,023
OTHER N.G.CONSUMERS	mN^3/h	422	INTERVALLS OF NEXT T			5
ANNUAL OPERATING HOURS	h/a	8 160	PRODUCT-QUANTITY	t/h		13,75
REFERENCE YEAR (=1986)		1986	PURE H2-QUANTITY	mN^3/h		8 420
YEAR OF START UP T		1990	INCREASE OF VALVE H2/N.G.F			1,40
PERIOD OF CONSTRUCTION	a	4	ANN.PRICE REDUCT.OF SOL.INVEST.Io			0,0125
LIFE IN YEAR	a	20	START OF REDUCTION			1991
RATE OF INTEREST	1/a	0,08	NUMBER OF YEAR FOR REDUCTION			20
RATE OF INFLATION R	1/a	0,05	PART OF INVESTMENT FOR REDUCT.			0,90

GENERAL RESULTS

$$v_E^I = 613,8 \ m_N^3 \ N.G. \ / t \ CH_3OH$$

TOTAL N.G.CONSUMPTION	mN^3/h	12 422
TOTAL PRODUCTION TIME	h	163 200
TOTAL PRODUCT QUANTITY	t	2 244 000

TOTAL N.G.CONSUMPT. UNDER CONSIDERATION OF H2-CREDIT mN^3/h 8 439,7

YEAR OF START UP	PRODUCT COSTS 86	HYDROGEN COSTS 86	REDUCED INVESTM. FOR T IN 1986	Io (t)	NAT.GAS PRICE g_1 (T) for 1986	CASH VALUE OF INVESTM. AT T(BT)	CASH VALUE FOR OPERAING COSTS AT T(CT)	CASH VALUE OF COSTS FOR N.G. T(DT)	TOTAL COSTS CASH VALUE AT T(CT)	CASH VALUE T/ CASH VALUE 86	COSTS OF N.G.Go AT T
	DM/t	DM/mN^3	10^6DM	10^6DM	10^6DM/GWh	10^6DM	10^6DM	10^6DM	10^6DM		10^6DM/a
1990	728.9	0.1886	790.0	790.0	0.0450	757.7	251.7	626.2	1635.6	1.0000	30.727
1995	773.3	0.2187	790.0	1008.3	0.0522	967.0	321.3	926.5	2214.8	1.2763	45.463
2000	800.2	0.2535	746.7	1216.2	0.0605	1166.5	387.5	1370.8	2924.8	1.6289	67.265
2005	836.7	0.2939	706.0	1467.6	0.0701	1407.6	467.7	2028.1	3903.4	2.0989	99.522
2010	884.2	0.3407	667.7	1771.7	0.0813	1699.3	564.6	3000.8	5264.6	2.6533	147.2
2015	944.0	0.3949	631.9	2139.7	0.0942	2052.2	681.8	4439.8	7173.8	3.3864	217.9
2020	1037.1	0.4578	631.9	2730.8	0.1092	2619.2	870.2	6569.0	10 058	4.3219	322.3
2025	1145.0	0.5308	631.9	3485.3	0.1266	3342.8	1110.6	9719.2	14 173	5.5160	476.9
2030	1270.0	0.6153	631.9	4448.3	0.1468	4266.3	1417.4	14 380	20 064	7.0400	705.6
2035	1415.0	0.7133	631.9	5677.2	0.1702	5445.0	1809.0	21 276	28 530	8.9850	1044.0
2040	1583.1	0.8269	631.9	7245.7	0.1973	6949.4	2308.8	31 480	40 738	11.467	1544.7
2045	1777.9	0.9586	631.9	9247.6	0.2287	8869.4	2946.7	46 576	58 392	14.636	2285.5
2050	2003.8	1.1113	631.9	11 803	0.2651	11 320	3760.8	68 912	83 992	18.679	3381.5
2060	2569.2	1.4935	631.9	19 225	0.3563	18 439	6126.0	150 855	175 419	30.426	7402.5

Tab. 5-17 A

COSTS OF PRODUCTS

METHANOL

TAB. 5 - 18

ALL PRODUCT COSTS IN DM/T CH₃OH
COSTS OF NATURAL GAS IN DM/MWH
PRICE OF HYDROGEN IN DM/M³N

All prices/costs related to 1986-prices. Standard conditions as described in text.

Year of start-up	1990	1995	2000	2005	2010	2015	2020	2025	2030	2035	2040	2045	2050	Year of Fossil Cost Parity
A - 6	489.8	546.0	605.4	675.5	758.0	854.8	972.6	1267.4			1663.6	2196.1		2066
A - 5	481.1	542.2	609.5	688.4	780.5	888.0	1016.0	1336.4			1767.1	2345.8		2015
A - 4.3	728.9	773.3	800.2	836.7	884.2	944.0	1037.1	1145.0	1270.0		1583.1	2003.8		2022
A - 4.2	632.0	684.4	729.7	785.6	853.6	935.4	1045.1	1172.4	1319.9		1689.1	2185.3		2024
A - 4.1	585.0	640.8	694.6	759.4	836.7	928.5	1045.2	1180.4	1337.2		1729.7	2257.2		2025
A - 3	536.7	595.2	655.6	727.3	811.9	911.4	1034.0	1176.1	1340.9		1753.2	2307.4		2025
A - 2.3	605.3	659.9	710.4	771.7	845.5	933.5	1047.8	1180.3	1333.9		1781.5	2235.3		2026
A - 2.2	543.4	601.4	660.8	731.5	815.0	913.3	1034.9	1175.8	1359.2		1748.2	2297.9		2024
A - 2.1	531.8	591.8	655.1	729.9	817.8	921.0	1046.5	1192.0	1360.7	1556.2	1782.9	2350.3		2031
A · 1	505.4	566.7	634.0	712.9	805.1	912.7	1041.0	1189.7	1362.2	1562.0	1793.7	2062.3	2373.7	2032
A - FOSSIL	449.5	513.5	587.7	673.7	773.4	888.9	1022.9	1178.2	1358.2	1567.0	1808.9	2089.4	2414.6	
PRICE OF HYDROGEN	0.1886	0.2187	0.2235	0.2939	0.3407	0.3949	0.4578	0.5308	0.6153	0.7133	0.8269	0.9586	1.1113	
COSTS OF NATURAL GAS	45.0	52.2	60.5	70.1	81.3	94.2	109.2	126.6	146.8	170.2	197.3	228.7	265.1	

COSTS OF PRODUCTS

AMMONIA BY SECONDARY REFORMING WITH AIR
TAB. 5 - 19

ALL PRODUCT COSTS IN DM/T NH$_3$

COSTS OF NATURAL GAS IN DM /MWH

Year of start-up	1990	1995	2000	2005	2010	2015	2020	2025	2030	2035	2040	2045	2050	Year of Fossil Cost Parity
B - 6	554.6	626.2	705.5	798.3	906.6	1032.9	1182.6		1557.5		2061.3	2738.3		2010
B - 5	542.1	615.4	698.3	794.8	907.2	1037.8	1191.3		1575.6		2092.0	2786.0		2009
B - 4.3	695.1	758.4	817.1	888.4	974.1	1076.3	1209.0	1362.8	1541.1		1987.4	2587.3		2022
B - 4.2														
B - 4.1														
B - 3	578.9	650.8	729.7	822.1	930.1	1056.3	1206.8	1381.2	1583.5		2089.8	2770.3		2025
B 2.3	622.4	691.9	764.5	850.4	951.7	1070.6	1216.1	1384.8	1580.3		2069.8	2727.6		2026
B - 2.2														
B - 2.1														
B - 1	560.0	633.5	716.1	812.3	924.5	1055.1	1209.0	1387.5	1594.4	1834.2	2112.2		2808.2	2030
B - FOSSIL	526.3	601.4	688.6	789.6	906.7	1042.4	1199.8	1382.2	1593.7	1838.9	2123.1	2452.7	2834.6	
Costs of natural gas	45.0	52.2	60.5	70.1	81.3	94.2	109.2	126.6	146.8	170.2	197.3	228.7	265.1	

ALL PRICES/COSTS RELATED TO 1986-PRICES. STANDARD CONDITIONS AS DESCRIBED IN TEXT.

COSTS OF PRODUCTS

AMMONIA BY NITROGEN ADMIXTURE TO HYDROGEN

TAB. 5 - 20

ALL PRODUCT COSTS IN DM/T NH$_3$
COSTS OF NATURAL GAS IN DM/MWH

	1990	1995	2000	2005	2010	2015	2020	2025	2030	2035	2040	2045	2050	Year of Fossil Cost Parity
C - 6	530.5	592.1	659.0	737.5	829.5	937.0	1066.0		1389.0		1822.9		2406.2	2013
C - 5	510.5	574.3	645.4	728.5	825.4	938.2	1071.6		1405.5		1854.3		2457.4	2011
C - 4.3	701.6	751.9	791.0	840.6	902.1	977.0	1082.2	1204.2	1345.7		1699.7		2175.5	2020
C - 4.2														
C - 4.1														
C - 3	554.3	616.2	682.0	759.6	850.8	957.7	1087.1	1237.2	1411.2		1846.7		2431.9	2022
C - 2.3	612.9	617.9	729.9	799.5	882.2	980.2	1103.7	1246.9	1412.8	1605.3	1828.3		2386.7	2026
C - 2.2														
C - 2.1														
C 1	531.4	595.6	666.5	749.5	864.4	959.5	1093.9	1249.7	1430.3	1639.8	1882.5		2490.2	2031
C - FOSSIL	487.8	554.2	631.0	720.1	823.4	943.2	1082.0	1243.0	1429.6	1645.9	1896.6	2187.3	2524.3	
COSTS OF NATURAL GAS	45.0	52.2	60.5	70.1	81.3	94.2	109.2	126.6	146.8	170.2	197.3	228.7	265.1	
YEARS OF START-UP	1990	1995	2000	2005	2010	2015	2020	2025	2030	2035	2040	2045	2050	

STANDARD CONDITIONS AS DESCRIBED IN TEXT.

ALL PRICE/COSTS RELATED TO 1986-PRICES.

COSTS OF PRODUCTS

OCTAMIX

TAB. 5 - 21

ALL PRODUCT COSTS IN DM/t FUEL METHANOL (45 % HIGHER ALCOHOLS)
COSTS OF NATURAL GAS IN DM/MWh
PRICE OF HYDROGEN IN DM/m^3N

All prices/costs related to 1986-prices. Standard conditions as described in text.

	1990	1995	2000	2005	2010	2015	2020	2025	2030	2035	2040	2045	2050	Year of Fossil Cost Parity
D - 6	862.5	965.0	1074.4	1203.4	1354.8	1532.1	1746.7			2283.7	3005.5	3975.5		2012
D - 5	830.0	936.8	1055.1	1193.5	1355.0	1543.3	1766.8			2326.3	3078.3	4088.9		2013
D - 4.3	1224.2	1305.1	1359.8	1431.9	1523.5	1637.3	1806.7	2003.1	2230.7		2800.6	3566.5		2022
D - 4.2														
D - 4.1														
D - 3	917.9	1021.2	1129.3	1257.2	1407.8	1584.9	1801.3	2052.3	2343.2		3071.5	4050.2		2024
D - 2.3	1032.5	1129.6	1220.9	1331.4	1463.9	1621.6	1824.9	2060.5	2333.7	2650.4	3017.5	3936.5		2026
D - 2.2														
D - 2.1														
D - 1	867.9	975.6	1093.3	1231.4	1392.9	1581.6	1807.1	2068.6	2371.7	2723.1	3130.5	4150.2		2030
D - FOSSIL	778.8	890.9	1020.8	1171.4	1345.9	1548.3	1782.9	2054.9	2370.2	2735.7	3159.4	3650.6	4220.0	
PRICE FO HYDROGEN	0.1886	0.2187	0.2535	0.2939	0.3407	0.3949	0.4578	0.5308	0.6153	0.7133	0.8269	0.9586	1.1113	
COSTS OF NATURAL GAS	45.0	52.2	60.5	70.1	81.3	94.2	109.2	126.6	146.8	170.2	197.3	228.7	265.1	
YEAR OF START-UP	1990	1995	2000	2005	2010	2015	2020	2025	2030	2035	2040	2045	2050	

Fig. 5 - 2

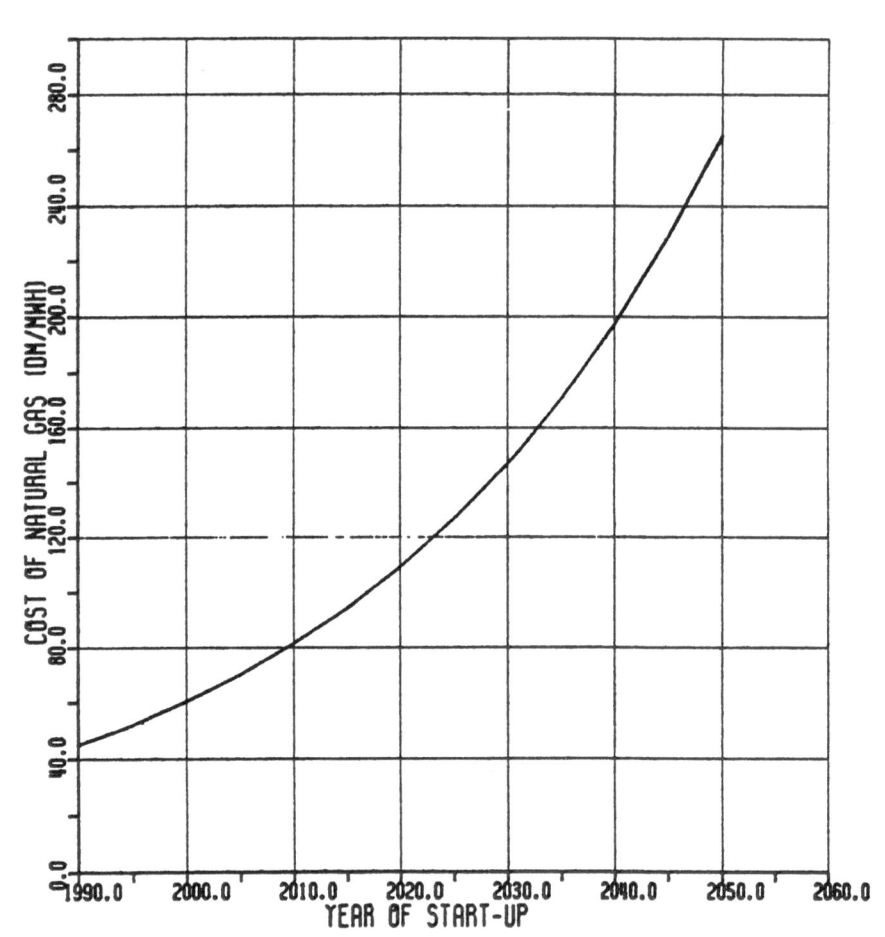

COSTS ARE ON BASIS OF 1986. THE H2-CREDIT IS BASED ON THE FOLLOWING RATIO HUH2/HUCH4=1.4

ANNUAL OPERATING PERIOD	8160 H/A	BASIC PRICE OF NAT. GAS	45.	DM/MWH
PERIOD OF CONSTRUCTION	4 A	LOWER HEATING VALUE	9915.	KWH/NM3
LIFE TIME	20 A	FRAC. OF OPERATING COSTS	.023	1/A
INTEREST RATE	.08 1/A	SOLAR COST DEPRECIATION	.01255	1/A
INFLATION RATE	.05 1/A	DEPRECIATION PERIOD	20	A
REAL INC. VALUE OF NAT.GAS	.03 1/A	START OF DEPRECIATION	1991	(YEAR)

Fig. 5 - 3

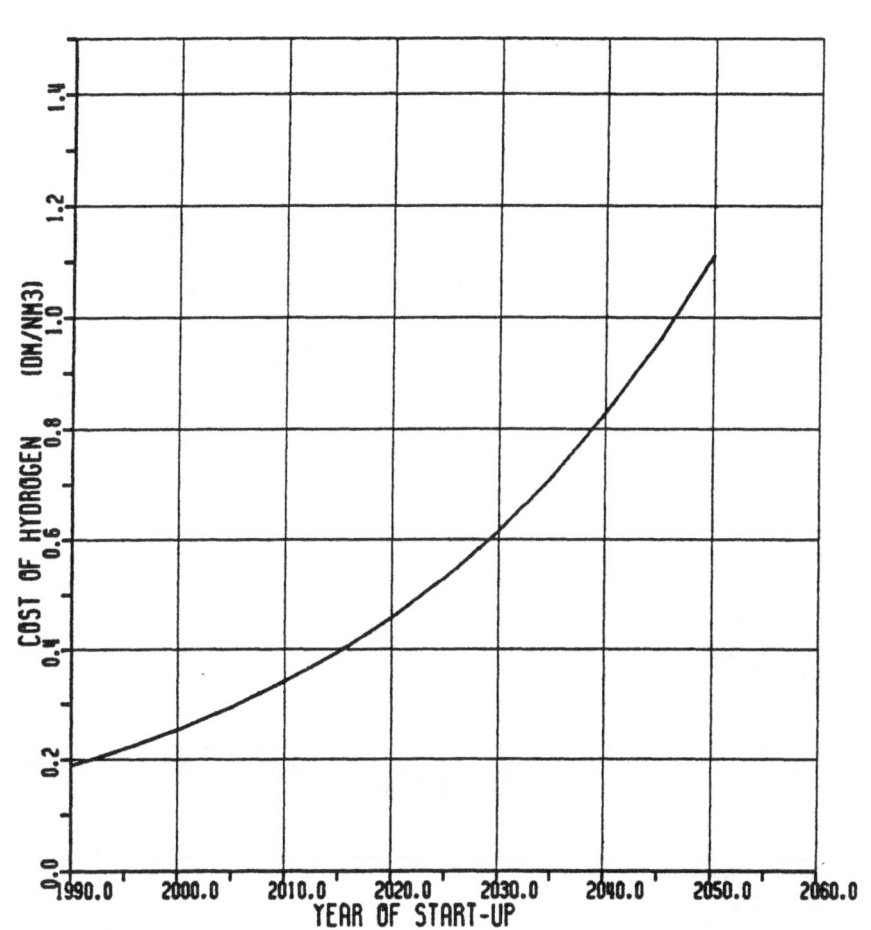

COST OF HYDROGEN (DM/NM3) vs YEAR OF START-UP

COSTS ARE ON BASIS OF 1986. THE H2-CREDIT IS BASED ON THE FOLLOWING RATIO HUH2/HUCH4=1.4

ANNUAL OPERATING PERIOD	8160 H/A	BASIC PRICE OF NAT. GAS	45. DM/MWH
PERIOD OF CONSTRUCTION	4 A	LOWER HEATING VALUE	9915. KWH/NM3
LIFE TIME	20 A	FRAC. OF OPERATING COSTS	.023 1/A
INTEREST RATE	.08 1/A	SOLAR COST DEPRECIATION	.01255 1/A
INFLATION RATE	.05 1/A	DEPRECIATION PERIOD	20 A
REAL INC. VALUE OF NAT.GAS	.03 1/A	START OF DEPRECIATION	1991 (YEAR)

LURGI	COST OF PRODUCTION	GKY6314

Fig. 5 - 4

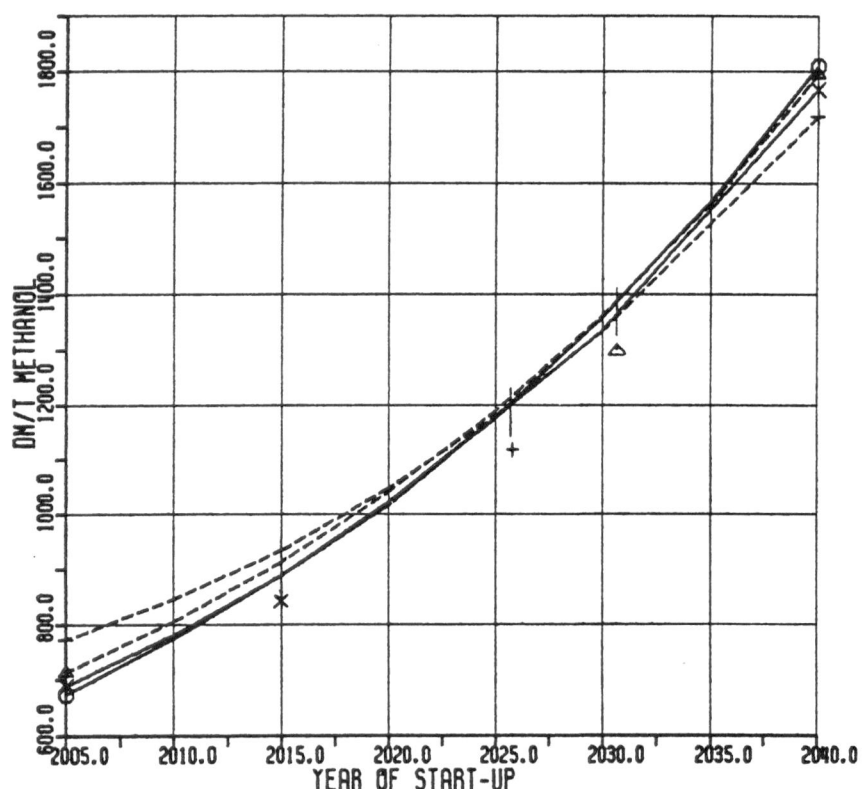

○ A-FOSSIL
△ A-1
+ A-2.3
× A-5

COSTS ARE ON BASIS OF 1986. THE H2-CREDIT IS BASED ON THE FOLLOWING RATIO HUH2/HUCH4=1.4
ANNUAL OPERATING PERIOD   8160 H/A   BASIC PRICE OF NAT. GAS   45. DM/MWH
PERIOD OF CONSTRUCTION   4 A   LOWER HEATING VALUE   9915. WH/NM3
LIFE TIME   20 A   FRAC. OF OPERATING COSTS   .023 1/A
INTEREST RATE   .08 1/A   SOLAR COST DEPRECIATION   .01255 1/A
INFLATION RATE   .05 1/A   DEPRECIATION PERIOD   20 A
REAL INC. VALUE OF NAT.GAS   .03 1/A   START OF DEPRECIATION   1991 (YEAR)

Fig. 5 - 5

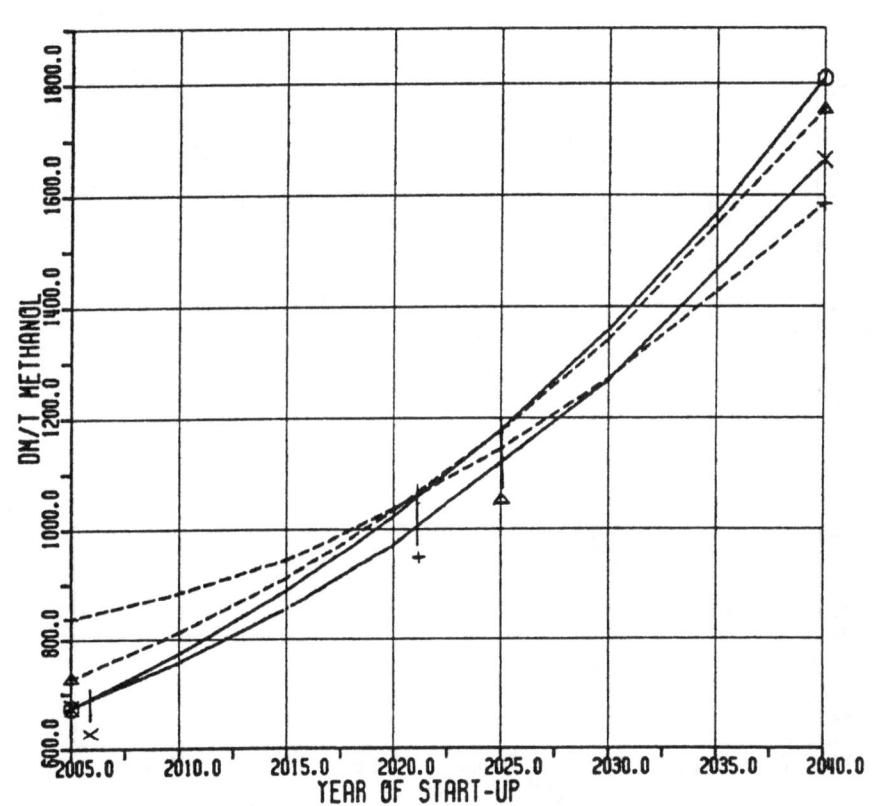

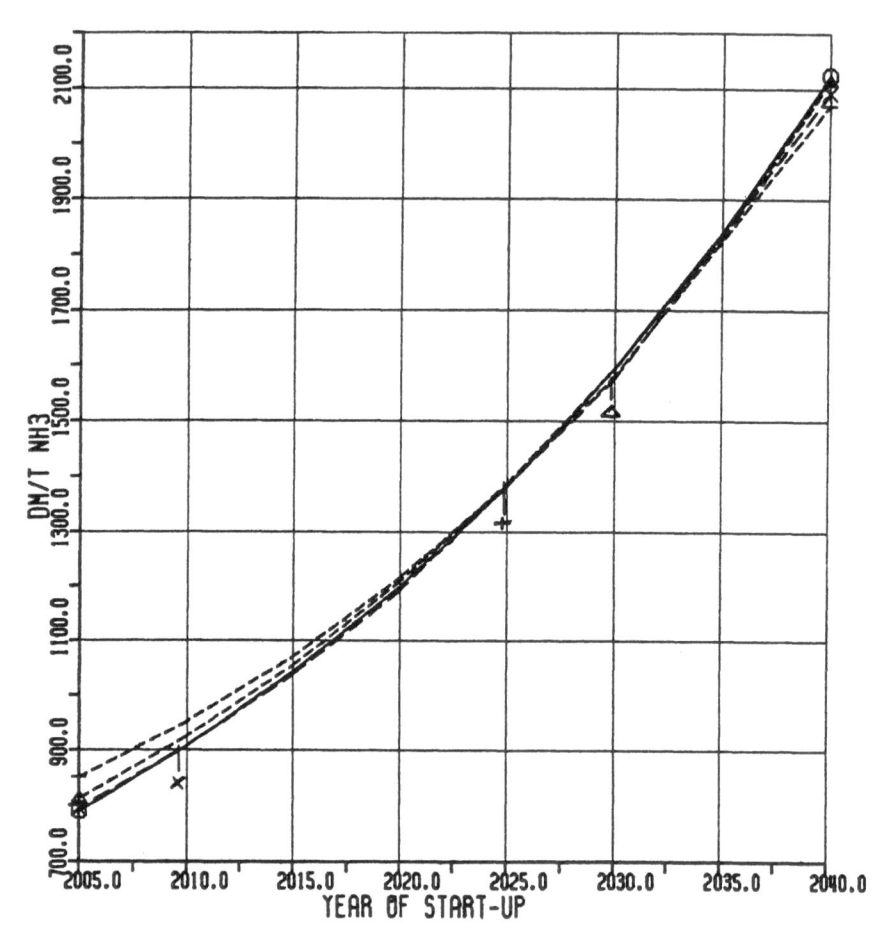

| | | COST OF PRODUCTION | | GKY63141 |

LURGI

Fig. 5 - 6

Legend:
- ⊕ B-FOSSIL
- △ B-1
- + B-2.3
- × B-5

COSTS ARE ON BASIS OF 1986.
ANNUAL OPERATING PERIOD   8160 H/A   BASIC PRICE OF NAT. GAS   45. DM/MWH
PERIOD OF CONSTRUCTION   4 A   LOWER HEATING VALUE   9915. KWH/NM3
LIFE TIME   20 A   FRAC. OF OPERATING COSTS   .023 1/A
INTEREST RATE   .08 1/A   SOLAR COST DEPRECIATION   . 01255 1/A
INFLATION RATE   .05 1/A   DEPRECIATION PERIOD   20   A
REAL INC. VALUE OF NAT.GAS   .03 1/A   START OF DEPRECIATION   1991 (YEAR)

Fig. 5 - 7

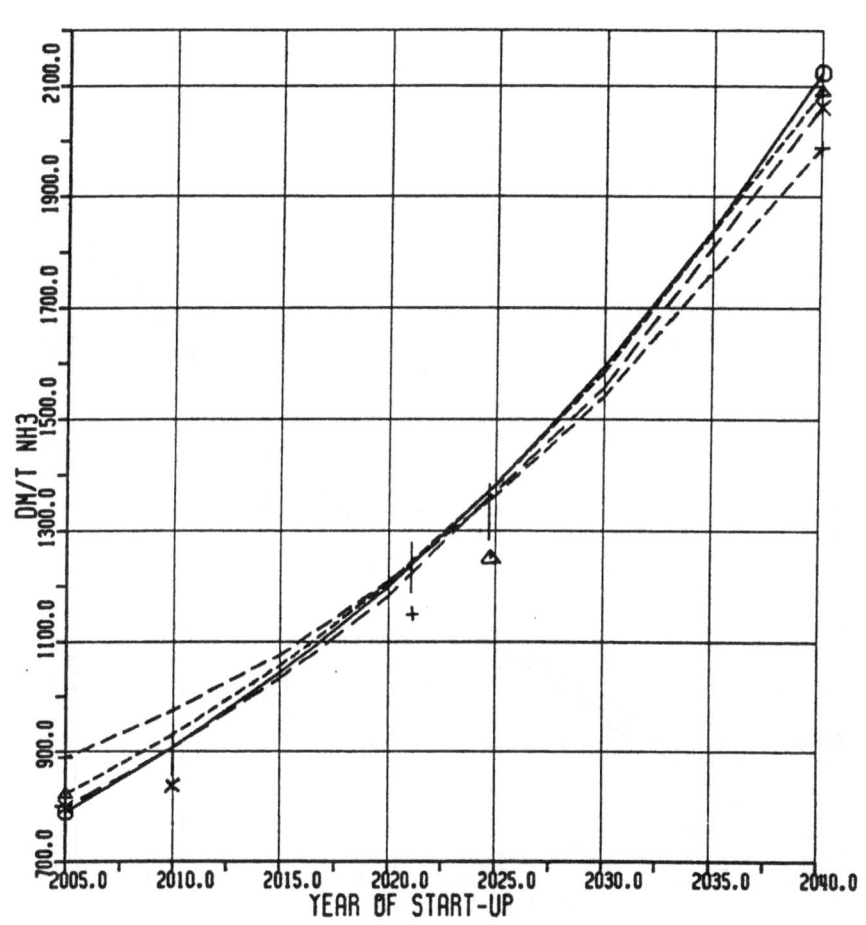

B-FOSSIL (⊙)
B-3 (△)
B-4.3 (+)
B-6 (✕)

COSTS ARE ON BASIS OF 1986.

ANNUAL OPERATING PERIOD	8160 H/A	BASIC PRICE OF NAT. GAS	45.	DM/MWH
PERIOD OF CONSTRUCTION	4 A	LOWER HEATING VALUE	9915.	KWH/NM3
LIFE TIME	20 A	FRAC. OF OPERATING COSTS	.023	1/A
INTEREST RATE	.08 1/A	SOLAR COST DEPRECIATION	.01255	1/A
INFLATION RATE	.05 1/A	DEPRECIATION PERIOD	20	A
REAL INC. VALUE OF NAT.GAS	.03 1/A	START OF DEPRECIATION	1991	(YEAR)

	COST OF PRODUCTION	
LURGI		

Fig. 5 - 8

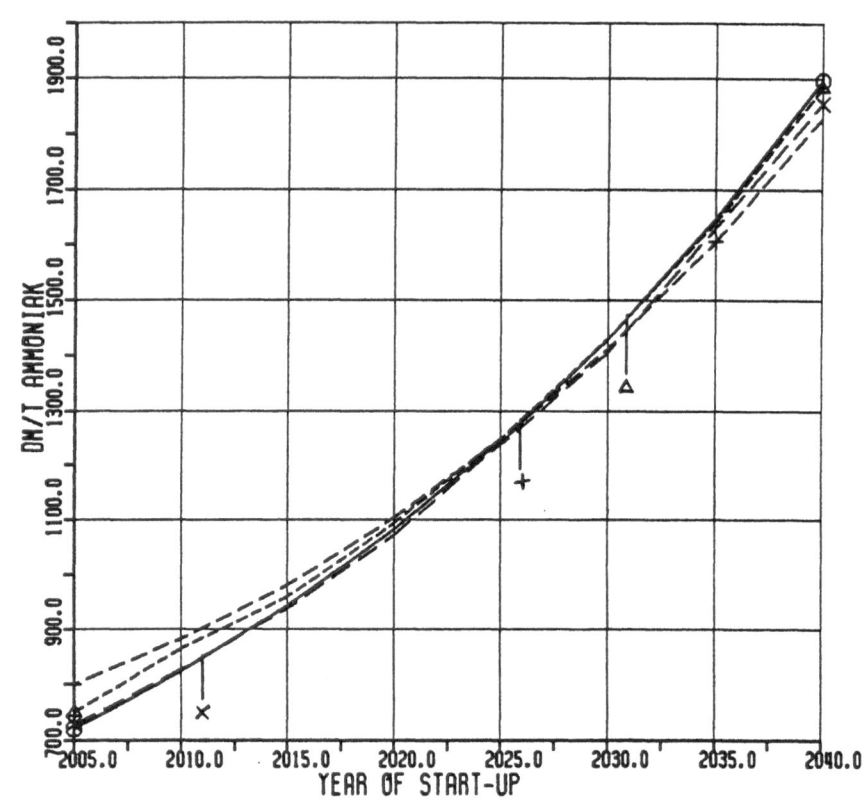

○ C-FOSSIL
△ C-1
+ C-2.3
✕ C-5

COSTS ARE ON BASIS OF 1986. THE H2-CREDIT IS BASED ON THE FOLLOWING RATIO HUH2/HUCH4=1.4
ANNUAL OPERATING PERIOD   8160 H/A   BASIC PRICE OF NAT. GAS   45. DM/MWH
PERIOD OF CONSTRUCTION   4 A   LOWER HEATING VALUE   9915. KWH/NM3
LIFE TIME   20 A   FRAC. OF OPERATING COSTS   .023 1/A
INTEREST RATE   .08 1/A   SOLAR COST DEPRECIATION   . 01255 1/A
INFLATION RATE   .05 1/A   DEPRECIATION PERIOD   20   A
REAL INC. VALUE OF NAT.GAS .03 1/A   START OF DEPRECIATION   1991 (YEAR)

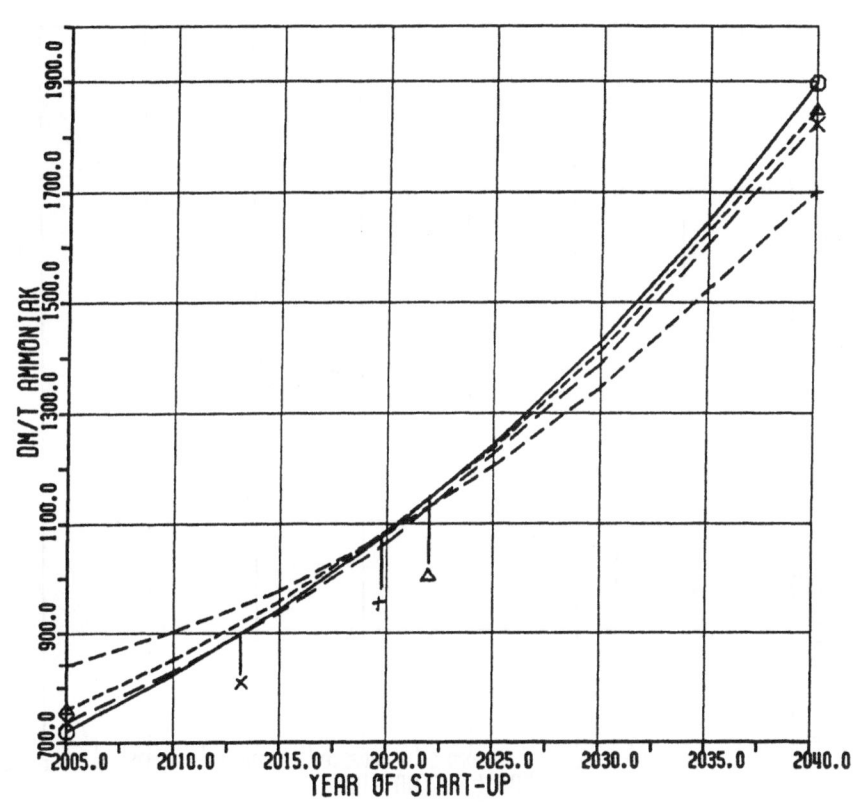

```
LURGI                    COST OF PRODUCTION
```

Fig. 5 - 9

○ C-FOSSIL
△ C-3
+ C-4.3
✕ C-6

COSTS ARE ON BASIS OF 1986. THE H2-CREDIT IS BASED ON THE FOLLOWING RATIO HUH2/HUCH4=1.4
ANNUAL OPERATING PERIOD   8160 H/A   BASIC PRICE OF NAT. GAS   45. DM/MWH
PERIOD OF CONSTRUCTION   4 A   LOWER HEATING VALUE   9915. KWH/NM3
LIFE TIME   20 A   FRAC. OF OPERATING COSTS   .023 1/A
INTEREST RATE   .08 1/A   SOLAR COST DEPRECIATION   . 01255 1/A
INFLATION RATE   .05 1/A   DEPRECIATION PERIOD   20   A
REAL INC. VALUE OF NAT.GAS .03 1/A   START OF DEPRECIATION   1991 (YEAR)

                              - 517 -
```
```

| (LURGI) | COST OF PRODUCTION | |

Fig. 5 - 10

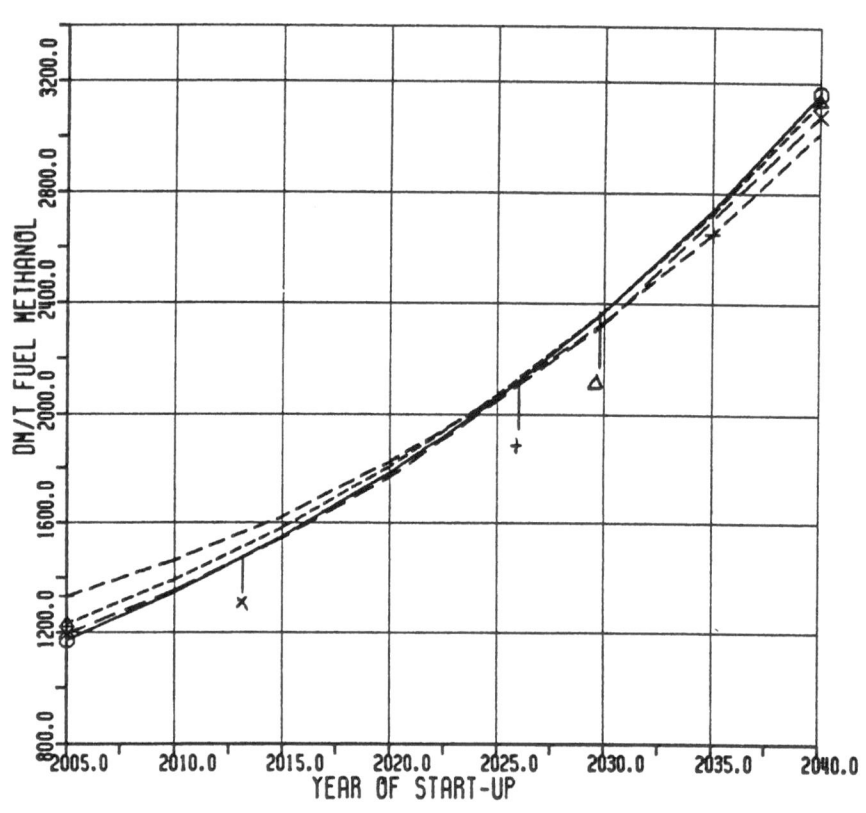

○   D-FOSSIL
△   D-1
+   D-2.3
✕   D-5

CÖSTS ARE ÖN BASIS ÖF 1986. THE H2-CREDIT IS BASED ÖN THE FÖLLÖWING RATIÖ HUH2/HUCH4=1.4
ANNUAL ÖPERATING PERIÖD   8160 H/A   BASIC PRICE ÖF NAT. GAS   45. DM/MWH
PERIÖD ÖF CÖNSTRUCTIÖN   4 A   LÖWER HEATING VALUE   9915. KWH/NM3
LIFE TIME   20 A   FRAC. ÖF ÖPERATING CÖSTS   .023 1/A
INTEREST RATE   .08 1/A   SÖLAR CÖST DEPRECIATIÖN   . 01255 1/A
INFLATIÖN RATE   .05 1/A   DEPRECIATIÖN PERIÖD   20   A
REAL INC. VALUE ÖF NAT.GAS   .03 1/A   START ÖF DEPRECIATIÖN   1991 (YEAR)

Fig. 5 - 11

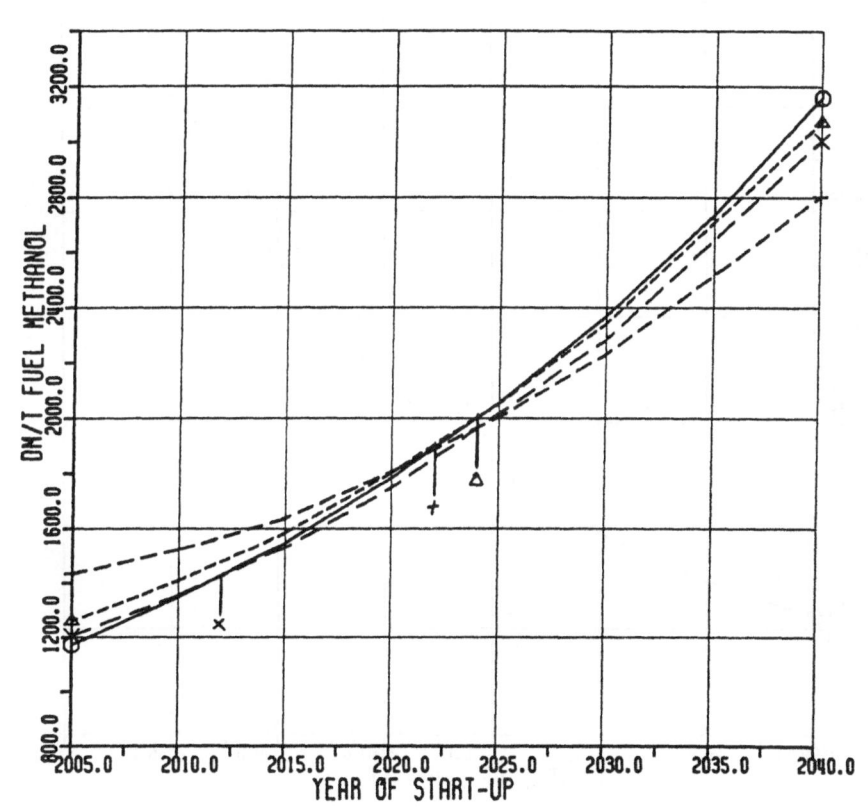

- ⊙   D-FOSSIL
- △   D-3
- +   D-4.3
- ✕   D-6

COSTS ARE ON BASIS OF 1986. THE H2-CREDIT IS BASED ON THE FOLLOWING RATIO HUH2/HUCH4=1.4

ANNUAL OPERATING PERIOD	8160 H/A	BASIC PRICE OF NAT. GAS	45. DM/MWH
PERIOD OF CONSTRUCTION	4 A	LOWER HEATING VALUE	9915. KWH/NM3
LIFE TIME	20 A	FRAC. OF OPERATING COSTS	.023 1/A
INTEREST RATE	.08 1/A	SOLAR COST DEPRECIATION	.01255 1/A
INFLATION RATE	.05 1/A	DEPRECIATION PERIOD	20 A
REAL INC. VALUE OF NAT.GAS	.03 1/A	START OF DEPRECIATION	1991 (YEAR)

COST OF PRODUCTION
- MAIN REACTOR -

Fig. 5 - 12

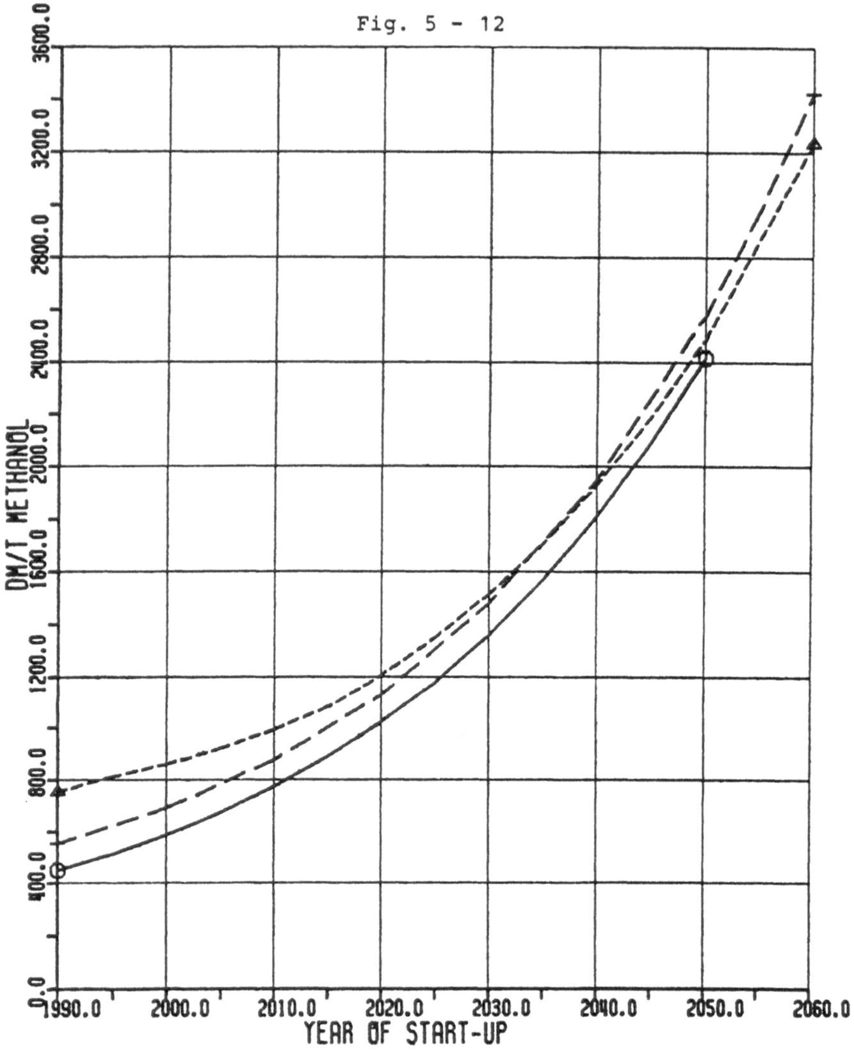

(Graph with Y-axis labeled "DM/T METHANOL" ranging from 0.0 to 3600.0, and X-axis labeled "YEAR OF START-UP" ranging from 1990.0 to 2060.0)

⊙	A-FOSSIL ,	
△	A-7 ; MAIN REACTOR SEPARATED	} at low pressure
+	A-8 ; MAIN REACTOR INTEGRATED	

COSTS ARE ON BASIS OF 1986. THE H2-CREDIT IS BASED ON THE FOLLOWING RATIO HUH2/HUCH4=1.4

ANNUAL OPERATING PERIOD	8160 H/A	BASIC PRICE OF NAT. GAS	45. DM/MWH
PERIOD OF CONSTRUCTION	4 A	LOWER HEATING VALUE	9915. KWH/NM3
LIFE TIME	20 A	FRAC. OF OPERATING COSTS	.023 1/A
INTEREST RATE	.08 1/A	SOLAR COST DEPRECIATION	.01255 1/A
INFLATION RATE	.05 1/A	DEPRECIATION PERIOD	20 A
REAL INC. VALUE OF NAT.GAS	.03 1/A	START OF DEPRECIATION	1991 (YEAR)

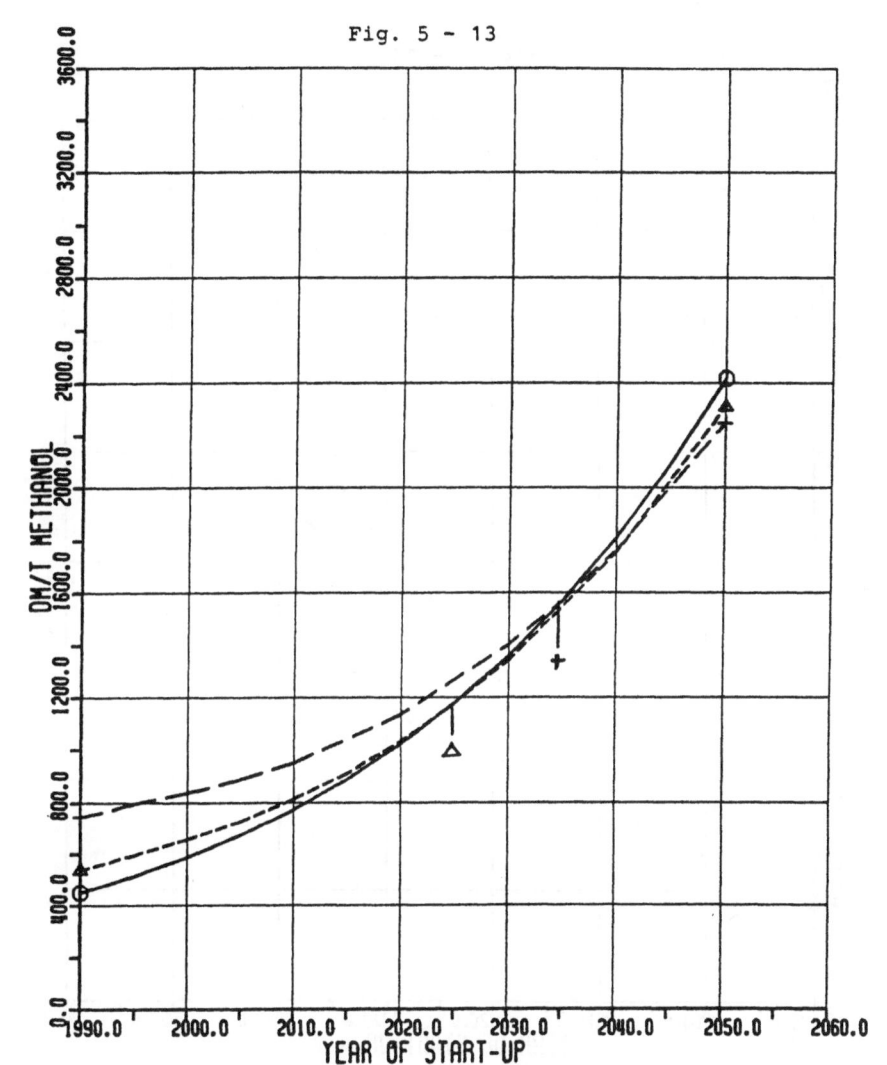

LURGI

DEVIATION FROM STANDARD STATE
- LOADING -

Fig. 5 - 13

Y-axis: DM/T METHANOL

X-axis: YEAR OF START-UP

○ A-FOSSIL , LOADING 100 %
△ A-3   ; LOADING 100 %
+ A-3   ; LOADING 15 % WITHOUT SOLAR

COSTS ARE ON BASIS OF 1986. THE H2-CREDIT IS BASED ON THE FOLLOWING RATIO HUH2/HUCH4=1.4
ANNUAL OPERATING PERIOD   8160 H/A   BASIC PRICE OF NAT. GAS   45. DM/MWH
PERIOD OF CONSTRUCTION   4 A   LOWER HEATING VALUE   9915. KWH/NM3
LIFE TIME   20 A   FRAC. OF OPERATING COSTS .023 1/A
INTEREST RATE   .08 1/A   SOLAR COST DEPRECIATION .01255 1/A
INFLATION RATE   .05 1/A   DEPRECIATION PERIOD   20 A
REAL INC. VALUE OF NAT.GAS .03 1/A   START OF DEPRECIATION   1991 (YEAR)

| LURGI | DEVIATION FROM STANDARD STATE |
| | - PRICE OF NATURAL GAS - |

Fig. 5 - 14

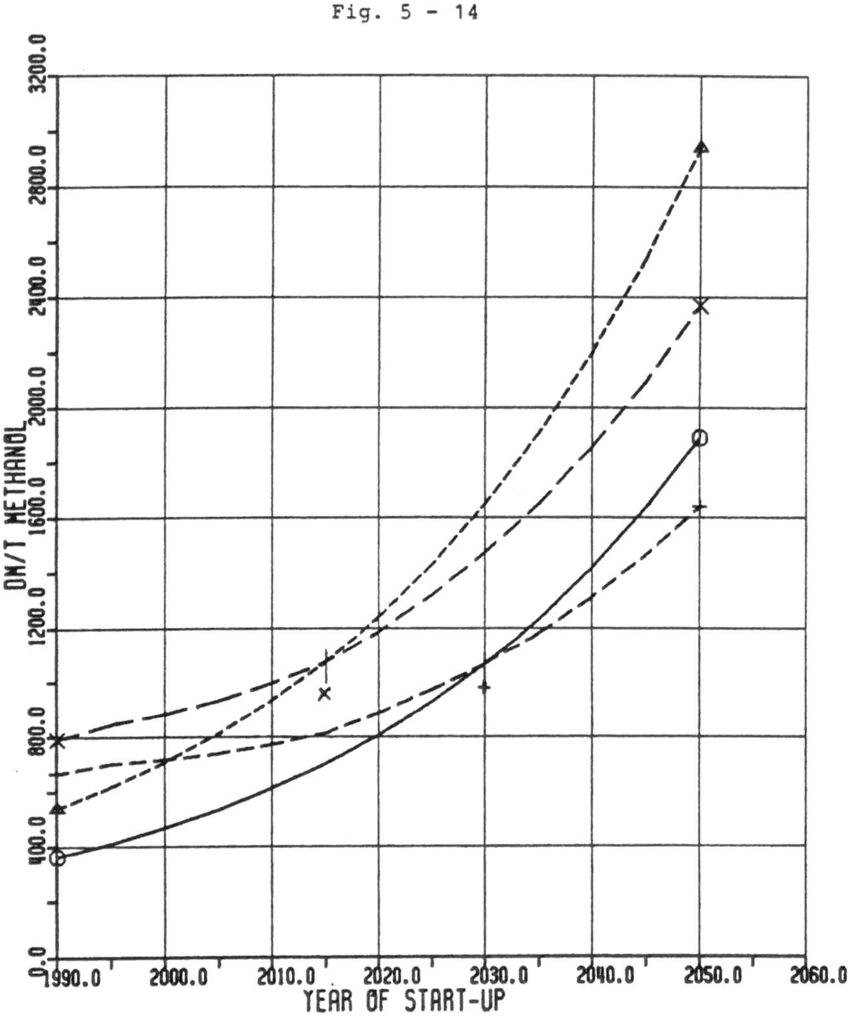

DM/T METHANOL

YEAR OF START-UP

○ A-FOSSIL , PRICE OF NATURAL GAS 35. DM/MWH
△ A-FOSSIL , PRICE OF NATURAL GAS 55. DM/MWH
+ A-4.3   ; PRICE OF NATURAL GAS 35. DM/MWH
X A-4.3   ; PRICE OF NATURAL GAS 55. DM/MWH

COSTS ARE ON BASIS OF 1986. THE H2-CREDIT IS BASED ON THE FOLLOWING RATIO HUH2/HUCH4=1.4
ANNUAL OPERATING PERIOD   8160 H/A   BASIC PRICE OF NAT. GAS   45. DM/MWH
PERIOD OF CONSTRUCTION   4 A   LOWER HEATING VALUE   9915. KWH/NM3
LIFE TIME   20 A   FRAC. OF OPERATING COSTS .023 1/A
INTEREST RATE   .08 1/A   SOLAR COST DEPRECIATION .01255 1/A
INFLATION RATE   .05 1/A   DEPRECIATION PERIOD   20   A
REAL INC. VALUE OF NAT.GAS .03 1/A   START OF DEPRECIATION   1991 (YEAR)

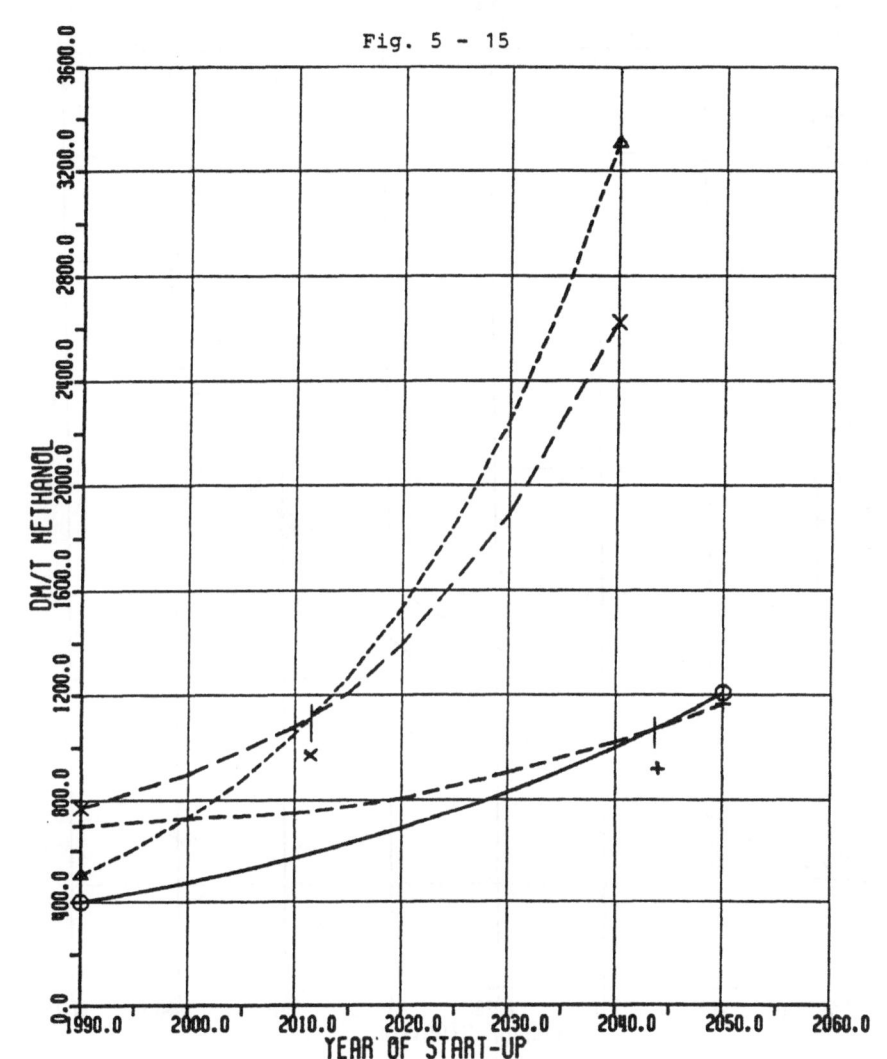

LURGI

DEVIATION FROM STANDARD STATE
-ESCALATION OF NAT.GAS-

Fig. 5 - 15

DM/T METHANOL

YEAR OF START-UP

⊙   A-FOSSIL , ESCALATION RATE OF NATURAL GAS 2 %
△   A-FOSSIL , ESCALATION RATE OF NATURAL GAS 4 %
+   A-4.3   , ESCALATION RATE OF NATURAL GAS 2 %
X   A-4.3   , ESCALATION RATE OF NATURAL GAS 4 %

COSTS ARE ON BASIS OF 1986. THE H2-CREDIT IS BASED ON THE FOLLOWING RATIO HUH2/HUCH4=1.4
ANNUAL OPERATING PERIOD   8160 H/A   BASIC PRICE OF NAT. GAS   45. DM/MWH
PERIOD OF CONSTRUCTION   4 A   LOWER HEATING VALUE   9915. KWH/NM3
LIFE TIME   20 A   FRAC. OF OPERATING COSTS   .023 1/A
INTEREST RATE   .08 1/A   SOLAR COST DEPRECIATION   . 01255 1/A
INFLATION RATE   .05 1/A   DEPRECIATION PERIOD   20 A
REAL INC. VALUE OF NAT.GAS .03 1/A   START OF DEPRECIATION   1991 (YEAR)

DEVIATION FROM STANDARD STATE
- OPERATING LIFE -

Fig. 5 - 16

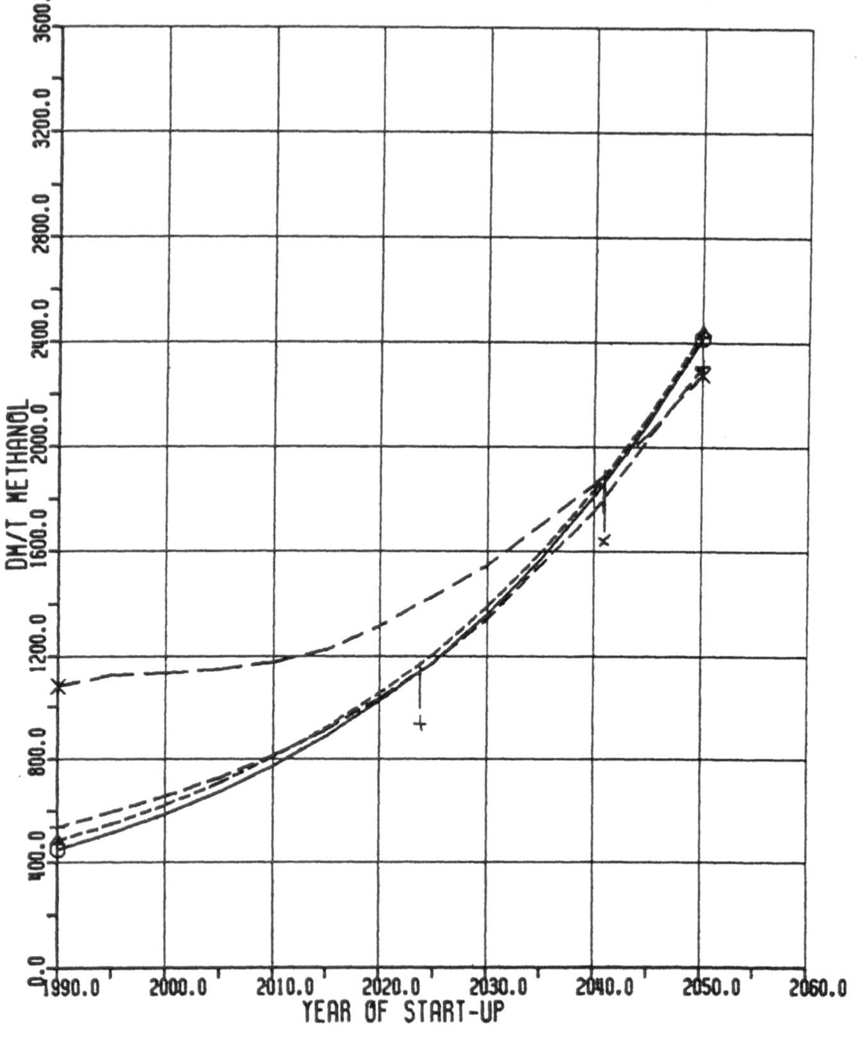

DM/T METHANOL

YEAR OF START-UP

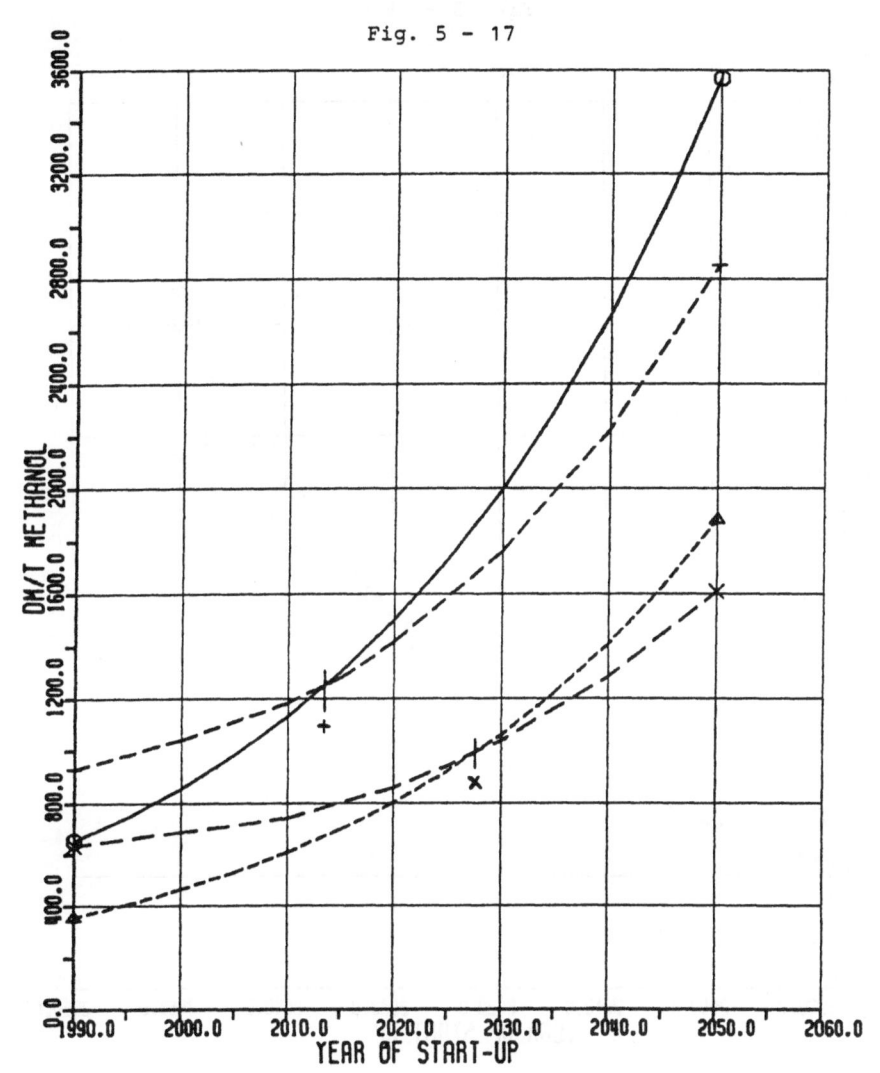

LURGI

DEVIATION FROM STANDARD STATE
-INFLATION AND INTEREST-

Fig. 5 - 17

⊙ A-FOSSIL , INFLATION RATE 6% , INTEREST RATE 6%
△ A-FOSSIL , INFLATION RATE 3% , INTEREST RATE 8%
+ A-4.3   ; INFLATION RATE 6% ; INTEREST RATE 6%
X A-4.3   ; INFLATION RATE 3% ; INTEREST RATE 8%

COSTS ARE ON BASIS OF 1986. THE H2-CREDIT IS BASED ON THE FOLLOWING RATIO HUH2/HUCH4=1.4
ANNUAL OPERATING PERIOD   8160 H/A   BASIC PRICE OF NAT. GAS   45.   DM/MWH
PERIOD OF CONSTRUCTION   4 A   LOWER HEATING VALUE   9915. KWH/NM3
LIFE TIME   20 A   FRAC. OF OPERATING COSTS   .023 1/A
INTEREST RATE   .08 1/A   SOLAR COST DEPRECIATION . 01255 1/A
INFLATION RATE   .05 1/A   DEPRECIATION PERIOD   20   A
REAL INC. VALUE OF NAT.GAS .03 1/A   START OF DEPRECIATION   1991 (YEAR)

| LURGI | DEVIATION FROM STANDARD STATE - INVESTMENT - | |

Fig. 5 - 18

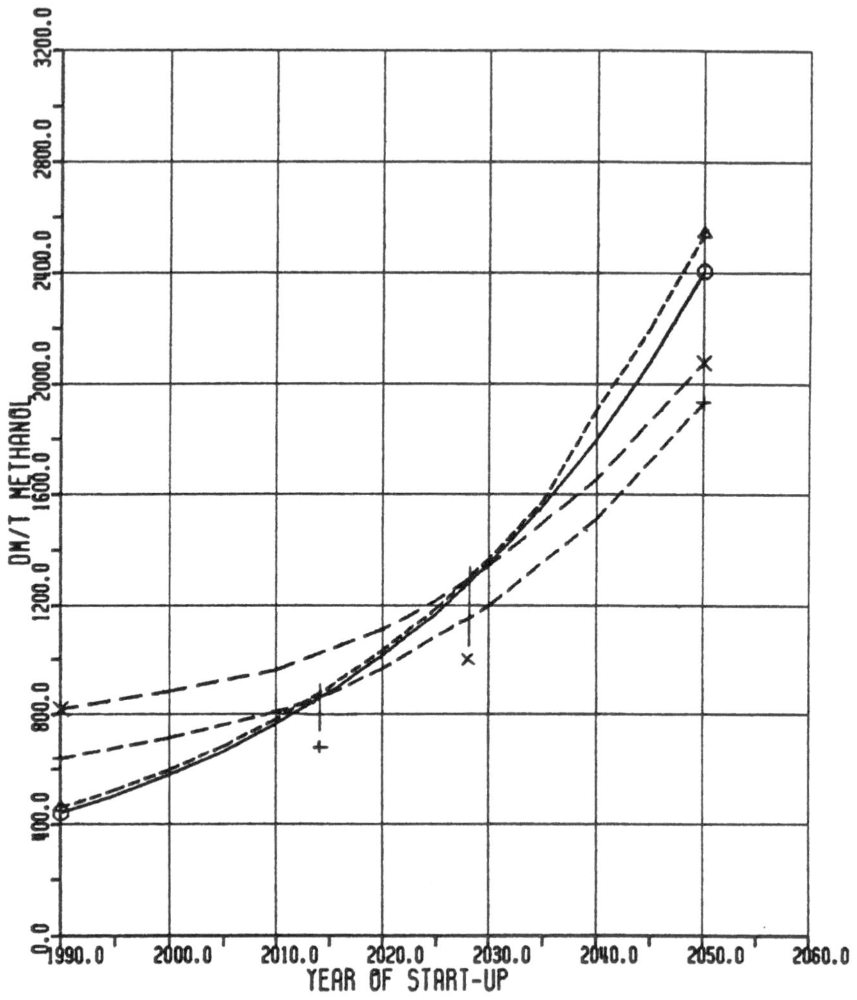

A plot with y-axis labeled "DM/T METHANOL" ranging from 0.0 to 3200.0, and x-axis labeled "YEAR OF START-UP" ranging from 1990.0 to 2060.0.

○ A-FOSSIL , INVESTMENT   80%
▲ A-FOSSIL , INVESTMENT   120%
+ A-4.3   ; INVESTMENT   80%
✕ A-4.3   ; INVESTMENT   120%

CØSTS ARE ØN BASIS ØF 1986. THE H2-CREDIT IS BASED ØN THE FØLLØWING RATIØ HUH2/HUCH4=1.4

ANNUAL ØPERATING PERIØD	8160 H/A	BASIC PRICE ØF NAT. GAS	45. DM/MWH
PERIØD ØF CØNSTRUCTIØN	4 A	LØWER HEATING VALUE	9915. KWH/NM3
LIFE TIME	20 A	FRAC. ØF ØPERATING CØSTS	.023 1/A
INTEREST RATE	.08 1/A	SØLAR CØST DEPRECIATIØN	. 01255 1/A
INFLATIØN RATE	.05 1/A	DEPRECIATIØN PERIØD	20 A
REAL INC. VALUE ØF NAT.GAS	.03 1/A	START ØF DEPRECIATIØN	1991 (YEAR)

- 526 -

6. Valuations of Solar Reforming Systems and Selection of a System for Development

6.1 Valuation of Solar Reforming Systems

6.1.1 Creation of Basic Valuation Data, Tab. 6 - 1

In general accord between the partners finally only four elements should be considered to be essential for the selection of the best suitable solar methane reforming system namely

  a) the solar energy rate in the product

  b) the specific product costs in the year 2030 (calculated with standard conditions)

  c) Specific investment in year 2030

  d) Risk of development for the system.

The figures, which can be found in table 6 - 1, have been derived from table 5 - 3, table 5 - 18 and table 5 - 12 or equivalent table 5 - 17 respectively. These figures are calculable, whereas the basic figure for the risks of development had to be fixed by mutual agreement, based on estimates on available techniques and known research programs.

6.1.2 Valuation of Systems, Tab. 6 - 2

Agreement was reached between the partners that the system with the highest interest with respect to a feature should get 5 points of merit, while that one with the lowest quote should get 1 point only. Therefore in each upper line of tab. 6 - 2 a case with 5 points and one with only 1 point can be found. A high risk got 1 point, a low risk 5 points. The intermediate points were calculated by the formulas given on the bottom of table 6 - 2.

Another agreement had to be reached between the partners,

what effectiveness should be given to each of the parameters. Finally it was accepted:

Solar energy rate	by	35 %
Specific product costs	by	35 %
Specific investment	by	10 %
Risk of development	by	20 %

Thus in table 6 - 2 the lower values of each line represent the product of the basic valuation figure of the upper line times the percentage of acceptance. The penultimate line of tabulation 6 - 2 gives the summation of the valuated figures for each solar reforming system.

These valuated figures determine the sequence of interest for the investigated systems.

One reads

         1   System A - 4.3

         2   System A - 6

         3   System A - 2.3

As was to be expected, the systems with high temperature (A - 4.3 and A - 6) are favoured, mainly by the high portion of solar energy in the product.

It should be recognized, if the same valuation procedure is applied for the reference year 2040 the sequence will change to

         1   System A - 4.3

         2   System A - 2.3

         3   System· A - 6

favouring also the high rate of solar energy in the product, but being made effective by the installation of storage systems. By the choice of a later reference date the sequence of product proces has got better established and there will be no later change in the basic figure-relation in line 2 of table 6 - 2 after the date 2040 with the adopted standard conditioins. But the revised valua-

tion shows that also separated prereactor systems with
high storage capacity are of interest.

6.1.3   Selection of Systems for Development

By the preceeding chapter the systems of highest interest
for solar steam reforming plants already have been determined. But, number one, the system A - 4.3 is the most
delicate one with the highest risk in development and also
the system A - 6 is thought to be at the limits of technical practicability with metallic tubes. Therefore the development should start for the demonstration of systems
similar to A - 2.3 and should proceed to the high temperature systems A - 6 and A - 4.3, when additional knowledge
about the behaviour of systems at $800^{\circ}C$, which is rather
limited still at present, will be available. As the
systems A - 2.3 and A - 4.3 reach parity in costs about
the same time under the assumptions of chapter 5.4 it may
well be that under different scenarios or aspects (politi-
cal, environmental, availability of natural gas) the solar
methane reforming systems may be interesting at an earlier
date and that it would be advisable to have operable
systems available at such earlier time.

Details for a preferred schedule for development and demonstration will be given in chapter 7.

VALUATIONS OF SYSTEMS/ BASIC DATA   TAB. 6 - 1

Number System Weight Code	Separated Prereactors A-1	A-2.1	A-2.2	A-2.3	Separated Main Reactors A-3	A-4.1	A-4.2	A-4.3	Integrated A-5	A-6
Item										
W 1 Solar Energy Rate in Product MWh/t CH_3OH	0.3387	0.5073	0.7464	1.1856	0.6833	1.0419	1.4628	2.8169	0.3915	0.7700
W 2 Specific Product Costs in Year 2030 DM/t CH_3OH	1362.2	1360.7	1339.2	1333.9	1340.9	1337.2	1319.9	1270.0	1336.4	1267.4
W 3 Specific Invest-ment in Year 2030 $10^{...}$ DM·D/t CH_3OH	0.5709	0.7076	0.7982	1.1485	0.7548	1.0427	1.3109	1.9148	0.4476	0.6130
W 4 Risks of Development Points	5	4	4	4	2.5	1	1	1	3	2

Points: 1 = worst case
         5 = best case

VALUATIONS OF SYSTEMS/RATING   TAB. 6 - 2

Number System Weight Code / Item	Separated Prereactors				Separated Main Reactors				Integrated	
	A-1	A-2.1	A-2.2	A-2.3	A-3	A-4.1	A-4.2	A-4.3	A-5	A-6
W1 35% Solar Energy Rate in Product	1.00 0.35	1.27 0.44	1.66 0.58	2.37 0.83	1.56 0.55	2.13 0.75	2.81 0.98	5.00 1.75	1.09 0.38	1.70 0.60
W2 35% Specific Product Costs in Year 2030	1.00 0.35	1.06 0.37	1.97 0.69	2.19 0.77	1.90 0.67	2.05 0.72	2.78 0.97	4.89 1.71	2.09 0.73	5.00 1.75
W3 10% Specific Investment in Year 2030	4.67 0.47	4.29 0.43	4.05 0.41	3.09 0.31	4.16 0.42	3.38 0.34	2.65 0.27	1.00 0.10	5.00 0.50	4.55 0.46
W4 20% Risks of Development	5 1.00	4 0.80	4 0.80	4 0.80	2.5 0.50	1 0.20	1 0.20	1 0.20	3 0.60	2 0.40
Summation	2.17	2.04	2.48	2.71	2.14	2.01	2.42	3.76	2.21	3.21
Sequence				3				1		2

$$W1 = 5 - 4 \cdot \frac{E_{MAX} - E}{E_{MAX} - E_{MIN}}$$

$$W2 = 1 + 4 \cdot \frac{K_{MAX} - K}{K_{MAX} - K_{MIN}}$$

$$W3 = 1 + 4 \cdot \frac{J_{MAX} - J}{J_{MAX} - J_{MIN}}$$

7.   Proposal for a System Experiment

The previous expositions of this study have shown that
natural gas can be substituted by coupling solar produced
process heat in a chemical high temperature process. The
system under examination here - heating a separate
reactor with a solar heated gas or the direct heating of
reactor pipes in the cavity - is not limited to the
application of methane reforming. This technology is also
of great interest for other processes already developed
or being developed, in particular for the storage of
solar energy, the transport of solar energy or for the
generation of hydrogen from non-fossil raw materials. The
prerequisite for examining such processes under realistic
solar conditions and with test plant parameters which can
be subsequently implemented in plant construction, is the
availability of the solar "process heat supplier" and the
experiences gained relating to its interaction with high
temperature processes which require process heat at a
constant temperature.

To test the solar section and the system function
together with a process plant, a system experiment is
therefore proposed for methane reforming on the
Plataforma Solar in Almeria. Methane reforming serves as
a reference process for all other processes since this
process is well known and reliably controlled. This
ensures that during the system experiment emphasis is
placed on the solar-specific development and not on a
process development which up to a certain development
stage should be carried out more effectively with
conventional heat sources.

The system experiment for solar methane reforming has the
following objectives:

  Verification of the operability of a chemical high
  temperature process with thermal energy produced by
  solar means under realistic insolation conditions.

. Verification of the function of a reactor integrated in
the cavity.

. Verification of the function of receiver and
storage, examination of design methods and testing
of design solutions.

. Verification of the function of an auxiliary cavity
firing system and of its design method.

For this purpose, the system experiment should be carried
out in two stages:

. An integrated receiver/main reactor orientated to the
plant concept 6;

. Test of a plant with separate heat transfer circuit
while using storage modules based on the plant concept
4.

7.1   System Experiment A for an Integrated Receiver/Main
Reactor (Plant Concept 6)

This system concept requires the installation of a
receiver in which process gas is heated to a temperature
of 860 °C. The construction of such a receiver is
considered as possible with the materials, semi-finished
products and production methods available today, whereby
the final selection of the material to be used must be
verified by material testing. The output of the receiver
should amount to approx. 3 MW; in certain cases,
restrictions may have to be taken into consideration
relating to the maximum size of receiver which can be
mounted on the tower of the CESA-1 plant.

The cavity of the receiver should already be planned and
designed to such a degree of flexibility that the structure
and the insulated inner walls as well as the aperture

area can be easily converted so that they can be used for
the test of the 1000 °C ceramic receiver required in
system experiment B (see below). The lines from the
receiver to the base should also be designed such that
they can also be used for the system experiment B with
1000 °C hot air as the heat transfer medium.
The main data of the receiver/reactor are:

Thermal capacity   3 MW
Inlet pressure   22 bar
Inlet temperature   350 °C
Outlet temperature   860 °C

The cavity is equipped with an auxiliary firing system
(refer to Fig. 7-1).

7.2   System Experiment B for a Separate Heat Transfer Circuit
        with Storage (Plant Concept 4)

This system experiment should follow the system
experiment A since the implementation of this system
eyperiment assumes the availability of a ceramic piping
system for the receiver with an outlet temperature of
1000 °C. The results of the tests envisaged in the
extended GAST program relating to a ceramic panel piping
system should therefore be evaluated before planning the
system experiment B. These results should be available up
to the end of 1987.
Fig. 7-1 shows the layout diagram of the system
experiment B, Tab. 7-1 contains the main data. The
receiver with a thermal output of approx. 3 MW feeds the
reactor with a thermal output of 1 MW. Two storage
modules each of approx. 7.5 MWh storage capacity must
enable a complete day's operation on several days
with solar energy only (similar to plant 4.3). The storage moduls are equipped with an auxiliary fossil fuel
heater which permits alternate charging and discharging
of the modules,   therefore enabling operation of the

plant even at times of inadequate insolation. At the same time, it should also be possible to carry out the tests as for normal storage modules. A recuperator is not included in the system experiment. The cooler must be activated when the blower inlet temperature is too high.

An auxiliary fossil fuel heater is also not included in the system experiment. Its function is carried out by the auxiliary cavity heater of system experiment A or by the firing system integrated in the storage moduls.

Depending on the planned test intensity, the supply of both system experiments with process heat can take place in the closed circuit as described in /1/ or in the open circuit in which the reformed gas is burnt off. The following cost estimations are based on a plant which is operated in the open process (see Fig. 7-2). It has the advantage of lower investment costs. The relatively high consumption of natural gas however renders necessary well prepared, limited test campaigns in natural gas consumption. In order to be able to prepare certain test procedures independent of the process and its fuel consumption with the solar section, an air cooler which can optionally be used in place of the reactor could be advisable for the system experiment B.

7.3   Cost Estimation and Time Scheduling

The estimated costs for the system experiments A and B are specified in Tab. 7-2. This estimation assumes that the process system used for the system experiment A can also be used for the system experiment B. Rough time scheduling is indicated in Fig. 7-3.

Bibliography to Section 7

/1/ Steam Reforming of Methane Utilizing Solar Heat,
    Final Report of MAN-Neue Technologie and LURGI
    GmbH, December 1984

Heat transfer medium	air
Receiver	
thermal power	3 MW
inlet pressure	8.5 bar
inlet temperature	500 °C
outlet temperature	1030 °C
Reactor	
thermal power	1 MW
Storage	
2 storage modules with additional combustion equipment	
storable energy/modul	7,5 MWh
dimensions	15 m x 2,5 m Ø
Blower	
power	200 kW
massflow	5 kg/s
Hot gas lines	
length	70 m
internal diameter	300 mm

Tab. 7-1:   : Preliminary main data of system experiment B
corresponding to plant concept 4.3

```
System experiment A                                     36 Mio DM

Design                                   8 Mio DM
Construction                            16 Mio DM
Assembly and start-up                    8 Mio DM
Test                                     4 Mio DM
                                        36 Mio DM


System experiment B                                     30 Mio DM

Design                                   6 Mio DM
Construction                            14 Mio DM
Assembly and start-up                    6 Mio DM
Test                                     4 Mio DM
                                        30 Mio DM

Total of system experiment A and B                      66 Mio DM
                                                        =========
```

Tab. 7-2: Cost estimation for system experiments

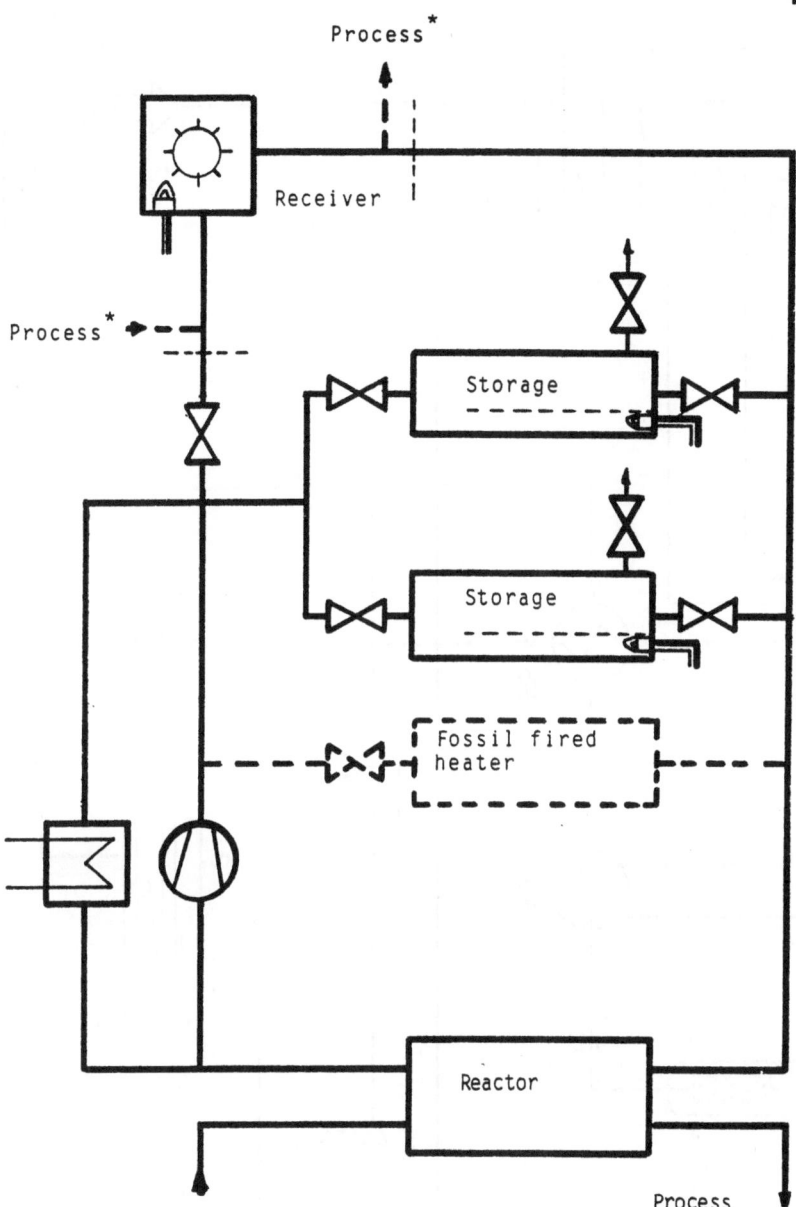

Fig. 7-1:   Simplified flow scheme of system experiment B
             *in case of system experiment A

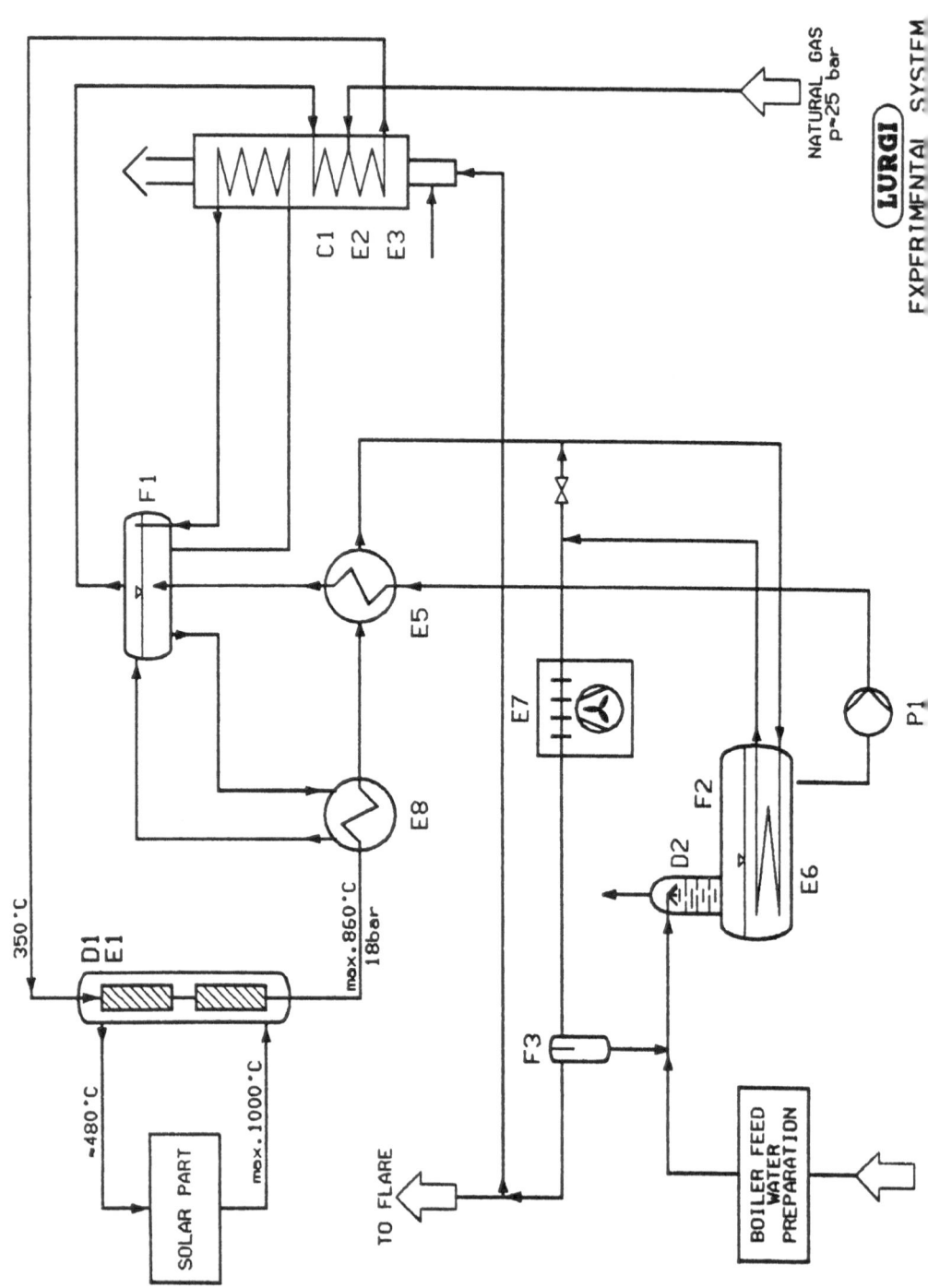

NATURAL GAS
p~25 bar

LURGI
EXPERIMENTAL SYSTEM

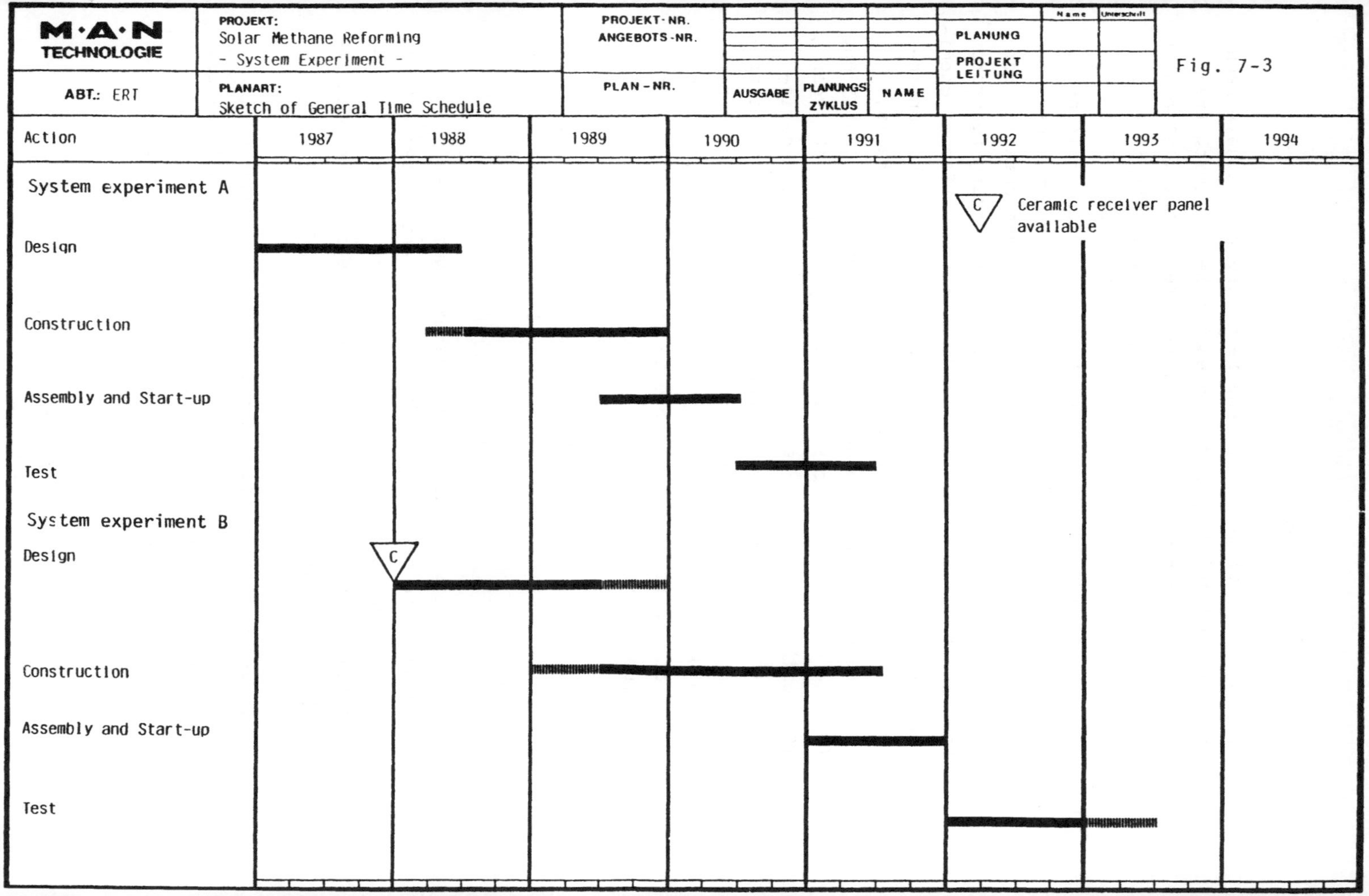

M·A·N TECHNOLOGIE

ABT.: ERT

PROJEKT: Solar Methane Reforming - System Experiment -

PLANART: Sketch of General Time Schedule

PROJEKT-NR.
ANGEBOTS-NR.
PLAN-NR.

	AUSGABE	PLANUNGS ZYKLUS	NAME
PLANUNG			
PROJEKT LEITUNG			

Name Unterschrift

Fig. 7-3

Action	1987	1988	1989	1990	1991	1992	1993	1994

System experiment A

Design

Construction

Assembly and Start-up

Test

System experiment B

Design

Construction

Assembly and Start-up

Test

C Ceramic receiver panel available

541

8. <u>Summary and Conclusions</u>

Various forms of solar methane reforming plants have been
investigated and compared with each other and with tra-
ditionally operated, purely fossil fired plants, also with
respect to a variety of chemical products. The investigations were extended for the differentiation of the solar plants with regard to construction, solar efficiency,
inclusion of storages, investment and secondary energy consumption. The investigations also have been detailed for
the construction of the solar operated reforming reactors
and the specific energy consumption of the different chemical's production as well as for investment of the non-solar parts of the plants. With the accumulated facts,
costs have separately been calculated for solar energy and
for the chemical products for a time span up to the year
2050 with a fixed cost related scenario and with deviations therefrom. Attempts also were made to vary the process and operating conditions of the solar plants to better adjust them to physically not changeable solar conditions and to present available technologies. By a scheme
of valuation the so found solar concepts were compared
another time with each other and those of highest interest
were found out by such method, thereby pretending programs
for further development of solar supported methane reforming plants.

As main results of this study one may summarize:
1) By adequate selection of the solar plant system at least
   theoretically it is possible to reduce the need of natural gas nearly down to the non substitutable quantit-
   ty for the process.

2) In solar operated plants for methanol production it
   will be possible to save up to 50 % of the natural gas
   for fuel with prereactor systems end up to 100 % of
   this item with main reactor systems, if adequate
   storage capacity will be provided.

- 542 -

3) The portions of saving of natural gas as fuel will be
   less with other chemicals for the same type of solar
   plant, but the portion can be increased with additional
   solar equipment, thereby nearly reaching again the favourable figures of methanol production. The actual
   volume of natural gas then saved by such enlarged solar
   plant, of course being much higher and the specific solar energy in the product being higher.

   Nevetheless the production of methanol claims a favourable position amongst the solar supported chemical production plants as it allows the high portions of solar
   substitution with relatively small solar plants.

4) Under the aspects of the chosen standard scenario the
   solar operated plants result in product costs, which
   are lower than with purely fossil operated plants,
   around the time span of the years 2020 to 2030. Though
   it is not easy to simplify the very complex, connexes
   between inflation, life time of plants, interest rates,
   specifically decreasing solar investments, etc., if
   looking at the most determining factor only, one may
   also put it to that way that solar aided plants of the
   studied nature will become economic, when the natural
   gas at the time of the start-up of the plant will have
   become about three times as valuable as to-day. So if
   the value increases more drastically the point of
   parity will be reached earlier and vice versa.

5) On the long run the savings and economic benefit, which
   can be achieved with solar operated plants, are the
   higher the more specific solar energy is handled. Therefore storage systems should be applied in any case and
   high temperature systems should be chosen, if the techniques are available. The development programs should
   be started with this principal aims, as changes in the
   energy price scenario and in the environmental aspects
   may demand for such plants earlier than anticipated
   with the standard conditions of this investigation.

The study clearly indicates that storage systems are
not a burden to solar plants, but anticipate the dates
of economic viability of "separated" solar plants. Besides this the profit on the long run will be higher
with storage systems than without them.

6) "Integrated" systems seem to become economic much
earlier than "separated" systems. Nevertheless the potential for saving money is small only in the time span
unless the separated systems become equivalent as a consequence of their higher investment and secondary energy requirements.

7) The present study was purposely based on ideal weather
conditions throughout the year, the influence of actual
weather disturbances therefore was eliminated. From
the results of this idealized study it may be concluded
that high temperature systems definitely should be
chosen for future applications. This may not turn out
to be correct, if real weather conditions are adopted
for such investigations as it is known that in reality
the effort will increase drastically to reach $1000^{o}C$
with the same factor of availability as to reach $800^{o}C$
This factor has not been considered in this study and
perhaps would have to be further investigated, when de-
velopments for far off periods have to be defined.

8) The study has shown that the effectiveness of a solar
system and the date of economic purity hardly are influenced by the sequential chemical product. As this
study has shown that solar supported steam reforming
plants are viable (even if only in the year 2020) it
must be assumed that also other high temperature processes reasonably may profit from such technology, if
available. Especially the generation of hydrogen,
which is often envisaged as the fuel and energy carrier
of the future, seems to be bound to high temperatures
(thermochemical cycles or hot-vapour electrolysis, Hot
Elly), to favourably provide it from water and solar

energy. It is therefore the strong feeling and the con-
viction of the group, having taken part in this study,
that solar supported methane reforming is a key process
to be further studied deligently by theory and practical demonstration to provide the techniques also for
analogous high temperature processes.

PROCESS SYNTHESIS OF A GASIFICATION
PROCESS MODIFIED FOR
HIGH SOLAR ENERGY INTEGRATION

G. BIRKE
R. REIMERT

LURGI, FRANKFURT

Contents

Summary

Solar energy can be integrated with chemical processes to transform solar energy into a more convenient form and to save natural ressources. A gasification process has been examined as an example of a solar process with advanced elements, including the combination of several energy absorbing steps, usage of solar-modified processes and solar-specific components.

The basic assumptions of this process are: the gasification in the Circulating Fluid Bed (CFB) to produce synthesis gas is the primary energy absorbing process. Solar energy is integrated in the reaction system via direct insolation. The raw materials are biomass and lignite. Solar gasification of biomass is an example of a scenario based on completely regenerable energy. The anticipated product is methanol, which is easily integrated in existing production and consumption schemes. Many aspects of the methanol synthesis route are exemplary for any synthesis. So, the CFB_TO_MEOH-solar process will serve as a reference process for studying problems of integration of solar energy into chemical processes, i.e. solar process synthesis.

A problem analysis and detailed breakdown leads to the basic concept: solar energy is integrated in an advanced step which combines specific features of solar energy (radiation) and the reaction system (energy absorption, distribution and consumption) in a nearly ideal manner. A further advantage is, that a combination of chemical reaction and power generation is possible. The problems deriving from the solar cycles are transformed into a problem of process synthesis of the CFB_TO_MEOH- synthesis route. The conventional synthesis route is therefore step by step modified to satisfy solar requirements. The main elements of process synthesis under consideration are: combination of energy absorbing steps, sequencing of process steps and usage of buffers of different physical and chemical quality.

A two-state-approach for the operation of the plant is the basis of all balance calculations made: one solar (day) state with constant solar energy input of 40 MW and one non-solar (night) state, during which the plant is switched to its alternate operation mode. Characteristic design parameters are evaluated to represent equipment size; this allows to compare similar process routes with respect to technical and (qualitatively) economic performance.

Solar CFB-gasification does not require oxygen as reaction agent and changes extensive and intensive properties of generated raw gas simultaneously: specific raw gas yield is increased by 31.8 - 33.6% for lignite and biomass and its composition is shifted towards higher H_2 + CO- contents. For biomass, the amount of power co-produced by the CFB-process is varied: up to 13.9 of 40 MW solar energy are used for power generation.

The upstream and downstream processes have to match the requirements of solar gasification: the oxygen plant is operated in a solar-modified manner using a buffer of liquid oxygen. The size of the oxygen plant is reduced and 85% of its energy consumption shifted to day operation. The shift conversion and gas purification processes work in dual modes and generate constant synthesis gas stoichiometry. For the lignite-cases, the operation of a shift gas buffer with H_2-rich shift gas produced during solar operation has been examined.

The design of the synthesis sequences has been evaluated by variation of throughput for all routes under consideration in order to find routes with the best compromise of load characteristics and integration potential. The key elements of process synthesis, which were examined, may have very different consequences. The co-production of synthesis gas and power by the CFB-process first of all increases integration potential for solar energy, but has only a negligible influence on the design of the synthesis sequence. There exists only a local design optimum for the CFB-process for small amounts of power co-produced. On the contrary, the shift gas buffer increases solar energy integration potential, too, but primarily it has a design effect for the whole synthesis sequence. Regarding load characteristics the synthesis routes with shift gas buffers look like a conventional plant.

Solar energy and the chemical process are coupled in a dual way: directly by insolation of solar energy into the CFB and indirectly, by a feedback path arising from process synthesis used for design of the solar CFB_TO_MEOH process. Process synthesis generates a balance scheme with demand for power, which is to some extent even synchronized with solar energy supply. In general, power demand for a synthesis plant cannot be supplied by co-production of power in the CFB-process alone. This option should therefore preferably be used as a means for optimizing energy scheme and steam system, for example for supply of a small "residual" power demand. As the most preferable option a dual solar plant is recommended with direct insolation and a separate process for power generation.

The integration potential for solar energy for different scenarios is estimated by the specific methanol yield. It may be increased by 14.3 (16.1) % for a process based on direct insolation only and by 28.6 (57.1) % for a process with dual solar plant for biomass (lignite).

The influence of load characteristics is more subtle, since it cannot be quantified in the same way as an integration potential. But when evaluating economics of a process or when comparing routes solely based on global design, there is a clearly defined limit in increase in the integration potential. Integration potential and load characteristics are closely interrelated for a

solar process route and adjusting them is one of the key problems
of design.

The fields of research will be determined by the basic design
concept: evaluation of chances and risks of direct insolation
becomes urgent. This requires, first of all, better insight into
the design of the interface between solar energy and reaction
system and the elemental physical and chemical processes in the
reaction system.

Regarding synthesis of process routes, simulation of solar process
plants is required, and some advanced tools for doing so have to
be provided. A larger number of processes, elements of process
synthesis and couplings between them should be examined, in order
to get a better overview of processes and routes. Taking into
account all relevant solar cycles plant modeling should be
refined. Preferable would be a modeling approach beyond the level
of dynamic plant simulation which exploits the special nature of
solar cycles.

1. The CFB_TO_MEOH-solar Process as Reference Process for solar Process Synthesis

Crude fuels, fossile or regenerable, require upgrading to produce clean fuels or industrial intermediates like synthesis gas. The energy consumption of these upgrading processes may amount up to 30 % of the feed. Replacing this energy by high-temperature solar energy will serve several purposes: solar energy is transformed into a well-manageable, storable energy form, natural resources are conserved and pollution is reduced. However, a solar upgrading process has to face some problems not relevant for conventional applications, but derive from the specific nature of solar energy: the low energy flux, the special form as radiation and the solar cycles.

A more conventional way to overcome two of the former problems is to concentrate solar energy by optical methods and heat a heat-carrier up to 900 °C. Solar energy is then brought into the reaction system by indirect heat transfer. For upgrading processes working at temperatures above 800 ° C this proved to be a process of low efficiency and poor performance.

A more advanced and, for energy reasons, preferable way to use high-temperature solar energy is the direct insolation of concentrated solar energy into the reaction system, a fluidized bed. In general, the reaction system has to be separated from atmosphere, requiring for example a quartz window for insolation. A large number of reactions and technically important processes can be considered for solar energy usage on the basis of this technique; however, neither is such a process available nor have its risks been evaluated.

Previous work (1) showed, that large effects of solar energy integration into the Circulating Fluidized Bed (CFB)-gasification process can be achieved using direct insolation. Potential for further optimization may be found, primarily in two fields:
- To increase the specific integration rate of solar energy by combination and modification of energy absorbing processes.
- To modify the overall process by nonconventional, solar-specific design techniques and components, for example buffers.

As mentioned, a very special solar-specific problem arises from the solar cycles. Especially when large integration effects can be realised in an energy absorbing process, modifications are required in the design of downstream processes. With respect to the economy of the overall process, this is one of the fundamental design problems of any solar process plant. Therefore, it is well justified to spend efforts in system design of solar process plants even at a very early stage of investigation.

Based on exemplary work on a suitable process, this study faces a dual purpose: to develop solar specific design techniques and to compare process routes, particularly with high solar energy integration potential with respect to technical and - partially - economic performance.

The basic assumptions of this study are:
- The primary energy absorbing step is gasification in the CFB to produce synthesis gas. Solar energy is integrated via direct insolation.
- The fuels considered are biomass and lignite. Each of these materials has its particular place in energy scenarios. Biomass plus solar gasification is an example of a scenario of completely regenerable energy, the biomass being produced in a biomass farm. Lignite is a fossil fuel, whose resources are available in geographic regions relevant for solar energy usage and so may be stretched by using solar energy for its conversion.
- The product is methanol, which is easily integrated in existing production and consumption patterns. Additionally, many aspects of the methanol synthesis route (raw gas conditioning and purification) are exemplary for any synthesis, such as ammonia or hydrogen production or oxo-synthesis.

Insofar, the CFB_TO_MEOH-solar process will serve as a reference process for studying problems of integration of solar energy into chemical processes, i.e. is solar process synthesis.

2. Solar-specific Design Topics

2.1 Problem Definition and Solving

The objective of this study and of research in high-temperature solar energy application as a whole is to integrate solar energy into process plants. This task obviously cannot be solved straightforward. Problem analysis was therefore the first step in the beginning of this study with a main effort towards problem decomposition, a well-known problem analysis and solving technique (2).

Problem analysis, as used here, includes the following steps:
(1) Define the problem
(2) Transform the problem into (several) subproblems of minor severity.
(3) Repeat step (2) with each generated subproblem, until the problem is reduced to existing technology or defines well research demands.
(4) Perform parameter variations and compare the results of the generated solution paths.

As a result of former work, the problem description graph presented with its main nodes in Fig. 2.1 was the starting point for this study. It is one important feature, that the problems stemming from solar energy are separated in two subproblems, which are then solved by different methods.

Solar energy is integrated by an advanced step, which combines characteristics of solar energy (radiation) and of the reaction system (energy absorption, distribution and consumption) in a nearly ideal manner. Then, arguing from a perception of the overall production scheme, the problems deriving from the solar cycles shall at first not be solved by buffering solar energy directly by a thermal buffer. Instead, they are transformed into a problem of process synthesis of the CFB_TO_MEOH- synthesis route. Its elements are: combination of energy absorbing steps and sequencing of process steps. In addition, there may be solar specific elements: buffers of different physical and chemical quality.

The main method to generate alternative process routes is to vary the relevant parameters. Because the number of parameters, and therefore the number of variations, might be very large a reduction strategy was applied:
- Elements which proved to be inefficient were not further investigated and dropped.
- The key parameters of the process were figured out by preliminary, semi-quantitative analysis. Only variations of these parameters were performed, each defining a case. Cases and subcases were generated by a hierarchical ordering. Formally, a case is denoted by a primary key (a number) and a secondary key (a character), for example 1.A.
- All other parameters, mainly pressures and temperatures, were fixed to representative or known values.
By this method, it is not possible - and it was not pretended to do so - to fix the optimum case, but to compare alternative processes and draw conclusions about the existence of favourable or optimum cases.

The goal node in Fig. 2.1 - integrated solar process plant - may be split into three nodes representing subgoals: economic and technical performance and high integration of solar energy. One important objective to be achieved by this study is, to obtain information about dependance or independance or about hierarchies of these subgoals.

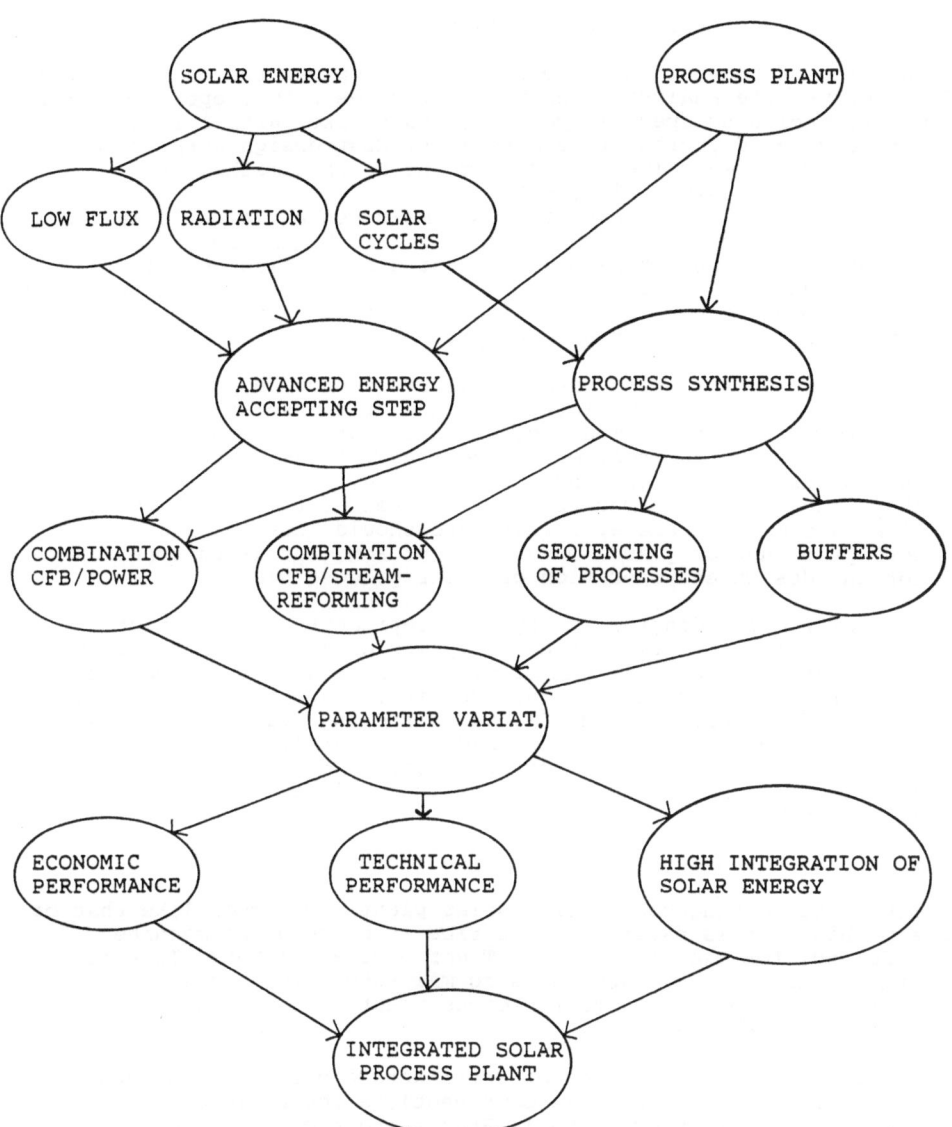

Fig. 2.1: Problem description graph for integration of
high-temperature solar energy into process plants

2.2 Modeling of solar Process Plants

Modeling of conventional process plants employs two main methods:
- steady-state-analysis, which investigates normal operation of a plant, including special operation modes like partial load or influence of different feed materials. Most design work is based on steady-state-analysis; modern means of this method are flowsheet simulators, like ASPEN.
- Dynamic simulation investigates time-dependant plant operation, including transients or shutdowns. Dynamic simulation is an extensive task and therefore the last step in plant simulation. For design purpose it is invoked for critical process steps or for processes which have to be designed for time-varying operation, for example an energy/steam system.

For solar plants, steady-state-analysis cannot be performed as usual, or it is not sufficient because of the solar cycles, but a third form of plant modeling which integrates several, at least two steady-states, may be designed. A basic property of this solar analogue of conventional steady-state-analysis is, that internal, sometimes transient states of the whole process plant are made discrete. Perhaps one should look at this perception not only as a modeling approach, but as an approach for the design and operation of solar process plants.

The solar cycle diagram in Fig. 2.2 represents the day/night cycle. It is simplified, because cycles of minor importance and transients (for example summer/winter and states with reduced solar flux) are not considered. But it represents the main solar cycle and at least two fundamental differing states will exist in a solar process plant. Solar steady-state-analysis has to investigate these states separately and then to integrate them. This two-state-approach was the basis of all calculations made in this study.

2.3 Formal Handling of solar Design Topics

Since the working scheme of a solar plant looks much like that of a machine, it is natural to describe it by means of machine modeling, for example Automata Theorie or Petri-Nets. In this study Petri-Nets are used as a formal scheme for state representation of the solar process plant; for a discussion of Petri-Nets see (3).

Generally, in the notation of Petri-Nets a circle represents a state or commodity (passive component), a rectangle a process or transition (active component). States and processes are combined by arcs, which denote flows of information, material, work and others. The arcs may be labeled and thereby the flow "through" the arcs defined. Fig. 2.2 is a simple example for a Petri-Net.

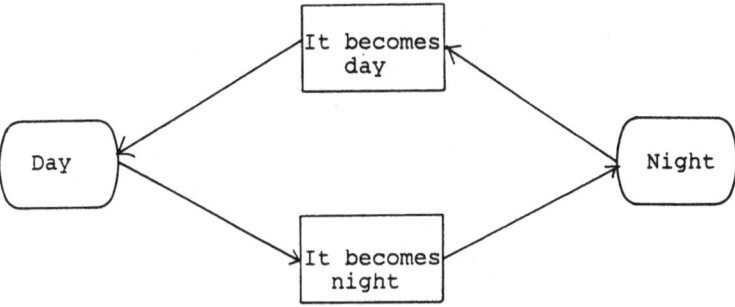

Fig. 2.2: Simplified solar cycle diagram

As a means for generation and representation of solar processes, a
number of logical processing units have been defined in terms of
Petri-Nets. Basis is the two-state-approach, so only two distinct
solar states (D/N) exist. These units define elemental state
transitions possible or relevant in solar plants; they are:

Process Unit: this is the general case, when a process accepts
varying input and produces varying output. See chapter 3.1 for a
discussion of gasification as an example.

Buffer Units: these are the classical solar units to manage the
solar cycles and they reverse the logical state of an input
stream. An example is a tank for liquid oxygen; see chapter 3.7
for a detailed discussion.

Switch Unit: this unit is realised by a process, which works in
different modes during different solar cycles and produces
constant output. In this context the attribute "constant" means
constant with respect to some quality of output; other properties
of output may vary as well. Consequently, by sequencing of process
units solar cycles may be mitigated or suppressed with respect to
their effect. An example is the shift conversion in the synthesis
chain, which generates in all cycles the correct
H2/CO-stoichiometry of the synthesis gas.

Network: This is a flexible process, perhaps outside the solar
plant, which accepts varying input and produces constant output.
An example would be a (public) electric grid, where
production/consumption-patterns could be optimized by combination
of the grid with the solar process plant. This topic was not
pursued further here.

NAME	FUNCTION	DESCRIPTION

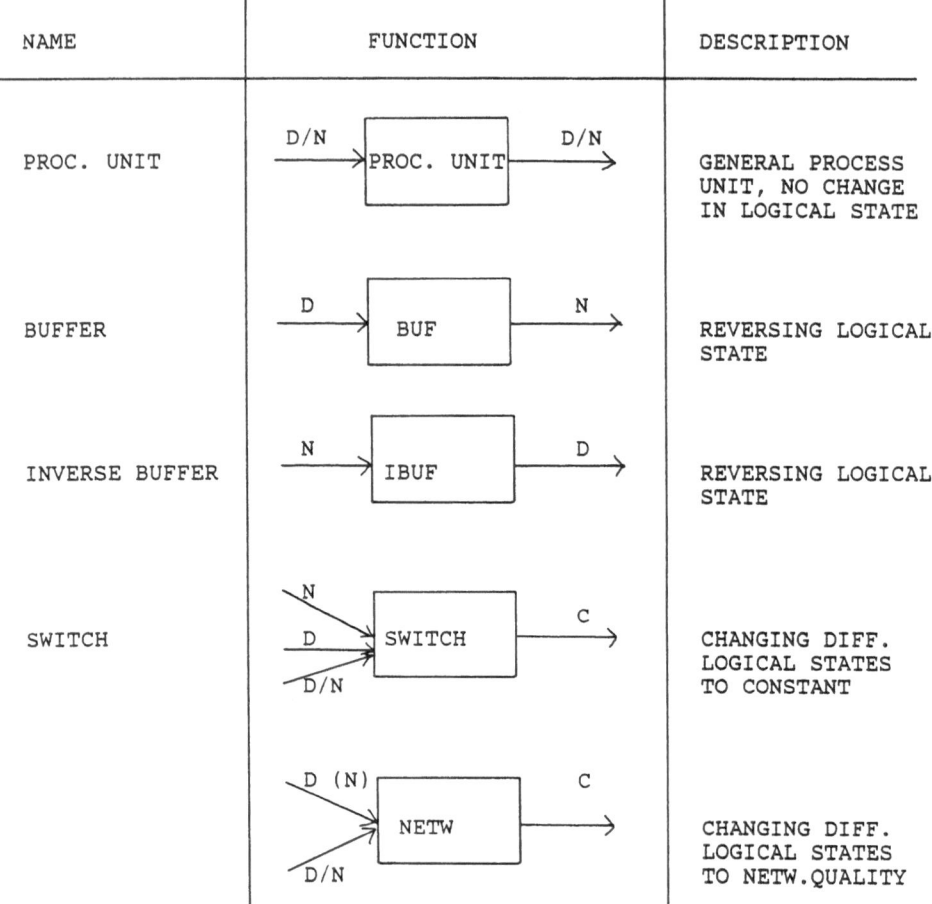

PROC. UNIT	D/N → PROC. UNIT → D/N	GENERAL PROCESS UNIT, NO CHANGE IN LOGICAL STATE
BUFFER	D → BUF → N	REVERSING LOGICAL STATE
INVERSE BUFFER	N → IBUF → D	REVERSING LOGICAL STATE
SWITCH	N / D / D/N → SWITCH → C	CHANGING DIFF. LOGICAL STATES TO CONSTANT
	D (N) / D/N → NETW → C	CHANGING DIFF. LOGICAL STATES TO NETW.QUALITY

Tab.2.1: Solar-specific logical process units
        Logical states:
                D   : Active only at day
                N   : Active only at night
                D/N : Active at day and night, but with different
                     physical state conditions
                C   : Active at day and night, but with same
                     state conditions with respect to some
                     physical quality

2.4 ASPEN-Model of the CBF_TO_MEOH solar Process and modeling
    Assumptions

Based on the two-state-approach the operation of a solar process
plant will be represented by the diagram in Fig. 2.3. Solar energy
input is fixed to 40 MW at day (solar) operation. As mentioned,
solar energy is not buffered as thermal energy on the level of the
energy absorbing step. The figure of 40 MW is chosen for two
reasons: the plant should reach industrial size, so that specific
plant data are representative for an actual plant and need not be
corrected for specialities of a pilot plant. Furthermore, 40 MW is
a reference figure coming from a former study (4). Assuming solar
energy input as constant, the size of processes with different
energy integration rate will vary. So, the question to be answered
is: What can be done best with 40 MW solar energy?

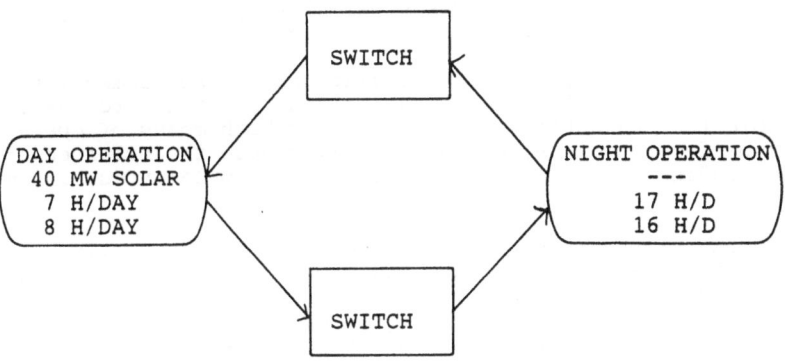

Fig. 2.3: Operation diagram for solar process plant

Balance calculations are made on the basis of 7 (8) hours solar
operation and 17 (16) hours non-solar operation. These cycle
time data are location-dependent; the 7/17-cycle as the base
value was chosen from data collected in Almeria (Spain). The
8/16-cycle was investigated in one case in order to check the
influence of solar operation periods.

Average balance results are then calculated by Eq. 2.1 from the
solar/non-solar results and operation periods:

Average = solar_op_hours * solar_result/24 (hours)

                                                          Eq. 2.1
         + non_solar_op_hours * non_solar_result/24 (hours)

The overall block flow diagram of the CFB_TO_MEOH solar process is shown in Fig. 2.4. It represents the synthesis chain with gasification, raw gas conditioning and compression, shift conversion, gas purification and MEOH-synthesis as its main steps. The balance scheme was constructed for the special solar design purpose; some process steps of minor importance in this context were omitted, for example gas water treatment. On the other side, some process steps had to be considered thoroughly, for example the oxygen plant, a "package unit" usually not described explicitly in a study.

Apart from solar gasification, only existing technology has been chosen and in so far the scheme looks like a conventional plant. But modifications are made on some processes with respect to their operation mode to adapt them to solar operation.

Gas purification, based on the well-known Rectisol process, is treated as a dummy and only the correct split factors are provided for separation of gases. Similarily, only the synthesis loop of the methanol synthesis unit is considered.

All mass- and energy balance calculations are performed with ASPEN. A special solar ASPEN-model of the plant is provided; it is suitable for evaluating plant states for both modes of operation. Solar buffer functions, which strongly couple the balances of corresponding operation modes, are not yet implemented, but made available by hand calculation.

Design and sizing is supplied by means of characteristic design parameters. The whole flowsheet is broken down into sections which correspond to the main process units of the plant. For each section, one or sometimes two characteristic flowsheet variables which determine the size of main equipment, were chosen for comparing alternative routes.These parameters are discussed in chapter 3 and were used to construct design diagrams for the whole plant. In this way, it is possible to draw qualitative conclusions about the economy of different routes, provided that they have the same or a related structure.

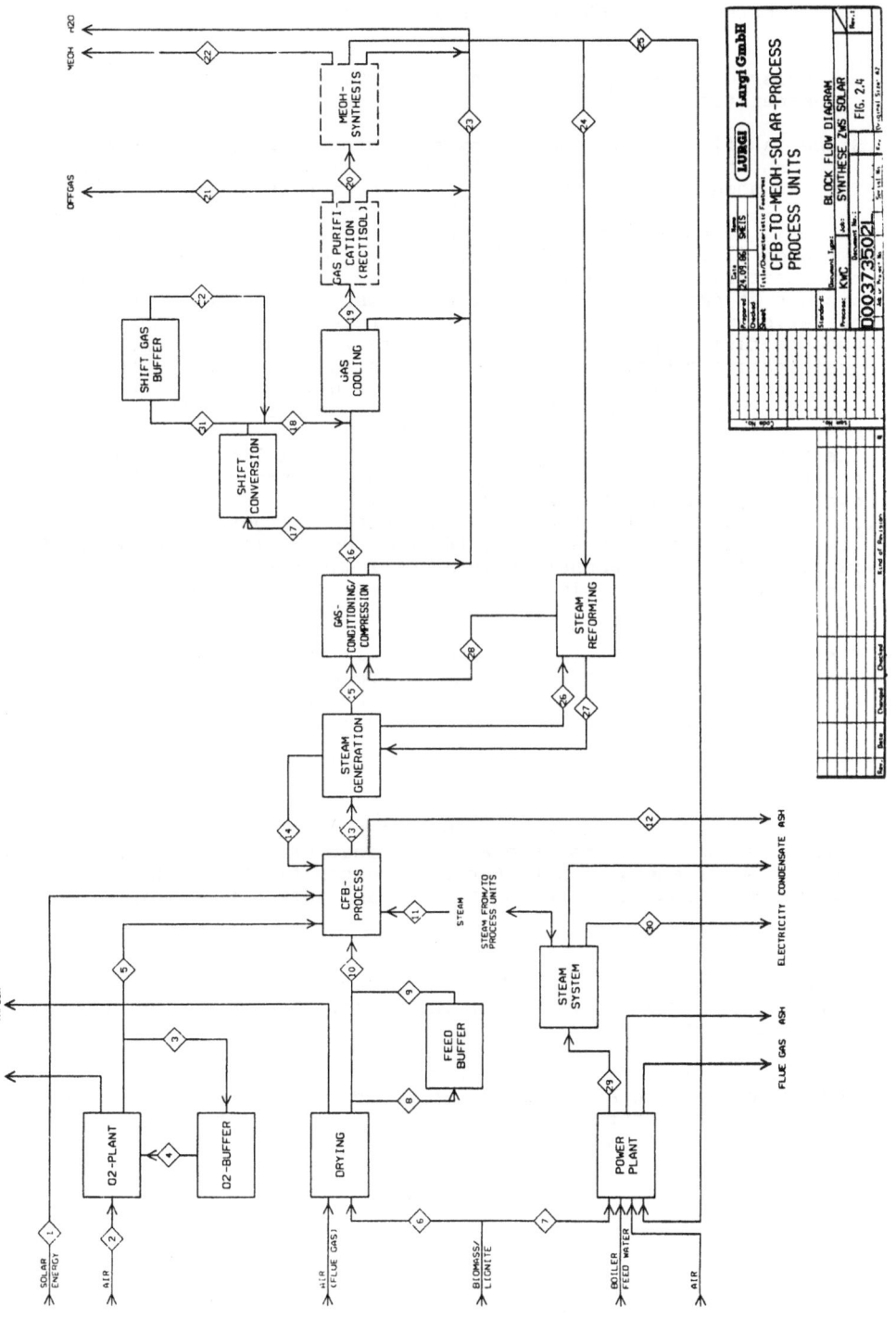

CFB-TO-MEOH-SOLAR-PROCESS
PROCESS UNITS

BLOCK FLOW DIAGRAM
SYNTHESE ZNS SOLAR

LURGI GmbH

FIG. 2.4

3. Elements of the CFB_TO_MEOH solar Process

3.1 Solar CFB-Gasification with integrated Power and/or Hot Gas
    Generation

Since gasification is the energy absorbing step, it is the most
important one whose characteristics determine the overall scheme.
The flowsheet for the solar CFB-gasification is shown in Fig. 3.1.

The main component is the CFB-reactor (1), which works as an
expanded fluidized bed. The fuels biomass or lignite are fed to
the reactor; via insolation solar energy drives the gasification
process at daytime. At night, the process is switched to standard
operation with oxygen/steam. Steam or recycle gas are additionally
used as fluidizing agents. The pressure is slightly elevated. The
gasification temperature is 850 °C, a standard value for the
gasification of highly reactive fuels. Ash is removed at the
bottom of the reactor with a (constant) C-content of 5%.

The raw gas and the circulating solid material leave the reactor
at the top and the solids are separated from the gas by the
cyclone system (2). The solids are recirculated to the reactor or,
if desired, through a fluidized-bed heat exchanger (3). The raw
gas is cooled down in a boiler (4) to 300 °C. The value of 300 °C
is fixed for two reasons: it is a lower limit for generation of
MP-steam and the recycle gas blower does not tolerate temperatures
well above 300 °C.

Steam for power generation is produced in the optional
fluidized-bed heat exchanger (3) and the boiler (4); both
components are commercially proven in the CFB-combustion process.
Power generation by solar energy can be increased, if the recycle
of raw gas via the blower (5) is adjusted. The recycle gas is used
as fluidizing agent for the heat exchanger (3) and the reactor;
therefore a minimum is required.

The amount of power co-produced by solar energy is chosen as a key
parameter and varied for biomass gasification. Tab. 3.1 lists the
data of solar energy input. For lignite gasification, this key
parameter is not investigated further and set to its minimum
value.

For a combination with steam reforming of methane, a heat
exchanger for the heat carrier gas replaces one unit for steam
generation in the upper part of the boiler. The heat carrier can
be heated up to 800 °C.

Gasification itself is a complex process, a large number of
coking, combustion and gasification reactions being involved. It
was modeled by restricted equilibrium of a set of three
independent reactions, Eq. 3.1-3.3, required to fix the amounts of

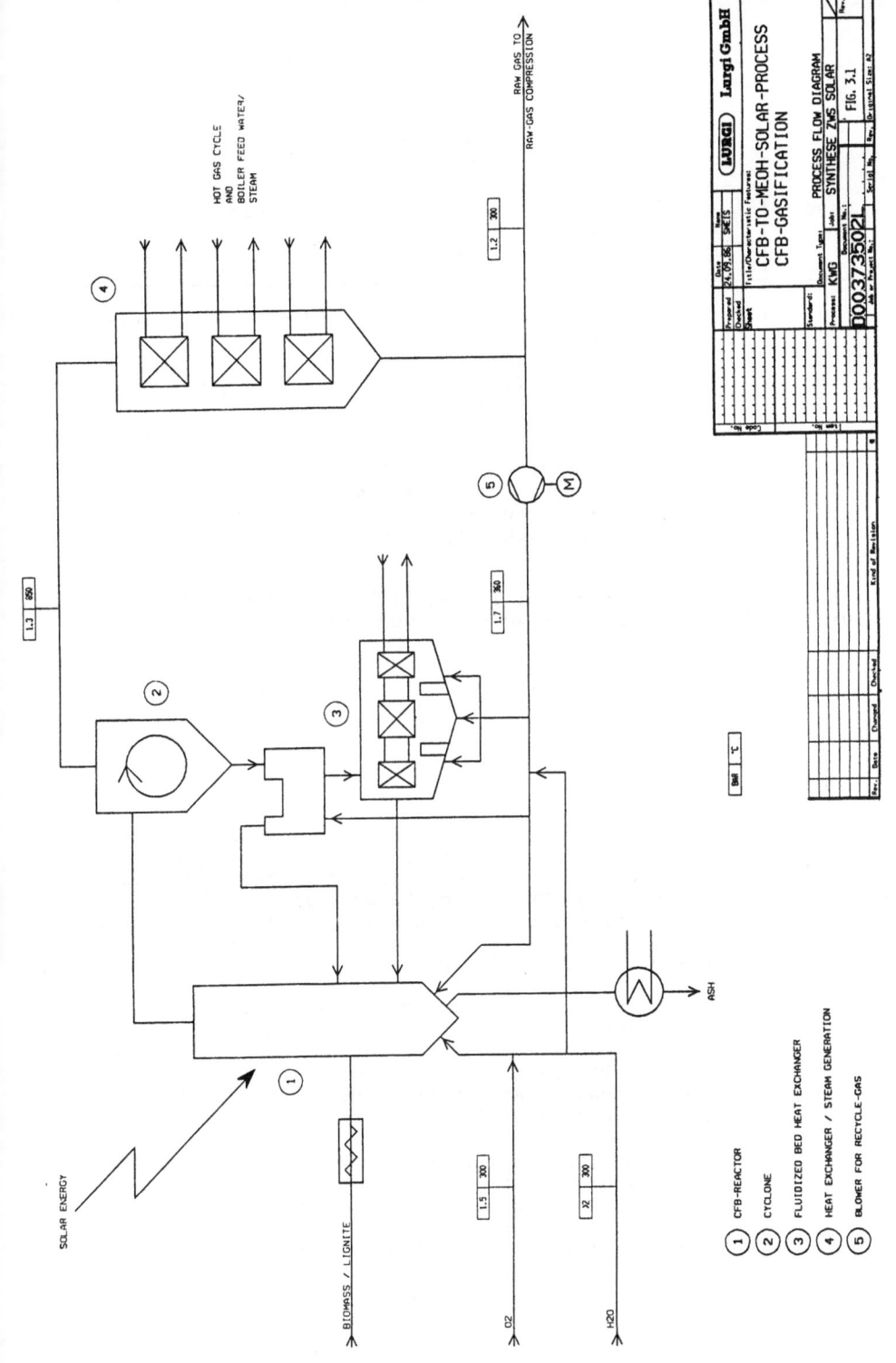

SOLAR ENERGY

HOT GAS CYCLE
AND
BOILER FEED WATER/
STEAM

RAW GAS TO
RAW GAS COMPRESSION

1.2 300

1.3 850

1.7 360

BIOMASS / LIGNITE

O2
1.5 300

H2O
32 300

ASH

① CFB-REACTOR
② CYCLONE
③ FLUIDIZED BED HEAT EXCHANGER
④ HEAT EXCHANGER / STEAM GENERATION
⑤ BLOWER FOR RECYCLE-GAS

Bar °C

LURGI Lurgi GmbH

Title/Characteristic Featured
CFB-TO-MEOH-SOLAR-PROCESS
CFB-GASIFICATION

PROCESS FLOW DIAGRAM
SYNTHESE ZWS SOLAR
FIG. 3.1

Date
24. 09. 86 SHEETS

Prepared
Checked
Sheet

Document Type:
Standard: KMG Job:
Process:

D00373502L
Document No.1

Job or Project No. Serial No. Rev. Original Size: A2 Rev.:2

Rev. Date Changed Checked End of Revision

- 565 -

Case	Biomass		Lignite	
	Chem (MW)	Power (MW)	Chem (MW)	Power (MW)
2	39.4	0.6	39.6	0.4
3	39.4	0.6	39.6	0.4
4	35.8	4.2	39.6	0.4
5	32.6	7.4		
6	26.1	13.9		

Tab. 3.1: Solar energy-input for gasification and power
generation. Total of solar energy input: 40 MW.
Chem: Solar energy input to gasification reaction
Power: Solar energy input to power generation

the main reaction products. The equilibrium restrictions and minor
yields of coking products were derived from experimental data.

$$C \;+\; CO2 \;\longrightarrow\; 2\;CO \qquad\qquad Eq.\ 3.1$$

$$CO \;+\; H2O \;\longrightarrow\; CO2 \;+\; H2 \qquad\qquad Eq.\ 3.2$$

$$CO \;+\; 3\;H2 \;\longrightarrow\; CH4 \;+\; H2O \qquad\qquad Eq.\ 3.3$$

The methane yield is determined by coking processes; the same is
true for C2H6 as a reference component for hydrocarbons CnHm (but
C2H6 was not integrated in the equilibrium calculations for
calculational reasons). The energy balance is closed by adjusting
oxygen flow or solar energy. Of course, the solar gasification
data are extrapolated from results of conventional gasification,
since no other data are available.

It is assumed, that biomass and lignite enter the gasification
process with a moisture content of 15 % and additional amounts of
water are fed as steam to adjust the raw gas moisture to an
appropriate level. The residual moisture in the raw material is
dried during gasification and their evaporation consumes a part of
the 40 MW high-temperature solar energy.

15 % moisture is a typical value for conventional gasification and
can be achieved by drying the raw material in a low-temperature
drying process, for example with low-value steam. From an
energy viewpoint it may be argued, that high-temperature solar
energy is "wasted" if a material with 15% moisture is fed to the
gasification. But lower moisture contents require sophisticated

drying at temperatures up to 400 °C. It is an optimization problem
to combine the moisture levels of the raw material and the
available solar energy at its temperature levels (see chapter 6).

Based on 15 % moisture, the maximum solar energy input is 6.7
MW/kg for biomass and 10 MW/kg for lignite (maf basis). As could
be estimated, lignite as the higher-ranking fuel has a greater
specific energy integration potential than biomass.

Tab. 3.2 compares the solar and non-solar gasification data. As an
effect of solar energy integration, the specific raw gas yield is
increased and its composition changed towards higher $H_2 + CO$ -
contents. For lignite, the H_2/CO-rate of the solar gas is 1.94, a
value, which comes near the required stoichiometry for methanol
synthesis.

A characteristic feature of solar gasification as a "free"
reaction system should be emphasized. The composition and yield of
the product gas are changed due to two independent facts: solar
energy directly substitutes one reaction component (O_2) and
the final state of gasification is not fixed by equilibrium. The
gasification data in Tab. 3.2 reflect the influence of
stoichiometry; this is not true for the influence of kinetics on
solar gasification. For a free reaction system intensive and
extensive properties of the product are changed simultaneously by
solar energy integration as a consequence . This is an important
difference to a catalytic reaction like the steam reforming of
methane and makes the design of a plant based on gasification more
difficult. On the other side, new possibilities are available.

Raw gas		Biomass		Lignite	
		Day	Night	Day	Night
H2	(Vol.%)	50.00	32.77	56.44	42.20
CO	(Vol.%)	30.25	32.45	29.04	32.26
CO2	(Vol.%)	14.91	27.85	29.04	32.26
CH4	(Vol.%)	3.23	4.30	1.67	2.20
H2S	(Vol.%)	0.04	0.06	0.23	0.30
N2	(Vol.%)	0.13	0.61	0.34	0.77
C2H6	(Vol.%)	1.44	1.96	0.17	0.23
Yield/feed (kMol/kg maf)		0.0795	0.0595	0.1373	0.1042

Tab. 3.2: Calculated results of CFB-gasification with maximum
solar energy input

Two characteristic design parameters are selected for this flowsheet section: total raw gas flow and the heat duty of the boiler for sizing the reactor and the boiler respectively (Tab. 3.3).

Section	CFB	
Parameter-Id.	REACTOR	BOILER
Parameter	raw gas flow(mf) (kmol/h)	Heat duty of boiler (MW)

Tab. 3.3: Characteristic design parameters for section CFB

3.2 Raw Gas Conditioning and Compression

The raw gas coming from the gasification section with 300 °C has to be cooled down and cleaned before compression (Fig. 3.2). This is done by a heat exchanger/scrubber sequence. By cooling down to 50 °C, the moisture and oily components are partially condensated; the scrubber washes out residual solid particles from the gas stream.

The raw gas is compressed by a three stage compressor with intercooling to 30 bar, the pressure used for the shift conversion and gas purification. Compression is one of the main power consuming steps. Optionally, the reform gas from the steam reforming section is integrated in the third compression stage which works at a pressure level of 10.4 bar. See chapter 3.6 for a detailed discussion. After compression, the raw gas is split and the appropriate part goes to the shift.

No solar specific design topics have to be considered here. The characteristic design parameter for COMPR-section is raw gas flow (maf).

3.3 Shift Conversion

Shift conversion is a process common to most synthesis plants, because the stoichiometry of primarily CO-rich synthesis gas has to be adjusted to the synthesis requirements. Fig. 3.3 shows the scheme of a standard raw gas shift conversion unit with two reactors. The shift reaction is given in Eq. 3.4; it is an exothermic catalytic reaction controlled by equilibrium.

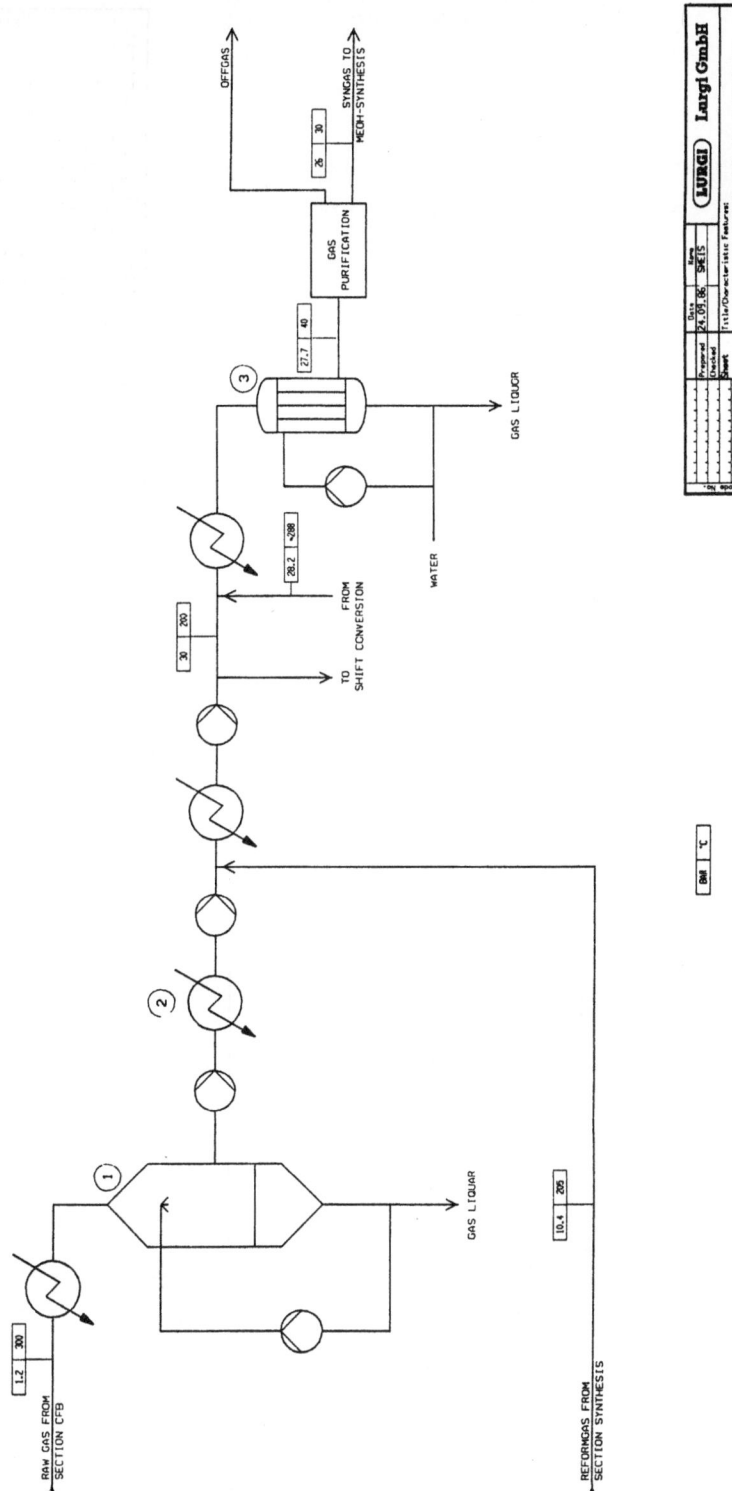

RAW GAS FROM SECTION CFB

REFORMGAS FROM SECTION SYNTHESIS

GAS LIQUOR

OFFGAS

SYNGAS TO MEOH-SYNTHESIS

GAS PURIFICATION

TO SHIFT CONVERSION

FROM SHIFT CONVERSION

GAS LIQUOR

WATER

1 GAS CLEANING / COOLING

2 COMPRESSOR (3 STAGES)

3 GAS COOLING

BAR °C

LURGI	Lurgi GmbH
Prepared	Date 24.09.86	SHEETS
Checked		
Sheet		

Title/Characteristic Features:
CFB-TO-MEOH-SOLAR-PROCESS
RAW GAS COMPRESSION AND
GAS PURIFICATION
PROCESS FLOW DIAGRAM

SYNTHESE ZWS SOLAR

D0037350ZL   FIG. 3.2

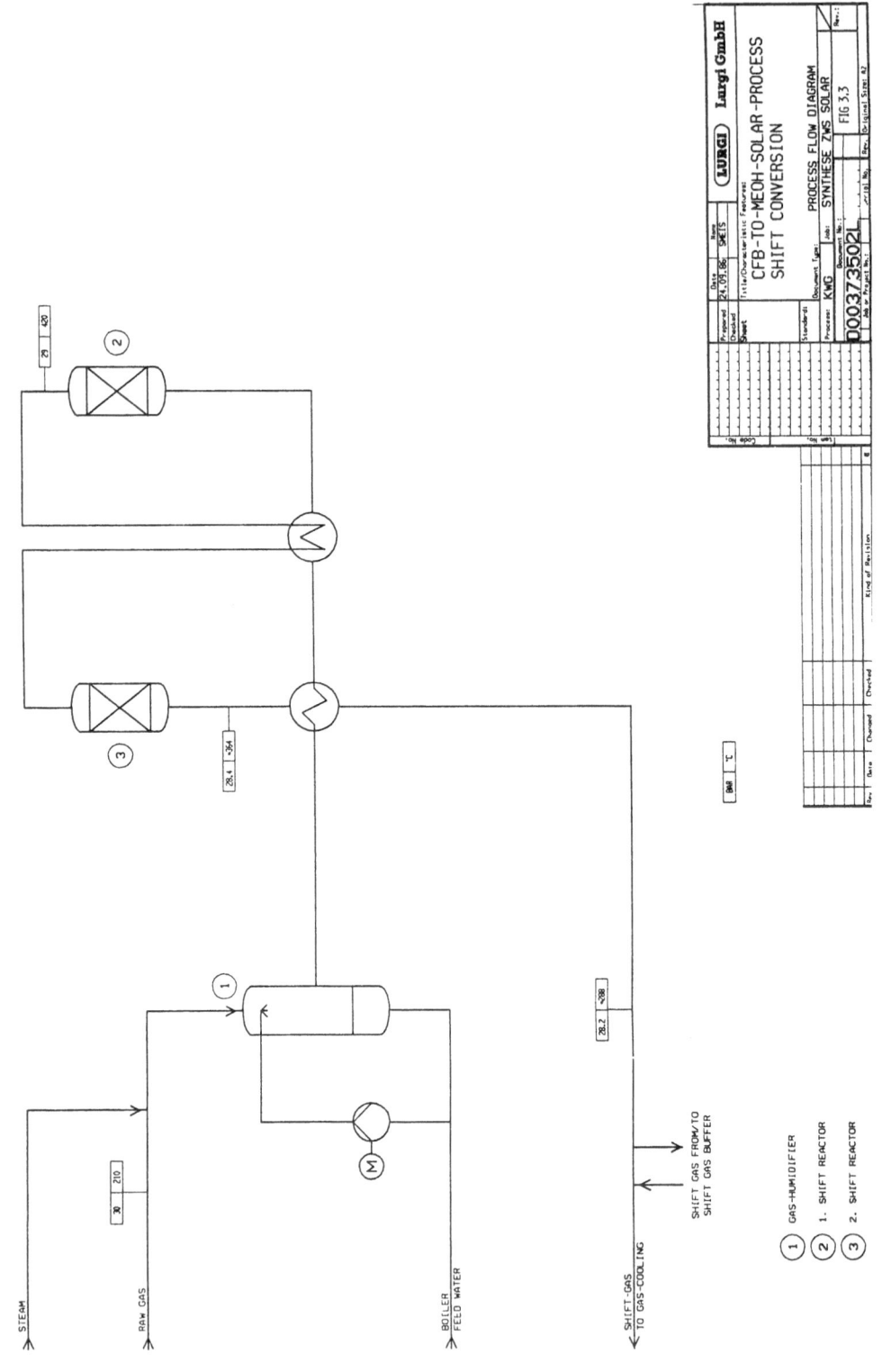

STEAM

RAW GAS

BOILER
FELD WATER

SHIFT-GAS
TO GAS-COOLING

SHIFT GAS FROM/TO
SHIFT GAS BUFFER

1 GAS-HUMIDIFIER

2 1. SHIFT REACTOR

3 2. SHIFT REACTOR

LURGI Lurgi GmbH

CFB-TO-MEOH-SOLAR-PROCESS
SHIFT CONVERSION

PROCESS FLOW DIAGRAM
SYNTHESE ZWS SOLAR

FIG 3.3

D0037335021

$$CO \quad + \quad H2O \quad \longrightarrow \quad CO2 \quad + \quad H2 \qquad Eq. \ 3.4$$

A side-effect of the shift conversion is, that unsaturated
hydrocarbons and some S-components are hydrogenated.

Because water is a reaction component, the water content of the
raw gas coming from the compression stage is adjusted by the
gas-humidifier. The gas is heated up by the two heat exchangers to
320 °C and is reacted in the reactors. The yield and the final
composition, i.e. above all the H2-content, can be adjusted up to
a maximum by appropriately setting the temperature levels and the
water content of the raw gas.

The shift conversion plays a special role in a solar plant, it
functions as a Switch Unit generating a "C"-stream with constant
H2/CO-stoichiometry. The final stoichiometry of the synthesis gas
is set by the gas purification process. See Fig. 3.4 for a cycle
diagram of the shift.

Raw gas with lower H2/CO-ratio is produced by non-solar operation
and the H2- "deficit" has to be compensated by the shift: either
the split is adjusted accordingly and a higher amount of raw gas
is fed to the shift, constant composition of shift gas being
assumed. Or the composition of the shift gas is changed towards
higher H2-content. In each case, in solar mode the shift operates
with partial load. The bypassed raw gas and the shift gas are
mixed to yield synthesis gas with the correct H2/CO-ratio, in Tab.
3.4 H2/CO = 2.38-2.39 for the mix gas. Tab. 3.4 lists the gas
compositions (mf) of raw, shift and mix gas of the shift for
biomass case 2.A, the reference case for biomass. For solar
operation the load factor is only 59.7%.

		raw gas		shift gas		mix gas	
		day	night	day	night	day	night
H2	(Vol.%)	50.0	32.8	55.4	41.1	53.0	40.6
CO	(Vol.%)	30.2	32.4	16.0	16.0	22.3	17.0
CO2	(Vol.%)	14.9	27.8	24.2	36.8	20.1	36.4
Rest	(Vol.%)	4.9	7.0	4.4	6.1	4.6	6.0

Tab. 3.4: Shift conversion operation characteristics
Case: Biomass.2.A, reference case for biomass

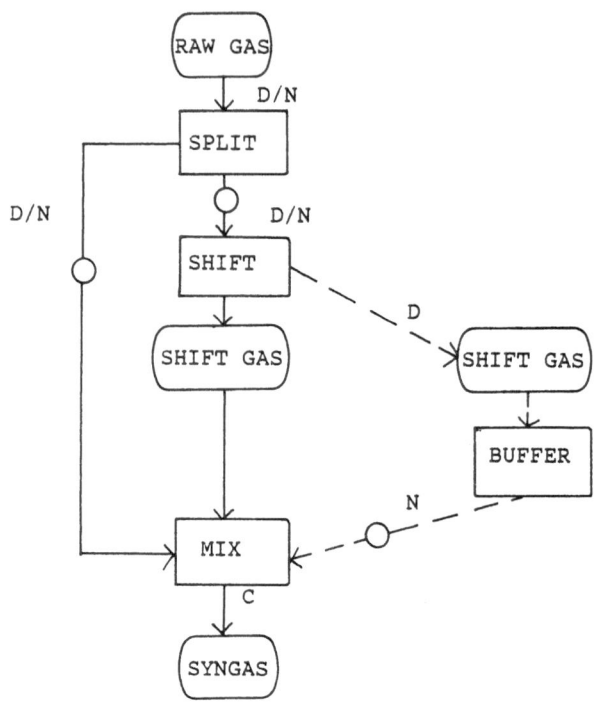

Fig. 3.4: Cycle diagram of the shift conversion
- - - : optional units
Definition of stream labels: see chapter 2.3

This basic scheme was used for the biomass cases. The lignite cases were calculated with an optional shift gas buffer. The buffer has a dual effect: it optimizes the design of the shift and the downstream units and effects higher solar energy integration. This topic is discussed in detail in the context of lignite results.

The shift gas buffer is a logical unit of the buffer type. Its function is: H2-rich shift gas is buffered during solar operation and removed from the buffer during non-solar operation. The corresponding buffer equations are:

BUFFER_SIZE = SOLAR_HOURS * SOLAR_BUFFER_STREAM   Eq.3.5

BUFFER_SIZE = NON-SOLAR_HOURS * NON-SOLAR_BUFFER_STREAM Eq.3.6

The variable BUFFER_SIZE (Kmol) represents the shift buffer size; it is the minimum value for a physical buffer, since the buffer should not be emptied to zero level. It is expressed as the product of the variables ..._BUFFER_STREAM (kmol/time), which denote the shift gas streams going to or coming from the buffer and the variables ..._HOURS, the daily operation period in each mode. In stationary operation, BUFFER_SIZE is equal in Eq. 3.5 and 3.6 and can be eliminated, Eq. 3.7:

NON-SOLAR_BUFFER_STREAM   =

                                                                            Eq.3.7

     SOLAR_HOURS/NON-SOLAR_HOURS * SOLAR_BUFFER_STREAM

Eq. 3.7 expresses the coupling of solar and non-solar balances. Fig. 3.5 gives the shift data for the lignite case 3.A as an example.

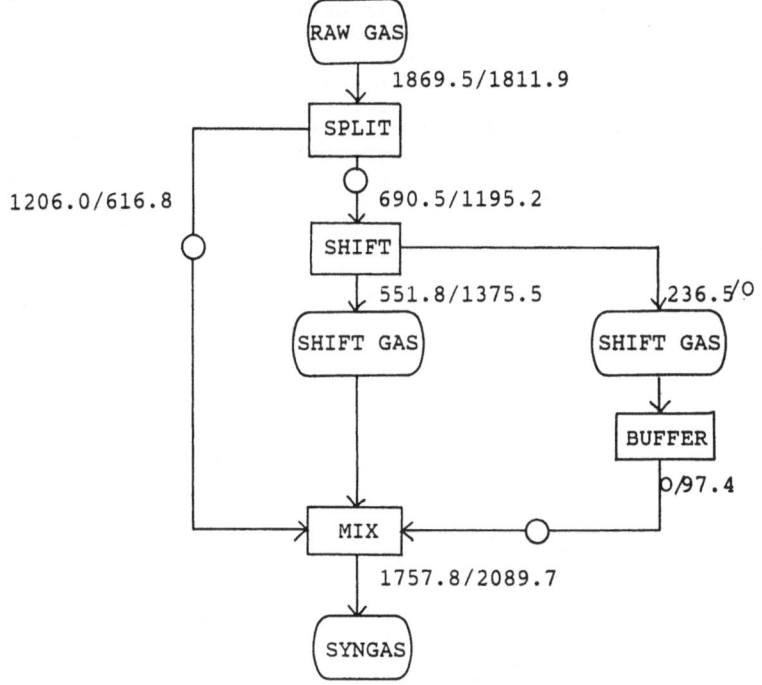

Fig. 3.5:   Operation characteristics of the shift conversion with
            buffer
            Case: lignite case 3.A
            Flow data in kMol/h on a moisture-free basis
            Data format: Day/night

The buffer may be realized for example by a spherical gas golder. In the balance calculations, it is assumed, that the buffer works at a maximum pressure of 100 bar. No design calculations for the buffer are made.

CO-conversion (kMol/h) was selected as characteristic design parameter for section SHIFT.

3.4 Gas Cooling and Cleaning

The mix gas coming from shift conversion section is cooled to ~40 °C by a sequence of heat exchangers. The gas liquor contains water, minor oily contents and physically dissolved gases. The internal gas-water recycle (3) in Fig. 3.2 removes solid components, which may crystallize from the gas.

The gas stream then enters gas purification, assumed here to be a Rectisol process shown as a black box. The Rectisol process is a physical gas purification process and has two purposes: it removes a number of components containing heteroatoms (first of all H2S) from the gas, which are poisonous for the methanol synthesis catalyst. Secondly, the CO2-content of the purified gas is adjusted finally to the required synthesis stoichiometry. The primary step of the Rectisol-process is the treatment of the raw gas with liquid methanol in a column, where the gas components dissolve in the liquid phase according to their vapour-liquid equilibria. Small amounts of the valuable components of the synthesis gas are removed here, too. In Fig. 3.2, the offgas from gas purification is withdrawn from the balance scheme in a very general manner; normally it has to be separated further and H2S, CO2 and a fuel gas containing H2, CO and CH4 can be produced.

Gas purification is a Switch Unit in the context of a solar plant, too. It handles raw syngas with different CO2-contents and produces constant stoichiometry. Tab. 3.5 shows as an example the gas compositions of raw and purified syngas taken from case lignite.2.

		Raw Syngas		Purified Syngas	
		Day	Night	Day	Night
H2	(Vol.%)	56.3	42.2	67.0	65.9
CO	(Vol.%)	29.0	32.2	27.3	26.8
CO2	(Vol.%)	12.1	21.8	3.5	3.5
Rest	(Vol.%)	2.6	3.8	2.2	3.8

Tab. 3.5: Calculated operating data of gas purification; gas compositions of raw and purified Syngas
Case: Lignite.2.A, reference case for lignite

Design of section GAS COOLING/PURIFICATION is based on two
parameters; total gas flow and CO2-flow. Total flow is
characteristic for sizing the cooling train and the first gas
cleaning column. CO2-flow is representative for the design of gas
cleaning and downstream CO2-handling units.

Section	GAS COOLING/PURIFICATION	
Parameter Id.	GAS	CO2
Parameter	gas flow (maf) to section (kMol/h)	CO2-flow to section (kMol/h)

Tab. 3.6: Characteristic design parameters for section
GASCOOLING/PURIFICATION

As can be estimated from the lower CO2-content of the solar raw
gas, the gas purification works in solar mode only with 54.5% of
load with respect to CO2-removal. For further discussion of these
design topics see chapters 4 and 5. But it should be mentioned
here, that turning down the Rectisol and downstream CO2-handling-
processes to 50% load is not that easy. The design of these steps
and possible savings of utilities (power for refrigerant
generation) as an indirect consequence of solar energy integration
have not been considered here.

3.5 Methanol Synthesis

Synthesis gas coming from gas purification enters the methanol
synthesis. The gas is first compressed to methanol synthesis
pressure of at least 50 bar. The balance calculations are based on
80 bar. In the flowsheet (Fig. 3.6) the synthesis loop is shown:
feed gas is mixed with recycle gas, methanol is produced in the
reactor (2) and condensed together with water in the condenser
(3). The purge gas is withdrawn from the synthesis, whereas the
recycle gas is recompressed by the compressor (4).

Methanol is produced by two reactions given in Eq. 3.8 and 3.9.

$$CO + 2\ H_2 \longrightarrow CH_3OH \qquad\qquad Eq.\ 3.8$$

$$CO_2 + 3\ H_2 \longrightarrow CH_3OH + H_2O \qquad Eq.\ 3.9$$

Both reactions are forced by a catalyst and are exothermic (ΔH_{298}
= 90.8 resp. 49.6 kJ/mol). The reactor is therefore cooled,
producing MP-steam to drive the recycle compressor. The

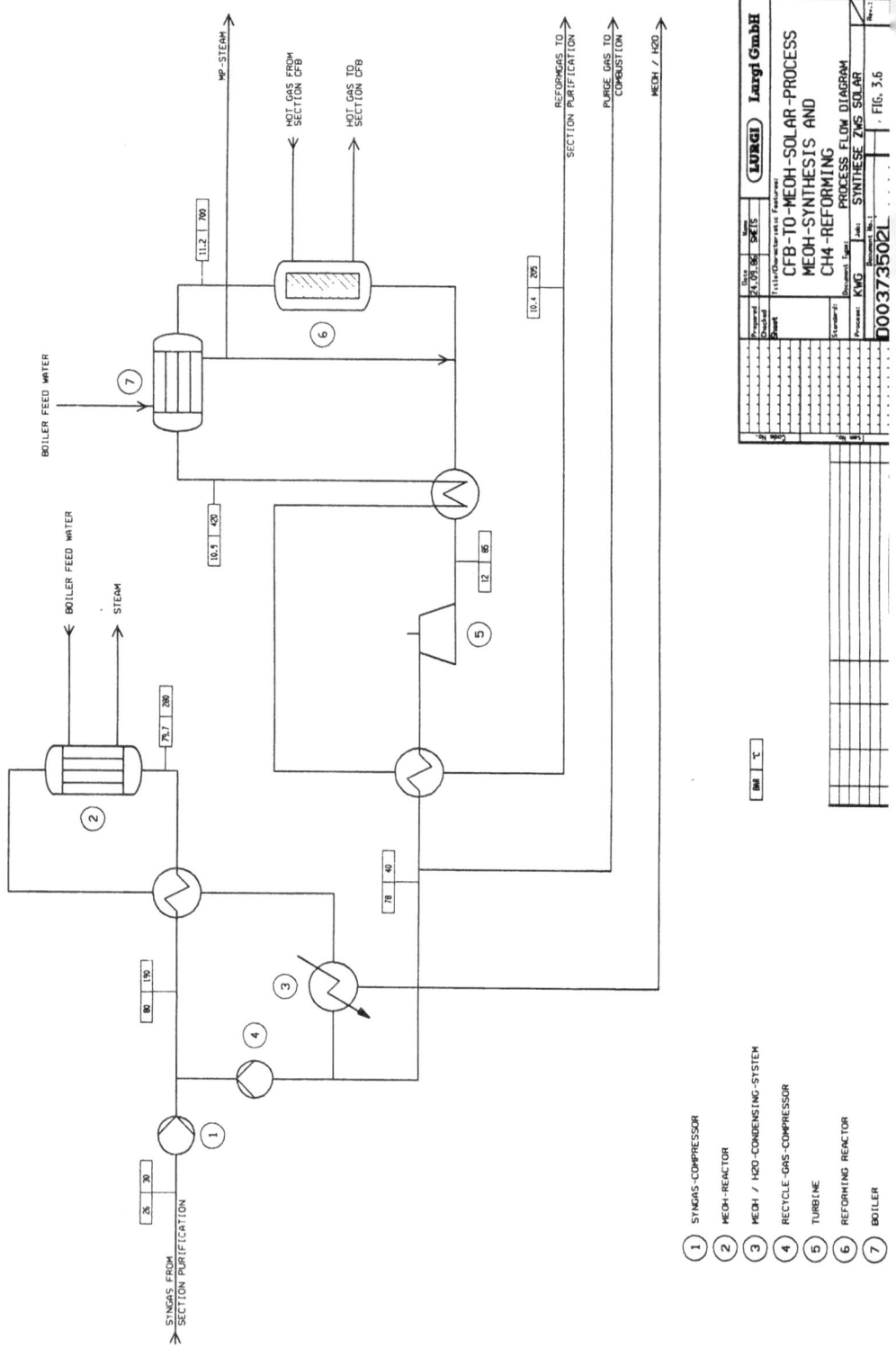

MP-STEAM

HOT GAS FROM
SECTION CFB

HOT GAS TO
SECTION CFB

REFORMGAS TO
SECTION PURIFICATION

PURGE GAS TO
COMBUSTION

MEOH / H2O

BOILER FEED WATER

BOILER FEED WATER

STEAM

SYNGAS FROM
SECTION PURIFICATION

(1) SYNGAS-COMPRESSOR
(2) MEOH-REACTOR
(3) MEOH / H2O-CONDENSING-SYSTEM
(4) RECYCLE-GAS-COMPRESSOR
(5) TURBINE
(6) REFORMING REACTOR
(7) BOILER

LURGI	Lurgi GmbH

CFB-TO-MEOH-SOLAR-PROCESS
MEOH-SYNTHESIS AND
CH4-REFORMING

PROCESS FLOW DIAGRAM
SYNTHESE ZWS SOLAR

, FIG. 3.6

D003735O2L

equilibrium composition of the reaction gas leaving the reactor, i.e. the composition of the purge gas, is determined by three factors: reaction temperature and pressure and the recycle rate of the looping gas. These parameters have to be adjusted to appropriate values in order to increase the contents of methane and higher hydrocarbons in the purge gas and hence to achieve better performance of the subsequent methane reforming unit. The pressure of the methanol synthesis is set to 80 bar for the balance calculations.

The solar and non-solar purge gases differ in their composition in a characteristic manner with respect to Nitrogen, see Tab. 3.7.

		Day	Night
H2	(Vol.%)	57.7	47.5
CO	(Vol.%)	5.4	4.4
CO2	(Vol.%)	6.9	5.8
CH4	(Vol.%)	23.4	28.9
N2	(Vol.%)	4.3	10.7
C2H6	(Vol.%)	2.3	2.7

Tab. 3.7: Calculated operating data of methanol synthesis;
          compositions of purge gas
          Case: Lignite.2.A, reference case for lignite

The inert components enriched in the purge gas originate from two sources: nitrogen is contained in the oxygen used as gasification agent and nitrogen and CH4/C2H6 are coking products from gasification. Solar gasification reduces both groups of inerts.

Methanol production (kMol/h) was used to characterize the size of the synthesis loop, section MEOH.

3.6 Hot Gas Loop and Methane Steam Reforming

As a result of the coking process of the raw material prior to gasification, the raw gas of the CFB-gasification always contains small amounts of methane and higher hydrocarbons. One idea at the beginning of this study was to increase the specific integration potential and simultaneously transform thermal energy from the raw gas into chemical energy by combination of two solar processes - CFB-gasification and steam reforming of methane.

Heat for steam reforming is conventionally integrated by a heat carrier loop; the helium carrier gas is heated up by raw gas in a heat exchanger placed on the hot side of the boiler (7) in Fig. 3.1. Depending on gasification temperature, a temperature of 700 °C may be reached. The carrier gas is circulated via the steam reforming reactor (6) in Fig. 3.6 and a blower.

The purge gas is removed from the methanol synthesis cycle after the condensing unit and split into two streams. One stream is going to steam reforming, the other to combustion; this stream cannot be zero in order to remove the inerts (N2) from the system. Purge gas going to steam reforming drives a turbine and is then preheated to reaction temperature and mixed with steam from the boiler (7). The steam reforming reactions are given by Eq. 3.10 and 3.11:

$$CH4 \;+\; H2O \;\longrightarrow\; CO \;+\; 3H2 \qquad\qquad Eq.\ 3.10$$

$$C2H6 \;+\; 2H2O \;\longrightarrow\; 2CO \;+\; 5H2 \qquad\qquad Eq.\ 3.11$$

The steam reforming reactions are catalytic and endothermic. As can be seen from the stoichiometric formula, the equilibria are pressure-dependent, shifting to the right as pressure is lowered. Steam is required for a dual purpose: it is reaction agent itself and must be set to an appropriate level in order to prevent formation of coke on the catalyst.

Reform gas is cooled in a boiler; the MP-steam produced is partially used fo the steam reforming reaction. The rest goes to the steam system. The reform gas is then conditioned and fed to the second cooling stage of the raw gas compressor.

The design conditions of the steam reforming reaction unit can be seen in Tab. 3.8.

	Pressure (bar)	Temperature (°C)
Carrier Gas		
Hot Side	11.5	800
Cold Side	11.2	350
Reaction Gas		
Hot Side	11.0	700
Cold Side	11.5	300

Tab. 3.8: Design conditions of the steam reforming reaction unit

Given pressure, thermodynamic equilibrium and therefore gas composition of the reform gas are determined by the temperature at the reactor outlet. 700 °C can be achieved by a carrier gas temperature of 800 °C. These temperature levels are relatively low to ensure good reaction yields in steam reforming. In order to achieve reasonable conversion rates at all, it is essential to set the pressure of steam reforming even lower then the level of shift conversion/gas cleaning. A pressure of 10.5 bar is therefore selected to match the operating pressure of the third stage of the raw gas compressor.

Considerable efforts are made to improve system performance by setting the pressure levels of methanol synthesis/steam reforming. But the exemplary compositions of purge gas and reform gas for solar operation in Tab. 3.9 indicate, that the reaction results are not satisfactory.

		Purge gas	Reformgas
H2	(Vol.%)	39.4	57.2
CO	(Vol.%)	4.0	7.9
CO2	(Vol.%)	5.3	9.5
CH4	(Vol.%)	35.3	24.5
N2	(Vol.%)	1.6	0.9
C2H6	(Vol.%)	14.4	

Tab. 3.9: Operating data of steam reforming unit: compositions of
          purge gas and reformgas
          Case: biomass.2.A, reference case for biomass

The major part of the methane cannot be reacted in a single loop. Ethane and higher hydrocarbons are reacted completely, but this is of no help: the data are taken from biomass results and C2H6-yield from biomass is higher than from lignite. Further, "C2H6" in this study is a reference component and stands for a mixture of ethane and minor amounts of higher hydrocarbons. In reality, the bulk of higher hydrocarbons is washed out from synthesis gas by the Rectisol unit. So the data for C2H6 in Tab. 3.9 represent the maximum.

For a discussion of energy efficiency, consider the formal cycle scheme for the steam reforming - methanol synthesis sequence in Fig. 3.7 which is used to check the efficiency.

Out of 100% energy supplied by the heat carrier, only 75% is transformed into chemical energy. To maintain the cycle, at least another 50% thermal energy is required, so the overall efficiency is only 50%. 24% energy used to produce MP-steam in the boiler of the steam reforming unit, are not lost, but they would be better produced in the CFB-boiler directly.

For this biomass reference case - solar operation - only 2.4 MW (6%) of total solar energy (40 MW) can be integrated at maximum into the steam reforming process and calculated with 50% efficiency. only 3% really transformed into chemical energy.

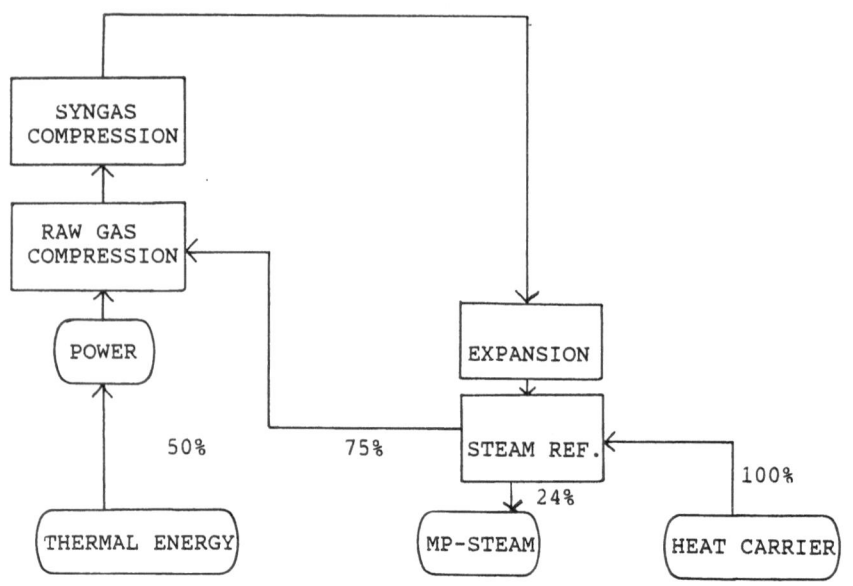

Fig. 3.7: Cycle scheme for a steam reforming - methanol synthesis
sequence
Case: Biomass.2.A, reference case for biomass

Consequently the solar energy integration potential is
rather poor and is so for two reasons: firstly, the temperature
level of steam reforming, coupled to gasification, is too low.
Rising the temperature level in gasification is of no help,
because it produces counteracting effects: performance of the
steam reforming process will be better, but the yield of methane
in gasification is reduced. Secondly, it is an effect of high
solar energy integration into gasification itself, which lowers
the specific yields of coking products and hence the potential for
a steam reforming unit. Naturally, for non-solar operation the
normal level of coking products is maintained, but this results in
a unfavourable production and consumption scheme: solar energy
available at day, integration potential at night.

The combination of CFB-gasification with steam reforming of purge
gas is not investigated further in this study due to the
availibility of more favourable elements of process synthesis.
However, it should not be disregarded altogether, but considered
in an new context, including:
- The direct heating of a methane reforming reactor by the raw
gas in order to increase the equilibrium temperatures. This is not

an existing technology and may cause problems because of the solid
contents of the raw gas.
- A conventional, non-solar steam reforming unit with
oxygen/steam. which works at a high temperature level. Solar
energy integration is in this case indirect by replacing purge gas
as fuel. A direct consequence is, that such a unit may be
installed only in a scenario where solar energy is available for
power generation.

3.7 Oxygen Plant

Oxygen is needed as gasification agent in high purity ($\geq$98% O2)
and is normally produced from air by low-temperature separation.
Oxygen consumption is determined by the gasification parameters,
mainly dependent on the raw material Eq. 3.12.

$$O2\_CON \quad = \quad O2/MAF \quad * \quad RAW\_MAT \qquad\qquad Eq. 3.12$$

O2_CON: throughput of oxygen plant (kMol/h)
O2/MAF: specific O2-consumption (kMol/Kg maf) referenced to raw
        material
RAW_MAT: Raw material throughput (kg maf/h)

The size of the oxygen plant is determined from Eq.3.12; the
oxygen plant is coupled to the gasification (apart from Eq. 3.12)
only by its power requirement, that is via the energy system. From
the viewpoint of a conventional synthesis plant the design of the
oxygen plant is therefore simple, being a "package unit". This is
not true for a solar gasification plant.

Consider the consumption diagram in Fig. 3.8; oxygen gas is
required during non-solar operation at night-time in the
CFB-reactor. No oxygen consumption takes place at day. This solar
consumption scheme can be generated on the production side by two
methods.
The first one is that oxygen is produced only during night by the
oxygen plant and no production takes place at day. The two
operation modes of the plant can then be characterized: normal
operation at night, at day the plant goes to a stand-by state with
zero production. This will in any case be a plant with low
performance, high energy losses, daily start-up and shut-down
phases and therefore high cost. An effect of solar energy
integration - as concerns the oxygen plant - would be saving a
part of utilities (power).

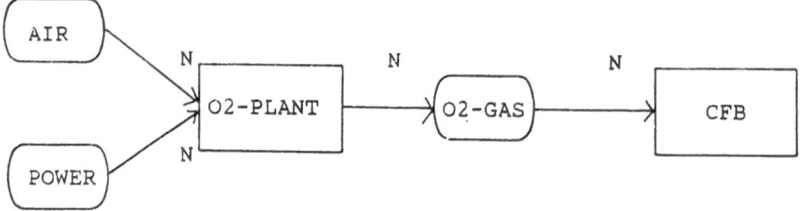

Fig. 3.8: Consumption diagram of an oxygen plant for solar gasification

A second possibility is to achieve constant operation of the oxygen plant by buffering the oxygen either in gaseous or liquid form. A general advantage of a buffer solution is that lower average O2-consumption of the gasification plant can really reduce the size of the oxygen plant. In the Fig. 3.9 the cycle diagram of a conventionally operating oxygen plant using a buffer is shown.

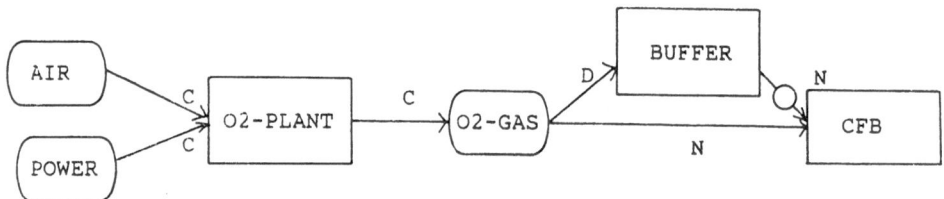

Fig. 3.9: Cycle diagram of an oxygen plant for solar gasification; conventionally operating plant with gaseous oxygen buffer

The plant in Fig. 3.9 produces gaseous oxygen, which is buffered during daytime, in steady-state operation just as a conventional plant. At night, currently produced and buffered oxygen is fed to the CFB-process. The buffer is realized for example as a gasometer. This solution minimizes energy consumption, because oxygen is produced in gaseous form. But the buffer will be large.

The corresponding scheme would be possible with a buffer for liquid oxygen. The primary product is liquid oxygen which has to be evaporated in a separate unit to be fed to gasification. The plant works in a conventional mode, but the extra energy for producing liquid oxygen is lost. The advantage of this solution is the small size and existing technology of the buffer, a low-temperature tank for liquid oxygen. But it is outweighed by far by the high energy consumption.

Finally, a scheme of a plant with solar-modified operation shall be discussed, Fig. 3.10. It works continuously, but in a dual mode: at day, liquid oxygen is produced and buffered in a tank. At night, the plant is switched to non-solar mode: liquid oxygen is recycled from the buffer to the plant, evaporated and mixed with currently produced gaseous oxygen. The advantage is the small size of the oxygen buffer and of course, that evaparation enthalpy of liquid oxygen can be used in night operation saving power. If power is produced by solar energy, this type of oxygen plant really works as an effective buffer for solar energy.

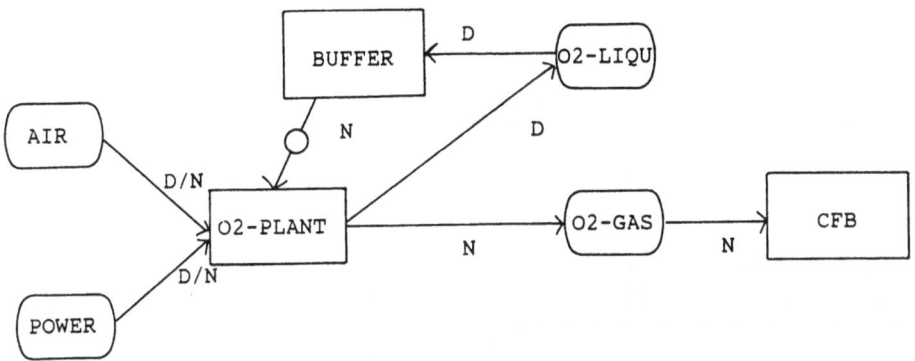

Fig. 3.10: Cycle diagram of an oxygen plant for solar gasification; solar-modified plant with liquid oxygen buffer

The problem is, that the plant works in dual mode; the required operating characteristics may be realized by switching the pressure level or by a recycle (or both). This is a nonconventional plant and therefore one with higher investment and operating cost. Based on the design of a conventional low-pressure oxygen plant with two columns, some design characteristics are investigated. See the flowsheet of the dual-mode oxygen plant in Fig. 3.11.

Air is compressed by the compressor (1), water and CO_2 are removed by (2) and the gas is cooled down by countercurrent heat exchangers. Air is then fed to the medium-pressure column shown at the bottom. Air separation occurs by distillation in a standard two-column design, see (5) for a discussion. Oxygen is removed from the bottom of the low pressure column in liquid form, nitrogen at the top, and a small portion of N_2/O_2 in the middle. The gases are heated up in the countercurrent heat exchangers. An independent air recycle is provided via the expansion turbine (3), a small part of the expanded gas going to the upper column.

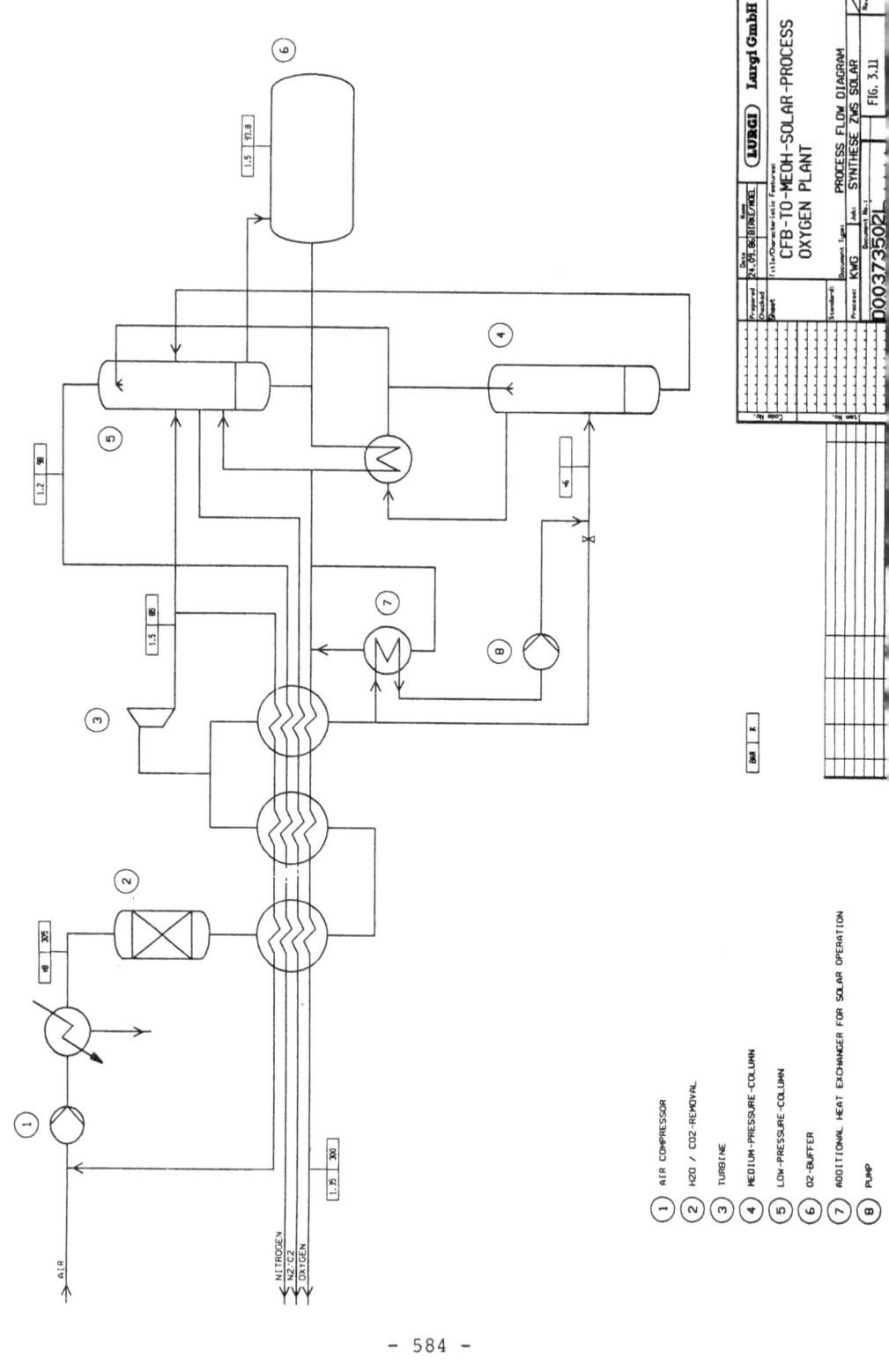

1	AIR COMPRESSOR
2	H2O / CO2-REMOVAL
3	TURBINE
4	MEDIUM-PRESSURE-COLUMN
5	LOW-PRESSURE-COLUMN
6	O2-BUFFER
7	ADDITIONAL HEAT EXCHANGER FOR SOLAR OPERATION
8	PUMP

CFB-TO-MEOH-SOLAR-PROCESS
OXYGEN PLANT

PROCESS FLOW DIAGRAM
SYNTHESE ZWS SOLAR
FIG. 3.11

LURGI Lurgi GmbH

In solar mode, the air is pressurized to 8 bar and then expanded as normal before entering the lower column. The recycle is adjusted, so that total oxygen can be removed in liquid form and buffered in tank (6). In non-solar operation, pressure is released to ~ 2 bar and the air is liquefied in the additional heat exchanger (7) by liquid oxygen drawn from the buffer. Liquid air is then fed by a pump to the lower column. Oxygen - currently produced and drawn from the buffer - leaves the plant in gaseous form at a pressure of 1.35 bar.

The following equations govern the operation of an oxygen plant with buffer; they are independent of the physical nature of the buffer material. The indices D,N in the variable names denote solar and non-solar operation.

Consumption side (see Eq. 3.12):

$$O2\_CON\_D \quad = \quad 0.0 \qquad\qquad\qquad\qquad\qquad\qquad \text{Eq. 3.13}$$

$$O2\_CON\_N \quad = \quad O2/MAF \quad * \quad RAW\_MAT \qquad\qquad \text{Eq. 3.14}$$

$$O2\_CON\_SUM \quad = \quad O2\_CON\_N * NON\_SOLAR\_HOURS \qquad \text{Eq. 3.15}$$

Production side:

$$O2\_P\_D \quad = \quad \alpha * O2\_P\_N \quad = \quad O2\_P\_N \quad (\text{for } \alpha = 1.0) \qquad \text{Eq. 3.16}$$

$$\begin{aligned} O2\_P\_SUM \quad = \quad & O2\_P\_D * SOLAR\_HOURS \\ + \quad & O2\_P\_N * NON\_SOLAR\_HOURS \end{aligned} \qquad \text{Eq. 3.17}$$

The load factor in Eq. 3.16 was fixed to 1.0 for the preliminary calculations here, but it would be subject to optimizing for a more sophisticated design. In steady-state operation, production and consumption of oxygen are equal:

$$O2\_CON\_SUM \quad = O2\_P\_SUM \qquad\qquad\qquad\qquad \text{Eq. 3.18}$$

Then, from 3.18, 3.17 and 3.15, we get the coupling and buffer equations:

$$\begin{aligned} O2\_P\_D \quad = \quad & O2\_P\_N \quad = \\ & NON\_SOLAR\_HOURS/24. * \quad O2\_CON\_N \end{aligned} \qquad \text{Eq. 3.19}$$

$$O2\_PUFF\_D = \quad O2\_P\_D \qquad\qquad\qquad\qquad\qquad \text{Eq. 3.20}$$

$$O2\_PUFF\_N = \quad SOLAR\_HOURS/NON\_SOLAR\_HOURS * O2\_P\_D \qquad \text{Eq. 3.21}$$

O2_PUFF_SUM =   Eq. 3.22

SOLAR_HOURS * NON_SOLAR_HOURS/24 * O2_CON_N

O2/MAF, RAW_MAT: see Eq. 3.12
α : load factor of the oxygen plant
O2_CON_X: O2-consumption in state X (kMol/h)
O2_CON_SUM: O2-consumption (KMol/day)
O2_P_X: current O2-production in state X (kMol/h)
O2_P_SUM: O2-production (kMol/day)
O2_PUFF_X: O2-stream to/from buffer in state X (kMol/h)
O2_PUFF_SUM: buffered O2 (minimum buffer size) (kMol)
(NON_)SOLAR_HOURS: daily operation hours in (non-)solar mode
          (h/day)

The size of the oxygen plant is determined by Eq. 3.19; it is an
effect of the buffer, that it is reduced by a factor of
NON_SOLAR_HOURS/24. The eqations 3.21 and 3.22 are the most
important buffer equations: the minimum buffer size is
determined from Eq. 3.22. Eq. 3.21 is especially important for a
solar-modified oxygen plant, since evaparation enthalphy and
thermal energy of the recycled buffer stream are used as an
additional cold stream to the plant in night operation.

	Day	Night	Avarage
Production data:			
Air throughput (kMol/h)	1581.9	1581.9	1581.4
O2_P_D (kMol/h)	324.6	-	94.7
O2_P_N (kMol/h)	-	324.6	229.9
O2-output to CFB	-	458.3	324.6
(kMol/h)			
Buffer data:			
O2_PUFF_D (kMol/h)	324.6	-	
O2_PUFF_N (kMol/h)	-	133.7	
Power consumption:			
Specific (J/kMol O2)	1.64E08	1.2E07	5.6E07
(% of total)	85.4	14.6	

Tab. 3.10: Calculated operation data of a solar-modified oxygen
          plant
          Case: Biomass.2.A, reference case for biomass
          Cycle: 7/17
          O2: 98% oxygen, 2% nitrogen
          Buffer data:
              Minimum size: 57.8 m3
              Spec. capacity: 1.6E10 J/m3 (minimum size)
              Conditions: liquid O2 (P=1.5 bar, T= 94 K,
              ζ =39.33 kMol/m3)

The operation data for a solar-modified plant in Tab. 3.10 are
taken from the biomass reference case. The flow data require no
further explanations, but pay attention to the power and buffer
data.

The buffer, realized as a tank for liquid oxygen, will have a
volume of ~80 m3, a minimum filling of 30% assumed. The energy
efficiency of the buffer was calculated as thermal energy
incorporated in the buffer material per volume unit of buffer.
Thermal energy estimation includes the specific power consumption
of the oxygen plant and the efficiency of power generation from
thermal energy.

The average power consumption estimated for a solar-modified
oxyygen plant is 5.6E07 J/kMol O2 and considerably higher than a
value of 4.E07 taken from literature (5) for a standard plant of
equal size. To some extent, this may be caused by poor design of
this solar-modified plant compared to a sophisticated standard
plant. But from the switching production scheme one can imagine,
that specific power consumption will really be higher for a
solar-adapted than for a conventional plant (buffer and exergy
losses).

As an effect of the liquid oxygen buffer, the bulk of the power
requirement - 85% - is shifted to day operation. Buffering oxygen
in liquid form has therefore a dual advantage for solar energy
usage: the size of the oxygen plant is reduced and a potential for
solar energy integration generated.

3.8 Drying Plant

Biomass and lignite are materials with moisture-content up to 60%
and require drying before gasification. Basis of the balance
calculations was drying from 40% down to 15% moisture. This drying
is accomplished at low temperature, for example with steam in a
drier with indirect heating.

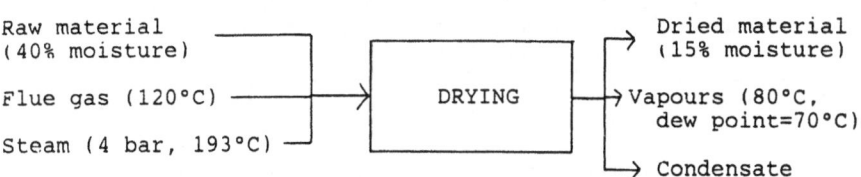

Fig. 3.12: Balance scheme for the drying process

The balance scheme in Fig. 3.12 results in a specific drying energy of 2.7E6 J/kg H2O to be stripped off. The drying process is operated with constant throughput, allowing minimum equipment size. The buffer of dried material matches the varying consumption of the CFB-process. Drying with low-pressure steam is favourable with respect to energy balance, but equipment is large and therefore expensive. Another drawback is, that low moisture contents cannot be achieved. Most preferable is a scenario with a separate solar plant for power generation and an integrated drying route based on a modern convective drier with drying temperatures up to 400 °C. For this option, moisture-contents down to 5% can be realized, minimizing the high-temperature energy requirements of the gasification.

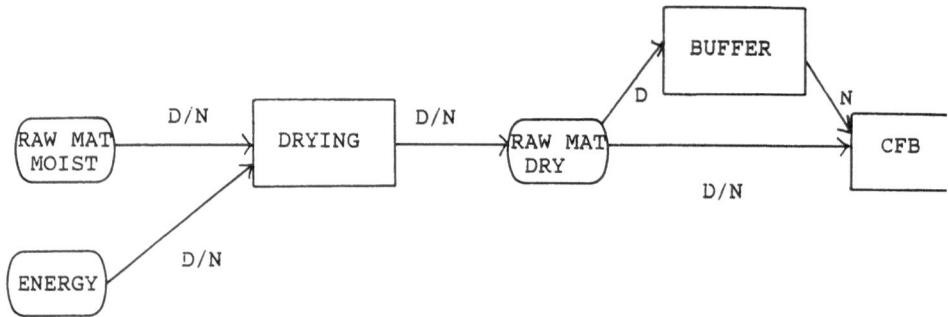

Fig. 3.13: Cycle diagram for a drying plant with a buffer of dried material

3.9 Steam System and Energy Scheme

Integral part of any chemical process plant is the steam system. In a gasification plant, steam is used for several purposes: as reaction agent, as medium for heating and for power generation. The main elements of a steam system are several steam-pressure and condensate levels, coupled by turbines and steam generation units. In Fig. 3.14 the simplified energy scheme for the CFB_TO_MEOH-solar process is shown. Steam is produced by a conventional boiler, in which purge gas of the methanol synthesis and biomass/lignite are burned. The design of a steam system has to be flexible in order to manage time-varying operation and insofar, a steam system for a solar plant will not differ much from a conventional unit.

Only global energy balance calculations are made in order to evaluate the energy requirements of the plant. These calculations are based on some (simplifying) assumptions: The balance scheme given in Fig. 2.4 is not complete for a whole synthesis plant and

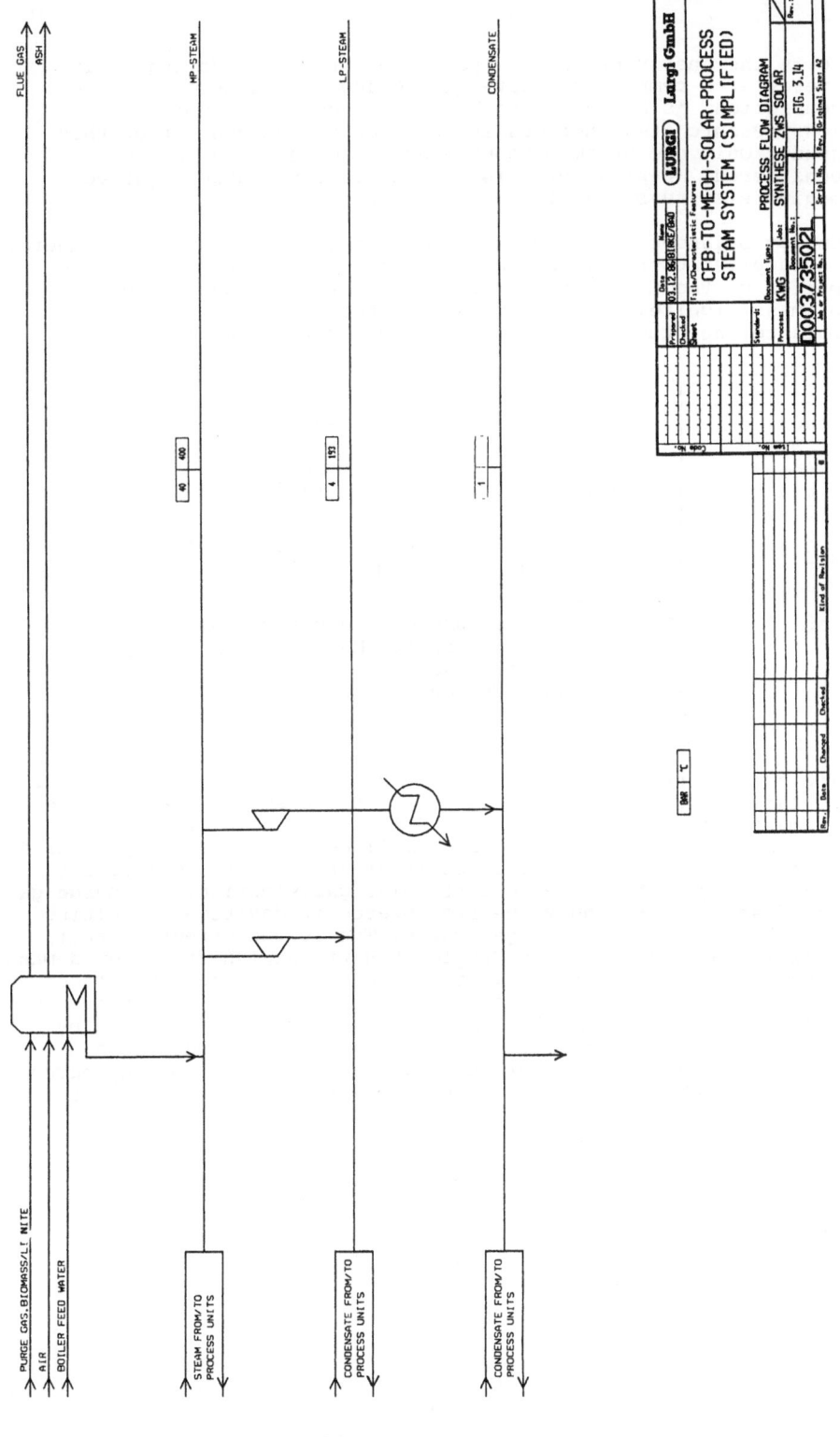

FLUE GAS
ASH
PURGE GAS,BIOMASS/LIGNITE
AIR
BOILER FEED WATER

NP-STEAM
40   400
LP-STEAM
4   193
CONDENSATE

STEAM FROM/TO
PROCESS UNITS

CONDENSATE FROM/TO
PROCESS UNITS

CONDENSATE FROM/TO
PROCESS UNITS

LURGI Lurgi GmbH

PROCESS FLOW DIAGRAM

CFB-TO-MEOH-SOLAR-PROCESS
STEAM SYSTEM (SIMPLIFIED)

SYNTHESE ZWS SOLAR

FIG. 3.14

Prepared | 01.12.06
Checked
Sheet

Title/Characteristic Features:

Standard:

Process: KWG

D0037/3502L

Original Size: A2

so is the energy balance. Apart from steam for drying, only
energy at a temperature level above 300 °C, as used for power
generation, is balanced. No balance is established for
low-pressure steam nor was it required in the context of this
study. Only one steam pressure-level (40 bar, 400 °C) is
considered; MP-steam of this quality is favourable to drive
medium-size turbines of process plants.

Energy balances are calculated for the two-state approach. Energy
is supplied partially by solar energy - steam produced in the
boiler of the CFB - and by combustion of purge gas and of
biomass/lignite. The latter is adjusted to close the energy
balance; no other source of solar energy was assumed.

Case	Day	Night	Average
2.A	70.2	0.3	20.5
2.B	63.4	0.3	18.5
2.C	57.1	0.3	16.9
2.D	50.3	-	14.7
2.E	44.0	-	12.8

Tab. 3.11: CFB_TO_MEOH-solar process on basis biomass
          Estimated energy deficit (MW) to be closed by
          combustion of biomass
          Case: biomass reference cases

The energy "deficit", which has to be closed by biomass/lignite,
is large at day, small at night. See the examplary data of the
biomass reference cases in Tab. 3.11. With increasing power
co-generation in the CFB even a small energy surplus may occur at
night. This phenomenon is a result of the efforts to integrate
solar energy. As side-effect of solar gasification, the purge gas
yield and its heating value are lowered at day-time, resulting
from higher specific raw gas yield. The second effect is that
buffers and design of the CFB_TO_MEOH process shift energy demand
from night to day. This is, of course, a highly desired effect and
offers a potential for solar energy on a thermal level suitable
for power generation. This potential. however. is not used by
solar energy available on the basis of the examined process
scheme. The conclusions drawn from this are one important outcome
of this study and lead to a scenario with a dual solar plant.

4. CFB_TO_MEOH-solar Process on Basis Biomass

So far, the elements of process synthesis - their function, solar-related topics, design and some results - have been presented. These elements have to be combined to synthesis routes on the basis of biomass and lignite as raw materials and examined as a whole. The main objective is not to compare different raw materials, but to evaluate process routes. In addition, the examination brought out new findings which were integrated in the further work. As a consequence, different elements of process synthesis are put in for both synthesis sequences. which are therefore not directly comparable. Because biomass was the first process to be examined, it serves as a base case and is discussed more thoroughly.

For the biomass sequence, two main key parameters are varied: firstly, the combination of CFB-gasification and power generation in order to find a favourable mix and secondly, the daily solar operation periods in order to check the influence of solar constants.

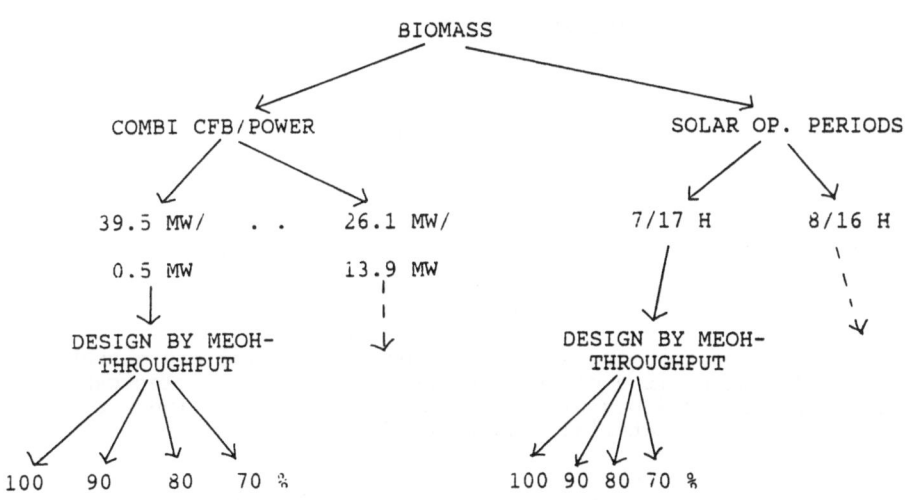

Fig. 4.1: Hierarchy of key parameters for biomass sequence

The secondary key parameter is the throughput of the synthesis

sequence, referenced to the methanol-throughput. The block
flowsheet for all biomass cases in Fig. 4.2 is identical with Fig.
2.4.: the units and commodities not used for biomass are removed.

4.1 Biomass

Biomass includes all growing material, like wood, bushes, grass or
newly developed fast-growing material. Beechwood was chosen as a
reference material, because gasification of this material in the
CFB was extensively investigated in the LURGI pilot plant. The
analytical data are listed in Tab. 4.1 and 4.2. the (gross)
calorific value of biomass is 1.774E7 J/kg on a ash and
moisture-free basis.

Moisture (%)	40.0
Vol. matter (%)	60.28
C-fix (%)	35.00
Ash (%)	4.72

Tab. 4.1: Proximate analysis of beechwood

Ash	4.72
C	46.94
H	5.48
N	0.29
S	0.10
O	42.47

Tab. 4.2: Ultimate analysis of beechwood

The low ash content is typical for biomass and may cause problems
in gasification. if a circulating bed from ash particles cannot be
built up. Either a bed with high coke-content or an inert solid as
an auxiliary bed material will be used.

Biomass is highly reactive and gasification takes place at a
reasonable rate even at 600 °C. But a gasification temperature of
800 °C is favourable in order to minimize yield of coking products
(hydrocarbons, phenols). Due to the character of biomass as a
low-ranking material, the specific O2-consumption for conventional
gasification and hence the solar energy integration potential is
lower than for any fossile material.

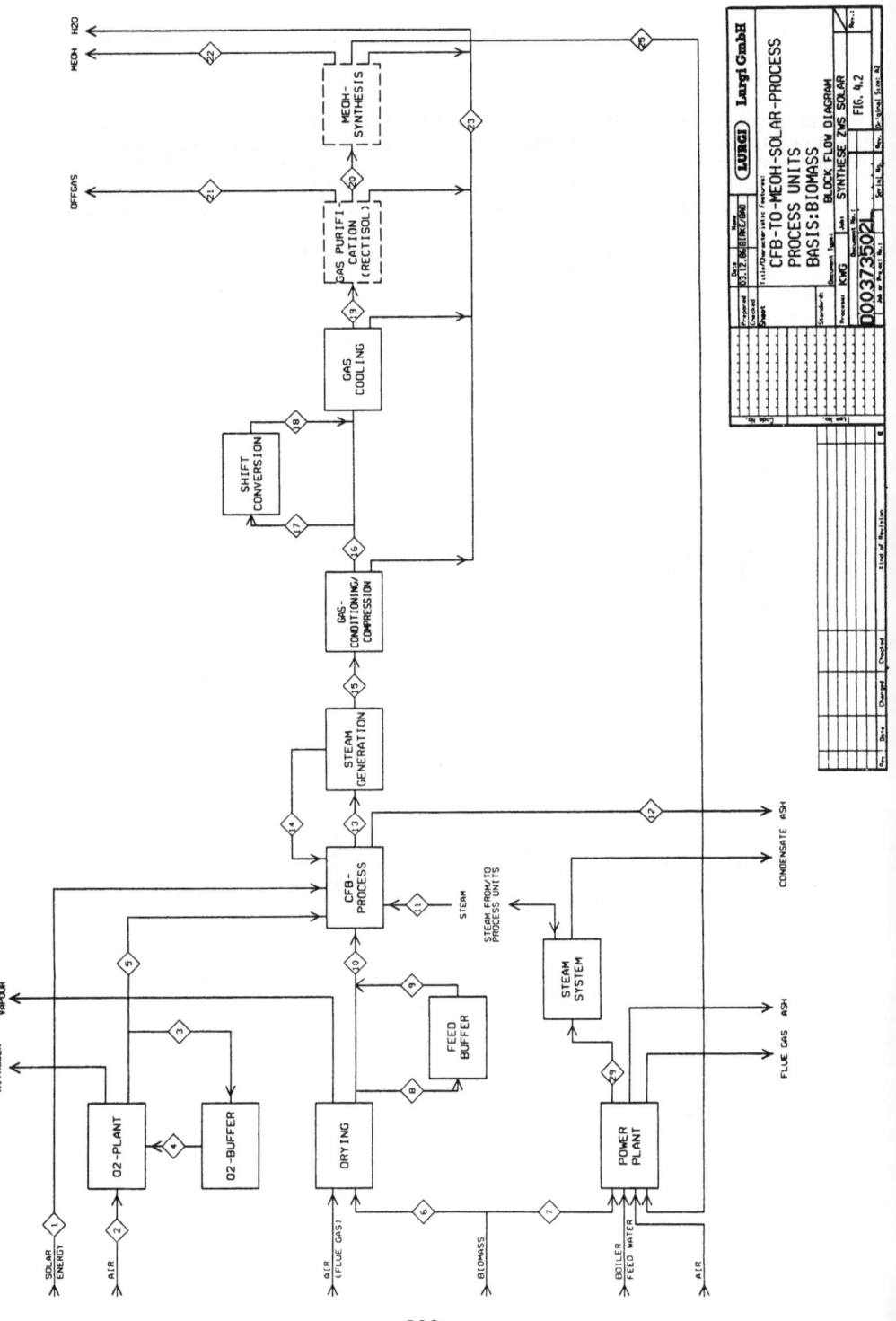

4.2 Combination of solar CFB-gasification with Power Generation

4.2.1 Definition of Cases

The primary key - the combination CFB-gasification and power generation - is varied on four levels; solar input for power generation ranging from 0.6 - 13.9 MW of 40 MW. These levels are chosen somewhat arbitrarily, because the states of synthesis sequences were unknown at the beginning. From the gasification results it becomes clear that for a favourable design, methanol throughput has to be reduced at night; it is reduced to 60% of its day-value to analyze the optimum design of the whole plant. See Tab. 4.3. for an overview of the balance cases. Case 1. is the case of conventional plant without solar energy integration and is calculated to compare the results.

CASE	SOLAR ENERGY (MW)		MEOH-THROUGHPUT (%)	
	CFB-GAS.	POWER	DAY	NIGHT
1.			100.	100.
2.A	39.4	0.6	100.	100.
2.B				90.
2.C				80.
2.D	↓	↓	↓	70.
2.E				60.
4.A	35.8	4.2	100.	100.
4.B				90.
4.C				80.
4.D	↓	↓	↓	70.
4.E				60.
5.A	32.6	7.4	100.	100.
5.B				90.
5.C				80.
5.D	↓	↓	↓	70.
5.E				60.
6.A	26.1	13.9	100.	100.
6.B				70.
6.C	↓	↓	↓	60.

Tab. 4.3: CFB_TO_MEOH- Solar process
        Biomass balance cases
        Cycle: 7/17

4.2.2 Design of Plant

The results of balance calculations in the design table Tab. 4.4 are based on the characteristic design parameters of the units. Due to the solar cycles, not all units can be operated with their design load in all states. Instead there exists an optimizing problem to generate a sequence with best load characteristics. Economically, a solar plant with maximum load is a plant with minimum specific investment cost. The decrease in specific investment cost with increasing plant size can be excluded here without loss of generality, because plant size is fixed solely by the assumption of 40 MW solar energy. The optimum size of a plant is not considered here.

Design and sizing is a different thing for the following units:

- Synthesis sequence (CFB-gasification - MEOH-synthesis): a conventional steady-state design cannot be achieved. Instead, design must be based on the data of that state - solar or non-solar - requiring larger equipment. In the opposite state the unit is then operated with partial load. Economically, design based on the non-solar state is favourable, because non-solar operation time is longer (17 hours) than solar (7 hours).

- O2-Plant: all balance cases are based on the solar-modified oxygen plant with liquid oxygen buffer. Since this plant works in dual state mode, but constant throughput, its size is only determined by oxygen consumption.

- Drying plant: the unit can be designed for constant throughput of raw material; the buffer matches different consumption of dried material by the CFB-process day and night.

- Energy system: data are not given in Tab. 4.4; Energy system is assumed to be designed flexibly as in a conventional plant.

For a detailed discussion of the design to find best cases, only the synthesis sequence shall be considered. Due to the buffers, oxygen and drying plant achieve 100% -load design and may be decoupled from further examination. The design diagrams (Fig. 4.3, 4.4) show the units of the synthesis sequence. The diagrams indicate the (partial)-load in one state as % of design load, based on the opposite state. This is expressed by Eq. 4.1 and 4.2:

$$\%\_LOAD\_X = DESIGN\_VAR\_X \ / \ DESIGN\_VAR\_X * 100. \qquad Eq. \ 4.1$$

$$= 100.$$

$$\%\_LOAD\_Y = DESIGN\text{-}VAR\_Y \ / \ DESIGN\_VAR\_X * 100. \qquad Eq. \ 4.2$$

$\%\_LOAD\_X(Y)$: load in state X (Y) in %
$DESIGN\_VAR\_X(Y)$: value of characteristic design variable in state X (Y)
X: Design state. may be Day or Night
Y: opposite state, may be Night or Day

CASE	CFB REACTOR KMOL/H		BOILER MW		COMPR. KMOL/H		SHIFT KMOL/H		GAS-COOLING/PURIFIC. GAS KMOL/H		CO2 KMOL/H		MEOH KMOL/H	
	DAY	NIGHT	DAY	NIGHT	DAY	NIGHT	DAY	NIGHT	DAY	NIGHT	DAY	NIGHT	DAY	NIGHT
1.	2797.2	2797.2	17.0	17.0	2140.9	2140.9	285.1	285.1	2435.0	2435.0	883.1	883.1	441.0	441.0
2.A	2139.8	2797.2	12.1	17.0	1748.0	2140.9	119.2	285.1	1872.4	2435.0	379.8	883.1	441.0	441.0
2.B	2139.8	2517.5	12.1	15.3	1748.0	1926.9	119.2	258.2	1872.4	2191.3	379.8	794.5	441.0	397.1
2.C	2139.8	2237.8	12.1	13.6	1748.0	1712.9	119.2	228.2	1872.4	1948.0	379.8	706.5	441.0	352.8
2.D	2139.8	1958.0	12.1	11.9	1748.0	1498.6	119.2	199.6	1872.4	1704.0	379.8	618.2	441.0	308.7
2.E	2139.8	1678.3	12.1	10.2	1748.0	1284.5	119.2	171.1	1872.4	1461.0	379.8	529.9	441.0	294.6
4.A	2677.3	2543.4	15.0	15.5	1588.8	1946.8	108.4	260.7	1702.1	2214.0	345.3	802.8	401.0	401.0
4.B	2677.3	2289.1	15.0	13.9	1588.8	1752.2	108.4	234.6	1702.1	1992.6	345.3	722.5	401.0	361.0
4.C	2677.3	2034.7	15.0	12.4	1588.8	1557.3	108.4	208.6	1702.1	1771.2	345.3	642.2	401.0	320.8
4.D	2677.3	1780.4	15.0	10.9	1588.8	1362.8	108.4	182.5	1702.1	1549.8	345.3	562.0	401.0	280.7
4.E	2677.3	1526.0	15.0	9.3	1588.8	1168.1	108.4	156.4	1702.1	1328.4	345.3	481.7	401.0	240.6
5.A	3171.6	2331.0	17.8	14.2	1454.6	1784.3	99.3	239.0	1560.2	2029.3	316.6	735.8	367.6	367.6
5.B	3171.6	2097.9	17.8	12.8	1454.6	1606.0	99.3	215.1	1560.2	1826.4	316.6	662.2	367.6	330.8
5.C	3171.6	1864.8	17.8	11.3	1454.6	1427.6	99.3	191.2	1560.2	1623.4	316.6	588.6	367.6	294.1
5.D	3171.6	1631.7	17.8	9.9	1454.6	1249.1	99.3	167.3	1560.2	1420.0	316.6	515.1	367.6	257.5
5.E	3171.6	1398.3	17.8	8.5	1454.6	1070.6	99.3	143.4	1560.2	1217.4	316.6	441.5	367.6	220.6
6.A	4110.5	1831.0	23.0	11.0	1137.3	1401.4	79.1	186.6	1220.0	1594.0	246.3	578.1	288.7	288.7
6.B	4110.5	1647.9	23.0	10.0	1137.3	1261.0	79.1	167.9	1220.0	1435.0	246.3	520.3	288.7	260.0
6.C	4110.5	1464.81	23.0	8.9	1137.3	1121.1	79.1	149.2	1220.0	1275.0	246.3	462.5	288.7	231.0

Tab. 4.4: CFB TO MEOH solar Process on basis Biomass
Design Data based on characteristic
design parameters
Data for case 3.A-3.E (8/16 cycle)
correspond to the data of case 2.

For example the load of the CFB-reactor - design variable REACTOR
- for case 2.A is calculated from Eq.4.1 and 4.2 (values taken
from Tab. 4.4: D:Day, N: Night); in case 2.A the size of the
CFB-reactor is based on night operation.

$*\_LOAD\_N$ = REACTOR_N / REACTOR_N = 100.   Eq. 4.3

$\%\_LOAD\_D$ = RECTOR_D / REACTOR_N * 100.   Eq. 4.4

                = 2139.8 / 2797.2 * 100. = 76.5

The diagrams characterize the design of a whole sequence.
An examination of alternative sequences is made first for cases
with the same primary key (= power input), for example cases
2.A-2.D.

Consider the reference cases 2.A-2.D and their design diagrams in
Fig. 4.3. The cases represent operable plants with the following
characteristics: the size of the methanol plant is constant, but
methanol production decreases due to partial load in cases B-D;
the size of all other units varies, solar energy integration
increases from A-D. What can be seen from the diagrams?
- In base case 2.A, the methanol plant has its design load in both
states, but the whole rest of the sequence is operated at daytime
with partial load. Two units, shift conversion and gas cleaning
(CO2-handling units) must be operated with load factors even lower
than 50%.
- Going from case 2.A to 2.B to 2.C. the methanol unit is operated
with partial load at night; but the rest of the plant is
increasingly better used.
- Going from case 2.C to 2.D, there is a switch in design: the
sections CFB, COMPRESSION and GAS-PURIFICATION are now operated at
night with partial load.
-As a consequence, there is an optimum case having the best load
characteristics. Apart from methanol synthesis. the size of all
other units is a minimum; case 2.C comes close to it.

A similar analysis can be made for the cases 4.-6.; as an effect
of increasing solar energy integration for power generation. the
CFB-section must be designed for day operation and is operated
with partial load at night. There is an optimum case, too, but
because of the countercurrent design effects of the sections CFB
and COMPR-GASCOOLING/PURIFICATION it is not possible to evaluate
an optimum case from each group in an exact manner. This would
require a sophisticated economical analysis. The optimum cases are
selected on the basis. that the units COMPR -
GASCOOLING/PURIFICATION are designed optimally and the CFB-unit
can be operated with reasonable partial load. These cases are

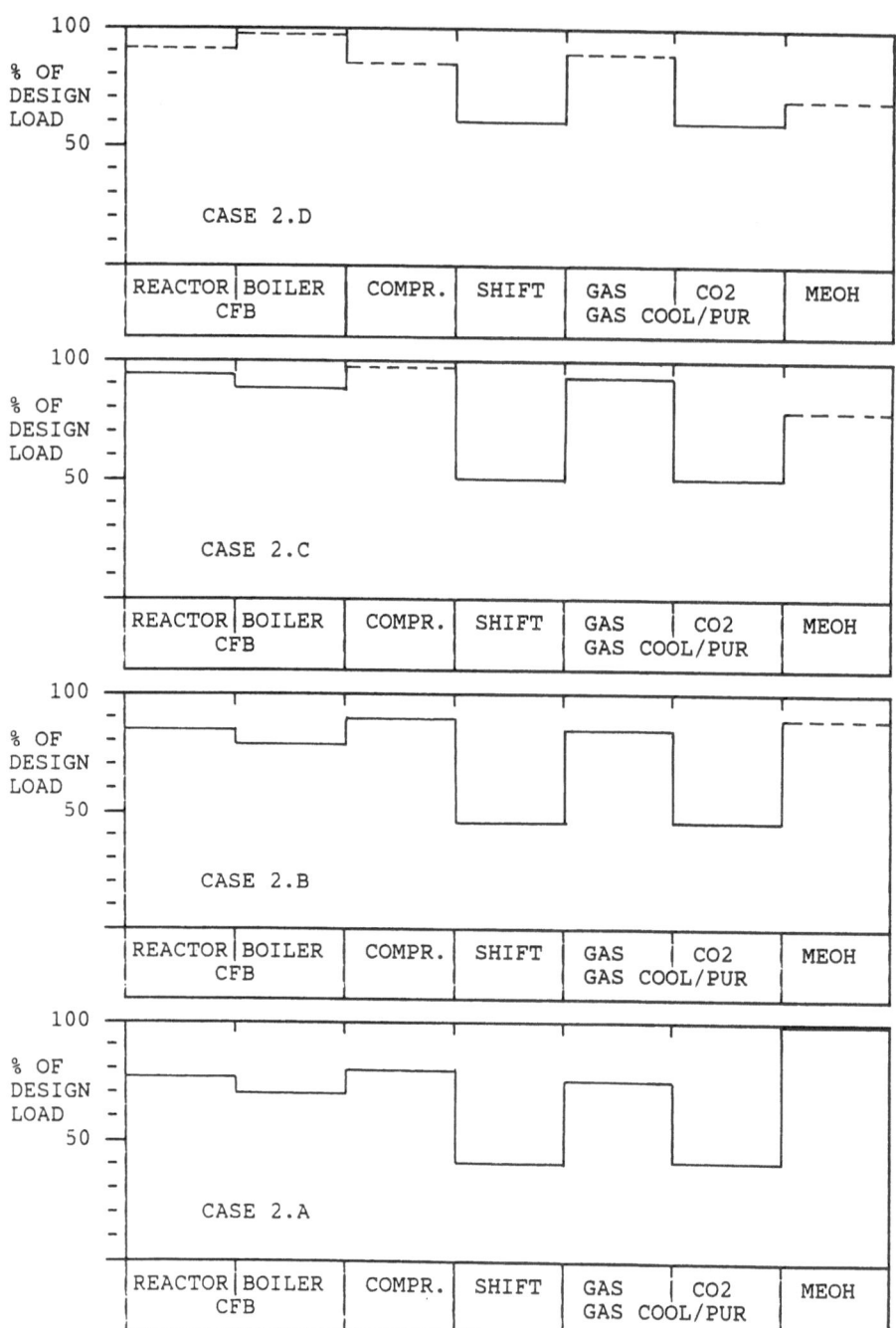

Fig. 4.3: CFB_TO_MEOH-solar process on basis biomass
Design diagrams for synthesis sequence
Cases: Reference cases for biomass
——— : Design based on night operation
----- : Design based on day operation

considered to represent the optimum good enough to allow further
discussion of design topics and an estimation of solar energy
integration potential.

See in Fig. 4.4 the diagrams for the cases 4.C, 5.C and 6.B,
chosen as optimal from their groups. Realize the effect of
increasing solar energy integration for power generation:
 - First, there is a switch in design of the CFB-unit going from
case 2.C to 4.C which defines a local optimum lying between these
cases. That is, with a relatively small part of solar energy used
for power generation, the CFB-unit can be operate with nearly
constant load. This local optimum results from the differing
specific gasification results for solar and non-solar operation.
 - With increasing solar energy integrated for power generation,
the CFB-section has to be designed larger and larger for day
operation. In the end - case 6.B - even no more optimization of
the whole synthesis sequence is possible, because one unit runs
out of range. To find a favourable case out of 2.C - 5.C, one has
to realize a compromise between increasing solar energy
integration and overall plant performance.

4.2.3 Results and Conclusions

For an assessment of different process alternatives individual
units need not to be looked at, but the data given at the battery
limits are sufficient. Hence, the flowsheet in Fig. 4.1. may be
compressed to an overall balance scheme representing solely input
and output streams, Fig. 4.5:

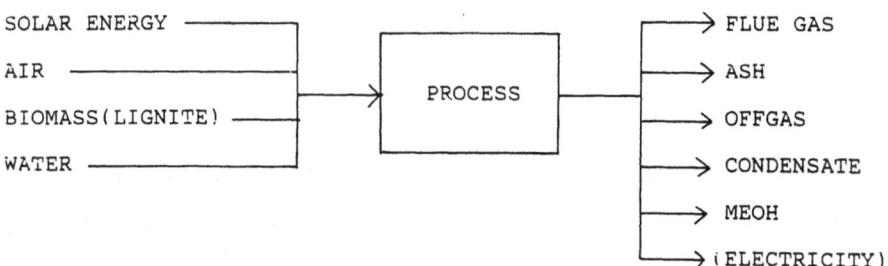

Fig. 4.5: CFB-TO-MEOH-solar process: overall balance scheme

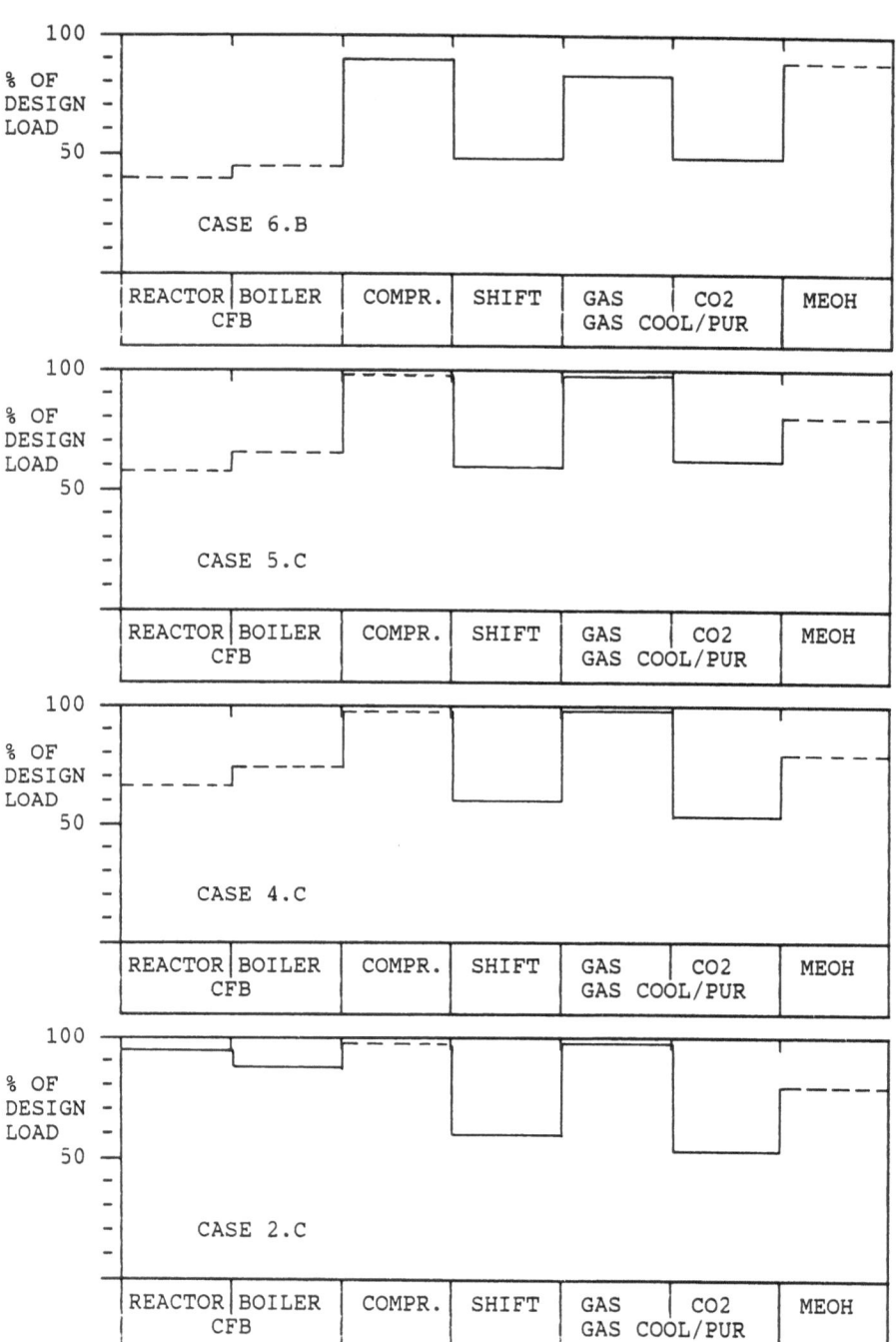

Fig. 4.4: CFB_TO_MEOH-solar process on basis biomass
Design diagrams of optimum cases
————— : Design based on night operation
- - - - : Design based on day operation

Moreover. all solar specifics shall be removed and data refer only
to the average of solar and non-solar states. Concentrated solar
energy, the raw material and the product methanol (the possible
by-product electricity was not produced in the cases under
consideration) are the main materials resp. energies o_ value. The
process alternatives shall be assessed in terms of characteristic
parameters derived from consumption/production of these three
production factors. Because there are three of them, analysis
cannot be a unidirectional process, but has to consider a scheme
of mutual dependences as it is represented in Fig. 4.6.

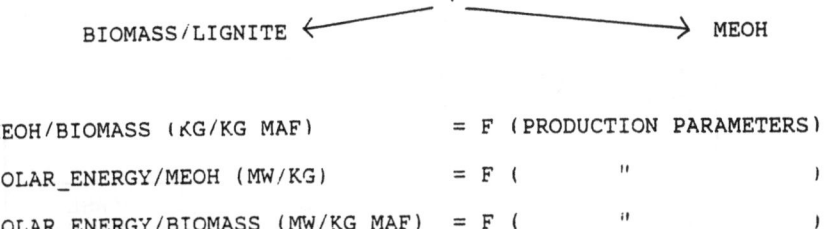

$$MEOH/BIOMASS\ (KG/KG\ MAF) = F\ (PRODUCTION\ PARAMETERS)$$

$$SOLAR\_ENERGY/MEOH\ (MW/KG) = F\ (\qquad " \qquad)$$

$$SOLAR\_ENERGY/BIOMASS\ (MW/KG\ MAF) = F\ (\qquad " \qquad)$$

Fig. 4.6: CFB-TO-MEOH-solar process: generation of characteristic
        parameters

The key parameter is the specific methanol yield MEOH/BIOMASS (in
kg/kg maf) which characterizes the degree of utilization of
biomass in general and the effect of solar energy integration in
particular. Then, there is a second independent parameter which
characterizes the efficiency of solar energy integration;
SOLAR_ENERGY/BIOMASS (in MW/kg maf) was selected for this purpose.

In Fig. 4.7 specific methanol yield is given for all biomass
cases; on the abszissa the subcases .A-.E are marked. Remember
that these cases are referenced to methanol-throughput, the
night-load of the methanol unit varying from 100-60% of design
load.

For conventional gasification (case 1), specific methanol yield is
0.357. As a result of solar energy integration methanol yield is
increased, at minimum by 9.5%, at maximum by 28%. The curves
reflect the influence of the two parameters, which have been
varied: by reducing throughput on night operation - following the

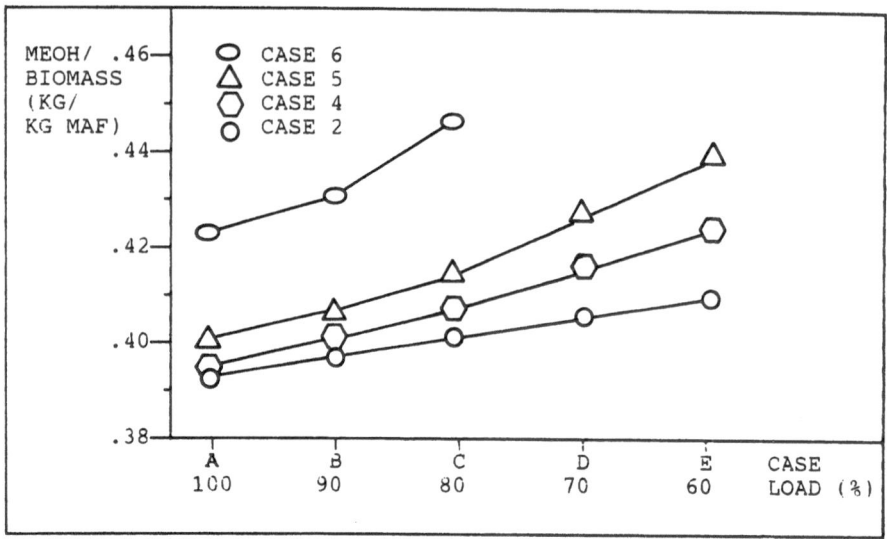

Fig. 4.7 : CFB_TO_MEOH-solar process on basis biomass
          Specific methanol yield

curves - a larger part of methanol is produced by solar operation.
The parallel curves represent increasing amounts of solar energy
integrated as power. In terms of energy, both effects are not
equivalent. This becomes clear from Fig. 4.8, where specific
methanol yield is presented as a function of the other
characteristic parameter, specific solar energy input.

The base case 2.A-D with a maximum of solar energy integrated as
chemical energy, has best efficiency. With increasing amount of
power co-generated, the curves shift to the right, indicating
decreasing efficiency of solar energy integration.

The design cases with optimal load characteristics (see 4.2.2 for
a discussion) are labeled with larger symbols in Fig. 4.8. Their
connecting line separates two fields of design: the region above
is characterized by the fact that increasing solar energy
integration is achieved with a simultaneous increase in investment
cost. This can be economically meaningful and depends on the
prices of solar energy and of the raw material biomass. The region
below combines decreasing solar energy integration with increasing
investment cost and is therefore not relevant.

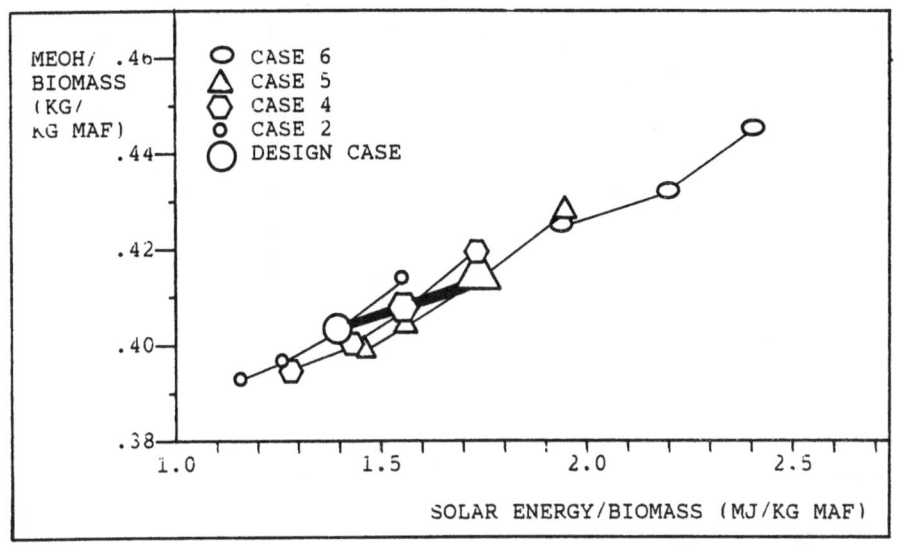

Fig. 4.8: CFB_TO_MEOH-solar process on basis biomass
          Specific process parameters

4.3 Influence of solar Operation Time on Integration Results

Most balance calculations are based on the 7/17 cycle, that is 7
hours solar and 17 hours non-solar operation time. Based on
location dependent solar flux data, the solar unit of the plant,
i.e. the mirror field has to be designed to match these cycle
data. But a design for other cycle data is possible. In addition,
the balance calculations are based on a two-state-approach which
idealizes transitions between day and night with their lower solar
flux. In a more accurate model these transition states and their
durations have to be considered with respect to their influence on
the design of the solar process plant. The cycle parameters are
varied in one case to get an impression of their importance. The
selected level is 8/16, that is more favourable for solar energy.
Tab. 4.4 lists the data; case 2 is the reference case for biomass,
case 3 just its equivalent for 8/16-cycle.

CASE	SOLAR ENERGY (MW)		PRODUCTION CYCLE (H)		MEOH-THROUGHPUT (%)	
	CFB-GAS.	POWER	DAY	NIGHT	DAY	NIGHT
2.A	39.4	0.6	7	17	100.	100.
2.B						90.
2.C						80.
2.D						70.
2.E	↓	↓		↓	↓	60.
3.A	39.4	0.6	8	16	100.	100.
3.B						90.
3.C						80.
3.D						70.
3.E	↓	↓	↓	↓	↓	60.

Tab. 4.4: CFB-TO-MEOH-solar process on basis biomass
Balance cases for 8/16-cycle

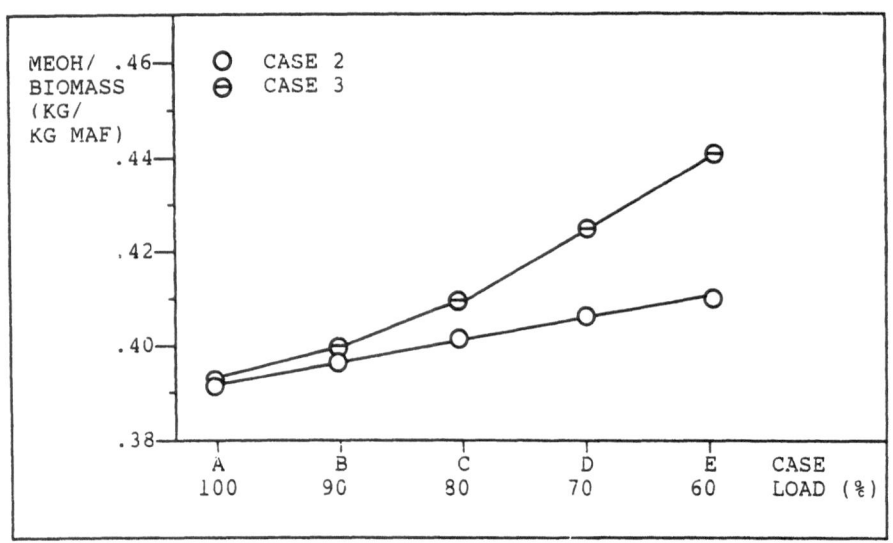

Fig. 4.9: CFB TO MEOH-solar process on basis biomass
Specific methanol yield as function of solar cycles
Case 2: 7/17-cycle
Case 3: 8/16-cycle

Design of cases 3. is the same as for their corresponding cases
2.; it is economically important, that the relative weight of the
solar state due to longer operating time is increased.

Comparing the results in Fig. 4.9 with the data in paragraph
4.2.3, one can see that the influence of cycle parameters on the
integration results of solar energy is of the same order of
magnitude as the effect of design modifications of the process
plant. As a consequence one should have in mind that there is a
coupling between the design (size, operation) of the mirror field
and that of the process units. Furthermore, this leads directly to
the problem to manage the transition states (for example between
day and night) of solar process plants.

5. CFB_TO_MEOH-solar Process on Basis Lignite

The handling of lignite cases follows the procedure outlined for
biomass and presentatioon can therefore be shortened. A new element
of solar process synthesis is introduced: the shift gas buffer. It
is the primary key parameter, which is varied for the lignite
cases; night throughput of the synthesis sequence is varied as a
secondary key parameter to evaluate design. Unlike the biomass
cases, variation of throughput in both directions is meaningful
for a sequence with shift gas buffer. The block flowsheet for the
lignite cases is shown in Fig. 5.2.

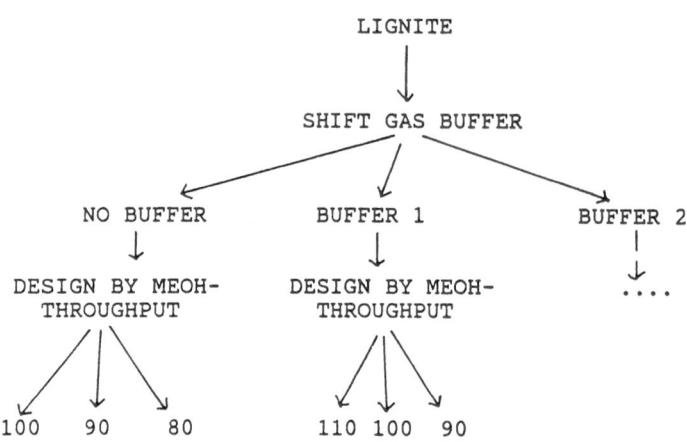

Fig. 5.1: Hierarchy of key parameters for lignite sequence

5.1 The Material Lignite

Lignite is a low-ranking coal, available world-wide and well
suitable for solar gasification. A North Dakota (USA) lignite was
chosen as a reference material ; results of conventional
gasification of this material in the CFB are available from tests
in the LURGI pilot plant. The analytical data of the lignite are
listed in Tab. 5.1 and 5.2; the (gross) calorific value of lignite
is 2.492E08J/kg on a ash and moisture-free basis.

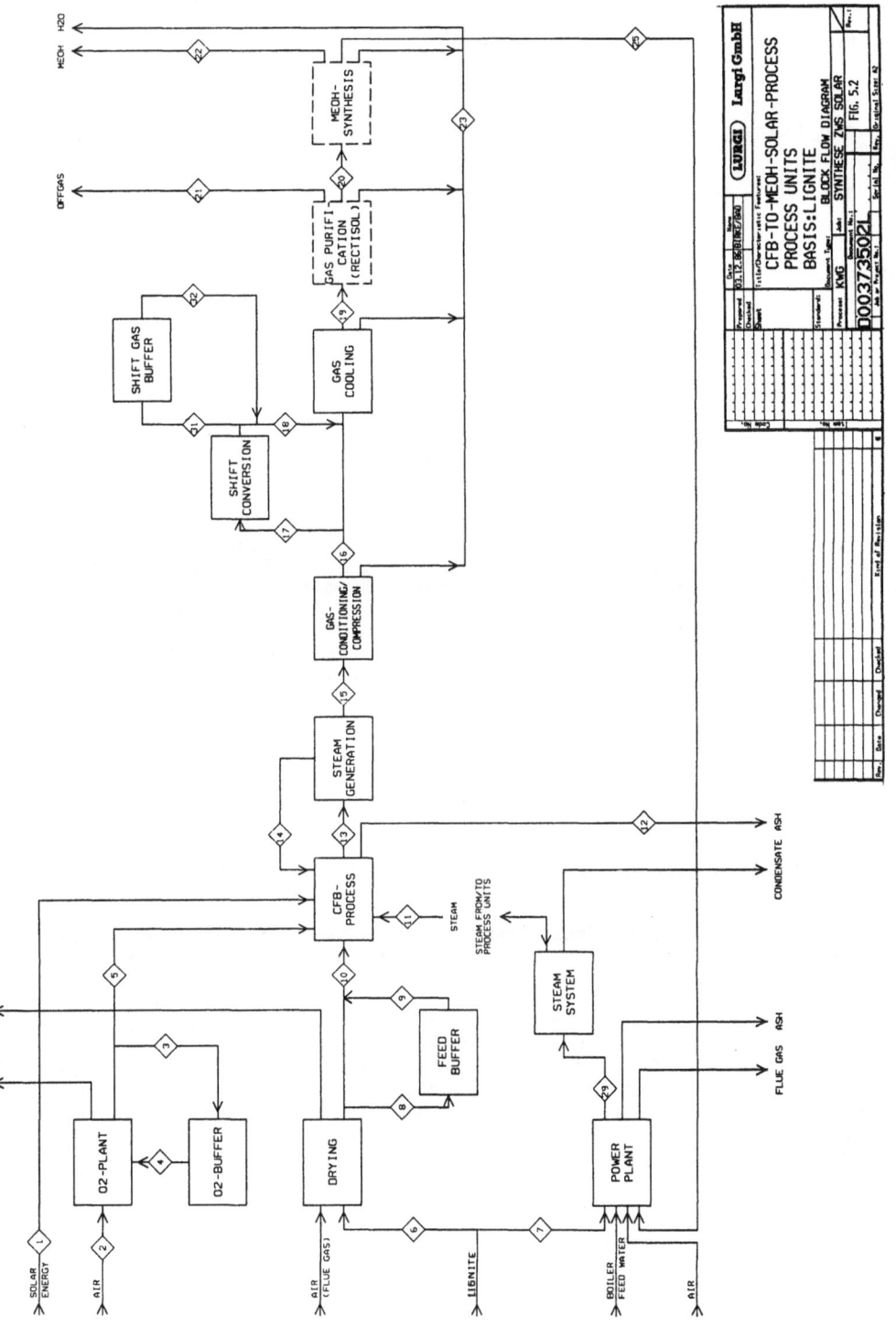

Moisture (%)	40.0
Vol Matter (%)	40.0
C-fix (%)	49.0
Ash (%)	11.0

Tab. 5.1: Proximate analysis of lignite (%)

Ash	11.0
C	63.9
H	4.2
N	1.0
S	0.9
O	19.0

Tab. 5.2: Ultimate analysis of lignite (%)

Compared to biomass, lignite is less reactive, but a gasification temperature of 800-850 °C is favourable as well. As can be seen from the analytical data (lower volatile matter and oxygen content), the solar energy integration potential is higher. Other specifics are: 11% ash content is considerably higher than for biomass. Further, the S-content of lignite is greater by an order of magnitude, resulting in a correspondingly higher H2S-content in the raw gas.

5.2 A solar Process with Shift Gas Buffer

5.2.1 Definition of Cases

The primary key, the shift gas buffer size - is varied on three levels. In Tab. 5.3 the ratio of shift gas going to either synthesis or to buffer is used to define the cases with increasing shift gas buffer size. Case 2 is the solar reference case without buffer. Because buffer gas acts as an additional source of synthesis gas at night, methanol throughput at night is varied from 90 -110% of its day-value to analyze optimal design of the plant. The labeling of cases A-D follows increasing specific solar energy integration. See Tab. 5.3. for a listing of the balance cases. Case 1. is the case of conventional plant without solar energy integration.

CASE	CFB REACTOR KMOL/H		CFB BOILER MW		COMPR. KMOL/H		SHIFT KMOL/H		GAS-COOLING/PURIFIC. GAS KMOL/H		CO2 KMOL/H		MEOH KMOL/H	
	DAY	NIGHT	DAY	NIGHT	DAY	NIGHT	DAY	NIGHT	DAY	NIGHT	DAY	NIGHT	DAY	NIGHT
1.	2582.6	2582.6	15.2	15.2	2173.3	2173.3	229.3	229.3	2410.6	2410.6	703.8	703.8	508.7	508.7
2.B	2190.8	2582.6	12.1	15.2	1896.5	2173.3	77.0	229.3	1978.6	2410.6	306.6	703.8	508.7	508.7
2.C	2190.8	2324.3	12.1	13.7	1896.5	1956.0	77.0	206.4	1978.6	2169.5	306.6	633.4	508.7	457.8
2.D	2190.8	2066.1	12.1	12.2	1896.5	1738.6	77.0	183.4	1978.6	1928.5	306.6	563.5	508.7	406.9
3.A	2190.8	2377.4	12.1	14.0	1896.5	2000.6	97.9	199.3	1762.6	2329.2	273.1	664.2	452.9	498.2
3.B	2190.8	2153.2	12.1	12.7	1896.5	1812.2	97.9	180.5	1762.6	2089.7	273.1	604.2	452.9	452.9
3.C	2190.8	1929.2	12.1	11.4	1896.5	1643.0	97.9	161.7	1762.6	1912.7	273.1	544.0	452.9	407.5
4.A	2190.8	1953.4	12.1	11.5	1896.5	1643.8	119.7	153.5	1538.3	2008.8	238.4	559.8	395.3	434.9
4.B	2190.8	1763.3	12.1	10.4	1896.5	1484.0	119.7	138.6	1538.3	1832.0	238.4	509.4	395.3	395.3
4.C	2190.8	1573.6	12.1	9.3	1896.5	1339.2	119.7	123.7	1538.3	1654.9	238.4	459.4	395.3	355.8

Гаb. 5.4: CFB TO MEOH solar Process on basis Lignite:
Design Data based on characteristic
design parameters

CASE	SHIFT GAS SPLIT		MEOH-THROUGHPUT (%)	
	SYNTHESIS	BUFFER	DAY	NIGHT
1.			100.	100.
2.B	1.00	0.00	100.	100.
2.C	↓	↓	↓	90.
2.D				80.
3.A	0.70	0.30	100.	110.
3.B	↓	↓	↓	100.
3.C				90.
4.A	0.50	0.50	100.	110.
4.B	↓	↓	↓	100.
4.C				90.

Tab. 5.3: CFB_TO_MEOH- Solar process
Lignite balance cases
Cycle: 7/17
Solar energy input:
CFB-Gasification: 39.6 MW
Power: 0.4 MW

5.2.2 Design of Plant

Design data of the lignite cases based on characteristic design
parameters are listed in Tab. 5.4. From each group. the case with
the best compromise of load characteristics and solar energy
integration is selected. i.e. as much as solar energy integration
is possible without affecting the load characteristics. These are
the cases 2.D, 3.C and 4.A and their design diagrams in Fig. 5.3
will be explained further.

The reference case 2.D looks just like the corresponding diagram
for biomass: by reducing night-throughput of the methanol plant it
is possible to adjust the load of the CFB/COMPR- section to nearly
100%. Although the loads of the shift conversion and CO2-handling
units can be increased, they cannot nearly reach design values.
Now consider the favourable effects of a shift gas buffer; there
are two:

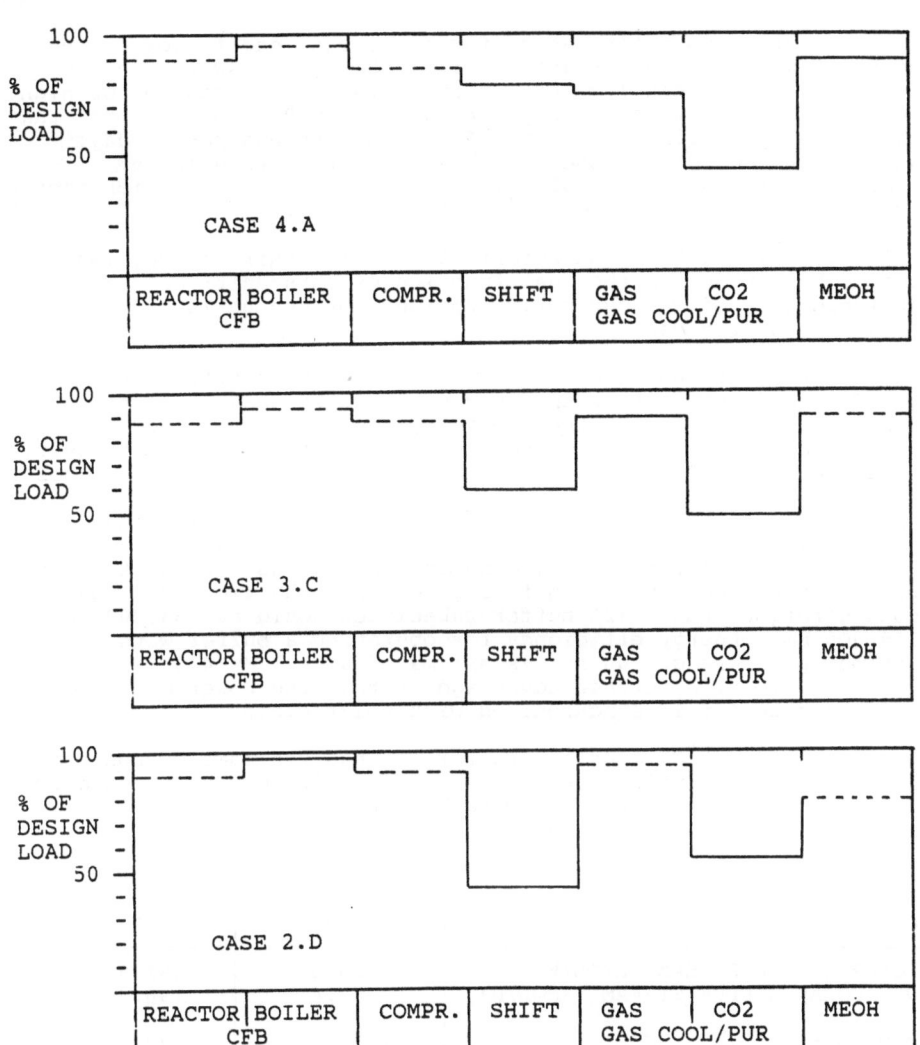

Fig. 5.3: CFB_TO_MEOH-solar process on basis lignite
Design diagrams of optimum cases
———— : Design based on night operation
----- : Design based on day operation

- With increasing shift gas buffer size, the load factor of the methanol plant increases and finally, there is a switch in the design of the methanol plant. In case 4.A it is designed for night operation and operated at day with 90.9 % (=100/110; remember that the cases are defined in terms of methanol-throughput at day, whereas the design is based on that state requiring larger equipment). This is a result of the shift gas buffer, which stores synthesis gas for the night-mode of the plant.

- Similarly, the load characteristics of the shift conversion become better due to the dual function of this unit: adjust stoichiometry of gas currently provided for synthesis and producing buffer gas.

For the rest of the plant, design differs not much from case 2. The CO_2-handling unit is the only unit, whose load characteristics cannot be improved with the applied production scheme. Compared to the biomass cases, the lignite cases with buffers have much better load characteristics, resulting in lower specific investment cost.

Shift gas buffering is therefore a means for optimizing the design of the plant. Managing the buffer requires an extra unit to handle the buffer gas streams, including at least a gas cooling train and pressurizing/depressurizing facilities. For the two cases under consideration, the shift buffer capacities would be 53,000 and 108,000 Nm^3. Energy efficiency for operating a buffer with a maximum pressure of 100 bar is estimated at 0.3 MW/4.3 MW, that is 7% energy loss. A further advantage is that the power demand of a shift gas buffer is synchronized with solar cycles.

For the calculated cases the buffer gas has the same composition as shift gas including 23-24% CO_2. Buffer size may be reduced by removing CO_2 by a simple Hot-Potash-washing process sequenced upstream the pressurizing unit.

Case	3.C	4.B
SOLAR_BUFFER_STREAM (KMOL/H)	236.5	481.8
NON-SOLAR_BUFFER_STREAM (KMOL/H)	97.4	198.4
MIN. BUFFER*SIZE (KMOL)	1655.5	3372.6
BUFFER SIZE (KMOL)	2365.0	4818.0
NET POWER DEMAND (MW)	0.3	0.65
BUFFERED SOLAR ENERGY (MW)	4.3	8.9

Tab. 5.4: Shift gas buffer data for lignite cases
Definition of buffer variables: chapter 3.3
*buffer size calculated with 30% reserve

5.3 Results

For a discussion of results it is important to see, that lignite cases do not reflect efforts to increase solar energy integration, but to optimize design. So the cases with shift gas buffers have much better design, but only moderately better methanol yield. For conventional lignite gasification, specific methanol yield is 0.566; for solar gasification it ranges from 0.638-0.671 for all cases in Tab. 5.5, from 0.652-0.663 for the cases with best design. For the design cases, there is an increase of 15.2-17.1%. For the lignite cases the energy balance deficit was large, during day and night. Since no source of solar energy for power generation was available, the balance had to be closed by combustion of lignite. Therefore, the data do not represent solar energy integration for lignite at its best; for an estimation see chapter 6.

| MEOH-THROUGHPUT | CASE | | | |
(%)	1	2	3	4
A 110			0.650	0.661
B 100	0.566	0.638	0.657	0.664
C 90		0.644	0.663	0.671
D 80		0.652		

Tab. 5.5: CFB_TO_MEOH-solar process on basis lignite
　　　　　Specific methanol yield

Another important feature arises from lignite balance data in Tab. 5.3: following the design cases considered optimal 2.D→3.C→4.A, the load characteristics improve. The corresponding values for specific methanol yield increase from 0.652 to 0.663 and then slightly decrease to 0.661. The case with small shift gas buffer has the same solar energy integration as with a larger buffer. A shift gas buffer as a means of process synthesis has therefore a dual function: in the process scheme under consideration the minor function is to increase solar energy integration, the main effect is to optimize design and hence economy of the plant. This shows very clearly, that integration of solar energy and design of the process route are closely interrelated.

6. Conclusions

6.1 Major Results of this Study

For a discussion of results one should come back to the problem description graph defined in the beginning and see how it evolved. A new version is established in Fig. 6.1; some nodes have been deleted, some replaced by others. As a result of the work done, the character of the whole scheme has changed to represent solution paths.

The major objective is, that solar energy and a suitable process are integrated; this is accomplished by two paths:

- Directly by insolation of solar energy into the CFB. For reasons discussed earlier, this is the most favourable integration step. As an option, power may be co-produced by the CFB-process.

- But there is another feedback path, arising from process synthesis used for the design of the solar CFB_TO_MEOH process. There are processes, which change the pattern of power demand; for example conventional purge gas reforming which lowers the amount of purge gas as a fuel for power generation. Solar energy on a thermal level suitable for power generation may substitute it; a thermal buffer for solar energy is required. Or there are buffer units which have a demand for power even synchronized with solar energy cycles. An example is a liquid oxygen buffer which shifts the major part of energy consumption of the oxygen plant to daytime.

As a result, a solar modified process plant has a large power demand, partially induced by integrating solar energy on a high-temperature level. Power generation by combustion of the raw material counteracts the objective of solar energy integration. Power has to be be generated by solar energy (or by another regenerable energy form). One option, co-production of power in the CFB, is examined in this study; the major advantage is, that it can be used for optimizing the CFB-process as a part of the synthesis route. But very large power supply or thermal buffering of solar energy with this option is not preferable.

As a consequence, a dual solar plant with direct insolation and a separate process for power generation, possibly based on another solar technology is recommended. Depending on the process scheme, the power generation plant may require a thermal buffer. The integration potential for solar energy on the basis of different scenarios was estimated for both raw materials, see Tab. 6.1. Biomass case with the largest amount of power co-produced nearly closes the energy balance. But only an exact evaluation of the design and economics will show, if this is a process with good

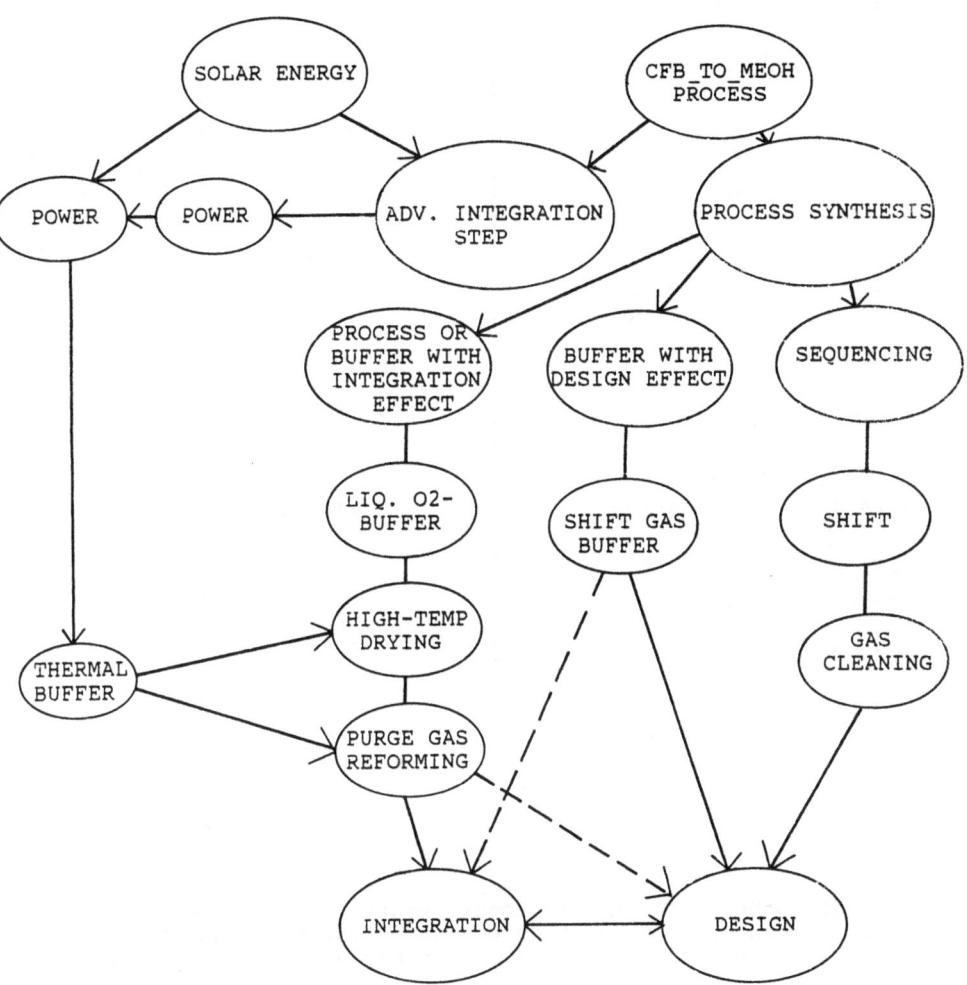

Fig. 6.1: CFB_TO_MEOH-solar process
Process synthesis graph

performance. Note the higher integration potential for lignite, which is not obvious from the balance data because of countercurrent effects.

	Biomass	Lignite
Dual solar plant plus shift gas buffering	≥ 0.50	≥ 0.91
Dual solar plant on basis direct insolation plus power generation		
with thermal buffer	0.45	0.88
without buffer	0.45	0.74
Co-produced power on basis direct insolation	0.43	
Reference case on basis direct insolation	0.40	0.65
Conventional gasification	0.35	0.56

Tab. 6.1: CFB_TO_MEOH-solar process
Solar energy integration potential for different scenarios, expressed by specific methanol yield from raw material (kg/kg maf)

There are other elements of process synthesis, which do not or not primarily generate integration potential for solar energy. Instead they influence the design of some process units. This design influence is a little more subtle, since it cannot be quantified in the same easy way as an integration potential. But when evaluating the economics of a process or - as has been shown here - when comparing routes solely based on a global design, it becomes evident and marks a well-defined limit to integration potential.

In terms of process synthesis, a fuel-upgrading process is represented by a solution path through the state space defined by the raw materials, products and the involved process steps. Solar cycles multiply the dimension of the state space. So new options are opened, but process synthesis becomes more difficult. In this study modeling efforts for solar process plants and an approach for formal handling have been provided. They showed to be essential tools for evaluating solar process routes.

6.2 Research required

In many aspects, the CFB_TO_MEOH-solar process is a good example for an advanced solar process; the basic concept - advanced integration step plus process synthesis - holds for a lot of processes. Insofar one may draw more general conclusions about perspectives of high-temperature solar energy application considering the results of this study. Of course, the basic concept will determine the key fields of research.

The integration step may be very generally defined as a system consisting of three elements: concentrated solar energy, the fluidized bed as a reaction system and the interface. An evaluation of chances and risks of direct insolation becomes urgent. This requires, first of all, a better insight into the design of the interface and reaction system, and into the elemental physical and chemical processes in the reaction system. Hand-shaking between experimental and theoretical work is of major importance. Contributions of theoretical work may be: modeling of direct insolation for example by a design model or by a sophisticated model of the reaction system based on a dynamic simulation of elemental processes.

In this study it is assumed, that direct insolation can be realized. If this assumption failed, this would be of no concern for the results regarding process synthesis. Any "free" reaction system like gasification, which integrates solar energy, generates the same or similar problems. Solar process synthesis - if successful - leads to a plant which looks like a conventional plant regarding its load characteristics; but of course it is not, because it integrates several states and their transitions. It is a plant consisting of strongly connected and coupled processes, and at the moment a high-cost plant too. Most of its units require sophisticated engineering: operation requires advanced control facilities and skilled personnel.

There is no need to be pessimistic about the solution of the aforementioned problems, because modern means of information processing and plant control - available in time - will solve most of them. But the special problem of managing the solar cycles will only be solved by solar research, at first theoretical work.

Simulation of solar process plants is required, and some advanced
tools for doing so have to be provided. Research should be divided:

-   in breadth: a larger number of processes, elements of process
    synthesis and couplings between them should be examined, in
    order to get a better insight into solar processes and
    routes.

-   in depth: refined plant modeling is required, which takes
    into account all relevant solar cycles. Of course, dynamic
    plant simulation can be used; but preferable would be a
    modeling approach beyond that level, because of the large
    size of the simulation problem and the early stage of
    investigation. A more qualitative modeling approach, which
    exploits the special nature of solar cycles and answers most
    design problems, would be the best at the moment. Perhaps a
    refined Petri-Net model or a cause-and-effect analysis (7)
    would do the job.

7. Literature

(1)   G. Birke, R. Reimert, Integrating high-temperature solar
      energy with fuel upgrading processes
      Proc. of the 3. Int. Workshop on solar-thermal central
      receiver systems (Vol. 2), Springer Verlag, 1986

(2)   J. Niellson, Principles of Artificial Intelligence
      (Springer Verlag)

(3)   For a review on process synthesis see for example:
      T. Umeda, "Computer aided process synthesis", Comp. and Chem.
      Eng., Vol. 7, Nr. 4, p. 279 (1983)
      V. Hlavacek, "Synthesis in the design of chemical processes",
      Comp. and Chem. Eng., Vol. 2, p. 67 (1978)

(4)   W. Reisig, "Petri-Netze - Eine Einführung" (Springer Verlag)

(5)   "Vergleichende Untersuchung und Bewertung verschiedener
      solarer Methanreformierungssysteme mit Folgeanlage" (Lurgi
      GmbH, MAN-Technologie GmbH)

(6)   "Ullmanns Enzyklopädie der technischen Chemie" (4. Ed.,
      Verlag Chemie)
      Vol. 20, "Oxygen and Ozone"
      Vol. 3, "Cryogenics"

(7)   T. Umeda, T. Kuriyama, E. Oshima, H. Matsuyama,
      Chem. Eng. Sci. Vol 35 (1980), pp. 2379-2388

DORNIER

Dornier-System GmbH - Friedrichshafen

UTILIZATION OF SOLAR ENERGY FOR
HYDROGEN PRODUCTION BY HIGH
TEMPERATURE ELECTROLYSIS OF STEAM

E. ERDLE
J. GROSS
V. MEYRINGER

DORNIER, FRIEDRICHSHAFEN

THIS STUDY WAS SUPPORTED
5o% BY THE EUROPEAN COMMUNITY

Contents

1. INTRODUCTION

Technologies for the utilization of solar energy are commonly envisaged to be of great importance for the world's energy supply in future. For that reason a variety of R&D programs have been and still are performed by many countries.

Two principally different concepts are pursued within these programs: the direct conversion of sunlight to electricity by means of photovoltaic cells and the use of solar energy by solar thermal plants. With the latter option not only electricity generation is possible but also the direct supply of high temperature heat for chemical processes.

Steam reforming of natural gas is a prominent example for such a process which could be demonstrated within the next few years.

Hydrogen could serve as a solar chemical or fuel, respectively, which will become important on a long term due to its specific advantages:

- Hydrogen is a clean and environmentally benign fuel: the product of its combustion is water.

- Hydrogen can be transported and stored easily.

- Hydrogen can be produced electrolytically the only raw material needed being water.

With the high temperature electrolysis of water vapor (Hot Elly) there is a new electrolysis technology under development, which has the potential for a substantial reduction of energy consumption as compared to state of the art conventional electrolyzers. Additionally, it is possible in this process to substitute electrical energy by high temperature heat in this process to a certain extent.

For the reasons quoted above it is worthwhile to investigate the possibilities for the coupling of solar thermal plants and high temperature electrolyzers. That is done within the present study for two different types of solar thermal plants, namely a distributed collector system (DCS) and a central receiver system (CRS) with operating temperatures of up to 300°C and 1100°C, respectively.

Within this study the design of a coupled system of an autothermal Hot Elly and a DCS-plant was performed (see chapter 3). The solar part of the system was taken to have essentially the same design data as the existing DCS-plant at Almeria (Spain), which was erected under the auspices of the IEA with the DFVLR acting as an operating agent.

For the investigation of an allothermal Hot Elly using high temperature heat and a CRS-plant no existing solar plant could be chosen as a basis, since they do not provide high enough process temperatures. As will be derived in chapter 4 a GAST-type plant represents a suitable solar system for that purpose.

Electrolysis of water vapor at high temperatures promises to be an attractive process as compared to conventional electrolysis of liquid water for a variety of reasons as can be derived from the thermodynamics of the water splitting process shown in Fig. 2.1.

- The total energy demand ΔH for water splitting is lower in the vapor phase than in the liquid phase. The energy for vaporization can be provided thermally instead of electrically.

- The minimum demand for electrical energy ΔG needed for electrolysis decreases with increasing temperature, i.e. one has the principal possibility to provide part of the splitting energy by thermal energy instead of electrical one thus achieving a higher total efficiency.

- The improved reaction kinetics at elevated temperatures are lowering overvoltages.

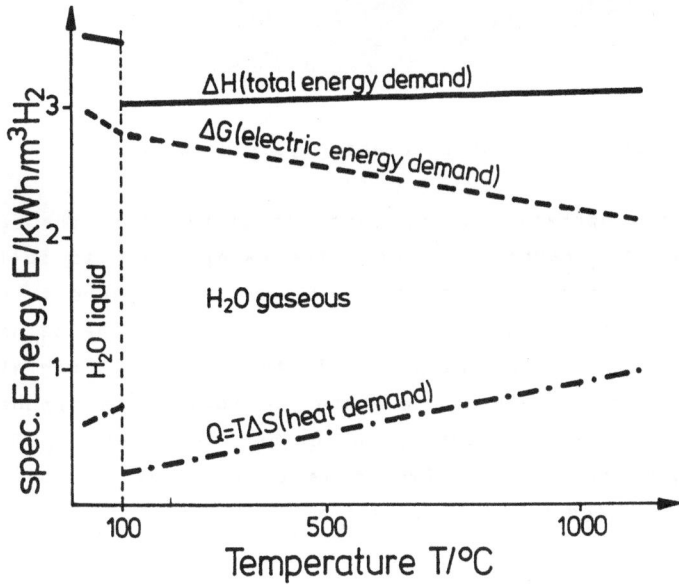

Fig. 2.1:   Thermodynamics of water splitting

The principal realization of the water splitting process at
high temperatures is illustrated by Fig. 2.2 which shows a
cross section of a real electrolysis membrane. It consists of a
gastight solid oxide electrolyte membrane which is coated with
porous electrode layers on both sides. The solid oxide elec-
crolyte is made of stabilized Zirconia, a material which exhi-
bits an ionic electrical conductivity with oxygen iones being
the charge carriers exclusively.

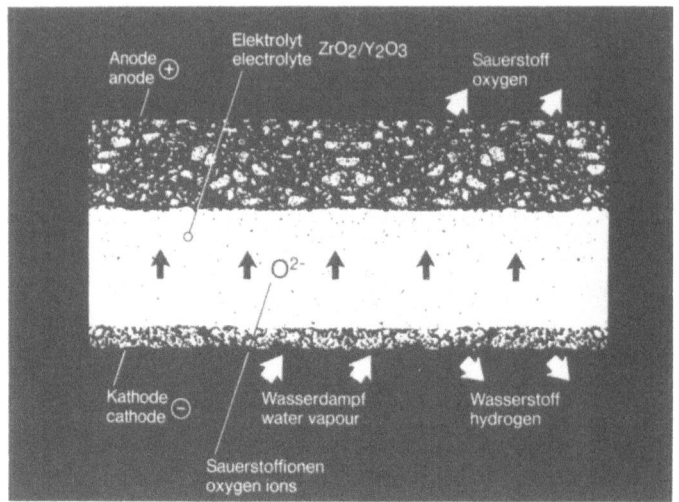

Fig. 2.2:   Cross-section of an electrolysis membrane

For electrolysis operation water vapor is supplied to the ca-
thodic side of the membrane. There, the water vapor is split
under the influence of an external voltage applied across the
membrane. The hydrogen molecules produced remain at the catho-
dic side while oxygen, following the electric field, is migrat-
ing in ionic form through the electrolyte towards the anode.
There, the ions are discharged and oxygen is released. The
products oxygen and hydrogen (which is enriched in the water
vapor) keep separated by the gastight electrolyte membrane.

The concept for a technical realization of this principal scheme which is pursued within the project Hot Elly is based on electrolytic elements which are mechanically stable by themselves. (Another possiblity would be the coating of a mechanically stable substrate with thin electrode and electrolyte layers.) Therefore the wall thickness of the electrolyte membrane cannot be reduced below about 0.3 mm.

Since the specific electric resistivity of the electrolyte material decreases with increasing temperature according to the curve shown in Fig. 2.3, an operating temperature of about 1000°C has to be foreseen in order to achieve low electric losses in the electrolyte.

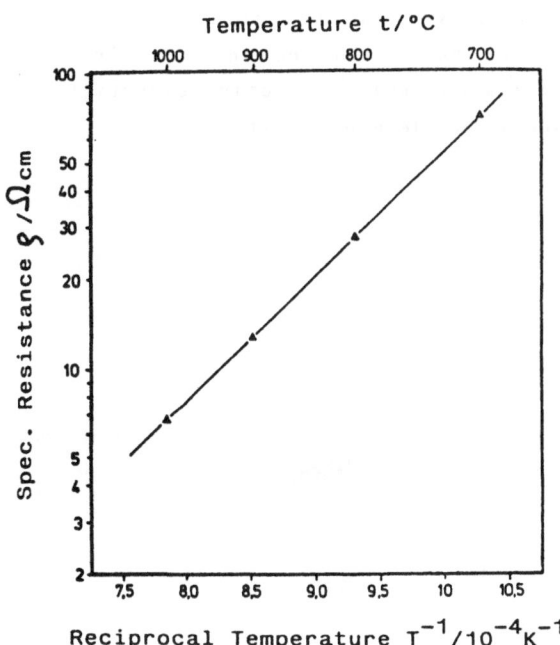

Fig. 2.3:   Specific electric resistivity of the electrolyte material (Z_rO_2 + 8 m/o Y_2O_3) as a function of temperature

With respect to mechanical stability as well as transport and separation of the gases involved a tubular shape of the electrolyte elements is the most favorable geometry. Due to electrical losses in the electrodes there is a rather low limit (~ 1 cm) for the length of the electrolysis cells, and for fabrication reasons their diameter is also restricted (Ø ~ 1.5 cm). Therefore, for a practical electrolyzer many of such single cells have to be integrated to larger aggregates. In a first step out of several single cells so-called electrolysis tubes are built with an integrated electrical series connection of the cells thus achieving high voltages at low currents already for these elements. This is done in the following way (see Fig. 2.4): Between two electrolyte cylinders an electrically conductive ring has to be placed which connects the cathode of one cell with the anode of the next one. For electrochemical reasons this conductive element has to be insulated with respect to the electrolyte.

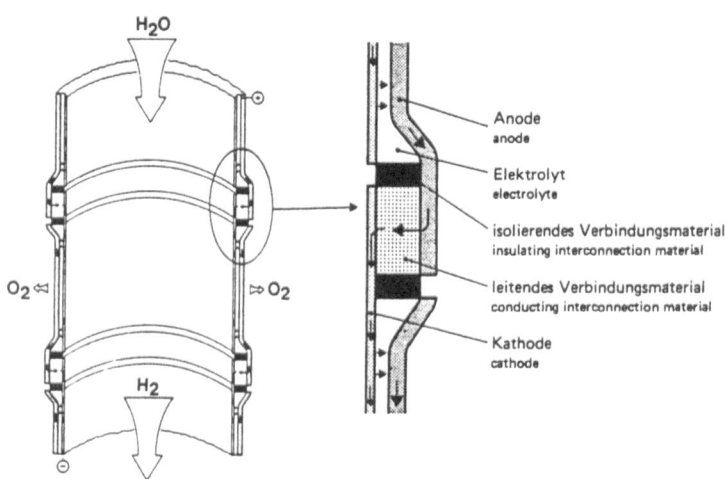

Fig. 2.4:   Setup of an electrolysis tube

As an example Fig. 2.5 shows an electrolysis tube with 20 cells as it was fabricated in the laboratory.

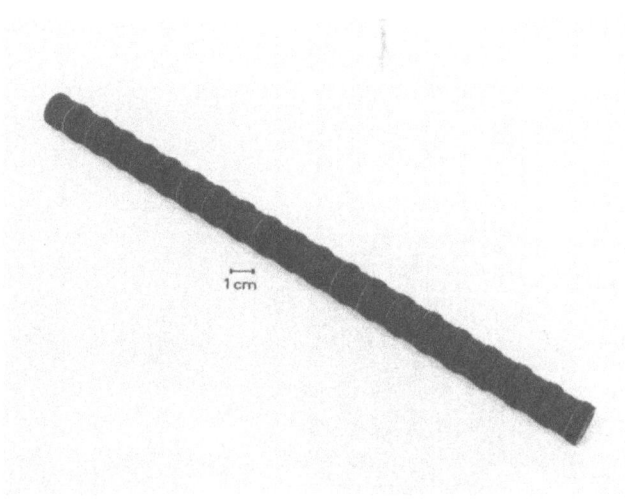

Fig. 2.5:   Electrolysis tube with 20 cells

In a second step many electrolysis tubes have to be assembled
within a so-called electrolysis module as is shown in Fig. 2.6.
It consists of so-called registers, each of them containing
several electrolysis tubes. Since the electrolysis tubes of
every register are to be connected electrically in series, they
have to be insulated from each other. Therefore, the register
body is made of Alumina. Many registers are combined to a module by two common metallic gas ducts, one for feed steam and the
other one for the outlet of the hydrogen produced inside the
electrolysis tubes. The steam enters the register through the
upper bore of the register body, wherefrom it passes through
the electrolysis tubes. There, the water vapor is widely
(~ 75 %) converted to hydrogen, which leaves through the central capillaries in the tubes and the lower channel of the
register body.

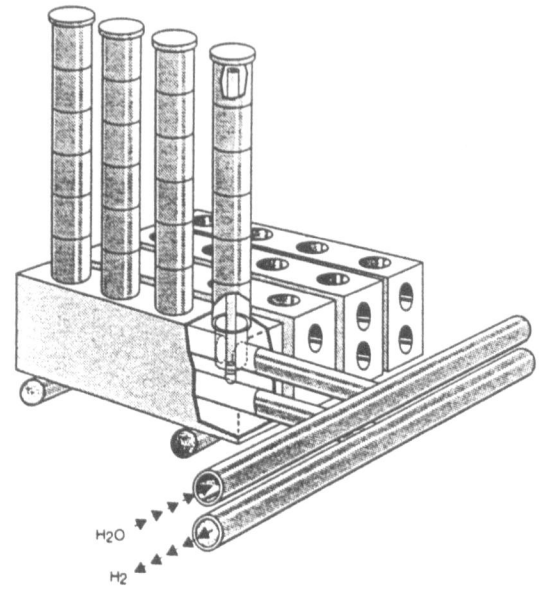

H₂O

H₂

Fig. 2.6:   Module-concept

As an example of cell testing Fig. 2.7 shows the current voltage characteristics of a single cell and for comparison one of a
commercially available conventional electrolysis. The potential
for an improvement with respect to energy consumption is clearly demonstrated by the result (note: the cell voltage is directly proportional to the specific electric energy consumption).

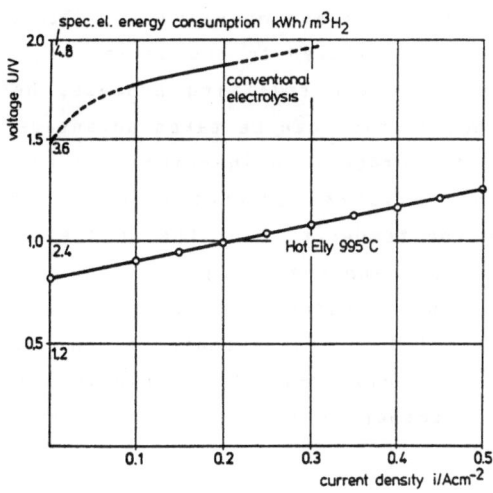

Fig. 2.7:   Current-voltage-characteristic of a high temperature electrolysis cell

With respect to the operation of high temperature electrolysis
two essentially different possibilities have to be distin-
guished:

At a cell voltage of 1.3 V (just corresponding to the splitting
energy demand of 3.1 kWh/Nm³ H_2, see Fig. 2.1) the process is
thermoneutral, i.e. the product gases hydrogen and oxygen leave
the cells at the temperature of the steam supplied to the
cells.

Below that voltage one has the principal possibility of coupling-in high temperature heat thus raising the efficiency of
the process. More details about that so-called allothermal
process version are explained within section 4.1 of this study.

- 633 -

If the electrolysis cells are operated at voltages above 1.3 V the electric energy delivered to the cells is greater than the one consumed by the water splitting process. Hence the cells produce "waste heat" which can be taken advantage of favourably with respect to the process engineering. Due to the exothermic operation of the cells the product gases can be superheated with respect to the temperature of the feed steam entering the electrolyzer. By the superheated product gases feed steam of low temperature can be heated up to a temperature necessary at the electrolyzer inlet in heat exchangers. The surplus voltages to achieve that are only 20-30 mV. A principal illustration of that so-called autothermal process version is given in Fig. 2.8.

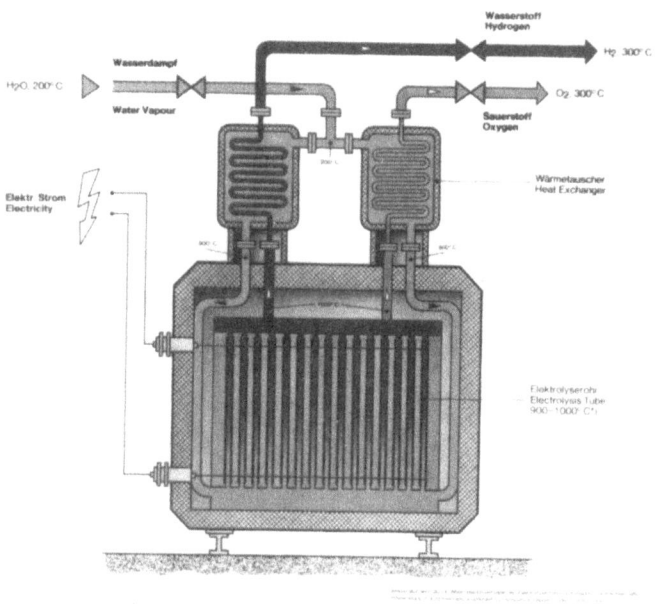

Fig. 2.8:   Autothermal process version

Due to the facts that this process version is only based on the availability of electricity and low temperature steam and that the current density in the cells is greater because of the greater cell voltage (i.e. electrolyzer size is smaller) the development performed so far was focussed on the autothermal and not on the allothermal process. Fig. 2.9 shows a picture of a model of an autothermal electrolyzer with a production capacity of 1000 Nm^3H_2/h which was built according to a detailed layout.

Its operating pressure is designed to be 4 bar. However, concepts for operation of high temperature electrolysis at elevated pressure have also been considered, and the feasibility of pressure operation has been demonstrated in the lab.

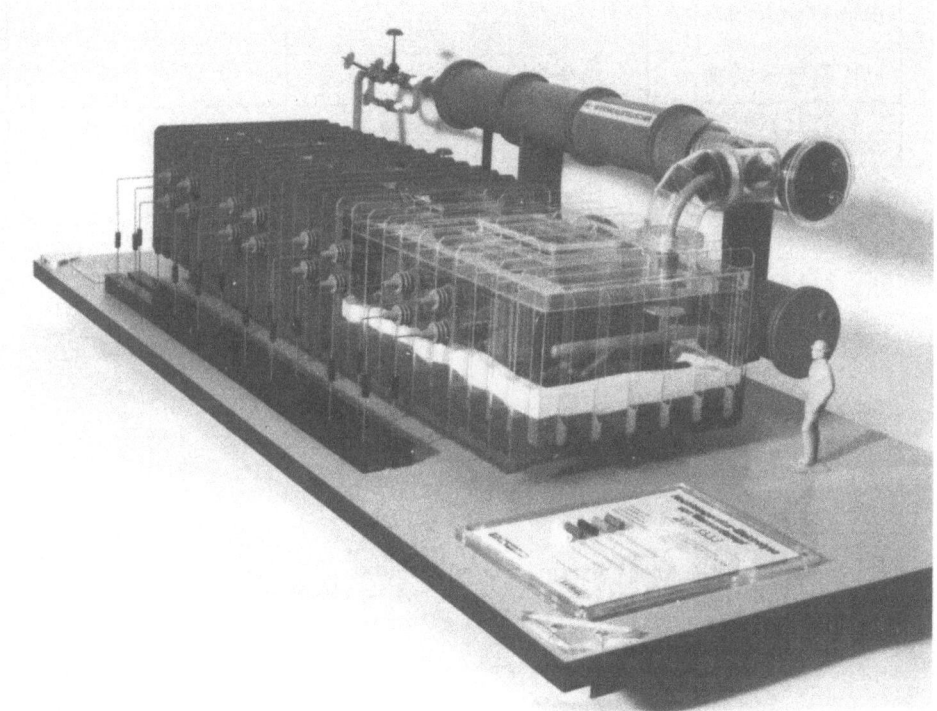

Fig. 2.9:   Model of an autothermal high temperature electrolyzer (1000 Nm^3H_2/h)

For the ease of obtaining a general view the characteristic energetic data of the two process versions of high temperature electrolysis - and for comparison of conventional electrolysis - are summarized at the end of this section in table 2.1.

Additionally, the total efficiency* is given based on the assumption that the generation of electric power from heat occurs at an efficiency of 40 %.

Energy Input ($kWh/Nm^3 H_2$)	High Temp. Electrolysis allothermal	High Temp. Electrolysis autothermal	Conv. Electrolysis
Electrical Energy	2.6 *(6.5)ₕₗ*	3.2 *(8.0)ₕₗ*	≳ 4.6
High Temp. Heat	0.5	-	-
Low Temp. Heat	0.6	0.6	-
Total Efficiency	47%	41%	31%

Tab. 2.1:   Characteristic Energy Data of Electrolysis Processes

---

* By convention the total efficiency is calculated on the
  basis of the higher heating value of H_2 (3.55 $kWh/Nm^3 H_2$)

3.   COUPLING OF A 'LOW TEMPERATURE' SOLAR THERMAL PLANT
AND AN AUTOTHERMAL HIGH TEMPERATURE ELECTROLYZER

Within this chapter the coupling of a 'low temperature' solar thermal plant (solar farm principle) and an autothermal high temperature electrolyzer is investigated in an exemplary manner. For that purpose the following two questions were to be treated:

- What does the coupled system look like?
- What operational data can be expected from the system?

The investigations were based on the existing DCS-plant at Almeria, a simplified flow sheet of which is shown in Fig. 3.1.

Obviously the interface for a coupling with a Hot Elly system has to be placed behind the thermal oil storage tank. Beginning there, a steam generation system, a power conversion system and the electrolyzer system would follow.

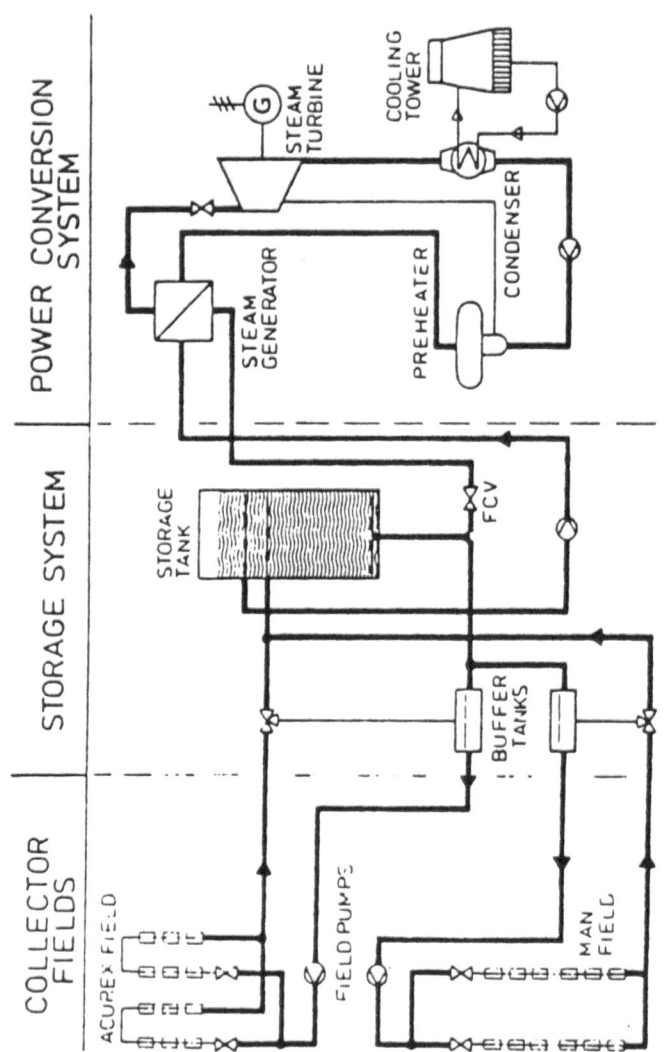

Fig. 3.1:   DCS-type solar thermal plant at Almeria

3.1   Brief description of the DCS-plant

The DCS-plant can be divided up into three subsystems (see Fig. 3.1):

- Accumulation of the insolated energy in the collector fields. There, a thermal oil is heated up and then transported to the

- storage system which decouples the varying insulation and the

- power conversion system (PCS) which is supplied with thermal oil out of the storage tank (net electrical design power 500 kW).

For the coupling with a high temperature electrolyzer mainly the features of the operational procedures with respect to the storage system are important. Therefore, they will be briefly reviewed here:

After sunrise the collector fields are firstly operated via their buffer tanks for warm-up until the circulating thermal oil has a temperature of 295°C. Then the three-way valves are opened in such a way that hot oil (295°C) is flowing into the upper part of the storage tank while simultaneously cold oil of a temperature of 225°C is pumped out of the lower part of the storage tank to the collector fields via their buffer tanks.

The power conversion system is operated at a constant flow rate of hot oil out of the storage. The oil is used thermally for the generation of steam in the course of which it cools down to 225°C; then it is pumped back to the storage tank.

The time of day for starting the power conversion system is defined by the following two requirements:

- The storage tank has to be empty of hot oil at the time the collector field stops operation in the evening.

- The mass of oil pumped into the storage tank from the collector field equals the one pumped out of the tank.

It can be deduced easily that these two requirements are equivalent to the request for minimal size of the storage tank.

An example for that operational strategy is given in Fig. 3.2 which shows the flowrates of the thermal oil from the collector field into the storage tank and out of the storage tank to the power conversion system for an equinox day.

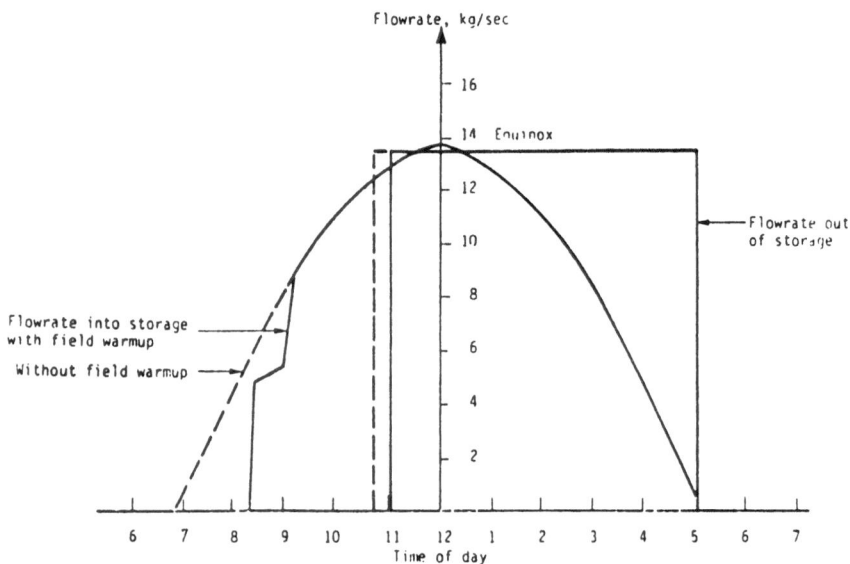

Fig. 3.2:   Flow rates of thermal oil into and out of the storage tank (equinox)

It should be noted that we are dealing with a short time storage system as can be seen from a mass of 83.1 t of hot oil when the storage tank is full and a design flow rate of about 13.5 kg/sec.

Assuming cloudless days all over the year about 2150 annual operating hours result from the layout of the DCS-plant. Their distribution over the year is shown in fig. 3.3. However, corresponding to experiences during operation of the plant a value of 1500 - 1800 h is a realistic one.

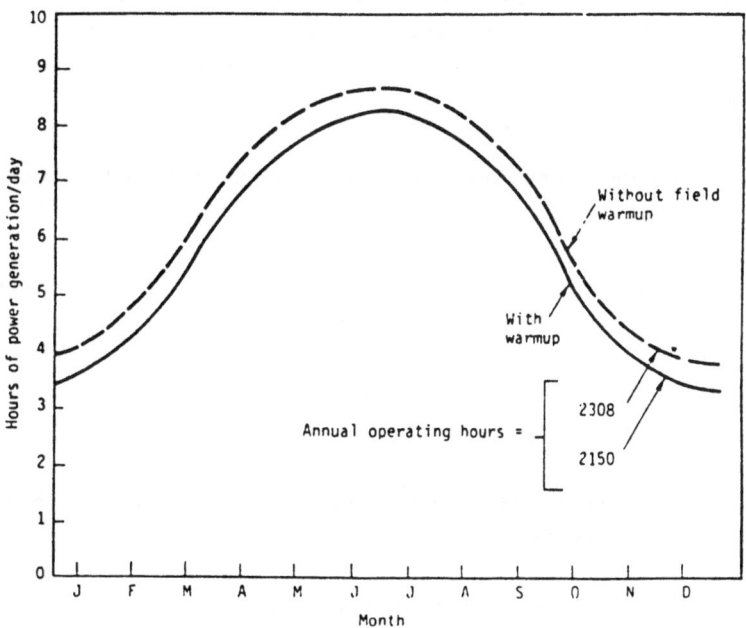

Fig. 3.3:   Daily operating hours of the DCS-plant during a year

3.2 Adjusted layout of an autothermal high temperature electrolyzer

3.2.1 Calculation of the modular gas and heat balance for one electrolysis tube

Within the project Hot Elly theoretical calculations of the voltage drop and of the temperature distribution of an autothermally operated electrolysis tube have been performed. Especially the thermal interaction processes of heat conduction and radiative heat exchange within the tube and with the surrounding were included in the model. Its results were taken as a basis for the layout of an autothermal Hot Elly to be coupled with the DCS-plant.

The calculations were based on the following boundary conditions:

- The feed steam is converted in every electrolysis tube up to a hydrogen concentration of 75 %

- The electrolysis tube is operated at a current of 1.505 A (= current density 0.35 A/cm²)

- The feed steam has a temperature of 950°C at the inlet of the electrolysis tube.

By Faraday's law a current of 1 A in a single electrolysis cell is causing a production of 7 Ncm³H_2/min and of 3.5 Ncm³O_2/min. Therefore, by the first two conditions noted above the gas flow rates to and from the electrolysis tube are fixed:

Input:

    $2.079 \cdot 10^{-4}$ mole H_2O/sec

Output:

    $1.559 \cdot 10^{-4}$ mole H_2/sec + $0.519 \cdot 10^{-4}$ mole H_2O/sec
    (inside the tube)

and

    $0.779 \cdot 10^{-4}$ mole O_2/sec (outside the tube).

Under these conditions the voltage of an electrolysis tube with 20 cells turns out to be 26.46 V corresponding to a power demand of $\dot{E}_{el}$ = 39.8 W. The resulting temperature distribution is shown in Fig. 3.4. The cell temperatures vary between about 970°C and 1040°C, the mean temperature is 1013°C.

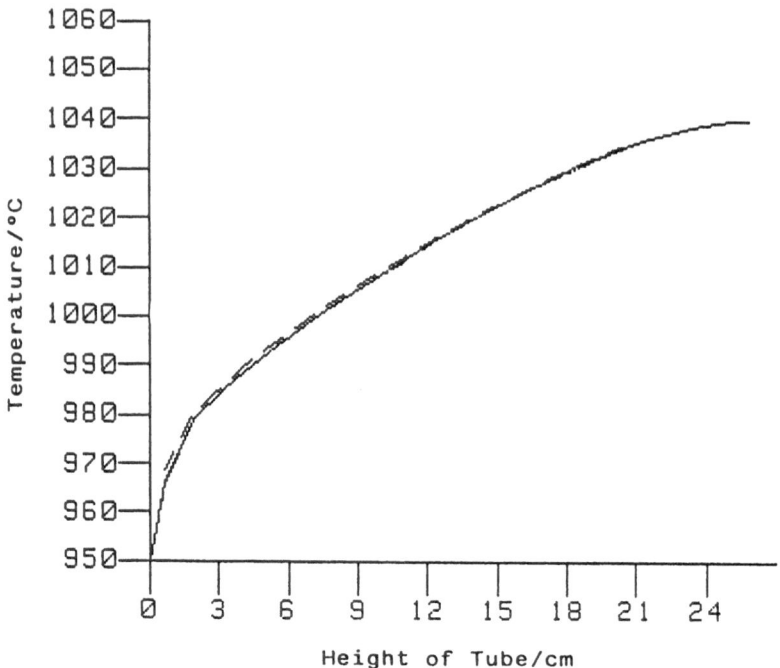

Fig. 3.4:   Temperature distribution along an electrolysis
            tube with 20 cells

Together with the assumption that the feed steam has a temperature of 900°C at the entrance of the electrolyzer the heat
balance of the process can be determined from the results of
the model calculations.

The mean cell voltage necessary to cover the energy demand of
the water splitting reaction at the calculated operating temperatures is 1.297 V.

Thus the electrolysis tube consumes a surplus of energy with respect to the energy demand of electrolysis, i.e. it produces a heat power $\dot{Q}_R$ corresponding to:

$$\dot{Q}_R = (26.46 - 20 \cdot 1.297) \cdot 1.505 \; W = 0.783 \; W \qquad (3.1)$$

By this heat the superheating of feed steam (from 900°C to 950°C) and of the product gases has to be accomplished and the heat losses of the electrolyzer to the surrounding have to be met. The latter amount to about 0.233 W per electrolysis tube (see section 3.3.1) thus remaining a heat power $\dot{Q}_g$ for the superheating of the gases of

$$\dot{Q}_g \approx 0.55 \; W. \qquad (3.2)$$

Superheating of the feed steam within the electrolyzer from 900°C at the entrance to 950°C at the first cell of the electrolysis tube consumes a heat power $\dot{Q}_{steam}$ corresponding to:

$$\dot{Q}_{steam}^{high} = c_{P \; steam} \cdot \Delta T \cdot \dot{m} \qquad (3.3)$$

$$= 43,512 \; \frac{Ws}{mole \; K} \cdot 50 \; K \cdot 2.079 \cdot 10^{-4} \; \frac{mole}{s}$$

$$= 0.452 \; W$$

Thus a heat power $\dot{Q}_{Product}$ of about 0.1 W remains (see equ. 3.2 and 3.3) for the superheating of the product gas stream ($H_2/H_2O + O_2$) with respect to a temperature of 950°C.

For the calculation of the temperature $T_{Product}$, at which the product gases leave the electrolyzer, the following equation is valid:

$$T_{Product} = 950°C + \frac{0.1 \text{ W}}{c_{product}^{high}} \approx 960°C \qquad (3.4)$$

with $C_{product}^{high}$ being determined by the molar specific heat of the product gases and their molar flow rates according to:

$$c_{product}^{high} = (1.559 \ C_{PH_2} + 0.519 \ C_{PH_{2O}} + 0.779 \ C_{PO_2}) \cdot 10^{-4} \ \frac{W}{K}$$

$$= 9.9 \cdot 10^{-3} \ \frac{W}{K} \text{ (at about 1000°C)} \qquad (3.5)$$

The heat content of the product gas streams is used for the superheating of the feed steam from a temperature of 120°C assumed at the outlet of a feed steam generator to 900°C at the entrance of the electrolyzer.

With $C_{PH_{2O}} = 38.67$ WS/mole · K being the mean specific heat capacity of steam in this temperature range superheating of the feed steam needs:

$$\dot{Q}_{H_{2O}}^{low} = 2.079 \cdot 10^{-4} \cdot 38.67 \cdot 780 \text{ W} = 6.27 \text{ W} \qquad (3.6)$$

Assuming heat losses of 5 % in the corresponding heat exchangers about 6.6 W have to be taken from the product gas streams for that purpose.

Using the mean specific heat of the H_2/H_2O and O_2 streams the cooling down of the product streams due to that heat exchange can be determined in an analogous way as is was done for their superheating before.

The resulting product temperature $T_{product}$ after heat exchange with the feed steam is:

$$T_{product} = 247 °C \qquad (3.7)$$

The remaining heat content of the products can partially be used for the preheating of the feed water. However, the heat demand for vaporization of the feed water cannot be met.

Hence, this energy has to be supplied by the solar system. The molar heat of vaporization of water being 40.603 Ws/mole and assuming 1.5 % losses in the evaporator 'steam making' needs a heat power $\dot{Q}_{vap}$ of:

$$\dot{Q}_{vap} = 8.6 \text{ W} \qquad (3.8)$$

The thermal oil used in the solar system has a mean specific heat C_{Poil} of 2.797 WS/g in the temperature range between 225 °C and 295 °C.

Thus a mass flow $\dot{m}_{vap}$ of hot oil is needed for vaporization corresponding to

$$\dot{m}_{vap} = \frac{\dot{Q}_{vap}}{70K \cdot c_{Poil}} = 4.4 \cdot 10^{-2} \, \frac{g}{s} \; . \tag{3.9}$$

As mentioned in section 3.1 13.5 kg/s of hot oil are needed for a power generation of 500 kW . Hence, for the mass flow $\dot{m}_{el}$ for supplying the electric power demand of 39.8 W results, if the same efficiency of power generation as in the DCS-plant is assumed:

$$\dot{m}_{el} = 1.07 \, \frac{g}{s} \tag{3.10}$$

Now we have determined the modular data with respect to one electrolysis tube with 20 cells of a high temperature electrolyzer which are listed in table 3.1.

- feed water (deionized) $3.74 \cdot 10^{-3}$ g/s
- electric power (DC)   39.8 W
- thermal oil ($\Delta T = 70°C$)
    for steam making $4.4 \cdot 10^{-2}$ g/s
    for power generation   1.07 g/s
- O_2 production   1.75 Ncm³/s at 247°C
- H_2 production   3.49 Ncm³/s at 247°C

Tab. 3.1:   Modular data of a high temperature electrolyzer
             with respect to one electrolysis tube with 20
             cells

As mentioned above, preheating of the feed water can be accomplished by the products. In fact, the H_2/H_2O gas stream can be used for that purpose before its water vapor content is condensated in a cooler to get pure hydrogen at a temperature of about 40°C. Use of the oxygen's remaining heat content for partially covering the energy demand of feed water vaporization is not worthwhile, since its further cooling down from 247°C to 120°C could only provide 4 % of it.

3.2.2   Adaptation of a high temperature electrolyzer and
         the power generation system of the DCS-plant

In the foregoing section it was derived that the heat demand
for the generation of the electrolyzer's feed steam is only
about 4 % of the one necessary for the corresponding generation
of electric power (see equ. 3.9 and 3.10). Hence, the easiest
way to incorporate an autothermal Hot Elly into the DCS-plant
is to reduce the power generation by about these 4 % thereby
providing the energy for the vaporization of the feed water.
The size of the storage tank as well as the flowrate out of it
have not to be changed.

Of course, the thermal oil flow has to be split into two
streams for the generation of feed steam and of electric power,
respectively. With the modular data derived in the section
before the electric power $\dot{E}_{el}$ of the electrolyzer can be determined to be:

$\dot{E}_{el}$ = (Power demand of one electrolysis tube)

$$(\frac{\text{total flow rate of thermal oil}}{\dot{m}_{el} + \dot{m}_{vap}})$$

    = 482 kW.   (3.11)

This figure corresponds to a hydrogen production rate $\dot{V}_{H_2}$ of

$\dot{V}_{H_2}$ = 152 Nm³ H₂/h .   (3.12)

These figures hold if the three-phase power generator of the power conversion system is substituted by a DC-generator which should have the same efficiency in the power range under consideration. Otherwise installation of an AC/DC-conversion system would be necessary which would reduce the efficiency of power generation and thus the size of the electrolyzer slightly. For example an efficiency of 95 % of the AC/DC-converter would change the layout of the electrolyzer to a nominal output power of 145 $m^3 H_2/h$.

With the data obtained so far in the sections 3.2.1 and 3.2.2 the principal design of a flow sheet for a combined system DCS-plant/autothermal high temperature electrolyzer can be made. It is given in fig. 3.5. On the left hand side the DCS-system is shown. However, as stated above, the thermal oil cycle behind the storage tank has to be split into two sub-streams by a three-way valve. A small oil stream of 0.53 kg/s corresponding to a thermal power of 105 kW ($\Delta T = 70°C$) is covering the heat demand of the feed steam generator.

The other stream of 12.97 kg/s corresponding to a thermal power of 2.57 MW is feeding the steam generators of the power conversion system generating 480 kW of electric power.

External feed water is initially combined with the condensated steam content of the H_2 product stream. Then it is preheated by the H_2/H_2O stream and vaporized in the feed steam generator. Subsequently the feed steam is split into two streams corresponding to the different heat contents of the H_2/H_2O and O_2 gas streams coming from the electrolyzer. They superheat the feed steam to 900°C in two recuperators before it enters the electrolyzer where ·it is split into hydrogen and oxygen. The product gases leave the electrolyzer at about 960°C and are

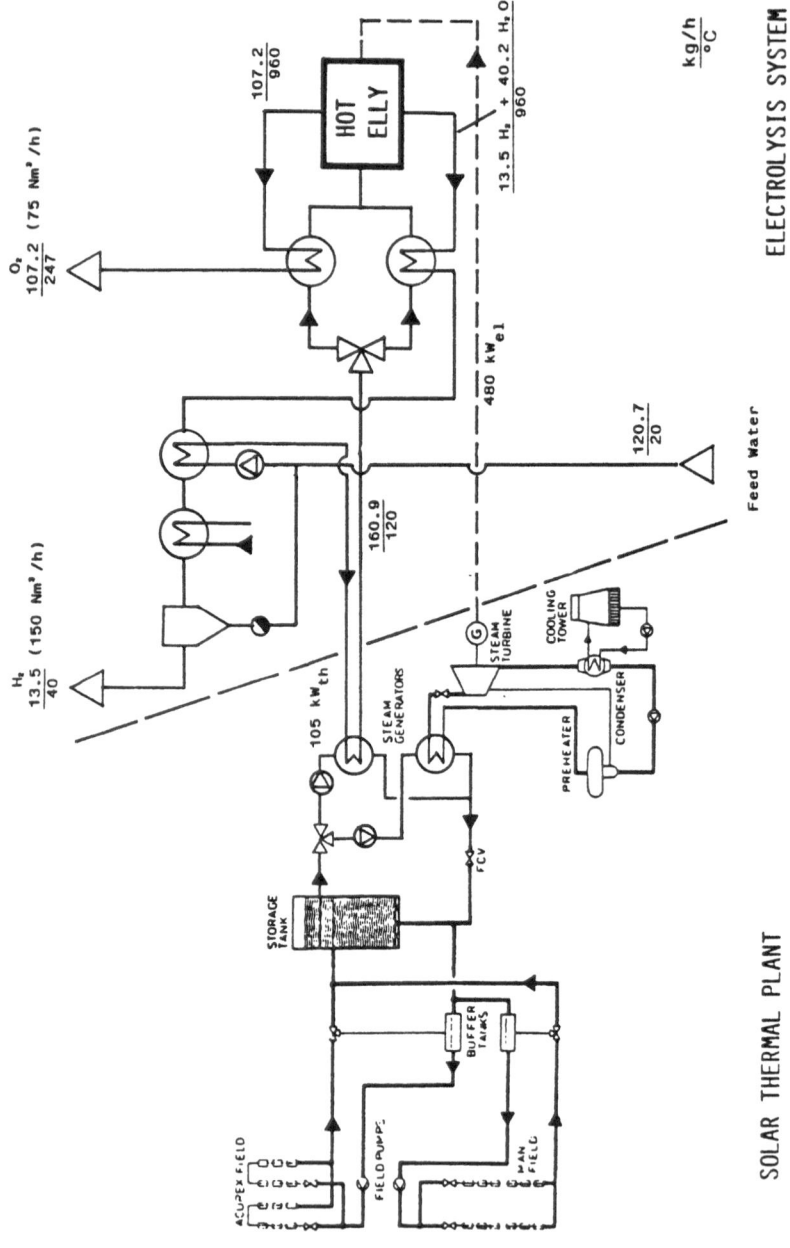

ELECTROLYSIS SYSTEM

$\frac{kg/h}{°C}$

O$_2$ (75 Nm^3/h)

$\frac{107.2}{247}$

$\frac{107.2}{960}$

HOT ELLY

13.5 H$_2$ + 40.2 H$_2$O

$\frac{}{960}$

480 kW$_{el}$

H$_2$ (150 Nm^3/h)

$\frac{13.5}{40}$

$\frac{160.9}{120}$

$\frac{120.7}{20}$

Feed Water

105 kW$_{th}$

STEAM GENERATORS

STEAM TURBINE

G

COOLING TOWER

PREHEATER

CONDENSER

FCV

STORAGE TANK

BUFFER TANKS

ACOUREX FIELD

FIELD PUMPS

MAN FIELD

SOLAR THERMAL PLANT

Fig. 3.5   Flow sheet for a combined system autothermal steam electrolyzer/DCS solar thermal plant

- 652 -

cooled down to about 250°C in the recuperators. The oxygen is vented (or cooled further if it shall be used), while the H_2/H_2O stream is preheating the feed water as mentioned above before its steam content is removed by condensation in a cooler.

The characteristic data of this layout - beginning at the interface of the storage tank - are summarized in the following Table 3.2.

feed water flow	:	120.7 kg/h
thermal oil flow to feed steam generator:		0.53 kg/s
thermal oil flow to power generation	:	12.97 kg/s
electric power generation	:	482 KW
H_2-production	:	150 Nm³/h
O_2-production	:	75 Nm³/h

Tab. 3.2: Characteristic data of the layout of an autothermal high temperature electrolysis system coupled to the DCS-plant

3.3 Operation of the coupled system DCS-plant/auto-
thermal Hot Elly

Within this chapter questions of the operational behaviour of
the system shall be discussed. These are mainly questions of
start-up and restart of the electrolysis system. Finally the
hydrogen yield of the total system will be estimated.

3.3.1 Thermal characteristics of the high temperature
electrolyzer system

Within this section an estimate of the thermal inertia of the
electrolyzer shall be evaluated. Again the design data for an
electrolysis tube given in section 3.2.1 are taken as a basis.
Furthermore it is assumed that 10 electrolysis tubes are integrated to a register on a common ceramic support structure.

For an electric power consumption of 1 kW_{el} about 2.5 registers
are necessary each having a mass of approximately 1 kg. The
specific heat of the ceramic materials used being 0.84 Ws/gK
the specific heat of the electrolysis modules per kW_{el} C_{kW} is
derived to be

$$C_{kW} \approx 2.1 \frac{kWs}{K \cdot kW_{el}} \qquad (3.13)$$

- 654 -

Hence the specific thermal energy Q_{kW} per kW_{el} for heating up the electrolysis modules from ambient temperature to their operating temperature ($\Delta T = 980°K$) is:

$$Q_{kW} = \Delta T \cdot C_{kW} = 2058 \text{ kWs/kW}_{er} = 0.57 \text{ kWh/kW}_{el} \qquad (3.13)$$

Therefrom one derives the heat content Q_{mod} of the electrolysis modules of the electrolyzer under consideration to be:

$$Q_{mod} \approx 275 \text{ kWh} \qquad (3.14)$$

That is one part contributing to the thermal inertia of the electrolyzer. Another one stems from the thermal insulation of the system.

In order to get an estimate for this figure we have to determine the volume of the electroysis modules, i.e. of the hot core of the electrolyzer at first. Assuming modules consisting of 10 registers with 10 electrolysis tubes each, which have approximately the shape of a cube with an edge length of 0.3 m, their specific dimensions including the other installations in the electrolyzer are estimated to be 0.4 m x 0.4 m horizontally and 0.5 m vertically. Taking into account that a single electrolysis tube consumes 39.8 W (see section 3.2.1), such a module has a power demand of 3.98 kW. Therefore, to meet the 480 kW electric power derived above, 120 modules are needed. They should be arranged in a rather compact way in order to achieve a low surface-to-volume ratio which is important for minimizing heat losses. In order to get an approximative figure an arrangement of the modules in three levels, each level consisting of 8 x 5 modules, was assumed.

That leads to a volume V_{hot} and a surface area A_{hot} of the hot core of the electrolyzers of:

$$V_{hot} = 9.6 \text{ m}^3, \quad A_{hot} = 28.4 \text{ m}^2 . \tag{3.15}$$

Of course, this specific arrangement was chosen arbitrarily. Nevertheless, it represents a reasonable basis for estimating these values, since other arrangements of the modules (e.g. 5x5x5) yield quite comparable data.

For the calculation of the heat losses $\dot{Q}_{loss}$ the mean thermal conductivity $\lambda = 0.1$ W/mK of a ceramic fibre insulation is taken and a thickness $d = 1$ m of the insulation is assumed.

Corresponding to

$$\dot{Q}_{loss} = \lambda \frac{A_{hot}}{a} \Delta T \tag{3.16}$$

with $\Delta T \approx 980°K$ the heat losses come out to be

$$\dot{Q}_{loss} = 2.8 \text{ kW} . \tag{3.17}$$

To a first approximation the volume V_{ins} of the insulation is

$$V_{ins} \approx A_{hot} \cdot d \approx 30 \text{ m}^3 . \tag{3.18}$$

With the density of the insulation of about 100 kg/m³ and its specific heat capacity of 1 kWs/kgK the heat content Q_{ins} of the insulation can be calculated assuming a linear increase of the temperature across the insulation, i.e. a mean temperature of 500°C for the insulation material.

The result is:

$$Q_{ins} \approx 400 \text{ kWh} \qquad (3.19)$$

Thus the total heat content of the electrolyzer Q_{tot} which has to be 'pumped' into the system before operation amounts to

$$Q_{tot} \approx Q_{ins} + Q_{mod} = 675 \text{ kWh} . \qquad (3.20)$$

From that figure and the heat loss $\dot{Q}_{loss}$ calculated in equation 3.17 it can be derived that once the system was heated up to operating temperature it has an hourly cooling rate $\dot{T}_{cr}$ after a stop of operation of:

$$\dot{T}_{cr} \approx \frac{\dot{Q}_{loss}}{Q_{tot}} (T_{op} - T_{am}) = 4°K/h \qquad (3.21)$$

with T_{op} and T_{am} being the operating and ambient temperature, respectively.

3.3.2   First start of operation and restart

Before the Hot Elly electrolyzer can be put into operation for
the first time it has to be warmed up to its operating temperature. The procedure to do this is not yet elaborated in detail,
but there are conceptual ideas to heat up the apparatus by
circulating a gas through it, for example air or nitrogen or
- at elevated temperature - steam. Of course, auxiliary equipment as an external heater (e.g. electrical or hydrogen fueled)
and a fan is necessary. Another possibility to warm up the
electrolyzer could be realized by the integration of electrically heating elements into the apparatus. The allowable temperature gradient for that heating up is - according to present
experimental results on electrolysis tubes - in the order of
about 200°C/h. Hence the apparatus can be 'warmed up' within
five hours.

Due to the solar based operation the electrolyzer has to be
restarted every day. From the layout data of the DCS-plant it
can be seen that the maximum time of interruption of operation
is about 20 hours. That causes a decrease of the mean temperature of the modules of 80°C at maximum since the hourly cooling
rate is about 4°C/h as was shown in section 3.3.1.

The tolerable temperature gradient for the modules of about
200°C/h shows that heating up again to operating temperature
can be accomplished within less than half an hour.

This could be done in the following ways:

    -   The auxiliary equipment necessary for starting the system
        first could be used.

- The apparatus could be heated electrically as soon as power generation starts. That would reduce the daily operating hours by about 30 minutes on the average.

- Since the electric resistance of the electrolysis cells increases with decreasing temperature, the modules could be operated at about nominal electric power, but at higher voltages and lower currents with respect to the nominal data. Higher voltages correspond to exothermic cell operation (see chapter 2) and thus would lead to an increase of the temperature of the modules. Of course, at lower currents the hydrogen production rate would be less than the nominal one, but reheating of the apparatus could be performed producing hydrogen simultaneously. From the experimental data available a rough estimate for such a procedure can be gained telling that for about half an hour to an hour the electrolysis modules would have to be operated at partial production capacity (starting at about 70-80 %). During this reheating procedure the voltage would have to be decreased and the current to be increased in a controlled way corresponding to the actual temperature of the modules. (If this is really a possible way for reheating, should be tested by means of a suitable prototype unit.)

- Finally, for reheating one could take into consideration to operate the modules in reverse, i.e. as high temperature fuel cells fueled by hydrogen. A high temperature fuel cell is only able to convert part of the heat content of the fuel to electricity. The other part is converted to high temperature heat which has to be taken out of the system by cooling. However, in the case of the system considered here the high temperature heat would be used to heat the cells directly.

3.3.3 Annual operation hours and hydrogen yield of the system

----

According to the design of the DCS-plant its power conversion system is completely decoupled from the collector fields, i.e. from the problems of varying insolation (e.g. cloud passages) by the storage tank acting as a buffer. This also holds for the modified layout containing an autothermal high temperature electrolyzer since the interface of the electrolysis system and the DCS-system is also behind the storage tank.

Thus the daily operation hours of the electrolyzer are the same as these of the power conversion system of the DCS-plant, except the time needed for reheating the electrolyzer to its operating temperature. Assuming half an hour every day for reheating the electrolyzer on the average the annual operation hours of the electrolyzer are about 180 hours less than these of the power conversion system. Taking 1800 hours as the annual operating hours of the power conversion system (see section 3.1) the annual operating time of the electrolyzer would be 1620 hours.

If one further assumes that no hydrogen is próduced during the daily reheating periods, the hydrogen yield per year is just the product of the electrolyzer's design output power times the annual operating hours. Thus an annual hydrogen yield of 243000 Nm^3 results.

However, if the part load procedure discussed in section 3.3.2 could be realized for the daily reheating of the electrolyzer, the production of 180 hours at about 70 - 80 % power would have to be added to the value given above. In that case the annual hydrogen yield would be about 262000 Nm^3.

3.3.4   Some final remarks

Operating pressure:

The system designed in section 3.2.2 is based on an electrolyzer working at ambient pressure. This is reasonable since - in
contrast to the system investigated in the next chapter - there
is no operational need for operation under elevated pressure.
The main advantage of this layout is that no control of the
differential pressure between the anode compartment of the
electrolyzer and the inside of the electrolysis tubes is necessary. That especially facilities the daily restarts. However,
the peripheral heat exchangers of the electrolyzer have to be
larger than they would be at elevated operating pressures.

Critical operation modes:

Besides the starting procedures no critical operation modes
seem to exist since the electrolyzer is not affected by variations of the insolation due to the storage system acting as a
buffer; i.e. the electrolyzer is operated in a steady state
mode.

Critical components:

Apart from the high temperature components of the electrolyzer
system, which have to be further developed and verified, no
critical components can be identified.

4.   COMBINATION OF AN ALLOTHERMAL HOT ELLY AND A "HIGH
      TEMPERATURE" SOLAR THERMAL SYSTEM

The prime "product" produced by solar thermal plants is heat
which can be converted to electricity consecutively. However,
it could also be used directly as process heat for the production of chemicals or fuels. In the case of high temperature
electrolysis one could take advantage of the heat for a reduction of the electric energy demand of the process by providing
part of the energy for water splitting thermally instead of
electrically.

Obviously thereby the overall efficiency, i.e. the hydrogen
yield of the electrolysis process can be increased.

Within this chapter the feasibility of such a process will be
investigated assuming that advanced "high temperature" solar
plants yielding process heat at a temperature level well above
1000°C will be available in future. This has to be requested
since the operating temperature of a high temperature electrolyzer, which shall be supplied with process heat, is about
1000°C.

Since a high temperature thermal plant of the kind required
here is not yet existing it seemed reasonable to evaluate modular data in a sense that they are based on an electrolyzer unit
with a production capacity of 1000 $Nm^3 H_2$/h. A system for a larger production capacity would simply contain several of these
electrolyzer units.

4.1   Concept for an allothermal operation of Hot Elly

The main question for the realization of an allothermal operation of high temperature electroysis is how high temperature
heat can be coupled into the process. The present concept for
the solution of that problem is a hot gas flowing along the
electrolysis tubes on their outside (anode side). That means an
allothermal Hot Elly is fed with two gas streams: the feed
steam which is converted to hydrogen inside the tubes and a gas
stream transferring heat to the outside of them.

The theoretical model for the calculation of the temperature
distribution and voltage drop of an electrolysis tube mentioned
in section 3.2.1 has been extended in order to treat that problem, i.e. heat exchange of the tube walls with a gas streaming
along the tubes has been included. However, before the corresponding results will be given, the essential features of allothermal operation shall be illustrated by some principal considerations.

A single electrolysis cell is considered which may be characterized by its resistance R. If U_o means the open circuit voltage
of the cell (which corresponds to the specific electric energy
demand ΔG; see Fig. 2.1) and I the operating current, the cell
voltage U_c is:

$$U_c = U_o + R \cdot I \qquad\qquad (4.1)$$

For an allothermal operation U_c must be smaller than the voltage U_H which corresponds to the total energy demand ΔH (see Fig.
2.1) of the vapor splitting reaction ($U_H \approx 1.3$ V). Only then
high temperature heat $\dot{Q}$ can be coupled into the process. In
total the following equation for the power (electric and thermal) consumed by the cell must be satisfied in order to cover
the total energy demand of the water splitting reaction:

$$U_H \cdot I = U_C \cdot I + \dot{Q} = U_0 \cdot I + RI^2 + \dot{Q} \qquad (4.2)$$

Writing ΔU for the difference of U_H and U_0 the thermal power absorbed by the cell can be written as

$$\dot{Q} = (\Delta U - R \cdot I) \cdot I . \qquad (4.3)$$

During a time Δt the thermal energy $Q = I \cdot \Delta t$ is transfered to the cell. Since the amount of hydrogen produced within Δt is directly proportional to $I \cdot \Delta t$, for the specific thermal energy Q_{spec} 'coupled into the product' the relation

$$Q_{spec} = A \cdot (\Delta U - RI) \qquad (4.4)$$

holds with A being a constant of proportionality

$$(A \approx 2.4 \ \frac{kWh}{Nm^3 H_2 \cdot V}).$$

The current I_{max} for which the cell is able to absorb a maximum $\dot{Q}_{max}$ of external heat power can be derived from equation (4.3). For I_{max} and $\dot{Q}_{max}$ holds:

$$I_{max} = \frac{\Delta U}{2R}; \quad \dot{Q}_{max} = \frac{\Delta U^2}{4R} \qquad (4.5)$$

Since for an autothermal operation, in which no high temperature heat is used, $\dot{Q}$ is approximately zero, the corresponding current has just to be twice the current I_{max}, as can be seen from equations (4.2) or (4.3). Thus a maximum of heat is transferred to the electrolysis cells in an allothermal operation at half the current of the autothermal process mode. At I_{max} a specific thermal energy Q_{spez} of $0.5 \cdot A \cdot \Delta U$ results, as can be seen by inserting the expression for I_{max} from equation (4.5) into equation (4.4). This corresponds to just one half of the total potential offered by thermodynamics which is $A \cdot \Delta U$. An exploitation of more than half of this potential for incoupling of high temperature heat can only be accomplished if the electrolysis cells are operated at currents below I_{max}. However, in this case the cells absorb less heat as they could in principle.

The relations between I, $\dot{Q}$ and Q_{spec} discussed above are illustrated in Fig. 4.1 which shows the heat power absorbed by one electrolysis cell and the specific heat energy used by the process as a function of I. The curves were calculated for the following set of boundary conditions:

Operating temperature:		1000°C
Cell resistance	:	0.3 Ω
Operating pressure	:	10 bar
Anodic atmosphere	:	air
H_2-concentration	:	37.5 % (= mean H_2-concentration of an electrolysis tube)

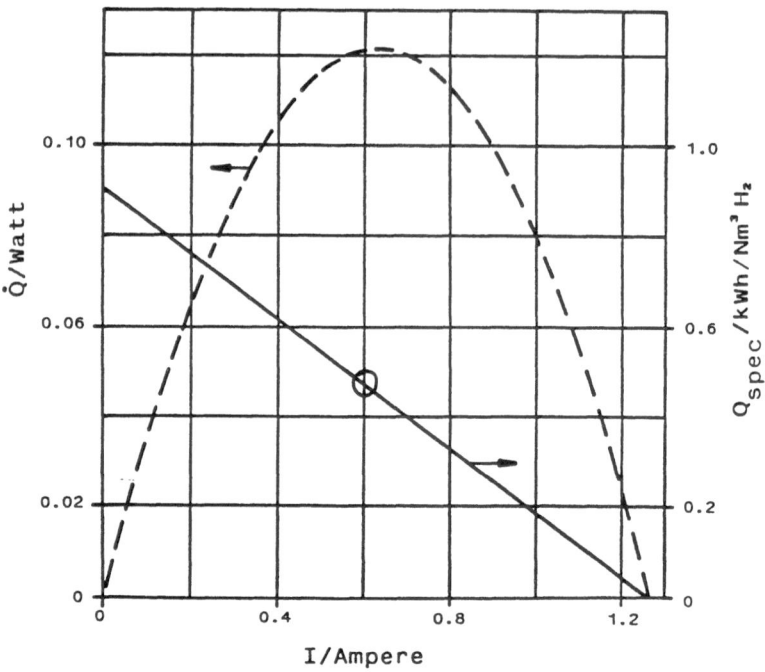

Fig. 4.1   Absorbed Heat Power Q̇ of an Electrolysis Cell and
Specific Heat Energy Use Q_{spec} as a Function of
Electric Current I

From Fig. 4.1 it can be clearly seen that it is reasonable to
operate an allothermal electrolysis at a current in the close
vicinity of I_{max} corresponding to a specific high temperature
heat energy use of about 0.45 kWh/Nm³H₂.

Otherwise the number of cells needed for a given production capacity would increase rapidly, if a substantially greater amount of specific thermal energy had to be coupled into the process. For example, the request for using 0.7 kWh/Nm³H₂ instead of 0.45 kWh/Nm³H₂ would require to reduce the current by more than a factor of two, i.e. the number of cells needed for the same production capacity would be more than doubled.

Now the thermal consequences for the heating gas flowing along the cells shall be considered briefly. Denoting by ΔT the temperature difference, by which the gas stream cools down due to heat transfer to the cell, and by $\dot{m}$ and C_p the mass flow and specific heat capacity of the gas, the following equation must hold:

$$\dot{m} \, C_p \, \Delta T = \dot{Q} \qquad\qquad (4.6)$$

Therefore, if the cell is operated at I_{max} for the temperature loss of the heating gas results:

$$\Delta T = \frac{\Delta U^2}{4 \, \dot{m} \, C_p \, R} \qquad\qquad (4.7)$$

The order of magnitude for that temperature loss can best be illustrated by an example. The mean cross section, through which the heating gas can flow along an electrolysis tube within a module, is about 2 cm². Keeping the boundary conditions given above and assuming a gas velocity of 0.1 m/s, a ΔT of about two degrees on the average results for every cell along the gas stream. Greater gas velocities would decrease ΔT and vice versa.

It should be stressed that the investigations made so far can only give trends since the problem was treated with a variety of simplifications. For example, the temperature dependence of the cell resistance R was neglected as well as radiative heat exchange between different cells and so on. All these effects have been included in the extension of the numerical model in order to gain detailled data. Of course the calculations were made in the neighbourhood of I_{max} as is recommended by the considerations made above.

Since concepts already exist for high temperature solar thermal plants using air as the heat transferring medium (gas cooled solar tower or GAST-type plant) the calculations were based on an air stream flowing through the anode compartment of the electrolyzer. According to the GAST-concept an operating pressure of about 10 bar was assumed.

The calculations were performed for an electrolysis tube with 20 cells. Several parameters have been varied, mainly the current, the mass flow (velocity) of the air stream and the temperature at the tube entrance, which is a boundary condition in the numerical model. The goal was to determine a set of parameters fulfilling the request that 0.5 kWh/Nm³H₂ are to be coupled into the process thermally. An example for the results obtained is given in Fig. 4.2.

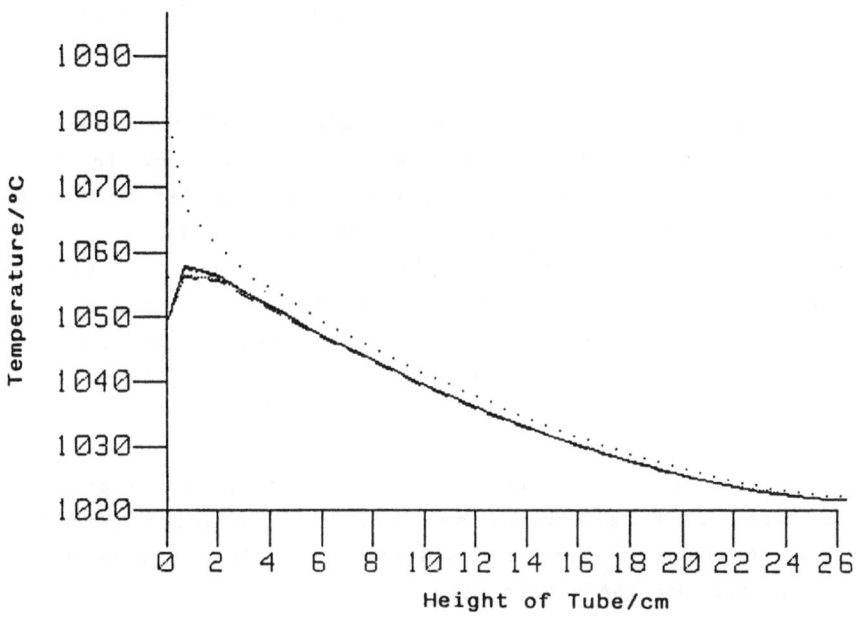

Fig. 4.2   Temperature Distribution of an Electrolysis Tube in
Allothermal Operation (air temperature ≙ dotted line,
tube temperature ≙ solid line)

The essential results delivered by the model calculations are the following:

- A coupling-in of thermal energy of the order of 0.5 kWh/-
  Nm^3H_2 can be achieved.

- The temperature of the heating air decreases from 1080°C
  to 1025°C, i.e. the temperature drop is 55°C.

Of course, the results of the calculations can only be used as approximate data for the feasibility of allothermal operation, because only the modelling of a single tube but not of a complete electrolyzer is possible.

Within an electrolyzer a great number of electrolysis tubes will be integrated laterally within modules and in several stages of modules one above another. A transfer of the data gained by the calculations to a complete electrolyzer can be made in the following way: The mass of hot air streaming through the electrolyzer is the one obtained in the calculations times the number of tubes integrated in the apparatus. Thus the gas velocity and therefore the mass flow $\dot{m}$ along a single electrolysis tube is n-times greater thán in the calculations with n being the number of stages within the electrolyzer. Corresponding to equation (4.7) the temperature drop along one tube will be n-times smaller, but the total temperature drop of the air along all stages of modules will be the one calculated for the single tube.

This interpretation of the results of the model calculations represents the basis for the layout considerations for a combined system allothermal Hot Elly/GAST-type solar plant given in the next two sections.

4.2   Thermal characterization of an allothermal
        Hot Elly unit (1000 Nm³ H₂/h)

A detailed design for an allothermal high temperature electrolyzer working at elevated pressures is not yet existing. Nevertheless some constructional features which are essential for
the layout of a combination with a GAST-type solar plant can be
derived approximately.

Assuming - corresponding to the model calculations - a design
current in the electrolysis cells for allothermal operation
which is approximately half the one of the autothermal case
about 3.4 million single cells have to be integrated in the
apparatus. The corresponding electrolysis modules can be arranged in many different ways. The geometrical features needed
here, namely volume and surface of the 'hot core' (modules) of
the apparatus are not sensitive with respect to that. Therefore
one arrangement was chosen arbitrarily for an estimate of these
data. It can be described as follows:

One module consists of 33 registers containing 47 electroysis
tubes with 30 cells each. Between the registers there is a gap
of about 1 cm in order to allow the air stream to flow along
the electrolysis tubes. The ground area of the module is a
rhombus with an edge length of about 1 m. Three modules are
arranged to fill a hexagonal area within a circle of about 2 m
in diameter. Twenty-four stages each containing three modules
have to be arranged above one another to reach the number of
cells required. The height of one stage is assumed to be 0.7 m
(0.4 m module height, 0.3 m height of the supporting structures) resulting in a total height of 16.8 m for all stages.
According to the elevated pressure this cylindrical volume
shall be closed by two semispherical domes at its ends. With
these data the volume V_{hot} and the surface area A_{hot} of the
electrolyzers hot core result to be

$$V_{hot} \approx 57 \ m^3; \quad A_{hot} \approx 120 \ m^2.$$

The total mass of the ceramic moduls is about 14 tons. Therefore the mass M_{hot} of the hot core including the support structures which are supposed to be ceramics, too, is estimated to be

$$M_{hot} \approx 20 \ t.$$

These ceramic elements have to be integrated into a pressure vessel with an insulation inside whose thickness was assumed to be 0.6 m. Hence the volume of the insulation is about 95 m^3 which leads - the density of ceramic fibre insulation being 98 kg/m^3 - to a mass M_{ins} of the insulation of approximately

$$M_{ins} \approx 9.3 \ t.$$

The heat content of the apparatus - once it is at operating temperature - and its thermal inertia can be estimated by analogy to the considerations made in section 3.3.1.

The important figures resulting are a heat loss $\dot{Q}_{loss}$ of

$$\dot{Q}_{loss} \approx 20 \ kW$$

and a total heat content Q_{tot} of the system of

$$Q_{tot} \approx 6000 \ kWh.$$

These two numbers tell that at least 6000 kWh of thermal energy have to be 'pumped' into the electrolyzer unit to heat it up to operating temperature and that the system is cooling down after a stop of operation at a rate $\dot{T}_{cr}$ of

$$\dot{T}_{cr} \approx 3,5 \ deg/h.$$

Thus the hourly cooling rate is similar to the one of the electrolyzer discussed in chapter 3 although the insulation thickness is less. The main reason for that lies in the lower surface to volume ratio of the hot core, i.e. per volume unit of the hot core there is less surface area through which heat losses can occur.

4.3 Basic flow sheet of a coupled system

With the considerations made in the previous two sections a first layout for a coupled system of an allothermal Hot Elly unit with a production capacity of 1000 Nm^3H_2/h and a GAST-type solar plant can be sketched.

According to the data of the model calculations the electrolyzer shall be supplied with 500 kW high temperature heat by a stream of hot air which enters the electrolyzer at a temperature of 1080°C and leaves it at 1025°C. The electric power demand of the electrolyzer is 2.6 MW, its operating pressure is supposed to be about 10 bar.

Firstly the gas and energy balance of the electrolyzer unit have to be considered. From the condition that the feed steam has to be converted to hydrogen by 75 % one deduces a feed steam demand $\dot{m}_{feed}$ of

$$\dot{m}_{feed} \cdot = 0.297 \ kg/s.$$

This feed steam has to be produced at a pressure of about 11 bar since the system is operating at 10 bar. Assuming 2 % losses in the evaporator, which has to work at 185°C, the heat demand $\dot{Q}_{vap}$ for vaporizing the feed water is:

$$\dot{Q}_{vap} = 606 \ kW.$$

Thereafter the feed steam has to be superheated to 1000°C. Assuming 5 % losses in the corresponding heat exchangers the heat power $\dot{Q}_{sup}$ required for that purpose is:

$$\dot{Q}_{sup} = 562 \ kW.$$

This heat demand can be partially covered by cooling down the H_2/H_2O-product stream from a temperature of 1025°C (of which it leaves the electrolyzer) to 210°C which is 25°C above the temperature of feed steam generation. That yields a heat power of 435 kW. Consecutively it is cooled down further to about 100°C thereby providing the heat necessary for preheating the feed water. Then the steam content of the product stream is already condensated. The remaining heat demand of 126 kW for superheating the feed steam is covered by the stream of hot air leaving the electrolyzer.

Thus the feed steam is superheated by means of two heat exchangers in series before it enters the electrolyzer. Since the cells therein are working at about 1040°C on the average additional heat is required to bring the feed steam up to that temperature. This heat requirement (29.5 kW) has to be met by the air stream as well as the electrolyzer's heat losses of 20 kW.

Hence the heating air stream has to provide a total of 550 kW heat power thereby cooling down from 1080°C to 1025°C. The corresponding difference of the specific enthalpy of air is 66.33 kJ/kg. Thus the mass flow $\dot{m}_{air}$ of hot air required is:

$$\dot{m}_{air} = 8.3 \text{ kg/s}$$

Due to the electrolysis process oxygen is produced at a rate of 0.198 kg/s (= 500 Nm³ O_2/h) which adds to the air stream flowing through the electrolyzer. Thus the mass flow of air $\dot{m}_{out}$ leaving the electrolyzer at 1025°C is:

$$\dot{m}_{out} = 8.5 \text{ kg/s}$$

As already mentioned above 126 kW of heat power have to be
extracted from this gas stream for the superheating of the feed
steam. That corresponds to a change of specific enthalpy of the
air of 14.8 kJ/kg. From that value the temperature of the air
having passed that heat exchanger can be calculated to be approximately 1010°C.

The gas and energy balance of the electrolyzer unit being
established now, the remaining part of the system has to be
considered next. It looks like a GAST-plant with the electrolyzer integrated in series between the receiver and the gas turbine as is shown in Fig. 4.3. Hence the way of the air through
the system comprises the following steps:

- compressor
- receiver
- electrolyzer + second feed steam superheater
- gas turbine
- waste heat boiler (feed steam generation for
  electrolyzer and steam turbine)

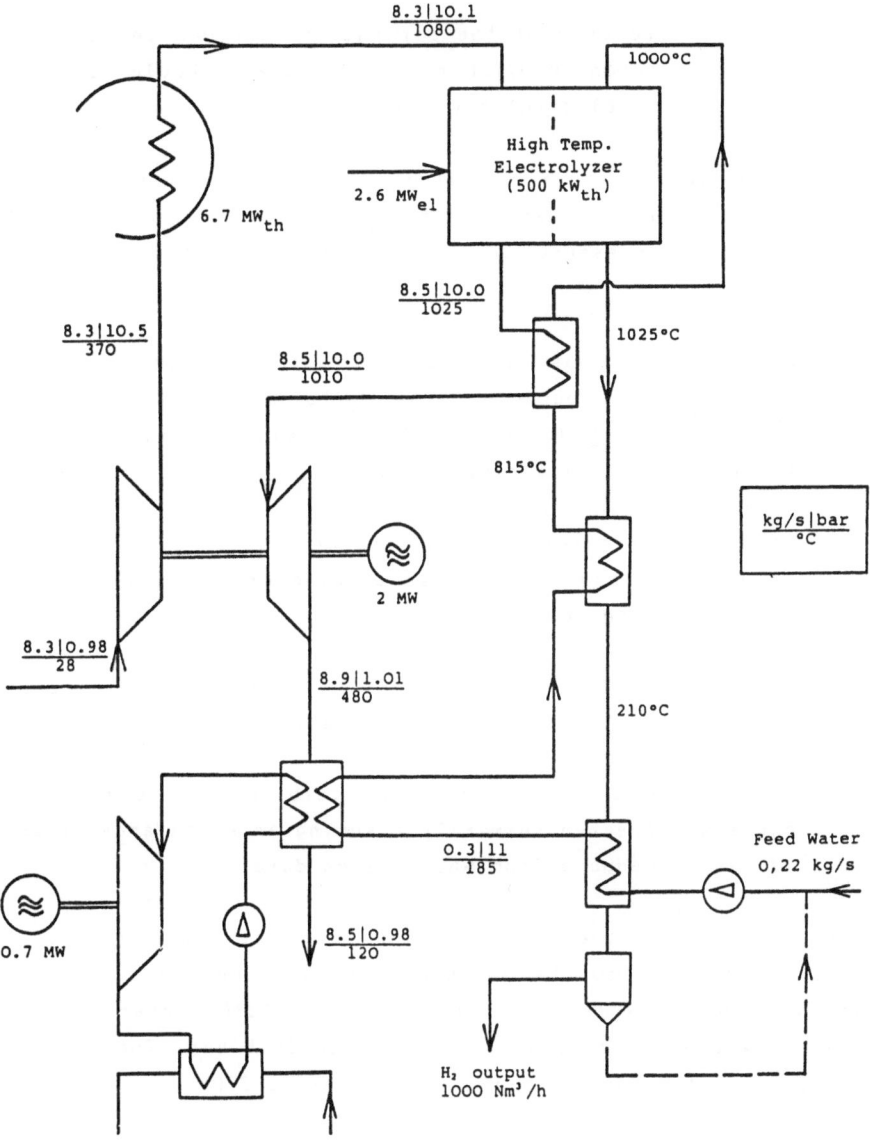

Fig. 4.3 Basic flow sheet of an allothermal high temperature
electrolyzer combined with a GAST-type solar thermal
central receiver plant

For the evaluation of the approximate figures characterizing
this system a number of assumptions had to be made:

- The pressure drop of the air passing through the electrolyzer was taken to 0.1 bar, a value which should be achievable for an electrolyzer unit.

- The inlet pressure of the gas turbine was supposed to be
  10 bar. In accordance with the design data of the GAST-
  project a pressure drop of 0.4 bar was assumed for the
  receiver.

- Pressure drops as well as heat losses in the air ducts
  were neglected because these ducts have to be short and
  therefore their losses should be negligible in comparison
  to the other system components.

- For the compressor which has to pressurize ambient air up
  to 10.5 bar and for the gas turbine efficiencies of 85 %
  each were assumed.

- The efficiencies of the generator at the gas turbine and
  of the rectifiers were taken to be 98 %.

- Power generation by the steam turbine in the bot…oming
  cycle was treated summarily assuming an efficiency of 26 %
  as can be deduced from GAST design data.

Corresponding to the considerations made at the beginning of
this section and to the calculations based on the assumptions
given above 8.3 kg/s of ambient air (T = 28°C) enter the compressor where the pressure is raised to 10.5 bar. The air temperature at the outlet of the compressor is 370°C.

Then the air passes the receiver wherein it is superheated to 1080°C which corresponds to a thermal power of the receiver of 6.7 MW. Consecutively the air passes the electrolyzer unit transferring 550 kW thermal power to it whereby it cools down by 55°C to 1025°C.

Within the electrolyzer 0.2 kg/s oxygen are added to the mass flow due to the electrolysis process. Thus an air stream of 8.5 kg/s is entering the second feed steam superheater in which it gets cooled down to 1010°C. The air stream is then entering the gas turbine at this temperature and at a pressure of 10 bar.

The turbine which drives the compressor yields a net mechanical power of 2.07 MW which is converted to an electric power of 1.989 MW by the generator and rectifier. Behind the turbine the air has a pressure of 1.01 bar and a temperature of 490°C. Cooling down the air further to 120° yields a thermal power of 3.3 MW which is used for feed steam generation and for a steam turbine bottoming cycle which yields another 0.688 MW of electric power. Thus a total of 2.68 MW electric power is generated which is enough to cover the power demand of the electrolyzer unit of 2.6 MW.

If the assumed efficiencies of the power generation components were to high, i.e. if the electric power would be lower, the system would have to be modified in the following way: The total mass flow of air had to be increased and split into two streams behind the receiver. One substream would still pass the electrolyzer and the second superheater whereas the other one would have to flow through a bypass joining the first one before it enters the gas turbine. Thus in total the heat supply for the electrolyzer would remained unchanged whereas the mass flow for the power generation system would be increased thus yielding greater electric power.

Another modification with respect to the flow sheet shown in Fig. 4.4 has to be foreseen, namely a burner between the receiver and the electrolyzer which has to compensate for temperature fluctuations at the receiver outlet due to fluctuations of the insolation. Since the system produces hydrogen which certainly has to be intermediately stored at least partially this hydrogen could resonably be used as a fuel for the burner. Thus the system would not depend on a fossile feed for support.

Additionally the burner allows to increase the operating hours of the system since operation can be started before the time in the morning the insolation reaches the level required and maintained beyond the time in the evening the insolation falls below this level.

Of course, the limit for the operation of the burner is to spend not more hydrogen than is produced simultaneously. Hence a maximum of only 24.8 g/s of hydrogen can be burned which corresponds to the production of 1000 Nm^3H_2/h. This figure is equivalent to 3.1 MW of thermal power the lower heating value of H_2 being 3.1 kWh/Nm^3.

Start and restart of the electrolyzer, i.e. warming up the electrolyzer to operating temperature, can be accomplished by heating it up using the air stream coming from the receiver. However, in order to avoid thermal shocks on the modules a bypass in parallel to the electrolyzer has to be foreseen through which the air can flow until the necessary temperature is reached.

For the first start-up again a temperature gradient with respect to time of about 200°C/h has to be obeyed. Since the nominal output power of the receiver is 6.7 MW and the heat required for warming up the electrolyzer is about 6 MWh (see section 4.2) no principal problem or bottleneck can be seen.

The same is true for rewarming the electrolyzer before it can
be restarted in the morning since the heat required for that is
small (in the order of 200 kWh). Hence it seems possible to
perform the rewarming during the timespan between sunrise and
the time the insolation reaches the level required for operation, within which the power generation components are also
brought to their nominal output power.

In total the basic flow sheet given in Fig. 4.4 has to be completed with respect to the modifications discussed above, i.e.
with respect to a bypass line in parallel to the electrolyzer
and a burner. The final flow sheet resulting is shown in Fig.
4.4.

In order to get an estimate of the benefits of coupling in high
temperature heat into the electrolyzer one has to compare the
system shown in Fig. 4.5 with a system consisting of an autothermal electrolyzer and a GAST-power plant which provides
electricity and feed steam. In other words: if the 500 kW thermal power are not used directly in the electrolyzer but for
power generation, what size of autothermal Hot Elly could be
supplied. Assuming the same efficiencies for the power generation components despite the fact that the gas turbine would
have to face higher inlet temperatures the corresponding calculation showed that in this case 920 Nm^3H_2/h instead of 1000
Nm^3H_2/h could be produced. Thus the potential for an improvement of efficiency by allothermal operation amounts to 8 %.

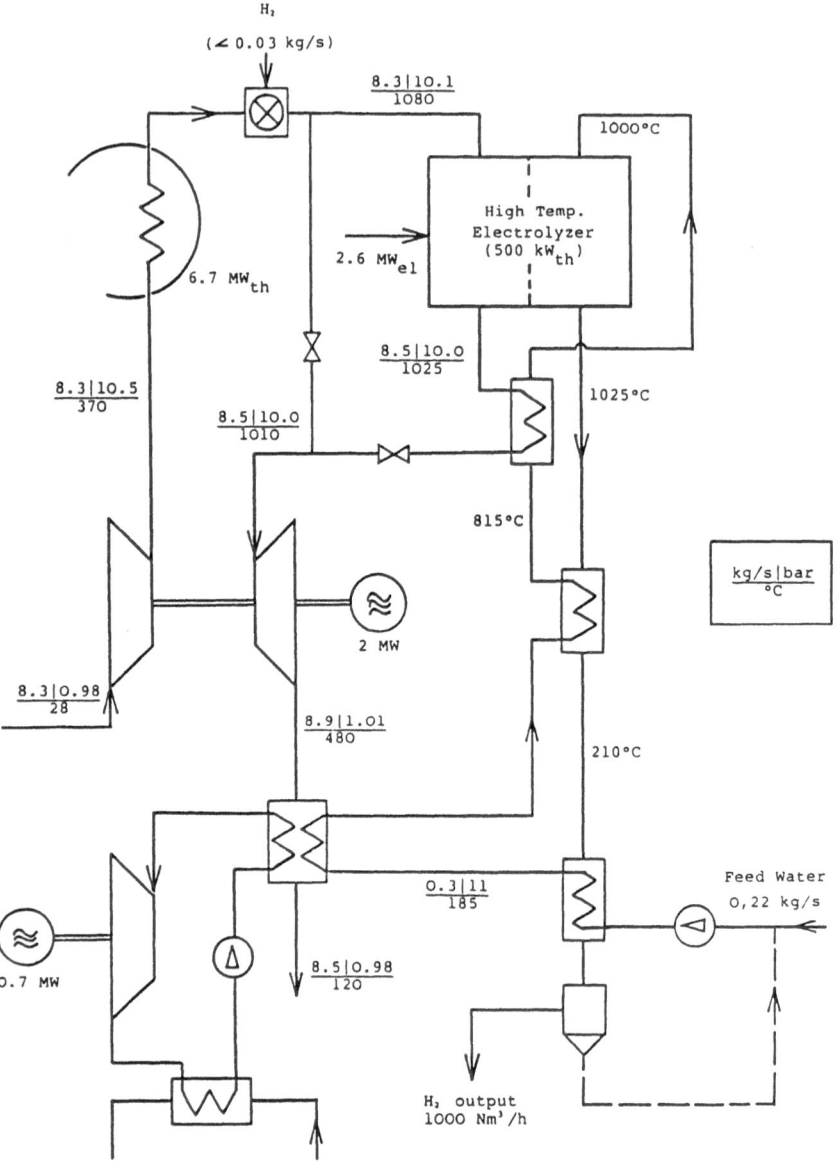

Fig. 4.4 Final flow sheet of an allothermal high temperature
electrolyzer combined with a GAST-type solar thermal
central receiver plant

4.4   Adaptation of the system to solar requirements and
estimate of the hydrogen yield

Within this section the question of the hydrogen yield of the
system is investigated. This is done mainly in terms of the
solar multiple, i.e. of the nominal field power (at noon of
June 21) in relation to the power demand of the system.

A detailed consideration of the system's behaviour at partial
load is rather difficult and would be beyond the scope of the
investigations possible within this study. Moreover the essential features can be already recognized assuming the system is
operated at constant power as it was done here.

In order to gain data which allow an estimate of the hydrogen
yield a numerical simulation program was developed which will
be described below. The calculations were based on field power
profiles given for the Almeria site as they are given in the
GAST study. They are shown in Fig. 4.5 for the representative
days of the year in a normalized form, i.e. the maximum power
at noon at summer solstice corresponds to a value of one.

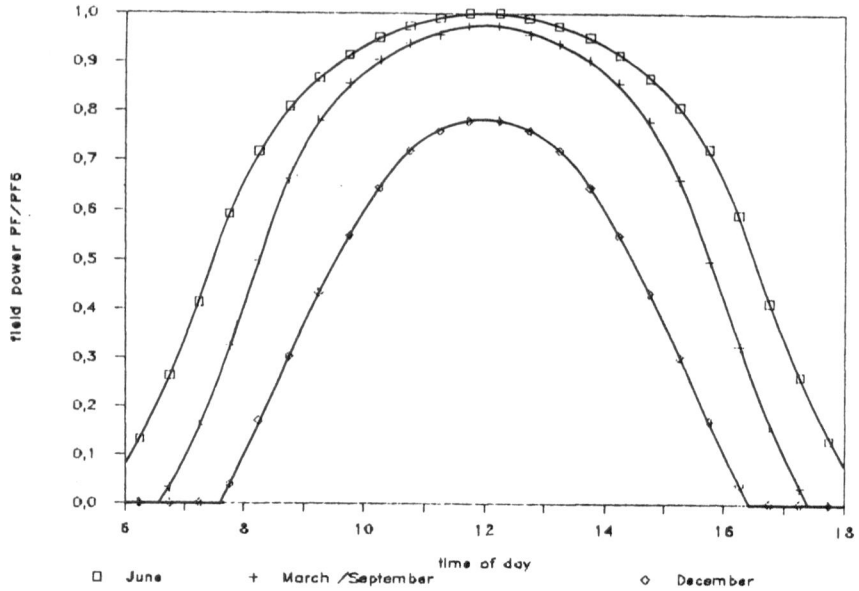

Fig. 4.5   Field power as a function of daytime for representative days of the year

It was assumed that the receiver is operated at a constant mass flow and at constant thermal power. The latter condition can be fulfilled by defocussing heliostats whenever the field power would exceed the level which is necessary to yield a receiver output power of 6.7 MW. The receiver efficiency was supposed to be 77 %. For the heliostat field an efficiency of 67 % was taken according to the data given in the GAST study; the burner was assumed to have an efficiency of 100 %.

By use of the hydrogen burner the operating time of the system can be extended beyond the time span within which the field power is great enough to cover the power demand of the system. This can be done as long as less hydrogen is burned than is produced. The nominal production rate of 1000 $Nm^3 H_2/h$ corresponds to a thermal power of 3.1 MW if the hydrogen is burnt (the lower heating value of H_2 is 3.1 kW/Nm^3). Thus the system can be operated whenever the field power is great enough to yield 3.1 MW less than the necessary 6.7 MW in the receiver, i.e. whenever the receiver power exceeds 3.6 MW. That corresponds to about 55 % of the nominal power of the receiver.

Taking into account that the mass flow is kept constant (and due to that the receiver output temperature is allowed to float) it can be seen from the efficiency characteristics for the receiver given in the Gast study and shown in Fig. 4.6 that the receiver works at constant efficiency between 55 % and 100 % power. This should also hold for the higher temperature range considered here.

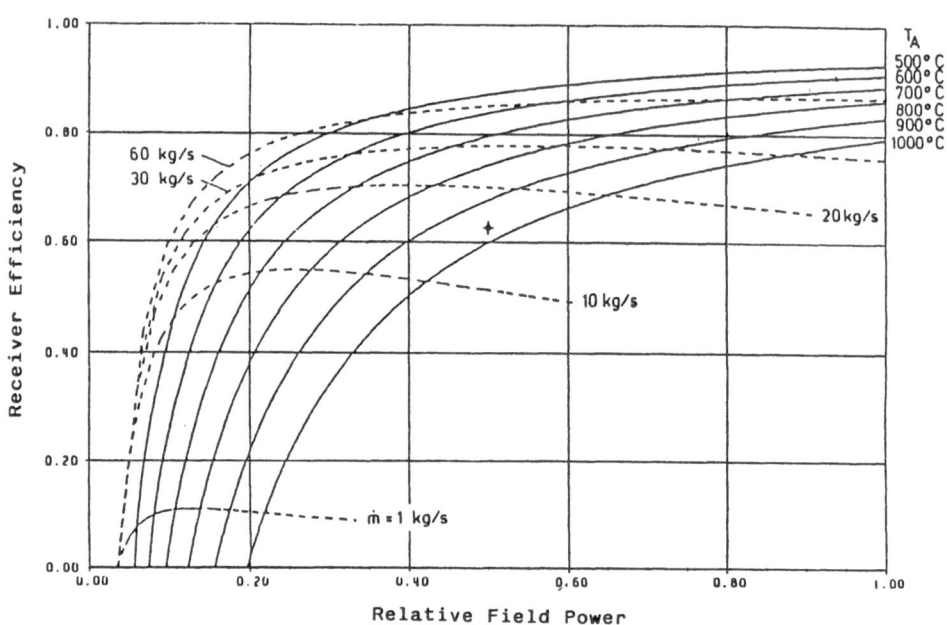

Fig. 4.6   Efficiency characteristics of a GAST-receiver

Based on these assumptions calculations were performed for the
winter solstice, summer solstice and the equinox days. For time
intervals of 30 minutes the receiver output power was determined. Thereby the receiver power was limited to the power demand
of the consumers (electrolyzer + power generation), i.e. surplus energy coming from the field in the case of solar multiples greater one was cut off. Thus the receiver output power
was approximated as a function of daytime for different solar
multiples. As an illustration the Figs. 4.7 to 4.9 show the
calculated receiver output curves for solar multiples of 1, 1.5
and 2.

- 686 -

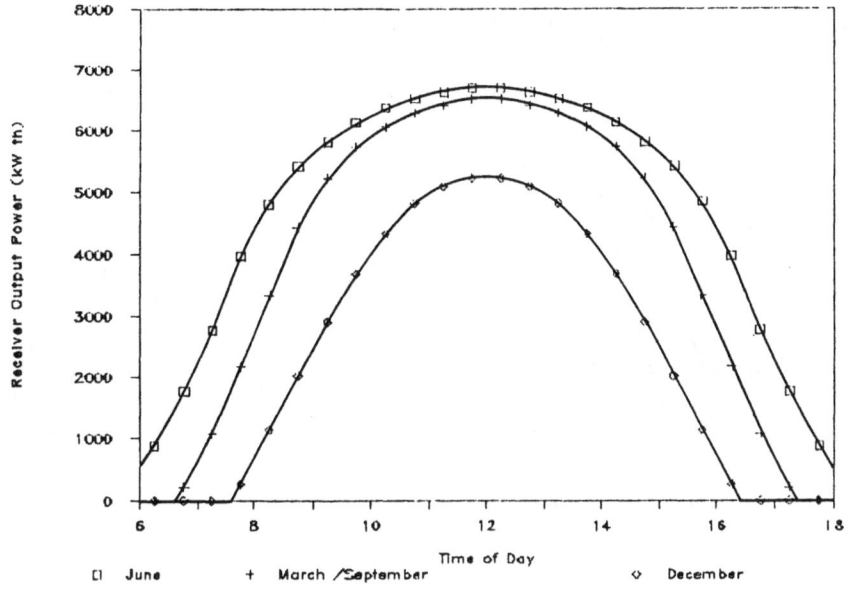

Fig. 4.7 Receiver output power as a function of daytime for the representative days winter solstice, equinox and summer solstice (Solar multiple = 1)

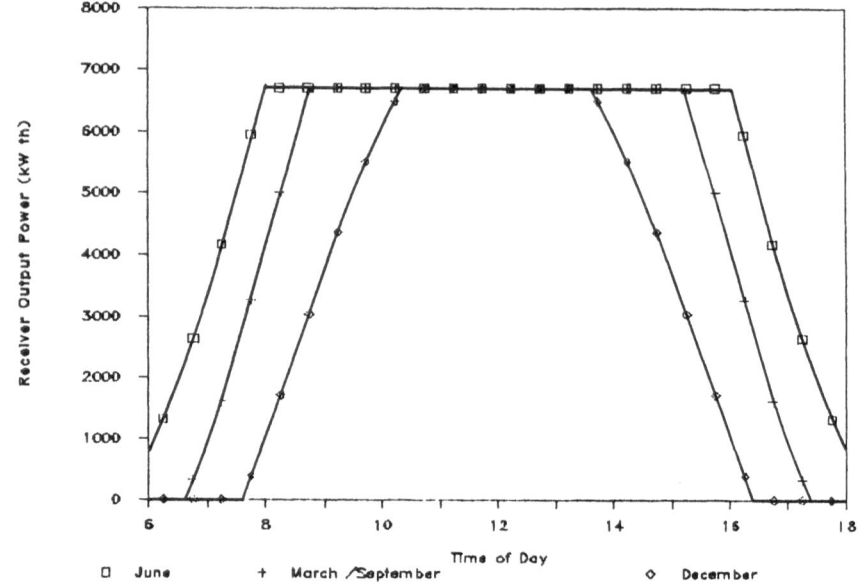

Fig. 4.8   As Fig. 4.7, but solar multiple = 1.5

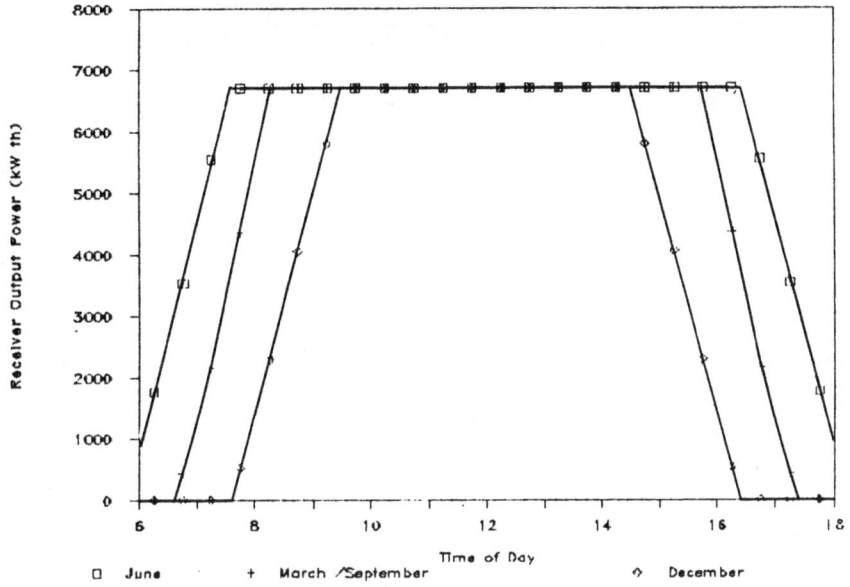

Fig. 4.9   As Fig. 4.7, but solar multiple = 2

Since one can assume that the system can be rewarmed in the
morning before the receiver output power reaches the requested
level of 6.7 MW the net daily hydrogen production (difference
of H_2 produced by the electrolyzer and H_2 spent in the burner)
could be directly calculated from these data. Simultaneously
the total energy insolated into the heliostat field was determined. Consecutively the annual average values were built for
the daily H_2-production and the insolation taking the results
for winter solstice, summer solstice and the two equinox days.

Sinally the total efficiency n_{tot}, which is another important characteristic of the system, was calculated according to the definition:

$$n_{tot} = \frac{\text{higher heating value of } H_2 \text{ produced}}{\text{solar energy insolated into the heliostat field}}$$

Sig. 4.10 gives an example for a typical data table calculated with the simulation program. The corresponding functions of the hydrogen production are shown in Figs. 4.11 to 4.13 for solar multiple of 1, 1.5 and 2.

SOLAR HOT ELLY

```
=================================================================================================
                         DESIGN PARAMETERS                              ABBREVIATIONS
field power (21.6. 12:00):  17403   Hot Elly rated power      6700 !  PF   field power actual      PHSol solar power for H2 gen.
efficiency receiver      :   0.77   heat.value H2 (kWh/m3) :   3.1  !  PF6  field power 21.6.       PHFue fuel power required
specif.H2 prod.(m3/kWhth):   0.149  solar multiple(PF6/PHR):   2.00 !  PR   receiver output power   HNet  H2 net prod.rate
                                                                     !  PHR  Hot Elly rated power

power(P..) in kW, energy (W..) in kWh, H2 (H..) in m3/h, time (T..) in h (decimal)
```

```
======== JUNE ========
Time    6.25  6.75  7.25  7.75  8.25  8.75  9.25  9.75  10.25 10.75 11.25 11.75 12.25 12.75 13.25 13.75 14.25 14.75 15.25 15.75 16.25 16.75 17.25 17.75
PF/PF6  0.132 0.263 0.414 0.592 0.718 0.809 0.868 0.915 0.951 0.974 0.990 0.999 0.990 0.974 0.951 0.915 0.868 0.809 0.724 0.592 0.414 0.263 0.132
PRec    1769. 3524  5548  6700  6700  6700  6700  6700  6700  6700  6700  6700  6700  6700  6700  6700  6700  6700  6700  6700  6700  5548  3524  1769
PHSol      0     0  5548  6700  6700  6700  6700  6700  6700  6700  6700  6700  6700  6700  6700  6700  6700  6700  6700  6700  6700  5548  3524     0
PHFuel     0  1152     0     0     0     0     0     0     0     0     0     0     0     0     0     0     0     0     0     0  1152     0     0     0
HNet       0   628   991  1000  1000  1000  1000  1000  1000  1000  1000  1000  1000  1000  1000  1000  1000  1000  1000  1000  1000   628     0     0

    *** AVERAGE DAILY PRODUCTION JUNE   9628 m3 H2 ****
```

```
======== SEPTE ========
Time    6.25  6.75  7.25  7.75  8.25  8.75  9.25  9.75  10.25 10.75 11.25 11.75 12.25 12.75 13.25 13.75 14.25 14.75 15.25 15.75 16.25 16.75 17.25 17.75
PF/PF6  0.000 0.161 0.325 0.498 0.662 0.780 0.868 0.856 0.905 0.938 0.957 0.974 0.957 0.938 0.905 0.856 0.780 0.662 0.498 0.325 0.161 0.032
PRec       0   429  2157  4355  6673  6700  6700  6700  6700  6700  6700  6700  6700  6700  6700  6700  6700  6700  6673  4355  2157   429
PHSol      0     0  2157  4355  6673  6700  6700  6700  6700  6700  6700  6700  6700  6700  6700  6700  6700  6700  6673  4355     0     0
PHFuel     0     0     0  2345    27     0     0     0     0     0     0     0     0     0     0     0     0     0    27  2345     0     0
HNet       0     0   244   991  1000  1000  1000  1000  1000  1000  1000  1000  1000  1000  1000  1000  1000  1000  991   244     0     0

    *** AVERAGE DAILY PRODUCTION SEPTE  8235 m3 H2 ****
```

```
======== DECEM ========
Time    6.25  6.75  7.25  7.75  8.25  8.75  9.25  9.75  10.25 10.75 11.25 11.75 12.25 12.75 13.25 13.75 14.25 14.75 15.25 15.75 16.25 16.75 17.25 17.75
PF/PF6  0.000 0.000 0.000 0.039 0.171 0.302 0.433 0.549 0.646 0.720 0.760 0.780 0.760 0.720 0.646 0.549 0.433 0.302 0.171 0.039 0.000 0.000
PRec       0     0   523  2291  4047  5802  6700  6700  6700  6700  6700  6700  6700  6700  6700  6700  5802  4047  2291   523     0     0
PHSol      0     0     0  2291  4047  5802  6700  6700  6700  6700  6700  6700  6700  6700  6700  6700  5802  4047  2291     0     0     0
PHFuel     0     0     0     0  2653   898     0     0     0     0     0     0     0     0     0     0   898  2653     0     0     0     0
HNet       0     0     0   144   710   898  1000  1000  1000  1000  1000  1000  1000  1000  1000  1000   898   710   144     0     0     0

    *** AVERAGE DAILY PRODUCTION DECEM  5855 m3 H2 ****

    YEAR AVERAGE DAILY PRODUCTION        7988 m3 H2
    Year Average Overall Efficiency     16.3 %
=================================================================================================
```

Fig. 4.10 Example of a data table calculated by the simulation program (solar multiple = 2)

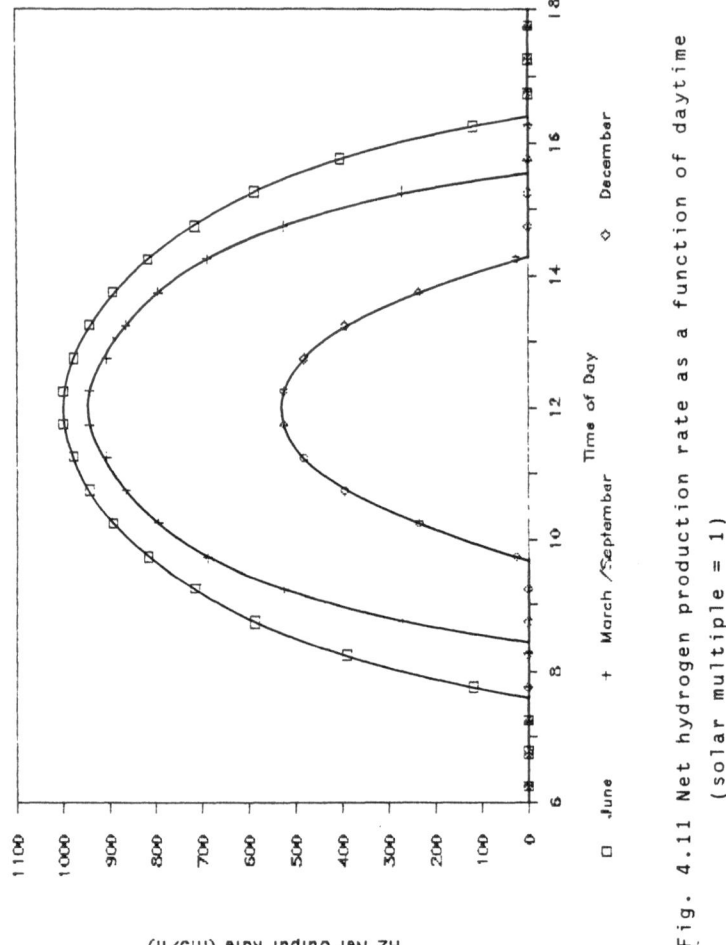

Fig. 4.11 Net hydrogen production rate as a function of daytime
(solar multiple = 1)

□ June   + March/September   ◇ December

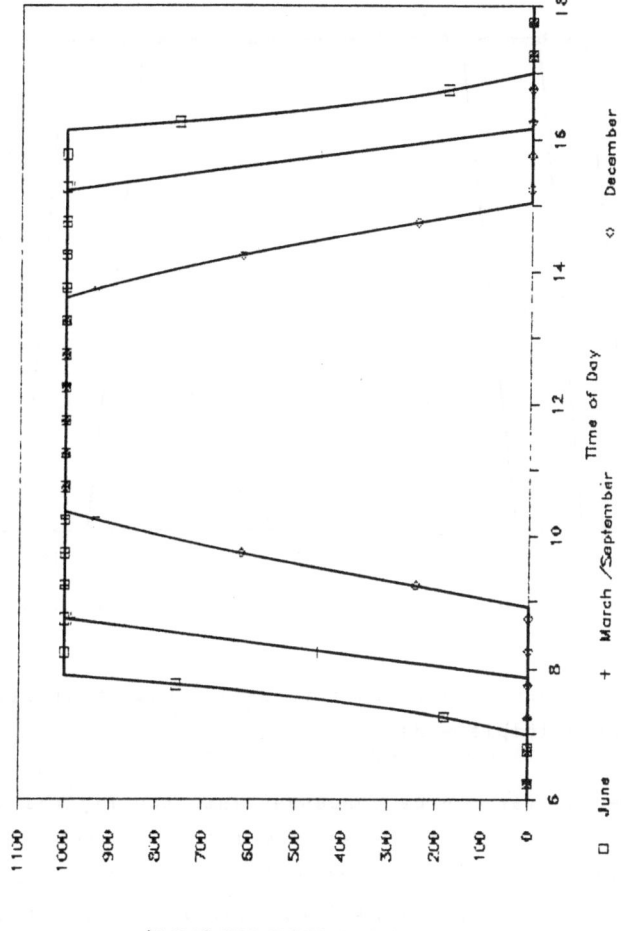

Fig. 4.12 Net hydrogen production rate as a function of daytime
(solar multiple = 1.5)

□ June   + March /September   ◇ December

H2 Net Output Rate (m3/h)

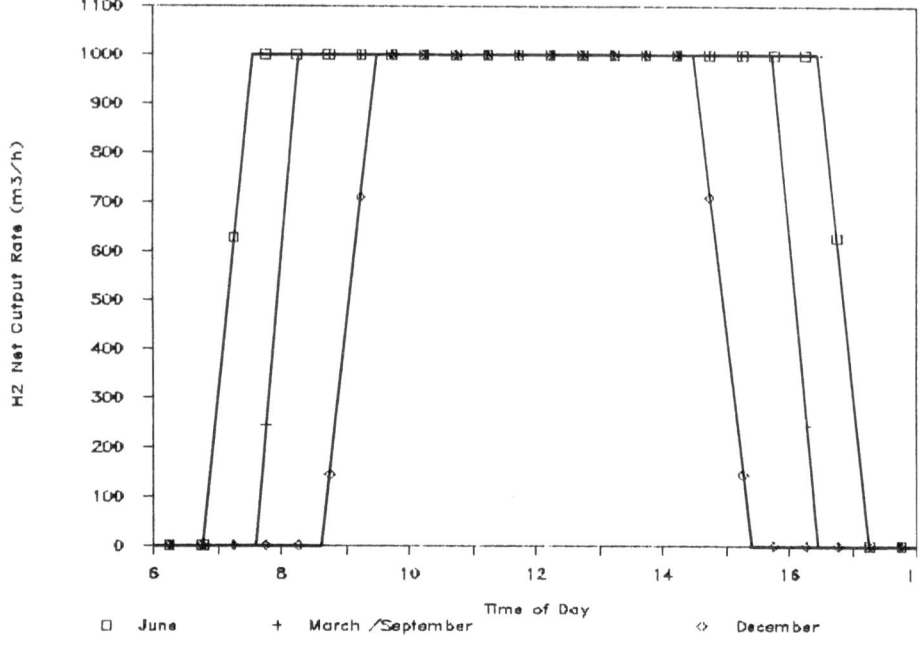

Fig. 4.13 Net hydrogen production rate as a function of daytime
(solar multiple = 2)

The data derived so far can be condensed into two figures which
contain the average daily hydrogen yield and the total efficiency as a function of the solar multiple. This is done in
Figs. 4.14 and 4.15, respectively. For comparison the results
of the same calculations for the case of a system without a
burner are also shown as dashed lines. However, the curves are
only given for solar multiples greater 1.2, since part load
behaviour was not investigated, i.e. hydrogen yield as well as
efficiency would be determined to be zero for a solar multiple
of 1 which is certainly not realistic.

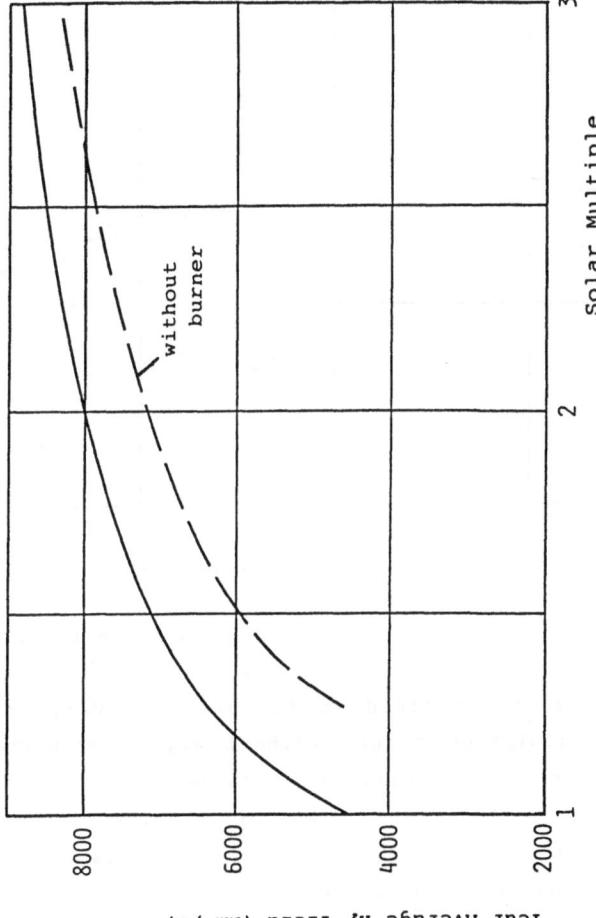

Fig. 4.14 Average daily H_2-yield as a function of solar multiple

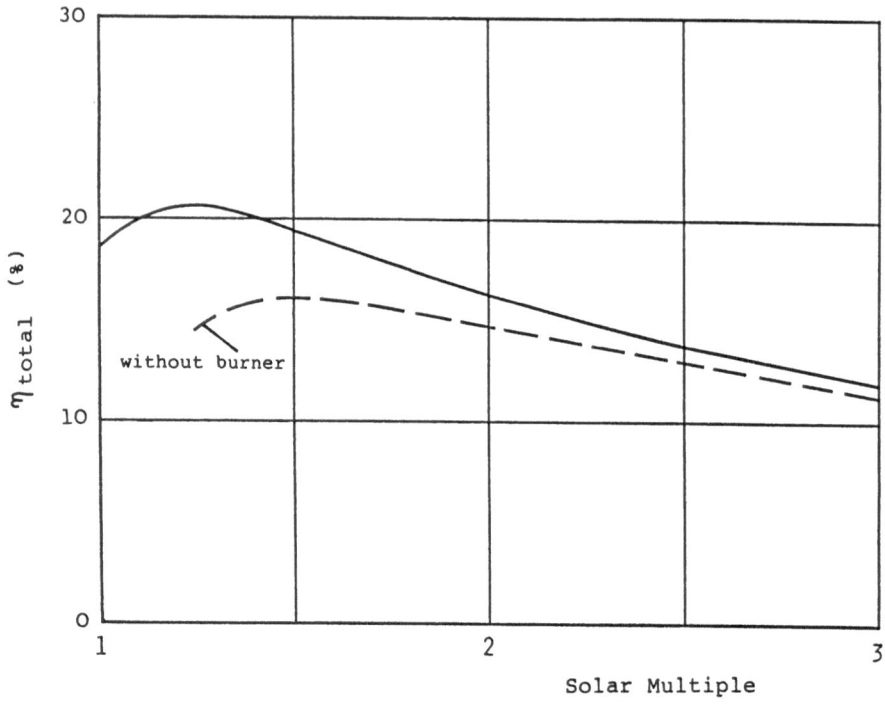

Fig. 4.15 Total efficiency η_{tot} as a function of solar multiple

Of course the results obtained so far are only valid on the condition that insolation occurs without any interruption or weakening (cloud passages, hazy or rainy days etc.) all over the year. However, in this early stage of investigation corresponding corrections are either possible nor necessary, since substantial conclusions can be drawn from the present data already.

Despite the fact that the burner spends part of the produced hydrogen the overall hydrogen yield and thus the efficiency of the system are cleary greater than in the case without a burner. Of course this advantage gets smaller .with increasing solar multiple and the curves will merge asymptotically.

At a solar multiple of one the system yields a mean hydrogen production of 4500 Nm³ per day. According to Fig. 4.14 this value can be substantially raised by enlarging the solar multiple. The maximum of the total efficiency lies in the region of a solar multiple of 1.3 at which the field covers the system's power demand just at winter solstice. At this solar multiple the hydrogen yield amounts to 6500 Nm³ per day which is 40 % more than at a solar multiple of 1. A further increase of the solar multiple leads to a further raise of the hydrogen yield at gradually decreasing efficiency of course.

Despite the fact that the total efficiency has its maximum at a solar multiple of about 1.3 greater solar multiples seem to be reasonable. Since increasing the solar multiple just means enlarging the heliostat field, at an additional expense for only that part of the whole system its hydrogen output can by improved. Though the determination of the optimal field size is only possible if the relative costs for the field and the rest of the system are known, Fig. 4.15 nevertheless recommends that a solar multiple of 1.5 at least should be foreseen. Presumably even higher values are reasonable as can be seen from a numerical example. At a solar multiple of e.g. 2 the system's hydrogen yield is nearly 80 % higher than at a solar multiple of 1. Assuming that the heliostat field amounts to 30 % of the system's total cost in the case of a solar multiple of 1, at an additional expense of 30 % for the field the product cost could be decreased by 45 %. Hence this measure seems worthwhile despite the fact that part of the insolated energy will not be used.

This portion E_{lost} of the total annual insulated energy which is not used for hydrogen production was also derived from the calculated data and is given in Fig. 4.16. Obviously a substantial part of the solar energy is lost (e.g. 35 % at a solar multiple of 2). Therefore one should try to improve the system by integration of a component capable of storing high temperature heat.

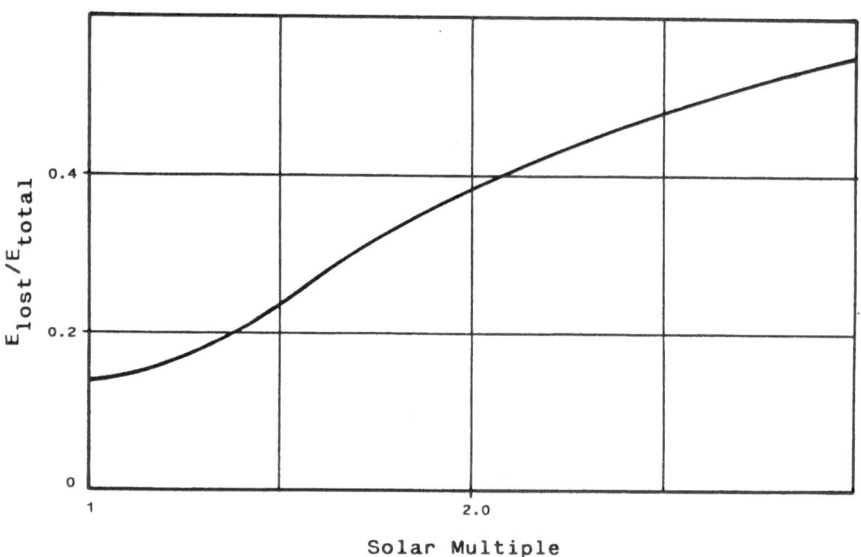

Solar Multiple

Fig. 4.16 Portion of <u>not</u> used insolation as a function of solar multiple

4.5   Critical components and need for development

The principal flow sheet of the system designed in section 4.3 contains a variety of critical components which are not yet available. The requirements which have to be fulfilled by these components can be derived from that flow sheet and thus the areas for further development activities can be identified.

Of course the electrolyzer unit itself is still a critical component. For the technology of the electrolysis modules rather detailed concepts exist, but the design of a complete 'allothermal unit is still an open question. The possibilities for the construction of such an apparatus should be studied in detail especially paying attention to the problem of insulation, support structures for the modules, distribution of gases as well as of electric current and not at least of accessibility for maintenance. These are topics which have to be treated within the future work on high temperature electrolysis.

Within the solar part of the system the critical components are obviously represented by the receiver, the hot gas ducts including the valves and eventually by the hydrogen burner and the gas turbine. The gas turbine is certainly not available as a standard component and need for further development exists, but there is no doubt about its principal feasibility using nowadays technologies. However, this does not hold for the other critical components mentioned above.

Outlet gas temperatures of approximately 1100°C have to be reached by the receiver. Therefrom wall temperatures of the order of 1200°C result for the side of the receiver tubes exposed to the sunlight. This requirement certainly can be fulfilled only by ceramic materials. Additionally the receiver as well as the hot gas ducts (which may be metallic tubes with a ceramic inside insulation) and the burner have to withstand an inside pressure of about 10 bar. Accordingly substantial development efforts are necessary in the field of ceramic technology. The most important topics to be treated thereby refer to questions of tensile strength, compensation of thermal expansion, thermal insulation, stability with respect to cyclic temperature loads and especially of technologies for the gastight joining of different components. With respect to the hydrogen burner a technology has to be developed which allows the controlled injection of H_2 into a hot air stream avoiding thermal damage due to the combustion which will take place spontaneously because of the high temperature.

Finally the two heat exchangers in which the feed steam is superheated have to be mentioned. Especially the second one, in which the steam is superheated from about 815°C to 1000°C by the air stream which cools down from 1025°C to 1010°C is a critical component due to the high temperatures and low temperature difference of the media. However, stress because of pressure drops accross the heat exchanging walls could be avoided by a suitable design. Therefore first of all it should be investigated if this heat exchanger could be built of metallic materials. Otherwise the development of a ceramic heat exchanger would be necessary.

5.   SUMMARY

Within this study the feasibility of combinations of solar
thermal plants and high temperature electrolyzers which are
favourable due to their superior efficiency has been investigated. For the autothermal as well as for the allothermal process
version of high temperature electrolysis combined systems could
be designed which are able to produce the solar fuel H_2 self-
sufficiently, i.e. without need for an existing infrastructure
as for esample a grid.

The combination of an autothermal high temperature electrolyzer, which has to be supplied with low temperature feed steam
and electricity, and a solar thermal plant would be possible
without any principal problem to be seen, if the technolgoy of
high temperature electrolysis were available on a technical
scale already today. This was shown by the layout of an autothermal electrolyzer adapted to the design of the existing
DCS-plant at Almeria (Spain). The corresponding flow sheet is
given again in Fig. 5.1.

The essential feature of the system is that the feed steam
generation for the electrolyzer and the power conversion system
are supplied with thermal energy from a short-time storage.
Thus they are not affected by variations of the insolation and
can be operated at nominal power.

The system's hydrogen production rate is 150 $Nm^3 H_2 /h$ at an
electricity consumption of 480 kW (the DCS-plant was designed
to yield 500 kW electric power).

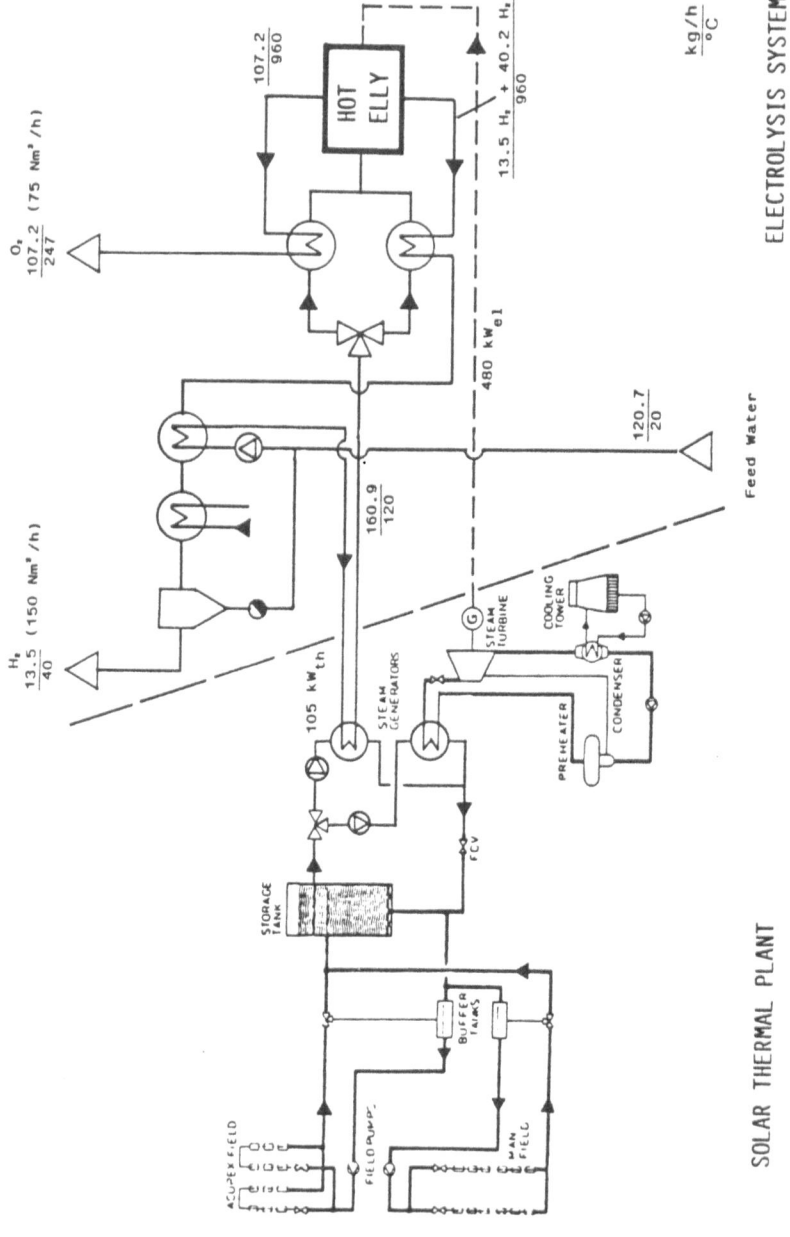

SOLAR THERMAL PLANT   ELECTROLYSIS SYSTEM

Fig. 5.1   Flowsheet for a combined system autothermal steam electrolyzer/DCS solar thermal plant

There is another, allothermal process mode for high temperature electrolysis in which part of the energy necessary for hydrogen production can be provided thermally instead of electrically. This could be achieved by a hot gas stream heating the electrolyzer. Based on model calculations on that problem a principal flow sheet of an allothermal high temperature electrolyzer combined with a GAST-type solar thermal plant has been sketched. It is given once more in Fig. 5.2.

The system contains a burner to cope with variations of insolation and to extend the operating time beyond the time insolation yields the nominal power. This burner is fueled by hydrogen produced on site, i.e. the system does not depend on assistance from outside (e.g. a fossil fuel).

The benefit of coupling in high temperature heat into the process turned out to be an improvement of efficiency of 8 % as compared to an autothermal electrolyzer.

Investigations on the hydrogen yield of the system in terms of the solar multiple revealed that values substantially greater than 1 should reasonably be foreseen.

Due to the high temperatures needed the feasibility of the system depends on the successful development of several components (receiver, gas ducts etc.) of its solar part and of allothermal high temperature electrolyzers.

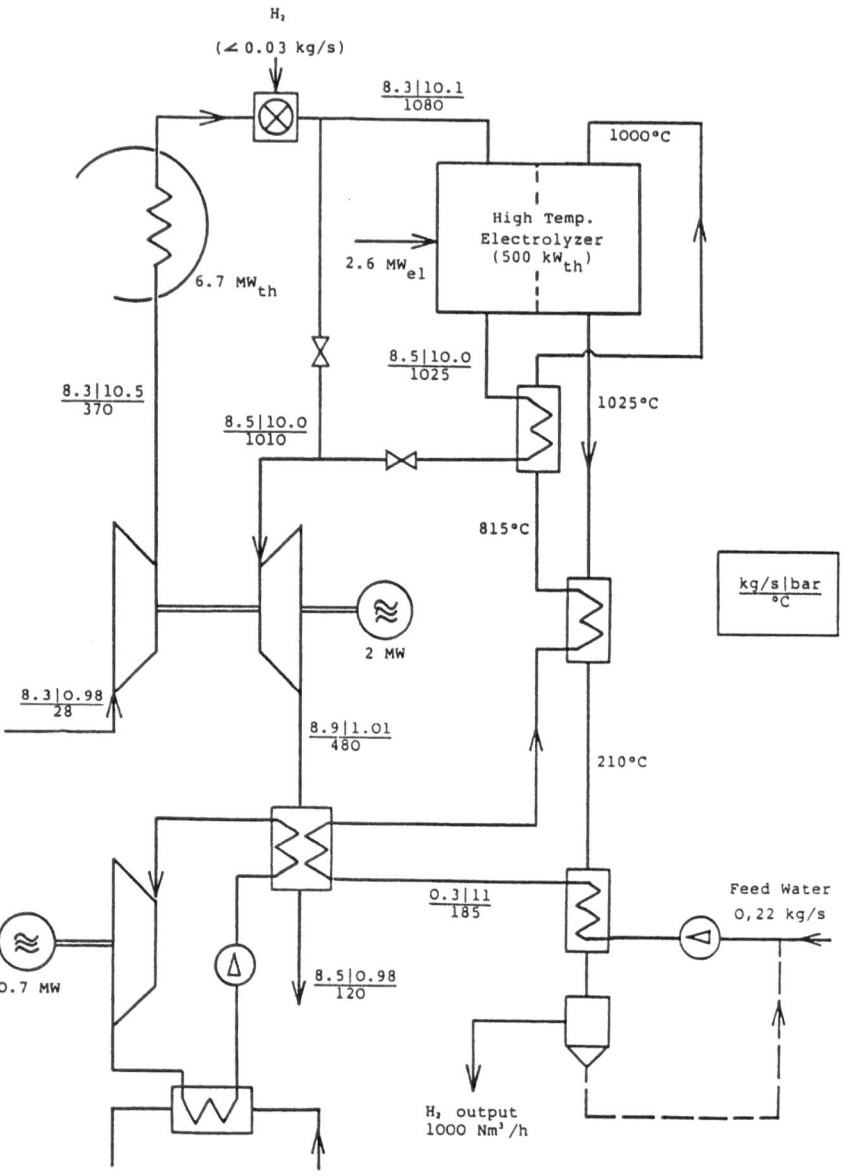

Fig. 5.2 Final flow sheet of an allothermal high temperature
electrolyzer combined with a GAST-type solar thermal
central receiver plant

6.   LITERATURE

For the present study use was made of the following reports:

- IEA Small Solar Power Systems Project SR1
  DCS Construction Report
  (Figs. 3.1, 3.2, 3.3 are taken from this reference)

- GAST-Schlußbericht Phase I

- GAST-Schlußbericht Phase II a

- Technolgieprogramm Gasgekühltes Sonnenturm-Kraftwerk (GAST)
  - Analyse des Potentials -

- Hochtemperatur-Elektrolyse von Wasserdampf
  Forschungsberichte BMFT-FB-T 80-051/84-032

Utilization of Solar Energy
For Hydrogen Production By High Temperature
Electrolysis of Steam

A N N E X

Contents

1. INTRODUCTORY REMARKS

In the report of February 1986* the principal feasibility of a coupled system combining an allothermal high temperature electrolyzer and a solar thermal central receiver system had been shown. According to the possibilities of heat transfer to the high temperature electrolyzer a GAST-type solar thermal plant turned out to be the natural choice for the solar system to couple with.

The present Annex to the report of February 1986 contains the results of further investigations of the following topics:

- Principal possibilities for part-load operation of the system

- Possible benefits from advanced cell technology of high temperature electrolysis with respect to the temperature level necessary for allothermal operation of the electrolyzer

- Elaboration of the possibilities for an improvement of the hydrogen yield by the integration of a high temperature storage system.

Storage of solar heat energy has turned out to be of particular relevance for the application of solar thermal central receiver systems and is hence facing growing interest as could be experienced on the Third International Workshop on Solar Thermal Central Receiver Systems at Constance in June 86. That corresponds exactly with our opinion and therefore the main emphasis of the investigations was dedicated to that topic which was found to be rather complex and demanding.

---

* In the following this report is reffered to as Final Report

2.   PRINCIPAL POSSIBILITIES FOR PART-LOAD OPERATION OF
        THE SYSTEM

In connection with the system designed in the Final Report in
which an allothermal high temperature electrolyzer and a GAST-
type solar thermal central receiver plant are combined the
question arose if and to what extend this system could be operated at part-load. The starting point for this discussion was
the finding that the hydrogen yield of this system which is for
convience shown once more in Fig. 2.1 approached zero at solar
multiples approaching one if the burner was not used. This
behaviour was a consequence of the assumption that the electrolyzer were only able to work at nominal load which is only
reached at noon of summer solstice in the case of a solar multiple of one. However, part-load operation should be possible
and therefore the possibilities for it shall be briefly discussed in the following.

Allothermal operation of the electrolyzer at partial load can
be achieved by reducing the electric current density of the
modules. Then - of course - less hydrogen is produced, i.e.
less electrical power is necessary. Additionally, the specific
amount of thermal energy coupled into the vapor splitting process increases (see chapter 4.1 of the Final Report) which
means that the specific electrical energy demand is reduced. In
principle one could take advantage of these characteristic
features for the realization of part-load operation, but in
view of the restrictions imposed by the 'solar part' of the
system a different way of part-load operation which will be
discussed now seems to be more reasonable.

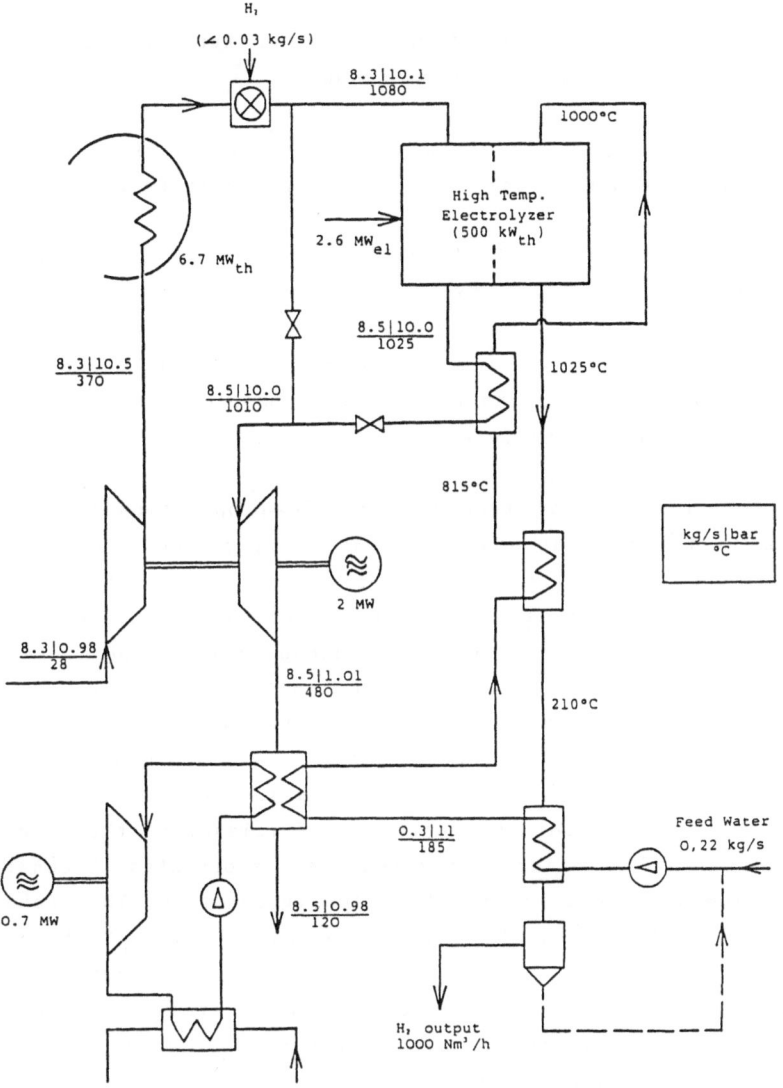

Fig. 2.1 Flow sheet of an allothermal high temperature electrolyzer combined with a GAST-type solar thermal
central receiver plant

With respect to the receiver part-load operation can be realized in two different ways:

- The receiver outlet temperature is kept at its design value and the mass flow is reduced.

- The rated mass flow through the receiver is kept constant and consequently the outlet temperature is reduced.

The first possibility is suited for processes which require thermal energy at a constant temperature level and which can complete with varying heat power. Especially it would have to be taken if the way of part-load operation explained above were to be realized, since the electrolysis modules must be kept of their operating temperature.

The second option of receiver part-load operation has to be taken in the case of power generation by an open gas turbine cycle as it is foreseen in the design of a GAST-plant. This is due to the fact that the gas turbine has to be operated at its nominal mass flow and hence at reduced inlet temperatures in the case of partial load. Since the lion's share of the thermal power delivered by the receiver has to be converted to electricity for the electrolysis process this part-load mode has to be taken. Besides this necessity there is another strong argument in favor of doing so: the part-load efficiency of the receiver is substantially better if the receiver is operated at constant mass flow instead of constant outlet temperature. This behaviour is illustrated by the examples given in Fig. 2.2 which are valid for a metallic tube receiver with a nominal outlet temperature of 800°C. It can also be derived from the receiver efficiency characteristics given in Fig. 4.6 of the Final Report or in Fig. 4.10 of this Annex, respectively.

The consequence of the requirement to reduce the receiver outlet temperature at partial load is that an allothermal operation of the electrolyzer is not possible in this case.

The reason for that is the fact that the electrical resistance
of the electrolysis cells strongly increases with decreasing
temperatures, i.e. only very small current densities and thus
hydrogen production rates would be possible in allothermal
operation at reduced temperatures.

However, one is not forced to operate the electrolyzer in the
allothermal mode all the time, and therefore part-load operation of the system could be realized in the following way: the
air coming from the receiver is not led through the electrolyzer but through the bypass parallel to it (see Fig. 2.1), which
has to be foreseen anyhow as was discussed in section 4.3 of
the Final Report. Thus the receiver output power is used exclusively for the generation of electrical power and of feed
steam. In the electrolyzer only part of the electrolysis modules are operated in the autothermal mode, in which the total
energy demand of the vapor splitting reaction is covered by
electrical energy, i.e. no thermal energy is coupled into the
process (see section 2 of the Final Report). Of course, depending on the electrical power available, only part of the electrolysis modules in the electrolyzer are to be operated.
Reasonably the electrolyzer is not only modified to allow the
operation of only part of its modules but also to enable a
supply of feed steam only to the modules in operation. By this
the feed steam flow could also be adjusted to the available
electrical power in order to keep the hydrogen concentration in
the product gas stream constant.

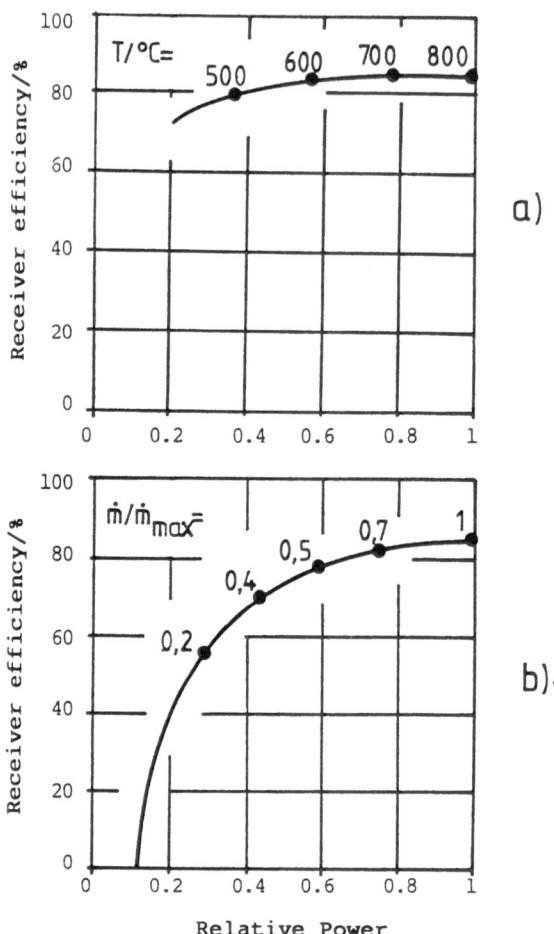

Fig. 2.2   Part-load efficiency of a GAST-receiver
            a) constant mass flow, varying outlet temperature
            b) varying mass flow, constant outlet temperature

By these measures part-load operation of the system would become possible in principle. Additionally one should notice that
the strategy of autothermal operation during part load periods
offers the great advantage that the electrolyzer is kept at its
high operating temperature.

For the sake of completeness it should be mentioned that another way of part-load operation with the electrolyzer working in
the allothermal mode seems to be feasible also. The receiver
would be kept at constant outlet temperature by reducing the
mass flow through it. The difference to the rated mass flow of
the gas turbine would be led through a bypass in parallel to
the receiver and the electrolyzer. Then it would join the hot
gas stream coming from the electrolyzer before it enters the
gas turbine. This strategy offers the advantage that the receiver temperature could be kept constant, but as stated already, it is certainly less efficient due to the part-load
characteristics of the receiver.

Finally it should be mentioned that the problem of part-load
operation is only of minor importance at this early stage of
investigation due to the fact that solar multiples substantially greater than one seem to be reasonable. As a consequence the
timespans during which nominal power is not available are short
with respect to the total operating time, i.e. their influence
or the overall hydrogen yield is small.

3.   BENEFITS FROM ADVANCED CELL TECHNOLOGY

The results presented in the Final Report were based on calculations for which the present data achieved by Hot Elly cells
in the laboratory were taken as an input. The corresponding
cell technology was described in chapter 2 of that report.

Within the Hot Elly project till now a concept of cells and
cell stacks has been pursued, which are mechanically stable by
themselves (self-supporting cells). Due to that a minimal wall
thickness of the electrolyte membrane is necessary ($\sim$ 0.3 mm).
However, there are other concepts for the realization of such
cells which allow a drastic reduction of the electrolyte thickness by a factor of about 10. This can be.taken advantage of in
two different ways:

-   Keeping the operating temperature the electrolyzer can be
    built smaller (using less cells) for a given output power,
    since the ohmic losses resulting from the electrolyte are
    reduced. That leads to a greater current density at a
    given voltage, i.e. to an increase of the specific hydro-
    gen production power of the cells.

-   The other possible advantage has to do with the fact that
    the electrolyte resistivity increases witn decreasing
    temperature (see Final Report, Fig. 2.3). That necessita-
    tes a certain level of the operating temperature in order
    to achieve practical current densities. This temperature
    level can be lowered due to the reduced thickness and the
    correspondingly reduced electrical losses of the electro-
    lyte whereby the current density self-supporting cells
    exhibit at higher temperatures can be kept. Thus with
    so-called thin film cells the same production power can be
    realized at lower temperatures, and that is the feature of
    interest in the context treated here.

Before the effects on the possibilities for the coupling of a high temperature electrolyzer and a GAST-type plant will be presented, a brief description of this alternative cell technology shall be given. The principal set-up of a thin film cell is shown in Fig. 3.1.

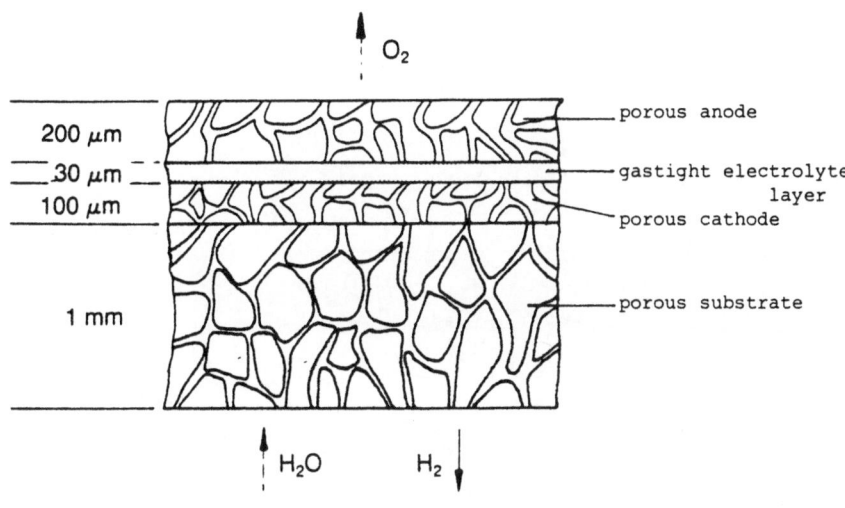

Fig. 3.1:   Schematic set-up of a thin film electrolyte cell

The essential features are the following: Mechanical stability is due to a porous substrate which has a considerable wall thickness (1-2 mm). Its porosity is necessary for reasons of gas transport. The substrate is coated with a porous electrode on which consecutively a dense electrolyte layer is deposited out of the gaseous phase by a special process called (electro-)chemical vapor deposition. By coating the electrolyte

layer with another electrode a complete cell is realized in
principle. Fig. 3.2 shows a micrograph of a cross section of
such a thin film electrolysis membrane which was prepared within a R&D-project sponsored by the European Community. It should
be mentioned that promising results have been achieved for that
alternative concept in the United States where the development
is mainly aimed on the reverse process, i.e. on using the technology for the realization of high temperature fuel cells.

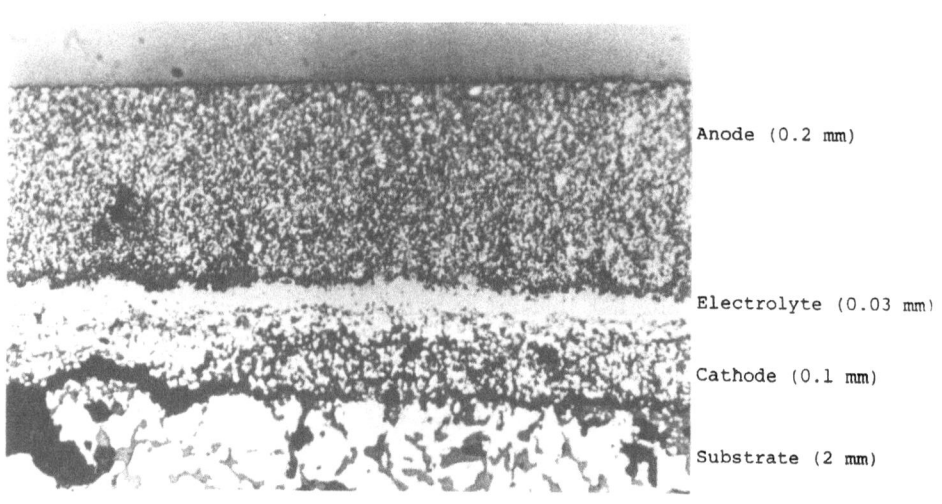

Anode (0.2 mm)

Electrolyte (0.03 mm)

Cathode (0.1 mm)

Substrate (2 mm)

Fig. 3.2:   Micrograph of a thin film electrolyte membrane

Under the assumptions that cells of the geometry chosen for Hot
Elly could be realized as thin film cells and that - compared
to the present state - polarization losses could be reduced by
half due to improvements of the electrodes, the cell resistance
would change according to Fig. 3.3 which shows the results of
corresponding calculations.

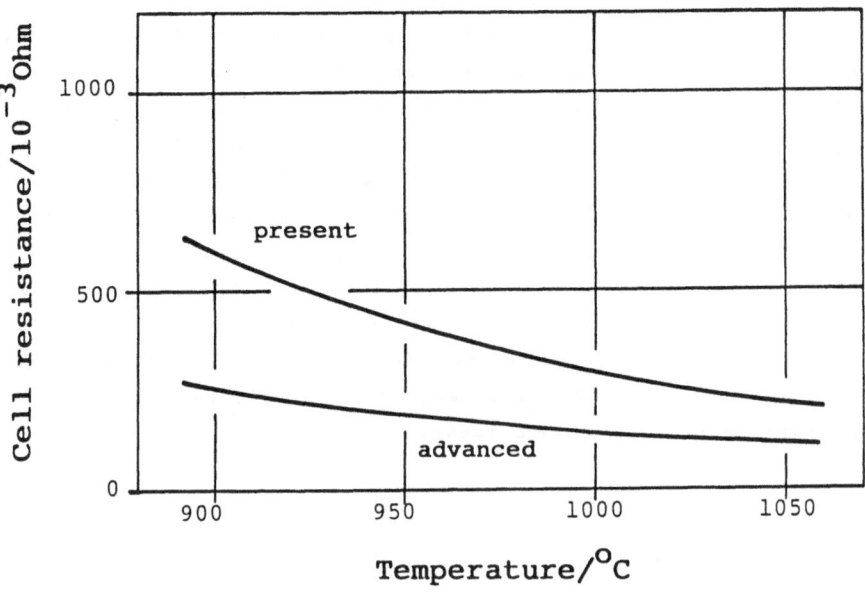

Fig. 3.3:   Possible improvement of cell resistance due to advanced technology

These data were taken as an input for the numerical program simulating allothermal operation which was already presented in section 4.1 of the Final Report.

However, due to the different set-up of the cell wall (porous substrate + layers) calculations of the thermal conductivity of the cells had to be performed in advance: one is dealing with a cell wall the pores of which are filled with gas (H_2, H_2O); thus the overall thermal conductivity depends on the porosity of the ceramic support and on the thermal conductivities of the ceramics as well as of the gases involved.

With this correction installed new computer runs were performed. In order to achieve proper comparibility with the former results the parameters current density, H_2-enrichment and pressure were taken to be the same and only the operating temperature and the mass flow of the heating air stream were varied. Fig. 3.4, which shows the temperature distribution along an electrolysis 'thin film' tube in allothermal operation, gives an example of the results. The shape of this new curve is quite similar to the one of former curves calculated for 'Hot Elly' tubes, but the temperatures are approximately 100°C lower. To be specific, at an inlet temperature of 980°C (instead of 1080°C) the coupling-in of 0.5 kWh/Nm^{3}H$_2$ of specific heat energy into the process came out to be possible at a temperature drop of about 50°C of the heating air. This could be expected from the principal considerations untertaken in chapter 4.1 of the Final Report since in the vicinity of 1000°C 'thin film' cells exhibit the same resistance as Hot Elly cells at a temperature level about 100°C lower.

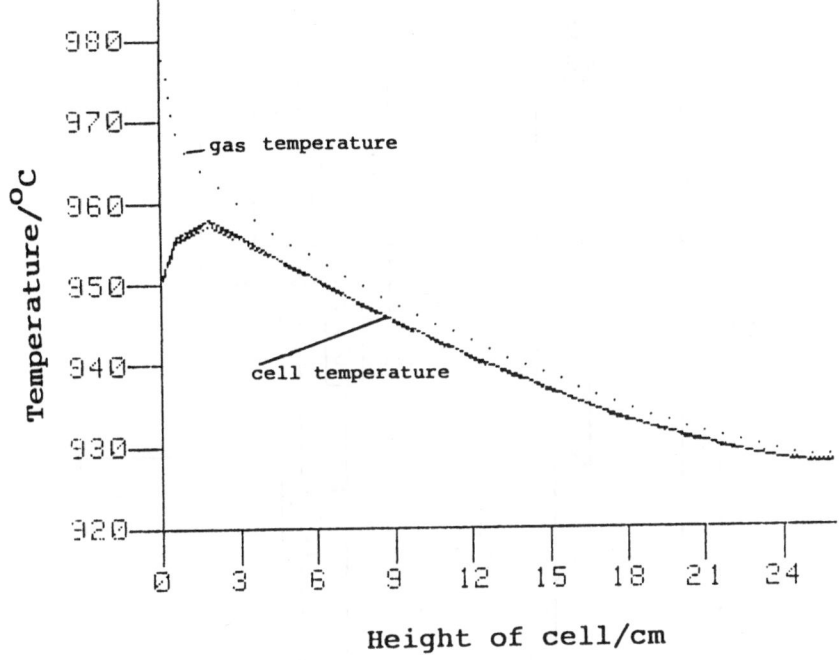

Fig. 3.4:   Temperature distribution of an advanced electroly-
             sis tube in allothermal operation (air temperature
             ≙ dotted line, tube temperature ≙ solid line)

With these results the calculation of new design data for a
flow sheet combining a GAST-type solar system and an allothermal high temperature electrolyzer could be performed in a next
step. In order to evaluate the benefits which are exclusively
due to the difference in electrolyzer technology all the characteristic data of the system's components (e.g. the efficiency of the gas turbine, losses of heat exchangers etc.) were
taken to be the same as in the Final Report (chapter 4.3). Of
course, lower heat losses of the electrolyzer due to the lower
operating temperature were accounted for. The results of these
calculations are presented in the flow sheet of Fig. 3.5.

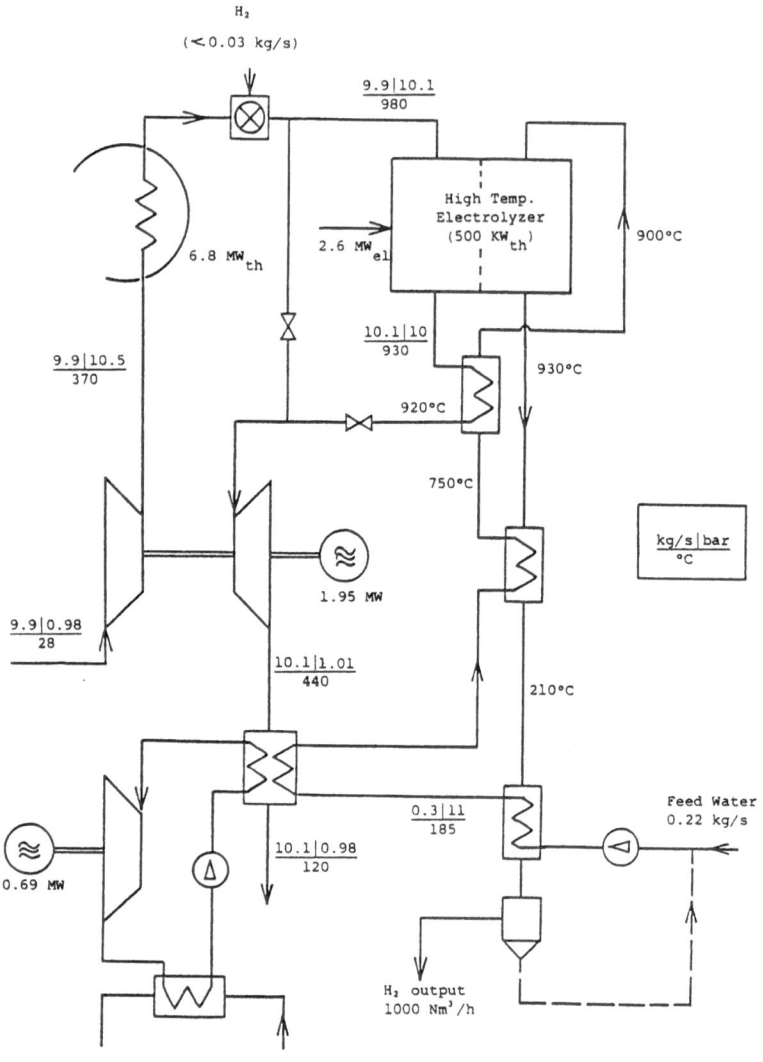

H₂
(<0.03 kg/s)

9.9|10.1
980

High Temp.
Electrolyzer
(500 KW_th)

900°C

6.8 MW_th   2.6 MW_el

10.1|10
930

930°C

9.9|10.5
370

920°C

750°C

1.95 MW

9.9|0.98
28

10.1|1.01
440

kg/s|bar
°C

210°C

0.3|11
185

Feed Water
0.22 kg/s

10.1|0.98
120

0.69 MW

H₂ output
1000 Nm³/h

Fig. 3.5:   Flow sheet of an allothermal high temperature electrolyzer with advanced cell technology combined
with a GAST-type solar thermal central receiver
plant

- 722 -

The essential differences of this flow sheet and the one elabo
rated in the Final Report are the following:

- A greater mass flow (9.9 instead of 8.3 kg/s) is necessary
  for an electrolyzer with a rated output power of
  1000 Nm^3 H_2/h.

- The thermal power needed in the receiver is slightly greater (6.8 MW instead of 6.7 MW) since the effect of the
  greater mass flow is only approximately compensated by the
  reduced superheating requirements (ΔT = 610°C instead of
  710°C across the receiver). I.e., a helistat field which
  is a few percent larger has to be foreseen.

For the sake of completeness it should be mentioned, that the
operational behaviour of the system corresponds exactly to the
one of the system designed in the Final Report. That means
especially that all the statements made there with respect to
the solar multiple hold also here.

Thus in summary thin film technology would allow to realize a
combined system of a GAST-type solar thermal plant and an allothermal high temperature electrolyzer at reduced requirements
with respect to the receiver outlet temperature. This would be
possible at approximately the same efficiency. However, the
temperature level of 980°C is still too high for metallic
solutions' of the receiver, i.e. ceramic components would still
be necessary.

INTEGRATION OF A HIGH TEMPERATURE STORAGE SYSTEM

In section 4.4 of the Final Report the hydrogen yield of the coupled system allothermal high temperature electrolyzer/GAST-type solar thermal plant was estimated. The corresponding flow sheet is given here once more in Fig.4.1. The data are to be understood as modular ones, since they are based on an electro-lyzer capacity of 1000 Nm^3 H_2/h. The investigations of the hydrogen yield of this system as a function of the solar multi-ple revealed that solar multiples substantially greater than one should be foreseen. Doing so, however, part of the insolat-ed solar energy cannot be used by the system. E.g., at a solar multiple of 2 almost 40 % of the annually insolated energy are to be rejected. For that reason investigations of possible improvements by the integration of a system capable of storing high temperature heat were performed, the results of which are presented in the following.

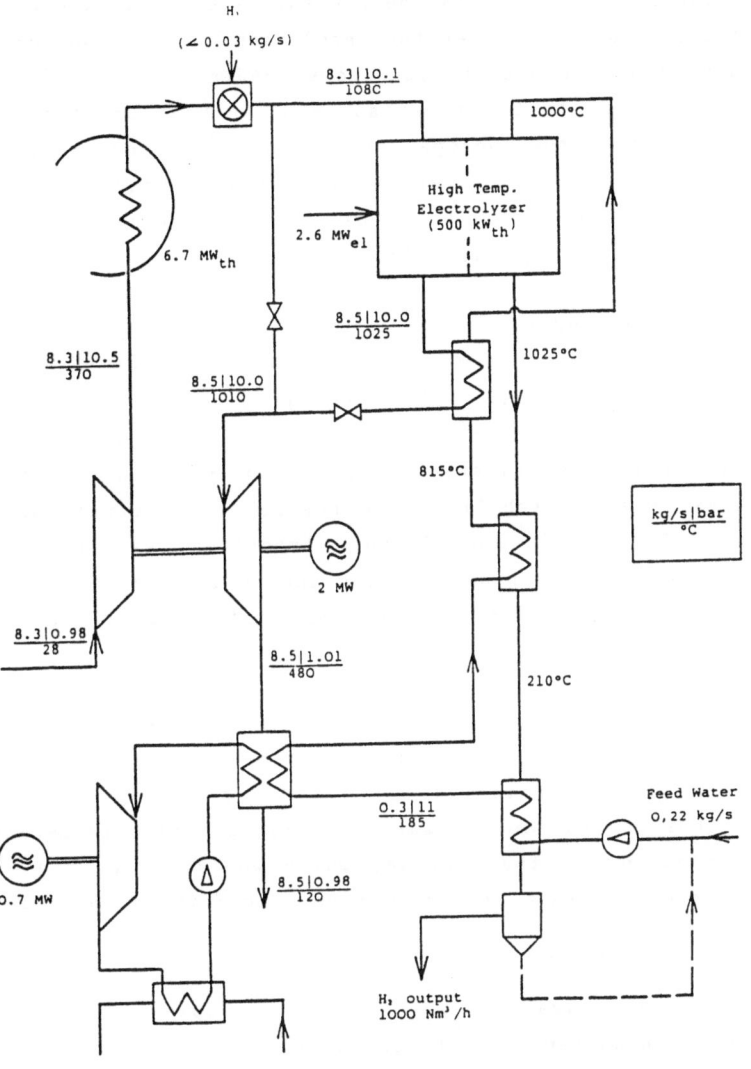

Fig. 4.1: Flow sheet of an allothermal high temperature electrolyzer combined with a GAST-type solar thermal central receiver plant

4.1 Description of the storage system

The storage system needed in the present case has to accomplish the following task: whenever the insolation exceeds the nominal power demand of the electrolyzer/power generation system, this excess energy has to be stored as sensible heat until insolation is less than the hydrogen producing system requires, i.e. storage is necessary for a short time of some hours. Then the stored heat energy has to be supplied to the process thus extending the hours of its operation.

For the system under consideration with its high operating temperature level (~ 1100°C) and air being the heat transporting medium a capacitive solid storage system using ceramic bricks as a storage medium is best suited. Such storage systems are state of the art in steel industry and offer a variety of advantages as e.g. stability with respect to frequent cycling and ease of construction. Another important feature is the capability of heat exchange between the ceramic storage bricks and air in direct contact without any need for a special heat exchanger equipment.

However, some special properties of this storage type have to be accounted for. E.g., stratification is rather poor compared to e.g. a low temperature (200 - 350°C) thermal oil storage system. Thus one has to be aware of varying outlet temperature during unloading of the storage, which would not be the case e.g. for a molten salt storage system comprising a 'hot' and a 'cold' tank.

The present investigations were based on studies performed on this type of high temperature heat storage by the DFVLR. In the following the main features of this solid storage will be reviewed briefly.

The storage consists of many ceramic bricks which have e.g. the
geometry of square stones or of so called 'Cowper' elements the
cross section of which is shown in Fig. 4.2. Favorite material
for the bricks in MgO whose specific heat capcity and density
are 1.26 kJ/kg and 3 g/cm³, respecively.

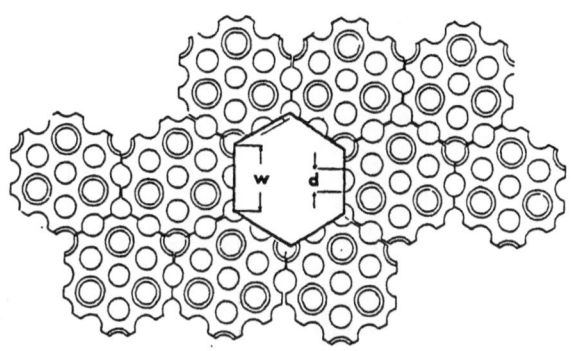

Fig. 4.2:   Example for geometry of ceramic storage bricks
            'Cowper' elements (the air is flowing normally to
            the plane of the picture.)

These bricks are stacked preferentially one above the other in
such a way that channels are formed through which air can flow.
Since heat has to be transported from the surface of the bricks
into their v<lume and vice versa by heat conduction, the thickness of the brick material between the channels has a typical
dimension of only some centimeters. The total volume of the
'storage block' is filled by the bricks to about 70 % the remaining 30 % being left for the gas channels. This stack of
ceramic bricks is integrated in a properly insulated container
of cylindrical shape which allows operation at elevated pressures. An example for this arrangement is given in Fig.4.3.

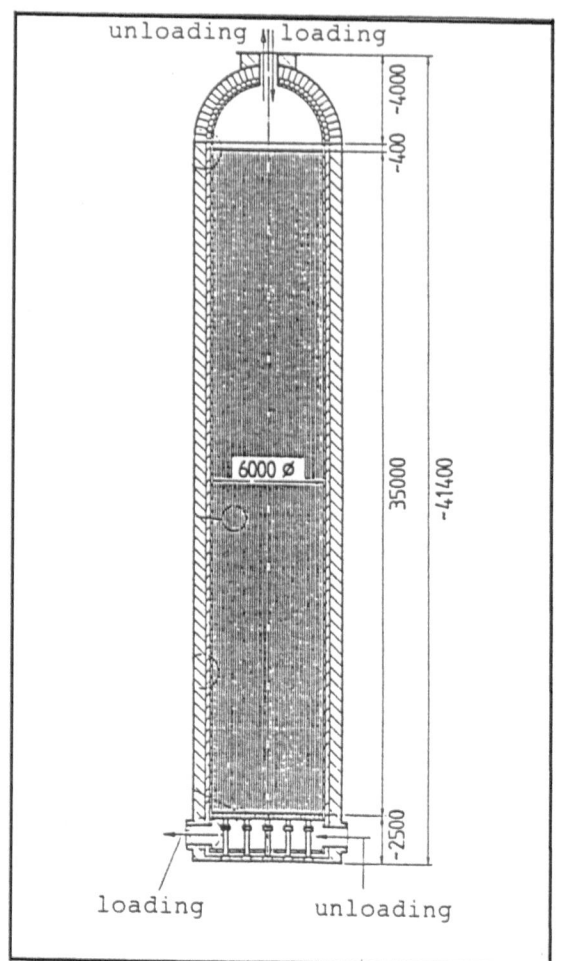

unloading ‖ loading

6000 ⌀

-4000

-400

35000

~41400

-2500

loading   unloading

Fig. 4.3:   Example of storage design

The storage is loaded with heat energy by hot air which flows through it from top to bottom and which is thereby cooled down since heat is absorbed by the bricks. For unloading, cold air is led through the storage from bottom to top whereby it is gradually heated up. The details of the storage behaviour during the first cycles of loading and unloading shall not be treated here, since after a few cycles steady state profiles of the temperature inside the storage are reached. For a storage which is loaded and unloaded with air of 1080°C and 370°C, respectively, Fig. 4.4 shows an example of such profiles, wich give the temperature of the bricks in the storage vessel as a function of the height at which the brick is located in the storage. The storage is 'empty' if the bricks have temperatures according to the 'unloaded profile'. During the loading phase the temperature profile gradually shifts towards the 'loaded profile' (intermediate profiles in Fig. 4.4) at which the stcrage is 'full' and back to the unloaded profile again during unloading. The storage capacity which can be used corresponds to the area between the loaded and unloaded temperature profile. It is of course only part of the total heat content in the storage since the heat 'stored' in the unlaoded profile cannot be used.

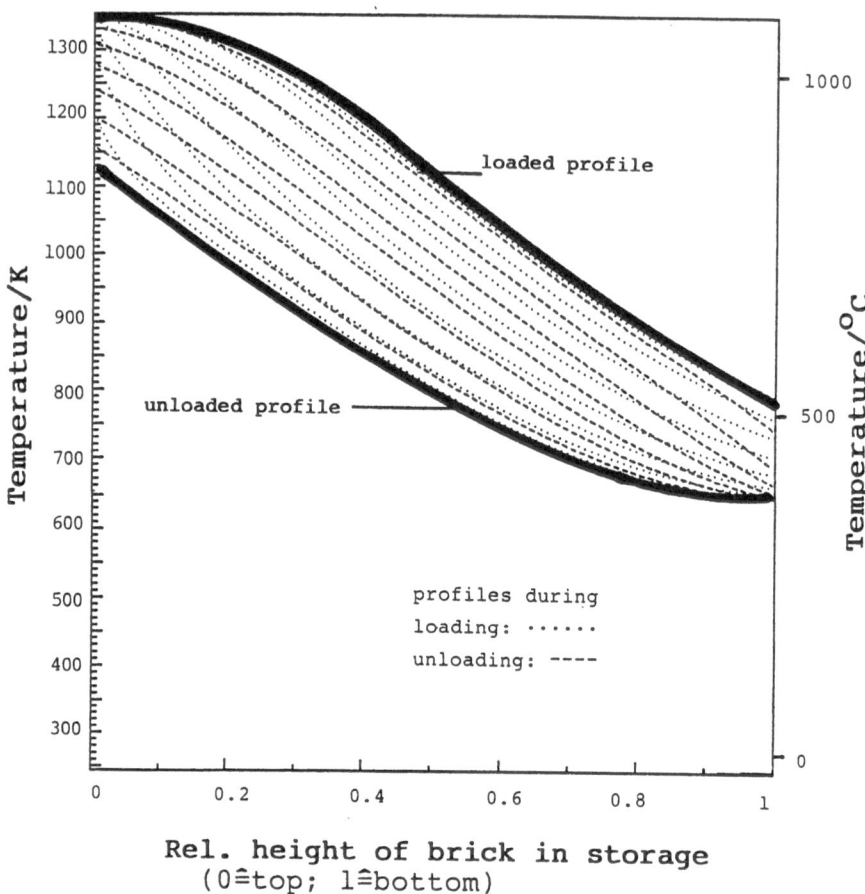

Fig. 4.4: Temperature profiles of a ceramic brick storage

As a consequence of these temperature profiles the gas temperature at the outlet of the storage varies with time during loading as well as during unloading. This property has to be taken
into account for the design of a system including a high temperature storage. Fig. 4.5 gives an example for this behaviour
for the storage whose profiles are shown in Fig. 4.4. It was
assumed that power during loading depends sinusoidally on time

(solar conditions) and that the storage is loaded at a constant inlet temperature and unloaded at a constant mass flow.

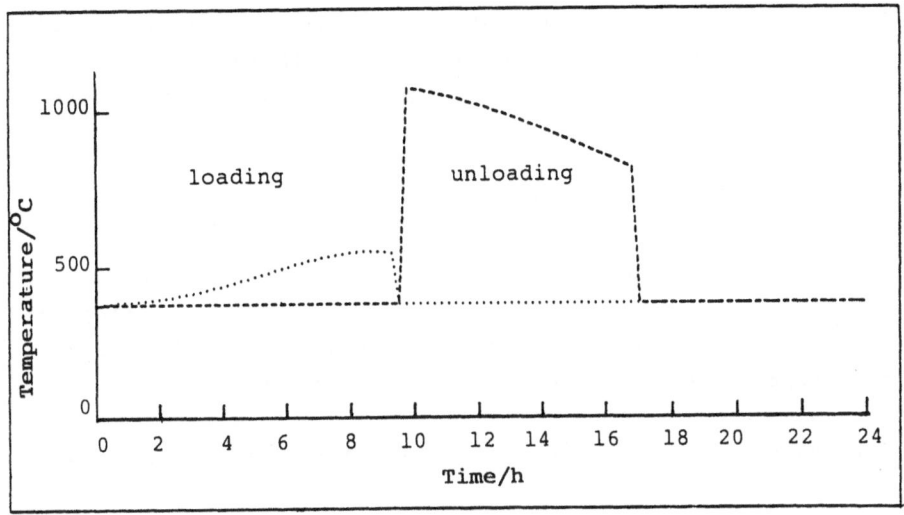

Fig. 4.5:   Gas outlet temperature of a ceramic brick storage as a function of time

At the end of the description of the essential features of this type of storage two further properties are to be mentioned which are important for the considerations of the following sections.

The first one is the fact that the pressure drop of the air passing the storage is of the order of less than 10 millibar, i.e. this pressure drop can be neglected in comparison to the others in the total system.

The second one concerns the heat losses, which amount to only 2 % of the useful stored energy within 24 hours. Therefore heat losses as well as pressure drops of the storage were neglected in the numerical calculations described in section 4.3.

4.2   Principal integration of the storage

Since the electrolyzer/power generation subsystem and the storage have to be operated simultaneously at times of high receiver output power the storage is to be incorporated in parallel.
For the times of low or no insolation there must be a possibility to supply the H_2-producing system with thermal energy by
the receiver and the storage or by the storage alone, respectively.

The resulting modified flow sheet of the system of Fig. 4.1 is
shown in Fig. 4.6. It is based on the assumption that the receiver is still only able to yield outlet temperatures of
1080°C. Thus the outlet temperature of the storage during unloading will be less, i.e. the burner is needed then to meet
the temperature requirements of the electrolyzer.

The system is operated in five different phases during a day:

1.   In the morning when insolation is too low to satisfy the
      H_2-generating system, energy could be stored in principle
      if there is a source of electricity to drive the blower.
      In this case the receiver would be operated preferentially
      at constant outlet temperature and varying mass flow cor-
      responding to the variation of insulation.

2.   At the time the receiver is able to yield at least 3.6 MW
      at a mass flow of 8.3 kg/s H_2-generation is started using
      the burner to cover the energy demand not provided by the
      receiver, which is now operated at constant mass flow and
      varying temperature.

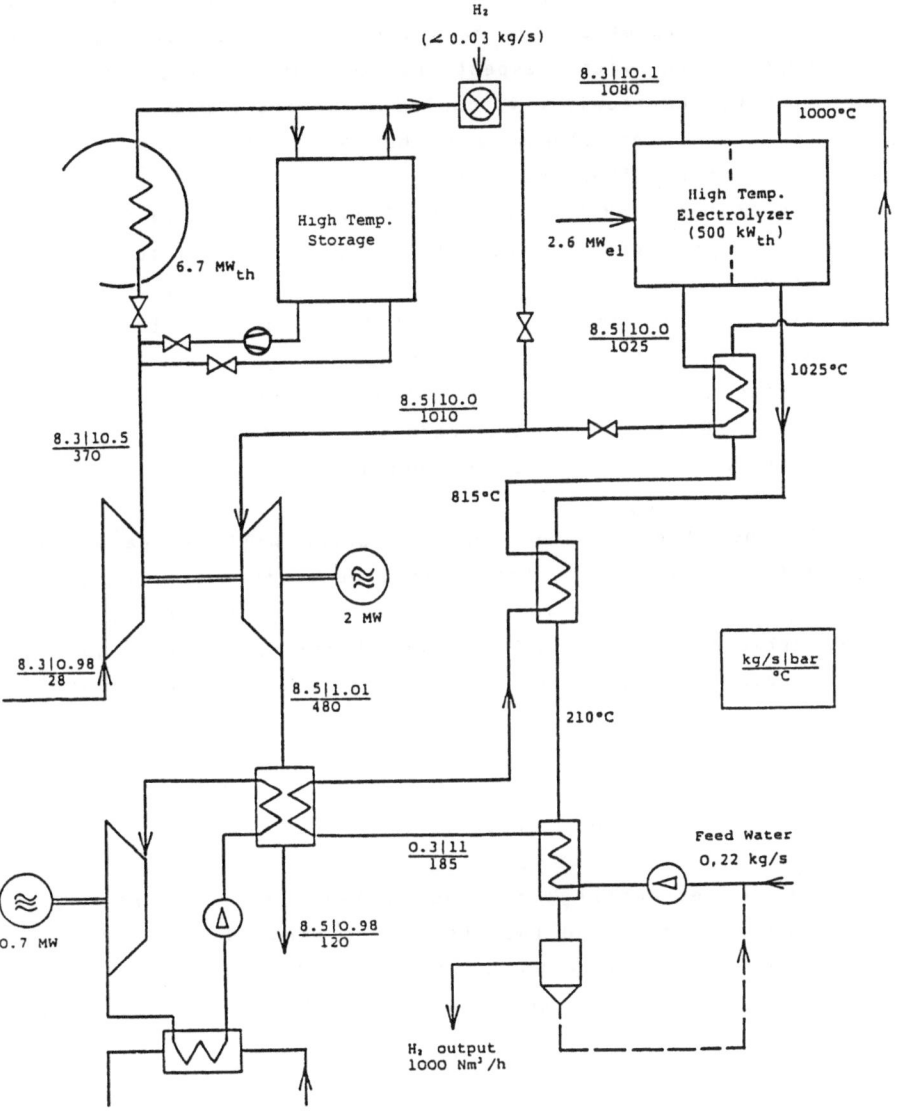

Fig. 4.6:   Flow sheet of an allothermal high temperature elec-
            trolyzer combined with a GAST-type solar thermal
            central receiver plant including a high temperature
            heat storage

3. When insolation exceeds the level necessary to yield
    6.7 MW receiver output power at 8.3 kg/s mass flow, i.e.
    keeping the mass flow receiver as well as air would be
    superheated with respect to an outlet temperature of
    1080°C, the receiver is cooled by an additional gas stream
    driven by the blower and thus kept at constant outlet
    temperature. Thus the receiver is now operated at constant
    temperature and varying mass flow of more than 8.3 kg/s.
    Behind the receiver the air stream is split into two sub-
    streams: 8.3 kg/s are provided for the H_2-generating sys-
    tem, and the excess flow is led through the storage. Be-
    hind the storage this substream is joint with the air
    coming from the compressor and led through the receiver
    again. Due to that the receiver inlet temperature will
    raise during this phase of operation.

4. In the afternoon the receiver power will decrease below
    6.7 MW at 8.3 kg/s. From this time on the air coming from
    the compressor is led partially through the receiver and
    partially through the storage. Thereafter these two sub-
    streams are joint and superheated by the burner to 1080°C
    until

5. after sunset the storage and the burner will be the energy
    sources until the storage is empty.

There is an essential difference between the present system
including a storage and the one of the Final Report without a
storage which should be emphasized here already because of its
important consequences which will become clear in the next
section:

In the former system one and the same receiver was used what-
ever solar multiple was foreseen. This receiver was operated at
nominal load and constant mass flow, respectively, and thus at
constant efficiency whereas the helistat field was operated at
'part load' (i.e. unused heliostats at high insulation). In the
present system the heliostat field is operated at 'nominal
load' (always all heliostats are used) whereas the receiver,
whose size now depends on the solar multiple, is operated at
part-load except at noon of summer solstice. Thus receiver
efficiency varies with time and is in general less than at
nominal power.

At the end of this section it should be mentioned that a different design avoiding the use of the burner during unloading
would be possible. In that case a receiver capable of higher
outlet temperature would be necessary. Then the storage could
be loaded at a higher inlet temperature which would ensure a
temperature of 1080°C or more during unloading. However, due to
the higher temperature the receiver would have less efficiency.
Additionally, the necessity to lead cold air through a bypass
in parallel to the receiver and the storage, respectively,
would arise in order to meet the electrolyzer's inlet requirement of 1080°C air temperature by mixing cold and hot air.
Because of these two disadvantages this alternate design was
not studied further.

4.3   Estimate of the system's H_2-yield

For the system considered in the Final Report a numerical code
had been developed for the calculation of the hydrogen yield.
For the present system this program had to be modified and
extended subtantially.

The main reason for that lies in the fact that the receiver
efficiency is no longer a constant since one is dealing with
varying mass flows through the receiver. Another problem was to
establish a reasonable set of data for the simulation of the
storage.

The latter task was treated in the following way: It was assumed that the storage size has to be adapted to the solar multiple in such a way that the storage capacity is totally used
just at summer solstice.

For the case of a solar multiple of 2 the energies to be stored
at summer solstice, equinox and winter solstice were estimated
as well as the time spans for the loading of these energies
into the storage. These data together with the boundary conditions

- that the power into the storage depends sinusoidally on
  time,

- that the storage is loaded at an inlet gas temperature of
  1080°C and

- that it is unloaded at a constant gas flow of 370°C inlet
  temperature

were given to DFVLR Stuttgart as input data for a computer code
existing there for the simulation of the high temperature solid
storage.

Starting with the layout of an optimized reference storage which was designed for a different purpose (i.e. different boundary conditions) the geometry of the storage was changed until storage profiles similar to these of the reference case were achieved. Additionally the behaviour of the outlet gas temperature during the loading and unloading phase were calculated. The essential results of these calculations are plotted in Figs. 4.7 to 4.9, which show the temperature profiles of the storage, the power into and out of the storage and its outlet gas temperatures for summer solstice, equinox and winter solstice. For the further calculations it was assumed that this behavior of the storage is valid for other solar multiples also since - as was mentioned already - the size of the storage was supposed to depend on the amount of energy to be stored, i.e. on the solar multiple.

In order to get a feeling for the size of the storage some approximate data for the case of a solar multiple of two shall be given: The masonry of storage bricks forms a cylinder which is 15 m high and 4 m in diameter. This block has a mass of about 400 t and is able to store 45 MWh of energy which is the surplus energy at summer solstice for a solar multiple of two.

The second modification of the computer code for the estimate of the system's H_2-yield concerned the efficiency characteristics of the receiver. Since all energy delivered by the heliostat field shall be used by the receiver its size is now dependent on the field size, i.e. on the actual solar multiple, in contrast to the system considered in the Final Report where the receiver size did not change with the solar multiple.

The calculations were based on the efficiency characteristics of a GAST-receiver already taken in the Final Report. However, for the calculations they had to be extrapolated and rescaled as is indicated in Fig. 4.10.

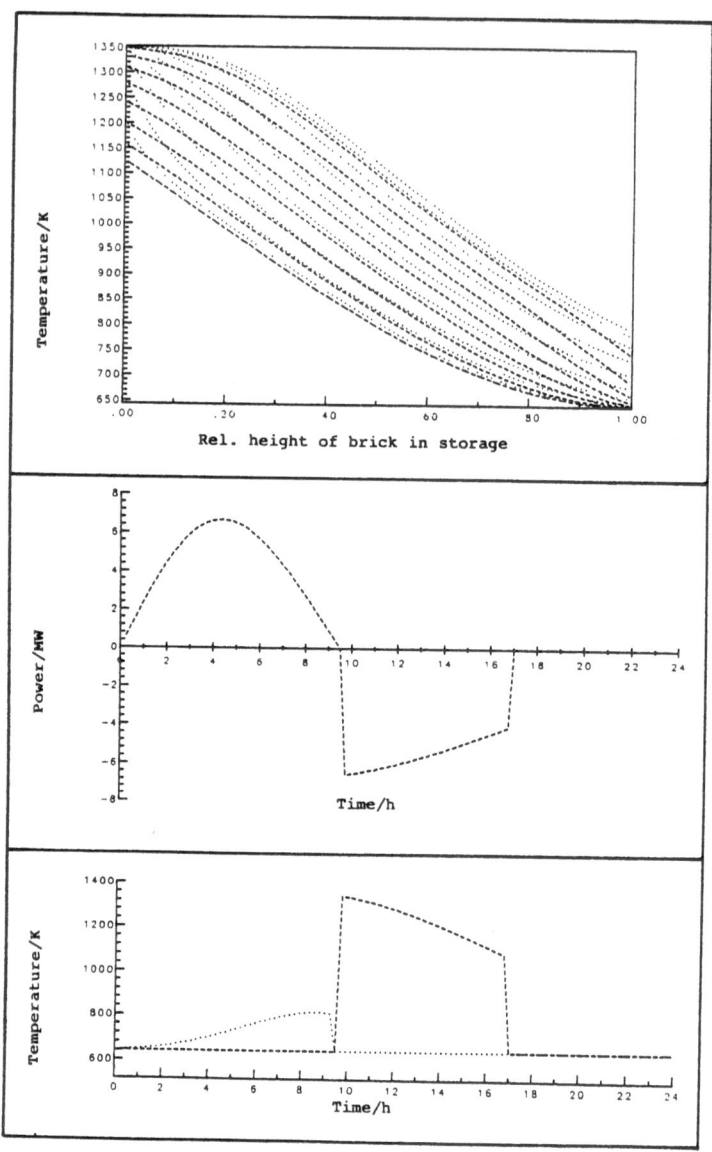

Fig. 4.7: Temperature profiles of the storage, power into and
out of storage and gas temperature at storage outlet during loading and unloading as a function of
time for a solar multiple of 2 (summer solstice)

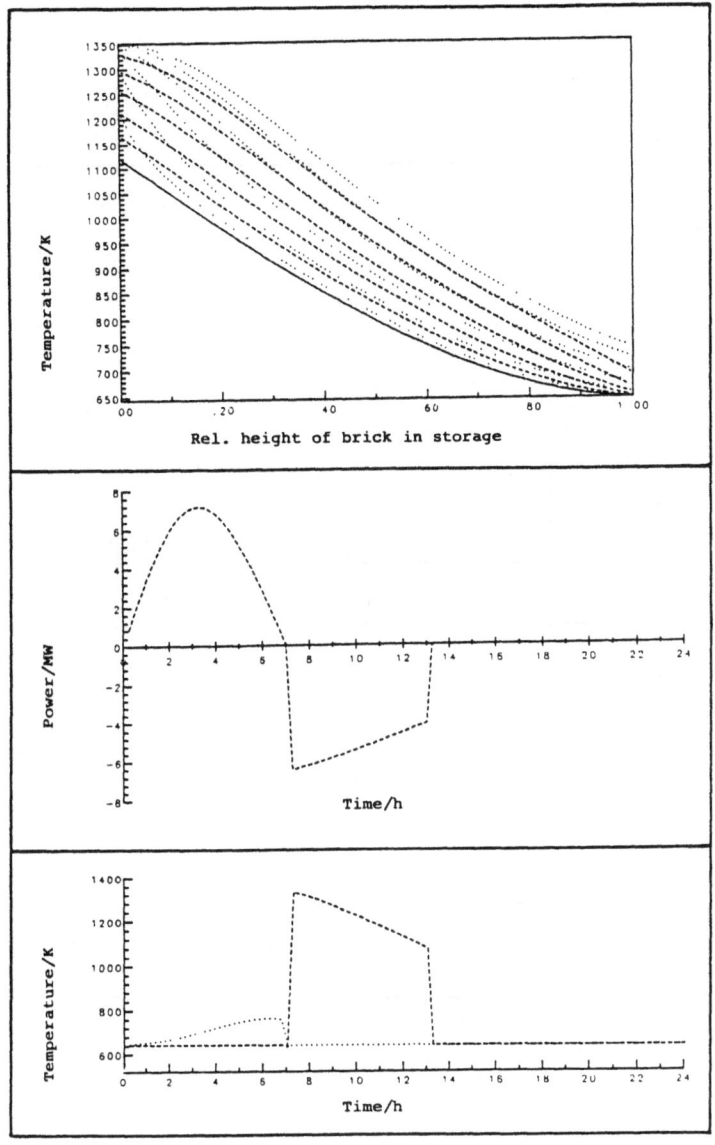

Fig. 4.8:   as Fig. 4.7, but at equinox

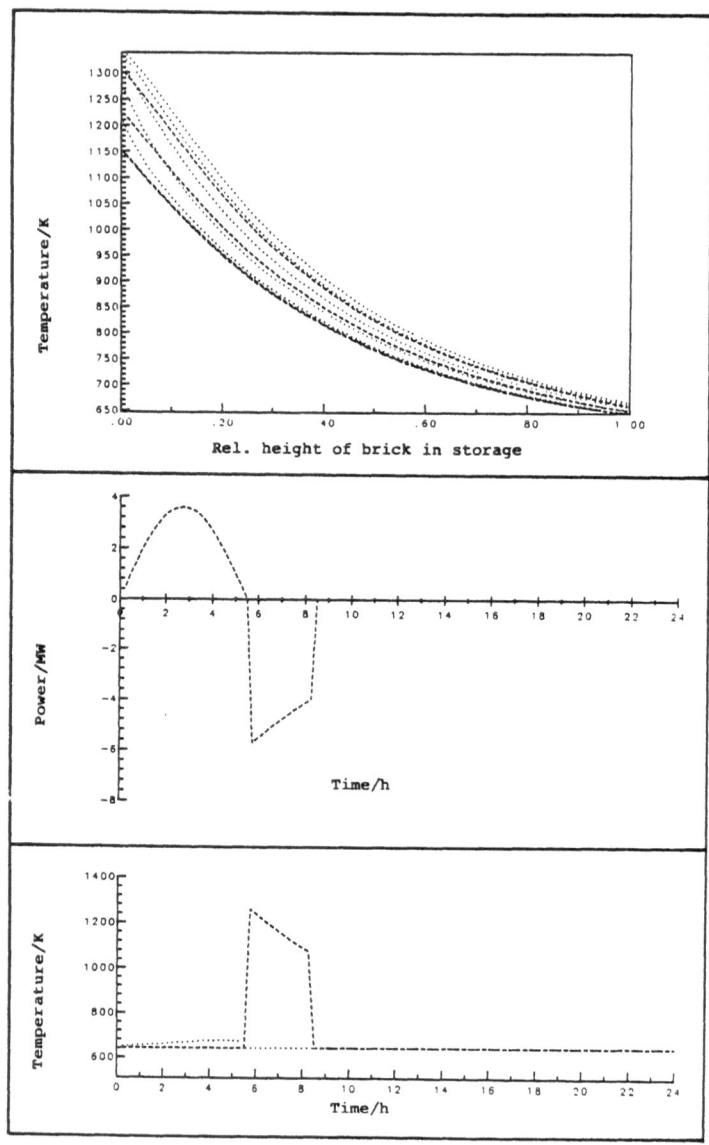

Fig. 4.9:   as Fig. 4.7, but at winter solstice

The nominal receiver load corresponds to a relative field power of 1 and a mass flow of $\dot{m}_{max}$ and thus to an outlet gas temperature of 1080°C (the relative field power is the actual field power divided by the maximum field power). This load is just reached at noon of summer solstice. At all other times the relative field power is less than 1 and thus the mass flow has to be smaller than $\dot{m}_{max}$ in order to fulfil the boundary condition of 1080°C gas temperature. For that reason the receiver efficiency during operation is - in general - less than the nominal value of 77 %.

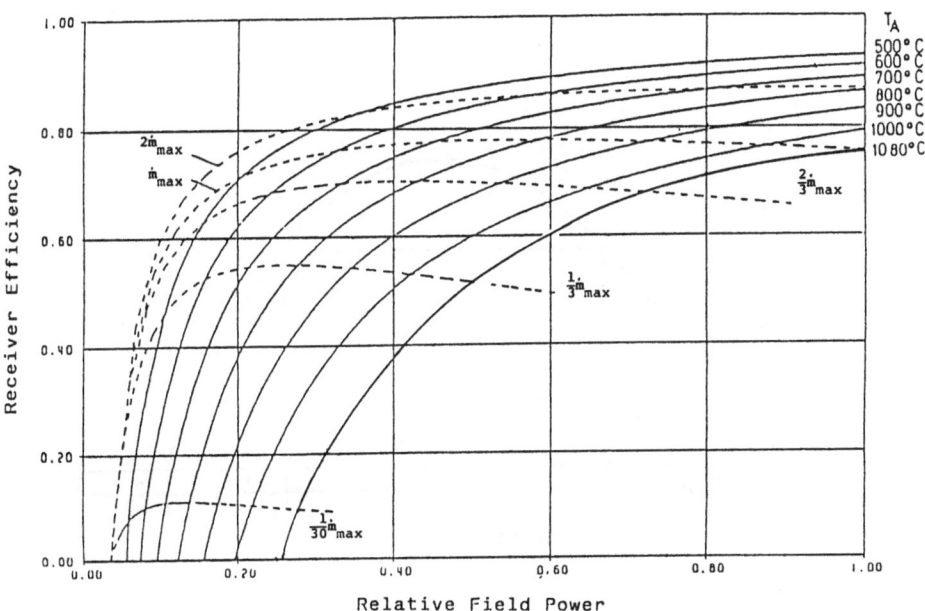

Fig. 4.10:   Efficiency characteristics of a GAST-receiver

The properties of the storage and the receiver explained so far were incorporated in the simulation program which will be described now. In this program the following operational scheme is pursued (see Fig. 4.11):

As can be seen from the receiver efficiency characteristics, the boundary condition of an outlet gas temperature of 1080°C can only be fulfilled at relative field powers exceeding 0.24. Additionally the following difficulty exists: If the mass flow $\dot{m}$ is small compared to $\dot{m}_{max}$ then the receiver efficiency is

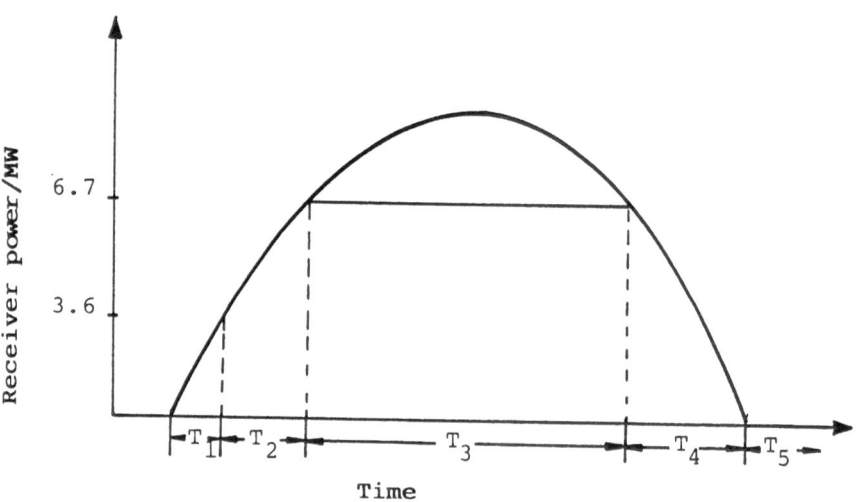

During T_1: unused solar energy
T_2: solar energy plus fuel energy from burner
T_3: solar energy only; storage loading with surplus
T_4: solar, fuel and stored energy used
T_5: stored and fuel energy used

Fig 4.11:   Qualitative operational scheme

(within the accuracy of the extrapolated receiver data) proportional to the mass flow. Hence for the receiver efficiency
η_{Rec}

$$\eta_{Rec} = \alpha \cdot \dot{m} \qquad (1)$$

is valid with α being a constant of proportionality. As long as
equ. 1 holds the outlet gas temperature is independent of the
mass flow and given by

$$T_{out} = T_{in} + \frac{\alpha \cdot FP}{C_p} , \qquad (2)$$

where

C_p = specific heat of the gas
FP   = field power
T_{in} = gas temperature at receiver inlet
T_{out} = gas temperature at receiver outlet.

According to equ. 2 there is only one value of field power
which meets the storage requirement of 1080°C gas temperature,
if $\dot{m}$ is small compared to $\dot{m}_{max}$. Thus during a period T_1 this
storage requirement is failed, i.e. during T_1 no energy can be
stored. However, one has to realize that T_1 is a very short
time span and its share of the total insulated energy is extremely small, especially at greater solar multiples.

From the time on at which the receiver power exceeds 3.6 MW at
$\dot{m}$ = 8.3 kg/s, the H_2-generating system is operated by the receiver and the burner during a period T_2 until the receiver
power passes the threshold of 6.7 MW (= power demand of the
electrolyzer/power generation system). During period T_2 the
receiver is operated at constant mass flow and variable outlet
temperature.

During a subsequent period T_3 the receiver power exceeds 6.7 MW; the surplus energy is loaded into the storage. In this period the receiver is operated at a constant outlet temperature and a variable gas flow (greater than 8.3 kg/s).

From the time on at which the receiver power decreases below 6.7 MW in the afternoon, the receiver is operated at 1080°C outlet gas temperature as long as possible (at a mass flow smaller than 8.3 kg/s). Simultaneouly the storage is unloaded and after mixture of the gas streams flowing through the receiver and the storage the gas temperature is raised to 1080°C by the burner.

This period T_4 ends when the receiver power is no longer great enough to yield 1080°C outlet gas temperature. From that time on the system is operated by unloading the storage and using the burner until the storage is empty (period T_5). As the calculations described below showed the storage can be completely unloaded before the burner - which superheats the air coming out of the storage with a temperature of less than 1080°C - would spend more hydrogen than is produced.

The equations governing the energy balance between the different subsystems which were used in the calculations are given in the appendix together with the nomenclature. What shall be presented here at the end of this chapter are the results.

The calculations with the modified program were performed by complete analogy to the ones of the Final Report. I.e., the same insulation data were used and values were also calculated at time intervals of 30 minutes for the representative days winter solstice, equinox and summer solstice. Therefrom mean values for the annual hydrogen production and for the total efficiency of the system were determined in the way described in section 4.4 of the Final Report. Of course, this was carried out for a variety of solar multiples.

The following figures give some examples for the output data
produced by the modified program for the case of a solar multiple of 2. In all these figures the respective curves are given
as a function of daytime. Figs. 4.12 and 4.13 show the relative
field power and the receiver efficiency respectively, for the
three representative days.

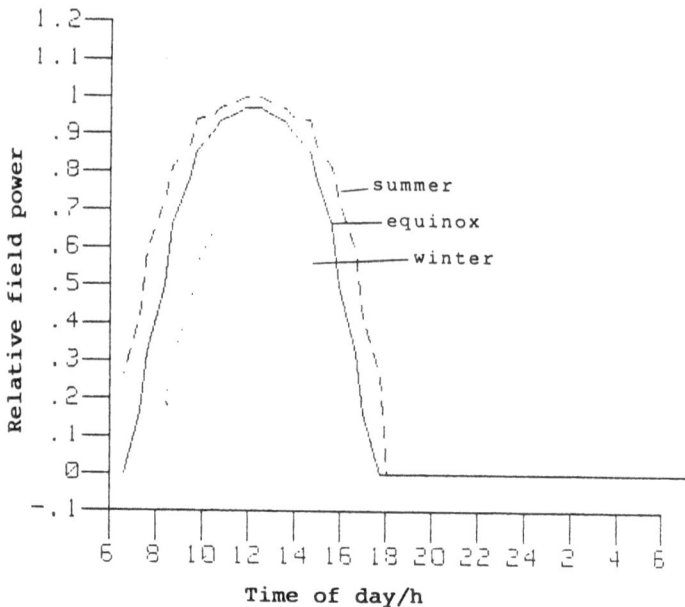

Fig. 4.12:   Relative field power

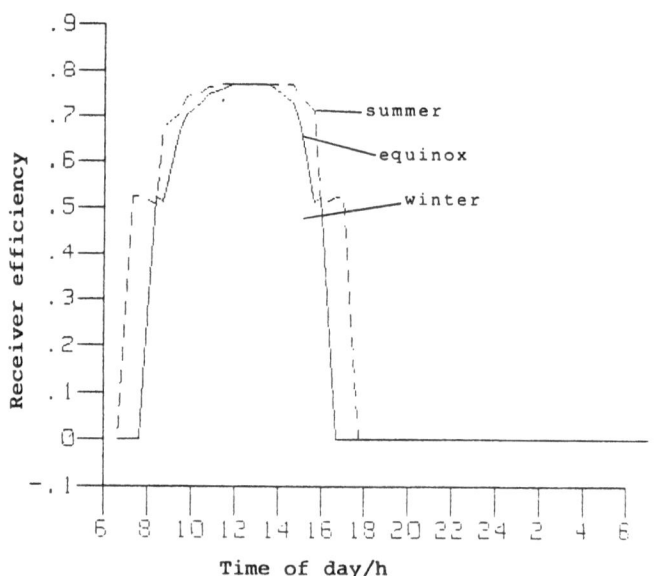

Fig. 4.13:   Receiver efficiency

From Fig. 4.13 it can be seen - as was already stated earlier - that the receiver is operated of part load efficiency ($n < 77$ % = nominal efficiency) most of the time. The various mass flows ($\dot{m}_1$ passes the receiver, $\dot{m}_2$ passes the storage during unloading, $\dot{m}_3$ passes the receiver and subsequently the storage during loading; see appendix) for the representative days are plotted in Figs. 4.14, 4.15 and 4.16.

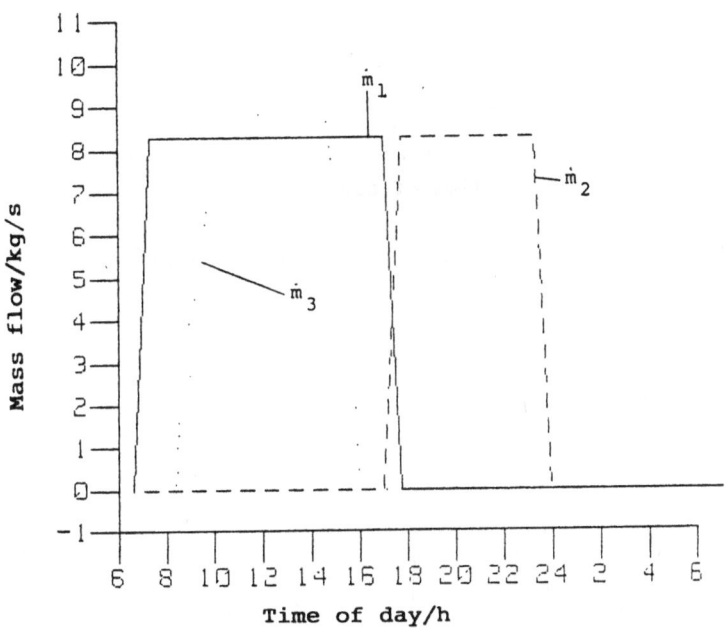

Fig. 4.14:   Mass flow at summer solstice (solar multiple = 2)

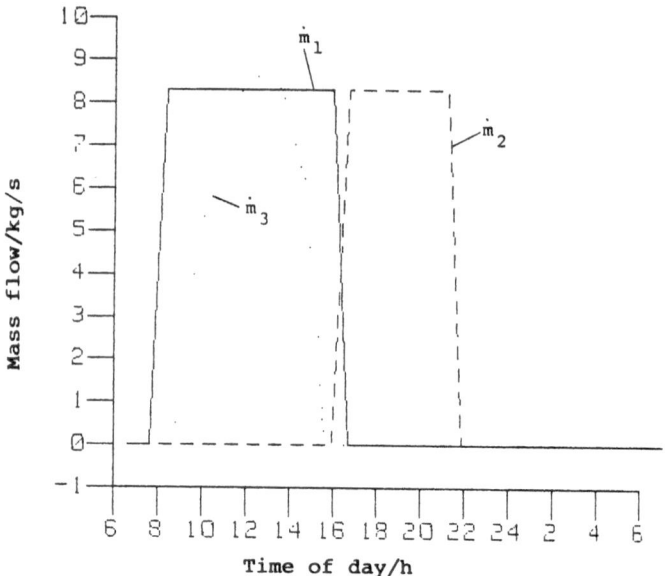

Fig. 4.15:   Mass flows at equinox (solar multiple = 2)

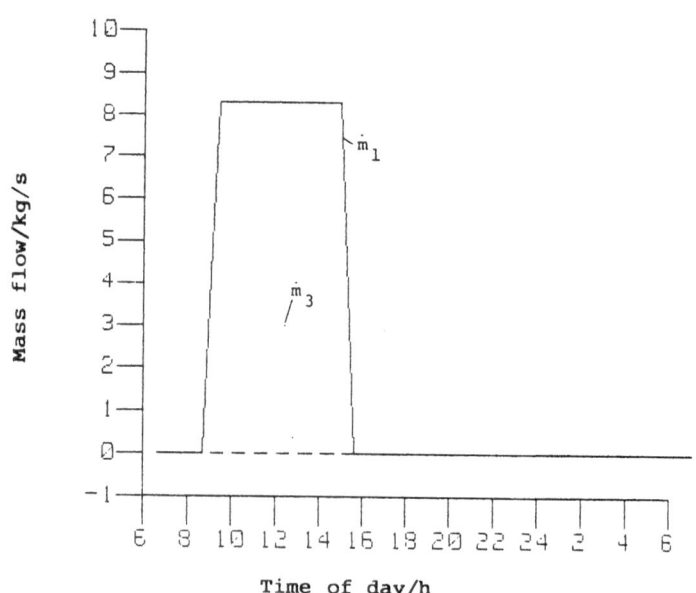

Fig. 4.16:   Mass flows at winter solstice (solar multiple = 2)

The power supplied by the receiver, the storage and the burner at summer solstice, equinox and winter solstice are shown in Figs. 4.17, 4.18 and 4.19 (negative storage power = power out of storage during unloading). From these figures the timespans during which operation is assisted by the burner can be seen as well as the sinusoidal loading of the storage and the slowly decreasing power from storage during unloading which is due to the decreasing storage outlet temperature. Furthermore Fig. 4.19 shows that due to the part-load properties of the receiver no useful energy can be stored at winter solstice.

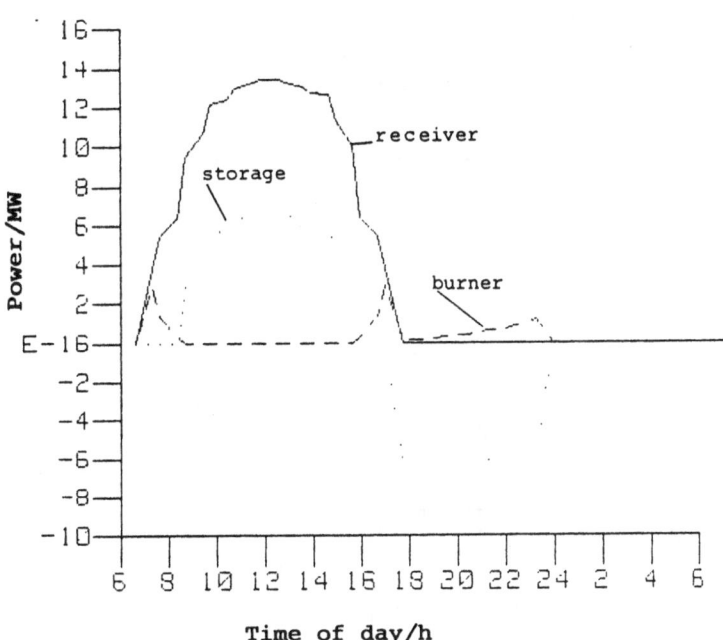

Fig. 4.17: Power of receiver, storage and burner at summer solstice (solar multiple = 2)

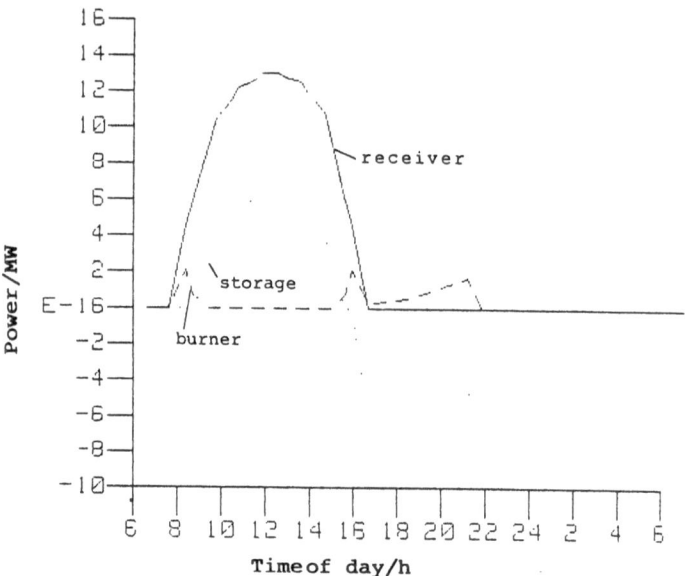

Fig. 4.18:   as Fig. 4.17, but at equinox

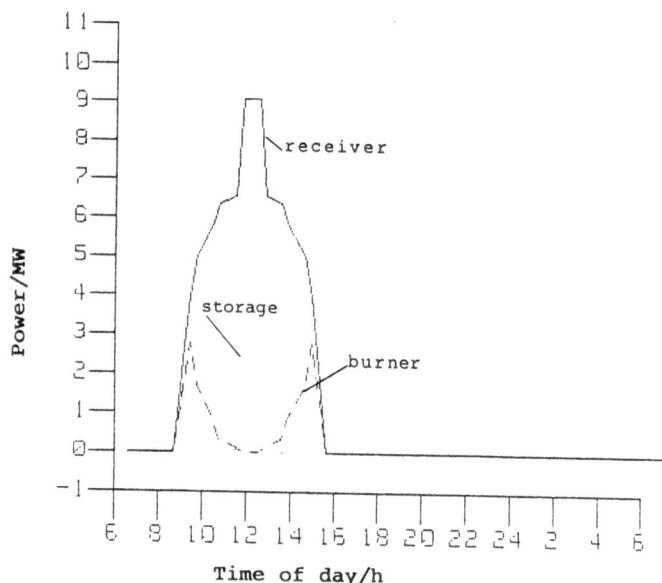

Fig. 4.19:   as Fig. 4.17, but at winter solstice

Finally at the end of these examples Fig. 4.2C shows the storage energy content for the representative days.

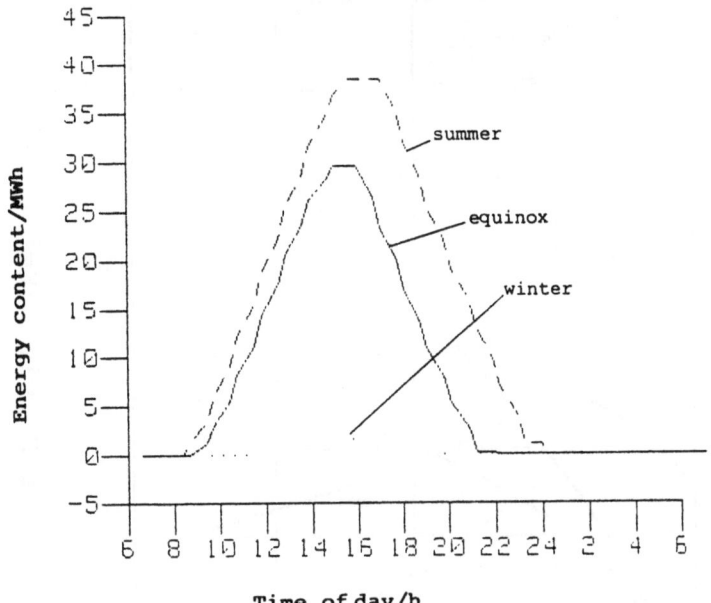

Fig. 4.20:   Energy content of storage for the representative
              days (solar multiple = 2)

Having demonstrated now the possibilities offered by the modified computer program the results on the hydrogen yield and the total efficiency of the system as a function of the solar multiple are presented in Fig. 4.21.

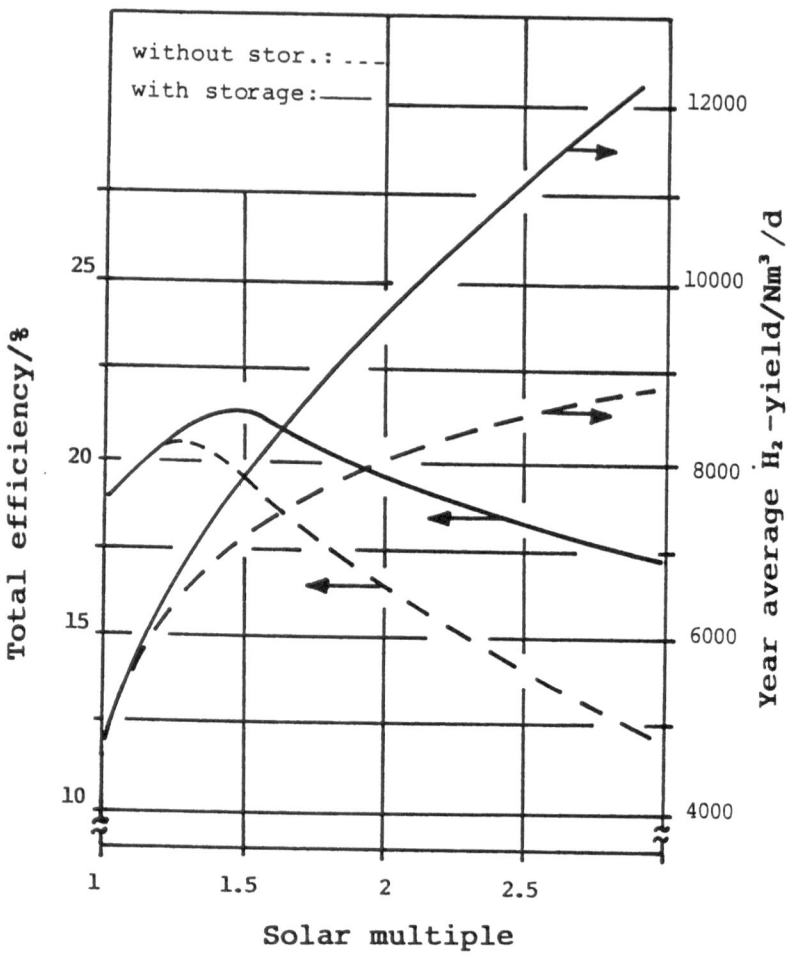

Fig. 4.21:   Mean annual H_2-yield and total efficiency as a
function of solar multiple

The benefits due to the integration of the storage into the
system can be clearly seen from the figure. The hydrogen yield
as well as the total efficiency are improved significantly. Of
course this effect is the greater the higher the solar multiple

is chosen since the former system without a storage could not use insolation exceeding its nominal power, i.e. the portion of unused energy increased with increasing solar multiple (see Fig. 4.16 of the Final Report).

Another effect that can be seen from Fig. 4.21 is that the maximum of the total efficiency is shifted towards a greater solar multiple (from 1.3 to about 1.5) if a storage is foreseen.

At the end of this section a comment should be made on what one could have expected to be the benefit of the storage and on what it turned out to be. For instance, at a solar multiple of 2 about 35 % of the insolated energy could not be used by the former system without the storage. Thus one could have expected that now for the system with the storage the hydrogen yield would be approximately 50 % greater (i.e. ~ 12 000 Nm³/d instead of ~ 8 000 Nm³/d). Obviously this is not the case. The reason for this is the fact that the receiver has to be adapted to the field size if the storage is used since it must be able to be operated at the peak power of the field. Therefore the receiver works at part-load efficiency most of the time. This is the distinct difference to the system without a storage where the receiver's size is determined by the electrolyzer/power generation system and not by the field size. Since then excess energy is rejected by defocussing of heliostats the receiver mostly operates at its rated power and efficiency, respectively. Or, in simple words: in the case of the system without a storage the receiver works at nominal load and the heliostat field works at partial load whereas in the case of the system with a storage the receiver works at partial load and the field works at nominal load.

What remains is the question if the part-load characteristics of the 'component receiver' can be improved. Probably some progress could be achieved for instance by the use of two small receivers instead of a single big one. Then both receivers could be operated in the vicinity of their rated power at times of high insolation whereas at low insolation levels only one receiver would have to be in operation in order to ensure high efficiency. Of course this strategy would require a heliostat field which can be focussed onto two different points.

Finally it should be mentioned that the results obtained so far only enlighten the 'technical side' of the problem. Which of the different feasible configurations one would have to take could of course only be decided if an economic assessment would be elaborated.

5.   SUMMARY

A system combining an allothermal high temperature electrolyzer
and a GAST-type solar thermal central receiver plant had been
designed in the Final Report from February 1986. Within this
annex the investigations were extended in three different directions taking the system of the Final Report in each case as
a reference system.

It was shown that part-load operation of the system is favora-
bly realized by using the receiver power for electricity generation exclusively and operating the electrolyzer in the autothermal mode. This procedure is recommended because of the
part-load characteristics of the receiver and the gas turbine.

Accounting for advances in the technology of high temperature
electrolysis by which the electrical cell resistance could be
reduced it was shown that a system almost identical to the
reference one could be realized at reduced temperature requirements. However, the temperature level required (980°C) turned
out to be still too high for the use of a metallic receiver.

Finally it was demonstrated that the system could be significantly improved with respect to its hydrogen yield by the
integration of a high temperature (short-time) storage. However, the improvement is limited due to the fact that the receiver has to be operated at part-load most of the time. Therefore strategies for improving the efficiency characteristics of
the 'component receiver' should be sought.

6.   APPENDIX

For the calculations the results of which are presented in section 4.3 a set of equations which are governing the energy balance between the subsystems was used. They are given here together with the nomenclature applied and a sketch of the mass flow scheme (Fig. A1) used.

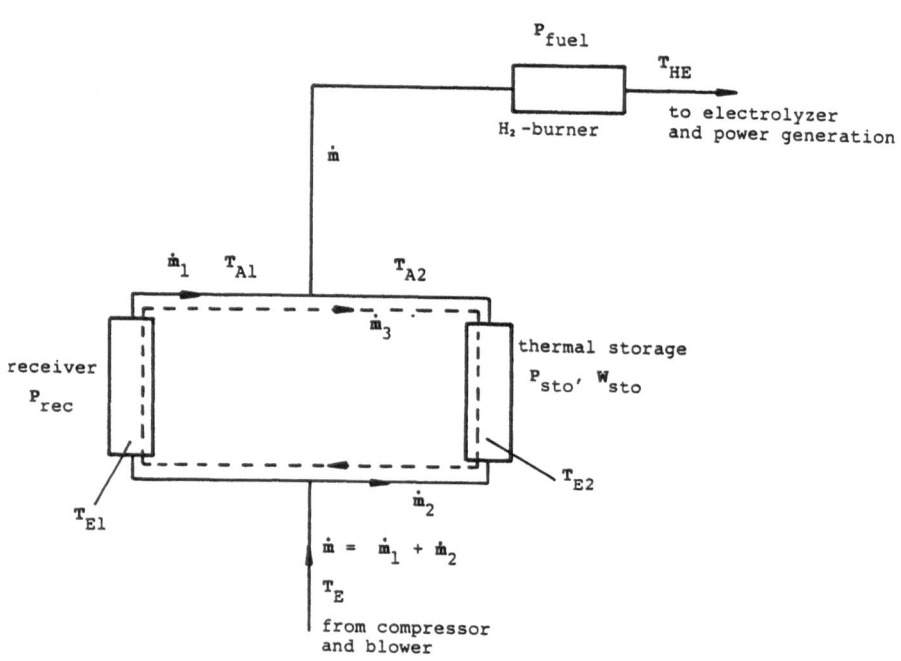

Fig. 6.1:   Mass flow scheme used in the calculations

Nomenclature:

T_{HE} gas temperature required by electrolyzer ($1080°C$)

T_E gas temperature behind compressor ($370°C$)

T_{E2} outlet gas temperature of storage during loading

T_{E1} gas temperature at receiver inlet not equal T_E since gas flows from compressor and from storage are mixed

T_{A1} gas temperature at receiver outlet

T_{A2} gas temperature at storage outlet during unloading

P_{rec} receiver output power

η_{rec} receiver efficiency

P_{fuel} power of burner

P_{sto} power to or from storage

W_{sto} thermal energy stored

$\dot{m}$ mass flow required by electrolyzer/power generation (8.3 kg/s)

$\dot{m}_1$ mass flow through receiver

$\dot{m}_2$ mass flow through storage during unloading

$\dot{m}_3$ mass flow through storage during loading

t time

t_0 time at which loading of storage starts

c_p specific heat of air

Equations:

$$\dot{m} = \dot{m}_1 + \dot{m}_2 \; ; \qquad \dot{m} = 8.3 \text{ kg/s} \tag{1}$$

$$P_{rec} (\dot{m}, FP) = n_{rec} (\dot{m}, FP) \cdot FP \tag{2}$$

Case a) $P_{rec} (\dot{m}, FP) < 3.6$ MW and $W_{sto} = 0$
The system is out of work (period T_1)

Case b) 3.6 MW $\leq P_{rec} (\dot{m}, FP) \leq 6.7$ MW and $W_{sto} = 0$

$$\dot{m}_1 = \dot{m} \tag{3}$$

$$\dot{m}_2 = \dot{m}_3 = 0 \tag{4}$$

$$P_{rec} = n_{rec} (\dot{m}, FP) \cdot FP \tag{5}$$

$$P_{sto} = 0 \tag{6}$$

$$W_{sto} = 0 \tag{7}$$

$$P_{fuel} = C_p \dot{m}_1 (T_{HE} - T_E) - P_{rec} = 6.7 \text{ MW} - P_{rec} \tag{8}$$

$$T_{A_1} - T_E = P_{rec} / (C_p \cdot \dot{m}_1) \tag{9}$$

(period T_2: receiver plus burner in operation)

Case c) $6.7 \text{ MW} < P_{rec} \, (\dot{m}_1, \text{FP})$

$$\dot{m}_1 = \dot{m} + \dot{m}_3 \tag{10}$$

$$P_{sto} = P_{rec} \, (\dot{m} + \dot{m}_3, \text{FP}) - \dot{m} \, C_p \, (T_{HE}-T_E) \tag{11}$$

$$P_{fuel} = 0 \tag{12}$$

The receiver inlet gas temperature T_{E1} is given by

$$T_{E1} = (\dot{m} \, T_E + \dot{m}_3 \, T_{E2})/(\dot{m} + m_3) \tag{13}$$

Since P_{rec} itself is a function of the mass flow,
$\dot{m}_3$ is a solution of the nonlinear equation

$$P_{rec} - C_{\dot{p}}(\dot{m}+\dot{m}_3) \cdot T_{HE} + C_{\dot{p}}(\dot{m} \cdot T_E + \dot{m}_3 \cdot T_{E2}) = 0 \tag{14}$$

$$W_{sto}^{(t)} = \int_{t_o}^{t} P_{sto} \, dt \tag{15}$$

(period T_3: H_2-production plus storage loading)

Case d) P_{rec} $(\dot{m}, FP) < 6.7$ MW and $W_{sto} > 0$

Case d1)   The receiver can be operated at a mass flow $\dot{m}_1 \leq \dot{m}$ such that the gas has an outlet temperature of 1080°C ($= T_{HE}$). Then:

$$\dot{m}_1 = P_{rec} (\dot{m}_1, FP)/[C_p \cdot (T_{HE}-T_E)] \tag{16}$$

$$\dot{m}_2 = \dot{m} - \dot{m}_1 \tag{17}$$

$$\dot{m}_3 = 0 \tag{18}$$

$$P_{sto} = - C_p \dot{m}_2 (T_{A2}-T_E) \tag{19}$$

$$P_{fuel} = 6.7 \text{ MW} - P_{rec} (\dot{m}_1, FP) - |P_{sto}| \tag{20}$$

$$W_{sto}^{(t)} = \int_{t_o}^{t} P_{sto} \, dt \tag{21}$$

Case d2) Case d1 is not satisfied (see problem discussed in connection with period T1 on page 35). Then the system is operated by unloading the storage and using the burner until the storage is empty.

$$\dot{m}_1 = \dot{m}_3 = 0 \qquad (22)$$

$$\dot{m}_2 = \dot{m} \qquad (23)$$

$$P_{sto} = - c_p \dot{m} (T_{A2} - T_E) \qquad (24)$$

$$P_{fuel} = 6{,}7 \text{ MW} - |P_{sto} \qquad (25)$$

$$W_{sto}^{(t)} = \int_{t_o}^{t} P_{sto} \, dt \qquad (26)$$

Operation is stopped in the evening if P_{fuel} exceeds 3.1 MW since otherwise the burner would consume more H_2 that is produced.

(periods T_4, T_5)

7.   LITERATURE

Beyond the reports already cited in the Final Report information of the following two sources were used:

- Entwicklung und Betrieb von Dünnschichtzellen für die
  Hochtemperaturelektrolyse
  Forschungsbericht BMFT-FB-T84-146

- Betriebsverhalten und Wirtschaftlichkeit eines modularen
  thermischen Gesteinsspeichers als Systemelement in Solar-
  termkraftwerken;
  draft of a thesis report by Michael Geyer, DFVLR,
  Almeria, Spain

Solar Thermal Energy Utilization
- German Studies on Technology and Application -

Index of Authors

Huber, P.E.:   Considerations   and   Proposals   for   Future
                    Research and Development of High Temperature
                    Solar Processes, (Motor Columbus, Stuttgart),
                    Vol. 1, p. 169.

Jäger, W.:   A Multistage Steam Reformer Utilizing Solar
                    Heat, (Interatom, Bergisch-Gladbach),
                    Vol. 2, p. 57.

Josfeld, F.J.:   Expert Opinion and Co-operation in the Devel-
                    opment Program High Temperature Storage Tank
                    (Uni-Essen GHS), Vol. 2, p. 211.

Kalfa, H.:   Layout of High Temperature Solid Heat Storages
                    (Didier, Wiesbaden), Vol. 2, p. 111.

Kalt, A.:   Solar Steam Reforming of Methane (SSRM)
                    Program Proposals, (DFVLR, Köln),
                    Vol. 3, p. 179.

Kappler, H.W.:   Considerations   and   Proposals   for   Future
                    Research and Development of High Temperature
                    Solar Processes, (Motor Columbus, Stuttgart),
                    Vol. 1, p. 169.

Karnowsky, B.:   Volumetric Ceramic Receiver Cooled by Open Air
                    Flow - Feasibility Study -
                    (Interatom, Bergisch-Gladbach), Vol. 2, p. 1.

Kaufmann, J.:   Literature Survey in the Field of Primary and
                    Secondary Concentrating Solar Energy Systems
                    Concerning   the   Choice   and   Manufacturing
                    Process of Suitable Materials
                    (NU-Tech-Neumünster), Vol. 1, p. 97.

Koepke, P.:   Yearly Yield of Solar CRS-Process Heat and
                    Temperature of Reaction
                    (Universität München), Vol. 1, p. 3.

Lammers, J.:   Considerations   and   Proposals   for   Future
                    Research and Development of High Temperature
                    Solar Processes, (Motor Columbus, Stuttgart)
                    Vol. 1, p. 169.

Lensch, G.:   Literature Survey in the Field of Primary and
                    Secondary Concentrating Solar Energy Systems
                    Concerning   the   Choice   and   Manufacturing
                    Process of Suitable Materials
                    (NU-Tech-Neumünster), Vol. 1, p. 97.

Leuchs, U.:   Solar Steam Reforming of Methane - Program
                   Proposals, Vol. 3, p. 195; A Multistage Steam
                   Reformer Utilizing Solar Heat, Vol. 2, p. 57,
                   (Interatom, Bergisch-Gladbach,

Lippert, P.:   Literature Survey in the Field of Primary and
                   Secondary Concentrating Solar Energy Systems
                   Concerning the Choice and Manufacturing
                   Process of Suitable Materials
                   (NU-Tech-Neumünster), Vol. 1, p. 97.

Meyringer, V.:   Utilization of Solar Energy for Hydrogen
                   Production by High Temperature Electrolysis of
                   Steam, (Dornier, Friedrichshafen),
                   Vol. 3, p. 621.

Müller, W.D.:   Steam Reforming of Methane Utilizing Solar
                   Heat, Vol. 3, p. 1; Comparative Investigations
                   and Ratings of Different Solar Systems Using
                   Tubular Steam Reformers (Lurgi, Frankfurt),
                   Vol. 3, p. 251.

Quenzel, H.:   Yearly Yield of Solar CRS-Process Heat and
                   Temperature of Reaction, (Universität München)
                   Vol. 1, p. 3.

Reimert, R.:   Process Synthesis of a Gasification Process
                   Modified for High Solar Energy Integration
                   (Lurgi, Frankfurt), Vol. 3, p. 547.

Siebert, W.:   A Multistage Steam Reformer Utilizing Solar
                   Heat, (Interatom, Bergisch-Gladbach),
                   Vol. 2, p. 57.

Sizmann, R.:   Yearly Yield of Solar CRS-Process Heat and
                   Temperature of Reaction, (Universität München)
                   Vol. 1, p. 3.

Streuber, Chr.:   Layout of High Temperature Solid Heat Storages
                   (Didier, Wiesbaden), Vol. 2, p. 111.

Werner, K.:   Expert Opinion and Co-operation in the Devel-
                   opment Program High Temperature Storage Tank
                   (Uni-Essen GHS), Vol. 2, p. 211.

Additional material from *Solar Thermal Energy Utilization. German Studies on Technology and Applications,*
ISBN 978-3-540-18032-6 (978-3-540-18032-6_OSFO1),
is available at http://extras.springer.com

Additional material from *Solar Thermal Energy Utilization. German Studies on Technology and Applications*,
ISBN 978-3-540-18032-6 (978-3-540-18032-6_OSFO3),
is available at http://extras.springer.com

Additional material from *Solar Thermal Energy Utilization. German Studies on Technology and Applications,*
ISBN 978-3-540-18032-6 (978-3-540-18032-6_OSFO5),
is available at http://extras.springer.com

Additional material from *Solar Thermal Energy Utilization. German Studies on Technology and Applications,*
ISBN 978-3-540-18032-6 (978-3-540-18032-6_OSFO6),
is available at http://extras.springer.com

Additional material from *Solar Thermal Energy Utilization. German Studies on Technology and Applications,*
ISBN 978-3-540-18032-6 (978-3-540-18032-6_OSFO7),
is available at http://extras.springer.com